# Environmental Science and Engineering

The ultimate goal of this series is to contribute to the protection of our environment, which calls for both profound research and the ongoing development of solutions and measurements by experts in the field. Accordingly, the series promotes not only a deeper understanding of environmental processes and the evaluation of management strategies, but also design and technology aimed at improving environmental quality. Books focusing on the former are published in the subseries Environmental Science, those focusing on the latter in the subseries Environmental Engineering.

Zhaosong Fang · Sheng Zhang · Zhang Lin ·
Xiwen Feng · Yuchun Zhang

# Outdoor Thermal Comfort and Heat Stress Mitigation

## Measurement Methods, Human Responses, Modeling and Design Strategies

Springer

Zhaosong Fang
Guangzhou University
Guangzhou, Guangdong, China

Sheng Zhang
Xi'an Jiaotong University
Xi'an, Shaanxi, China

Zhang Lin
City University of Hong Kong
Hong Kong, China

Xiwen Feng
Guangzhou University
Guangzhou, Guangdong, China

Yuchun Zhang
City University of Hong Kong
Hong Kong, China

ISSN 1863-5520 ISSN 1863-5539 (electronic)
Environmental Science and Engineering
ISBN 978-981-95-8898-5 ISBN 978-981-95-8899-2 (eBook)
https://doi.org/10.1007/978-981-95-8899-2



This Springer imprint is published by the registered company Springer Nature Singapore Pte Ltd.
The registered company address is: 152 Beach Road, #21-01/04 Gateway East, Singapore 189721, Singapore

# Preface

With the acceleration of urbanization and the continuing impacts of global climate warming, outdoor thermal environments in cities have become increasingly concerning. Under the national carbon-neutrality agenda and the pursuit of climate-resilient urban development, providing safe, comfortable and healthy outdoor spaces has become a shared priority in urban planning, building environmental studies and public health. This issue is especially critical in hot and humid regions, where outdoor thermal comfort affects not only the usability of public spaces but also the heat risks encountered during commuting, walking, labor and physical training. Motivated by these challenges, this book presents a systematic investigation into the assessment and improvement of dynamic outdoor thermal comfort.

Current research on outdoor thermal comfort faces multiple challenges. Outdoor environments are highly dynamic; solar radiation, disturbed wind around pedestrians and ground-surface temperature vary rapidly with time, increasing the complexity of environmental measurement and human heat-balance analysis. Human metabolic rate and skin temperature also display pronounced temporal characteristics during outdoor exposure, which traditional steady-state models cannot adequately capture. Psychological expectations, seasonal acclimatization and behavioral adjustments further complicate thermal perception. Classical indices such as PMV, PET and UTCI show noticeable deviations under dynamic conditions, while real-world scenarios such as commuting, outdoor labor and intensive physical training impose higher demands on heat-safety assessment. These challenges not only limit the explanatory power of existing theories but also introduce uncertainties in practical heat-safety judgment, creating a dual gap between research and application. Therefore, it is necessary to establish a more comprehensive research framework to effectively address the key issues related to environmental variability, human responses and model applicability in dynamic outdoor thermal comfort.

Against this backdrop, the book is organized into six interlinked parts, each addressing a key component while collectively forming a coherent progression. The first part reviews the development of outdoor thermal comfort research and provides the conceptual basis for the following chapters. The second part focuses on methods for measuring outdoor environmental parameters and addresses how to accurately capture actual exposure conditions, including wind fields, ground temperature and radiative components. The third part examines the influences of metabolic rate, seasonal variability and thermal pleasure on thermal perception, offering physiological and psychological foundations for model improvement. The fourth part evaluates the applicability of classical models and develops calibrated and adaptive approaches, thereby establishing a systematic methodological framework. The fifth part applies the models to typical scenarios such as commuting, outdoor work and military training, demonstrating their explanatory strength and relevance for heat safety. The sixth part integrates experimental and modeling findings and proposes practical improvement strategies, including tree shade, building shadows, ventilated umbrellas and overhead structures, which support planning practice and microclimate design. Together, these six parts form a logical chain from identifying challenges to obtaining accurate measurements, understanding human responses, improving models and supporting practical applications.

Thank you to the collaborators involved in the relevant research in this book, including Prof. Jianlei Niu, Prof. Zhengrong Li, Prof. Weiwei Liu, and Prof. Jianlin Liu. We also thank Xinyu Li, Jiantao Gan, Zhan Li, Jiaying Chen, Qinghao Meng, Futao Liu and Changyun Wang for their assistance with figures and proofreading. We also express our sincere appreciation to the team members who contributed to experimental design, data collection, model validation and case studies, as well as to the institutions and experts who supported field coordination, project funding and data sharing. Portions of this work were supported by national and local research grants, for which we extend our gratitude.

We hope that this book will provide researchers, graduate students and practitioners in urban climatology, building environmental science, public health and urban planning with a systematic theoretical framework, methodological guidance and practical case references. It aims to address the gaps in theory and application of dynamic outdoor thermal comfort and to support the development of more refined and resilient urban outdoor environments.

Guangzhou, China
December 2025

Zhaosong Fang

# Contents

# Abbreviations and Nomenclatures

## Abbreviations

| | |
|---|---|
| *ar*PTS | Adaptive-rational predicted thermal sensation |
| BMI | Body mass index |
| CBE | Center for the built environment |
| CFD | Computational fluid dynamics |
| DHI | Diffuse horizontal irradiance |
| DNI | Direct normal irradiance |
| GHI | Global horizontal irradiance |
| GHM | Globe thermometer method |
| HR | Heart rate |
| HVAC | Heating, ventilation, and air conditioning |
| MAPE | Mean absolute percentage error |
| MEMI | Munich energy balance model for individuals |
| MMSV | Mean air movement sensation vote |
| MRT | Mean radiant temperature |
| MTHM | Modified two-hemisphere method |
| MTSV | Mean thermal sensation votes |
| MWC | Maximum working capacity |
| PET | Physiological equivalent temperature |
| PMV | Predicted mean vote |
| PPD | Predicted percentage of dissatisfied |
| PTS | Predicted thermal sensation |
| RH | Relative humidity |
| RMSE | Root mean square error |
| SDAE | Standard deviation of absolute errors |
| SDIM | Six-direction integral method |
| SET* | Standard effective temperature |
| THM | Two-hemisphere method |

| | |
|---|---|
| TSV | Thermal sensation vote |
| UTCI | Universal thermal climate index |
| WBGT | Wet bulb globe temperature |

## Nomenclature

| | |
|---|---|
| $A$ | Age in years; Body surface area, $m^2$; Temperature factor |
| $A_1$ | Efficiency |
| $A_{Be}$ | Surface area of the clothed body, $m^2$ |
| $A_{Du}$ | Surface area of the skin, $m^2$ |
| $A_{eff}$ | Effective radiative area, $m^2$ |
| $A_{nude}$ | Nude human surface area, $m^2$ |
| $A_p$ | Direct solar radiation projected area, $m^2$ |
| $A_{sk}$ | Surface area of the skin, $m^2$ |
| $\alpha$ | The body mass fraction in the skin compartment; Improved PET model via experimental data, °C; Improved UTCI model via experimental data, °C; Improved SET* model via experimental data, °C; Improved UTCI, °C; Improved PET, °C; Improved SET*, °C |
| $\alpha_{LW}$ | Human surface absorptivity for longwave radiation, $\alpha_{LW} = 0.7$ |
| $\alpha_{SW}$ | Human surface absorptivity for shortwave radiation, $\alpha_{SW} = 0.97$ |
| $\beta$ | Efficiency; PET model output by BIO-met, °C; UTCI model output by BIO-met, °C; SET* model output by BIO-met, °C |
| $C$ | Convection, W; Convective heat transfer, $W/m^2$ |
| $c$ | Tissue heat capacitance, J/(kg·°C) |
| $C_{p,b,\ ck}$ | Specific heat capacity of body, kJ/kg |
| $C_{res}$ | Dry respiratory heat loss per unit area, $W/m^2$ |
| $D$ | Globe diameter, mm; Globe diameter, m |
| $D$(x) | Deviation |
| $d$ | Days |
| $E$ | Sum heat lost due to evaporation, W; Sum evaporating heat transfer, $W/m^2$ |
| $E_D$ | The latent heat lost due to diffusion of water vapor, W |
| $E_{res}$ | Rate evaporative heat loss from respiration, $W/m^2$ |
| $E_{SW}$ | The latent heat lost due to evaporation of sweat, W |
| $E_{sk}$ | Total rate of evaporative heat loss from skin, $W/m^2$ |
| $\varepsilon$ | Surface emissivity |
| $\varepsilon_g$ | Globe emissivity |
| $\varepsilon_p$ | Human emissivity coefficient, $\varepsilon_p = 0.97$ |
| $\eta$ | Mechanical efficiency |
| $E_B$, $E_{Bi}$ | Sky diffuse radiation from background part, $W/m^2$ |
| $E_C$, $E_{Ci}$ | Sky diffuse radiation from circumsolar part, $W/m^2$ (with the background radiation), $W/m^2$ |

| | |
|---|---|
| $E_{HD}$ | Total horizontal sky diffuse radiation, W/m$^2$ |
| $F_i$ | View factor |
| $F_B$ | Portion of the irradiance of the background part on the total horizontal diffuse irradiance |
| $F_{cs}$ | Energy flux from the body core to the skin surface, W |
| $f_{cl}$ | Clothing area factor; Clothing area coefficient |
| $f_{eff}$ | Effective radiative area factor |
| $f_p$ | Direct solar projected area factor |
| $f_{rad}$ | Correction coefficient of the effective surface area |
| $G$ | The body mass, kg |
| $H$ | Internal heat produced by the metabolism, W; Height, m |
| $H_{sk}$ | Heat loss from the skin, W/m$^2$ |
| $HR_0$ | Heart rate at rest, bpm |
| $h_c$ | Convective heat transfer coefficient, W/(m$^2$·°C) |
| $h_e$ | Latent heat transfer coefficient, W/(m$^2$·kPa) |
| $h_r$ | Radiation heat transfer coefficient, W/(m$^2$·°C) |
| $h_s$ | Standard heat transfer coefficient, W/(m$^2$·°C) |
| $I/I_{cl}/I_{clo}$ | Thermal resistance of clothing, m$^2$·°C/W; Total clothing insulation, clo |
| $K$ | Clearness index; Conductivity, W |
| $K_i$, $L_i$ | Shortwave and longwave radiation from environmental surfaces onto a human surface, respectively, W/m$^2$ |
| $K_{abs}$, $L_{abs}$ | Shortwave and longwave radiation absorbed by a human body, respectively, W/m$^2$ |
| $K_{abs,\ UH}$, $L_{abs,\ UH}$ | Shortwave and longwave radiation absorbed by a human body from the upper hemisphere, respectively, W/m$^2$ |
| $K_{abs,\ LH}$, $L_{abs,\ LH}$ | Shortwave and longwave radiation absorbed by a human body from the lower hemisphere, respectively, W/m$^2$ |
| $K\downarrow$, $L\downarrow$ | Shortwave and longwave radiation from above, respectively, W/m$^2$ |
| $K\uparrow$, $L\uparrow$ | Shortwave and longwave radiation from below, respectively, W/m$^2$ |
| $K_{hor-i}$, $L_{hor-i}$ | Shortwave and longwave radiation from four horizontal cardinal directions, respectively, W/m$^2$ |
| $L$ | Sensible heat lost due to respiration, W |
| $M$ | Internal heat produced by the metabolism, W; Metabolic rate, W/m$^2$ |
| $M_0$ | Metabolic rate in the inactive state, W/m$^2$ |
| $M_{shiv}$ | Metabolic heat generation by shivering, W |
| $P_a$ | Ambient water vapor pressure, kPa |
| $P_{sk}$ | Vapor pressure of water on the skin surface, kPa |
| $q_m$ | Metabolic heat production, W/m$^3$ |
| $q_{m,sh}$ | Metabolism of shivering, W/m$^3$ |
| $q_{m,w}$ | Metabolism of exercise, W/m$^3$ |
| $\Delta q_m$ | Any additional heat gain, W/m$^3$ |

| | |
|---|---|
| $\Delta q_{\mathrm{m,bas}}$ | Changes in the local basal metabolism, $W/m^3$ |
| $R$ | Sensible heat loss from skin by radiation, W; Radiant heat transfer, $W/m^2$ |
| $R + C$ | Sum sensible heat loss from skin, W |
| $R_{\mathrm{ecl}}$ | Wet permeability coefficient of the clothing |
| $R_{\mathrm{str,g}}$ | Ground radiation, $W/m^2$ |
| $RM$ | Increase in the heart rate per unit of metabolic rate, bpm/($W/m^2$) |
| $\rho_{\mathrm{b}}$ | Density of blood, $kg/m^3$ |
| $S$ | Heat content in the body, W; Heat storage, $W/m^2$ |
| $S_{\mathrm{cr}}$ | Rate of heat storage in core compartment, $W/m^2$ |
| $S_{\mathrm{ck}}$ | Rate of heat storage in the skin node, $W/m^2$ |
| $S_{\mathrm{str}}$ | Total radiation, $W/m^2$ |
| $S_{\mathrm{sk}}$ | Rate of heat storage in skin compartment, $W/m^2$ |
| $skbf$ | Peripheral blood flow, L/(h·$m^2$) |
| $SW$ | Sweat rate, kg/s |
| $T_{\mathrm{a}}$ | Air temperature, °C; Ambient temperature, °C |
| $T_{\mathrm{bla}}$ | Arterial blood temperature, °C |
| $T_{\mathrm{bs}}$ | Surface temperatures of the body sector, °C |
| $T_{\mathrm{cl}}$ | Clothing surface temperature, °C |
| $T_{\mathrm{g}}$ | Globe bulb temperature, °C; Globe temperature, °C; Black-bulb temperature, °C |
| $T_{\mathrm{i}}$ | Tissue temperature, °C |
| $T_{\mathrm{M}}$ | Mean body temperature, °C |
| $T_{\mathrm{mrt}}$ | Mean radiant temperature, °C; MRT in six-direction integral method, °C; MRT in two-hemisphere method, °C |
| $T_{\mathrm{op}}$ | Operative temperature, °C |
| $T_{\mathrm{sk}}$ | Mean skin temperature, °C; Skin temperature, °C |
| $T_{\mathrm{w}}$ | Wet bulb temperature, °C; Natural wet bulb temperature, °C |
| $T_{\mathrm{cr}}/t_{\mathrm{cr}}$ | Core temperature, °C |
| $T_{\mathrm{sk}}/t_{\mathrm{sk}}$ | Mean skin temperature, °C |
| $\Delta T$ | Differences in auditory canal temperature, °C |
| $TI$ | Turbulence intensity |
| $u$ | Average wind velocity, m/s |
| $V_{\mathrm{a}}$ | Air velocity, m/s; Wind speed, m/s |
| $V_{\mathrm{i}}$ | Wind speed of test point, m/s |
| $V_{\mathrm{s}}$ | Surrounding air velocity, m/s |
| $V_{\mathrm{w}}$ | Walking velocity, m/s |
| $V_{\mathrm{b}}$ | Blood flow from the core to the skin, $m^2/s$ |
| $V_{\mathrm{Pa}}$ | Ambient water vapor pressure, kPa |
| $W$ | Heat lost due to mechanical work, W; Mechanical work, $W/m^2$; Weight, kg |
| $W_{\mathrm{d}}$ | Wind direction, ° |
| $\omega$ | Skin wittedness; Moisture index of skin |
| $X$ | Vertical distance, m |

| | |
|---|---|
| $\chi$ | Angle between the sun and a sky element with the center of the sky hemisphere as the vertex, degree |
| $\gamma$ | Solar altitude angle, degree; Angle between the sun and a sky element, degree |
| $\theta$ | Solar zenith, degree; Time, s |
| $\sigma$ | Stefan–Boltzmann constant, $\sigma = 5.67 \times 10^{-8}$ W/(m$^2$·K$^4$) |

# Overview of Outdoor Thermal Comfort Research

# Introduction to Outdoor Thermal Comfort Research

**Abstract** This chapter provides an overview of research progress in outdoor thermal comfort, focusing on the challenges associated with dynamic and complex outdoor environments. The growing importance of outdoor thermal comfort is first introduced in the context of urbanization, climate change, and increasing outdoor activities. The evolution of thermal comfort research from indoor-oriented frameworks to outdoor-specific indices and field-based approaches is then reviewed, highlighting advances in environmental measurement, human physiological responses, and psychological and behavioral adaptation. Key methodological developments in assessing wind, radiation, and surface temperatures are summarized. Major unresolved issues are subsequently identified, including measurement uncertainties, human variability, limitations of existing thermal comfort models, and heat-related safety concerns. Based on these challenges, the chapter outlines the structure of the book, covering environmental measurements, human-centered factors, model evaluation and improvement, safety-oriented applications, and design strategies for outdoor thermal mitigation. This chapter establishes a conceptual foundation for the subsequent chapters and frames future research directions in outdoor thermal comfort.

## 1 Introduction

Urbanization, climate change, and the intensification of the urban heat island effect have made outdoor thermal comfort a pressing issue worldwide. People increasingly spend time outdoors, engaging in recreational activities, commuting, or working in semi-open environments. Consequently, outdoor thermal comfort has evolved into a critical research domain that intersects environmental science, urban planning, physiology, and public health [1–3].

Unlike indoor environments, outdoor thermal conditions are dynamic and strongly influenced by solar radiation, fluctuating wind fields, and surface heat exchange. These complexities introduce new challenges for thermal comfort research, ranging from accurate measurement of environmental variables to the modeling of dynamic human responses [4–6]. The study of outdoor thermal comfort thus not only informs



Z. Fang et al., *Outdoor Thermal Comfort and Heat Stress Mitigation*, Environmental Science and Engineering, https://doi.org/10.1007/978-981-95-8899-2_1

individual health and well-being but also supports sustainable design strategies and urban resilience under extreme climate scenarios.

In recent years, outdoor thermal comfort has become a focal point for sustainable city design and adaptation to global warming. Research has shown that thermal discomfort in outdoor spaces discourages walking and cycling, reduces public space utilization, and increases heat-related morbidity [7, 8]. Meanwhile, the demand for livable urban spaces underscores the importance of developing reliable assessment tools and mitigation strategies for outdoor thermal environments.

## 2 Development of Outdoor Thermal Comfort Research

Thermal comfort research has traditionally been dominated by indoor studies, leading to well-established standards such as ISO 7730 [9] and ASHRAE 55 [10]. These frameworks rely on static assumptions regarding metabolic rate and clothing insulation. However, their predictive power is limited in outdoor contexts, where both environmental factors and human activities fluctuate considerably.

Over the past two decades, attention has shifted toward outdoor and semi-open environments, motivated by urban design needs and public health concerns. Advanced indices such as the Physiological Equivalent Temperature (PET) [11] and the Universal Thermal Climate Index (UTCI) [12] were specifically developed to address outdoor contexts. In parallel, a growing body of field studies has investigated outdoor activities, revealing the importance of wind speed variability, solar radiation, and ground heat flux in shaping thermal perception [13, 14].

At the human level, metabolic rate has emerged as a pivotal factor in determining thermal comfort [15–17]. Unlike static environmental variables, metabolic rate is inherently dynamic and directly tied to activity intensity. Incorporating this variability into thermal comfort models remains a major research challenge.

Further developments have also examined psychological and behavioral adaptations, such as seasonal thermal expectations, cultural differences, and personal thermal preferences [18, 19]. These findings highlight that outdoor thermal comfort cannot be fully explained by physical models alone but requires integration of physiological and psychological perspectives.

Moreover, methodological advances have improved the accuracy of environmental measurement. Techniques such as computational fluid dynamics (CFD) for disturbed wind flow [20], infrared thermography for surface and radiant temperature [21], and big-data or GIS-based approaches for large-scale thermal comfort assessment [22, 23] provide more realistic data inputs. These developments underpin the methodological sections of this book.

## 3 Current Challenges in Outdoor Thermal Comfort Research

Despite decades of progress, several challenges remain unresolved:

Accurate measurement of outdoor environmental parameters: Disturbed wind speed near the human body, ground temperature variations, and radiant heat exchange are difficult to capture in real-world settings, yet they are essential for understanding thermal perception [11–14, 20–23].

Human physiological and behavioral variability: Dynamic metabolic rate, clothing insulation, seasonal adaptation, and psychological responses introduce significant uncertainty in thermal comfort evaluation [15–19].

Model limitations and calibration needs: Classical indices such as PMV, PET, and UTCI often fail to represent dynamic outdoor conditions, requiring correction, calibration, or new adaptive frameworks to improve their applicability [9–12].

Safety considerations in diverse scenarios: Commuting, outdoor labor, and intensive physical training highlight the close link between thermal comfort and human health risks under heat stress conditions. Recent case studies have reported increased risks of heat-related illness during construction work, student training, and large-scale public events [24].

Design and planning strategies: There is still a gap between theoretical models and their practical translation into urban design measures, such as shading systems, cooling interventions, and the use of overhead structures for thermal mitigation [4–6, 25, 26].

These challenges also define the thematic structure of this book. Part II focuses on environmental measurement methods (wind, ground, radiation). Part III addresses human physiological and psychological factors. Part IV evaluates and improves thermal comfort indices. Part V applies these frameworks to outdoor safety scenarios (commuting, labor, training). Finally, Part VI translates findings into practical design strategies for urban cooling and resilience.

## 4 Contributions and Outlook

This book aims to provide a comprehensive framework for understanding and evaluating outdoor thermal comfort under dynamic conditions. By integrating environmental measurements, human-centered factors, model refinement, and real-world applications, it contributes to bridging the gap between theoretical advances and urban design practice.

In particular, this work emphasizes the multi-dimensional nature of outdoor thermal comfort research: measurement precision, human physiological and psychological dynamics, model adaptability, and practical design strategies. By systematically organizing research across these domains, the book offers both a consolidated knowledge base and an agenda for future research.

The future of outdoor thermal comfort research lies in dynamic, adaptive, and personalized approaches, supported by real-time monitoring technologies, cross-disciplinary collaboration, and climate-responsive urban planning. The subsequent chapters collectively address these directions, offering both theoretical foundations and practical strategies to guide future work.

## References

1. ASHRAE. ASHRAE Handbook—Fundamentals, *American Society of Heating, Refrigerating and Air-Conditioning Engineers, Atlanta, GA, USA* (2017)
2. M. Santamouris, Heat Island research in Europe: the state of the art. Adv. Build. Energy Res. **1**(1), 123–150 (2007)
3. T.R. Oke, The energetic basis of the urban heat island. Q. J. R. Meteorol. Soc. **108**(455), 1–24 (1982)
4. M. Nikolopoulou, K. Steemers, Thermal comfort and psychological adaptation as a guide for designing urban spaces. Energy Build. **35**(1), 95–101 (2003)
5. T.P. Lin, A. Matzarakis, Tourism climate and thermal comfort in Sun Moon Lake, Taiwan. Int. J. Biometeorol. **52**(4), 281–290 (2008)
6. S. Oliveira, H. Andrade, An initial assessment of the bioclimatic comfort in an outdoor public space in Lisbon. Int. J. Biometeorol. **51**(5), 323–334 (2007)
7. D. Lai, W. Liu, T. Gan, K. Liu, Q. Chen, A review of mitigating strategies to improve the thermal environment and thermal comfort in urban outdoor spaces. Sci. Total Environ. **661**, 337–353 (2019)
8. E. Johansson, M.W. Yahia, I. Arroyo, C. Bengs, Outdoor thermal comfort in public spaces in warm-humid Guayaquil, Ecuador. Int. J. Biometeorol. **62**(3), 387–399 (2018)
9. ISO 7730, *Ergonomics of the Thermal Environment—Analytical Determination and Interpretation of Thermal Comfort Using Calculation of the PMV and PPD Indices and Local Thermal Comfort Criteria* (International Organization for Standardization, Geneva, 2005)
10. ANSI/ASHRAE Standard 55, *Thermal Environmental Conditions for Human Occupancy.* (American Society of Heating, Refrigerating and Air-Conditioning Engineers, Atlanta, GA, USA, 2017)
11. P. Höppe, The physiological equivalent temperature—a universal index for the biometeorological assessment of the thermal environment. Int. J. Biometeorol. **43**(2), 71–75 (1999)
12. D. Fiala, G. Havenith, P. Bröde, B. Kampmann, G. Jendritzky, UTCI—Fiala multi-node model of human heat transfer and temperature regulation. Int. J. Biometeorol. **56**(3), 429–441 (2012)
13. L. Chen, E. Ng, Outdoor thermal comfort and outdoor activities: a review of research in the past decade. Cities **29**(2), 118–125 (2012)
14. S. Thorsson, T. Honjo, F. Lindberg, I. Eliasson, E.M. Lim, Thermal comfort and outdoor activity in Japanese urban public places. Environ. Behav. **39**(5), 660–684 (2007)
15. P.O. Fanger, Thermal comfort, in *Analysis and Applications in Environmental Engineering* (McGraw-Hill, New York, 1970)
16. ISO 8996, *Ergonomics of the Thermal Environment—Determination of Metabolic Rate* (International Organization for Standardization, Geneva, 2004)

17. B.E. Ainsworth, W.L. Haskell, S.D. Herrmann et al., 2011 compendium of physical activities: a second update of codes and MET values. Med. Sci. Sports Exerc. **43**(8), 1575–1581 (2011)
18. N. Kántor, J. Unger, Benefits and opportunities of adopting GIS in thermal comfort studies in resting places: an urban park as an example. Landsc. Urban Plan. **98**(1), 36–46 (2010)
19. T.P. Lin, R. de Dear, R.L. Hwang, Effect of thermal adaptation on seasonal outdoor thermal comfort. Int. J. Climatol. **31**(2), 302–312 (2011)
20. B. Blocken, J. Carmeliet, Pedestrian wind environment around buildings: Literature review and practical examples. J. Therm Envelope Build Sci. **28**(2), 107–159 (2004)
21. M. Taleghani, M. Tenpierik, A. van den Dobbelsteen, Heat in courtyards: field measurements and CFD analysis of radiation, air and surface temperatures. Sol. Energy **103**, 108–124 (2014)
22. A. Middel, J. Lukasczyk, S. Zakrzewski, M. Arnold, R. Maciejewski, Urban form and composition of street canyons: a human-centric big-data and deep-learning approach to thermal comfort. Landsc. Urban Plan. **183**, 122–132 (2019)
23. N. Kántor, A. Kovács, Á. Takács, et al., Human-biometeorological significance of shading in urban public spaces—summertime measurements in Pécs, Hungary. Landsc. Urban Planning **177**(article) (2018)
24. J. Xiang, P. Bi, D. Pisaniello, A. Hansen, Health impacts of workplace heat exposure: an epidemiological review. Ind. Health **52**(2), 91–101 (2014)
25. R. Emmanuel, A. Loconsole, Green infrastructure as an adaptation approach to tackling urban overheating in the Glasgow Clyde Valley Region, UK. Landsc. Urban Planning **138**, 71–86 (2015)
26. L. Shashua-Bar, D. Pearlmutter, E. Erell, The influence of trees and grass on outdoor thermal comfort in a hot-arid environment. Int. J. Climatol. **31**(10), 1498–1506 (2011)

# Measurement of Outdoor Thermal Environment Parameters

# Measuring Disturbed Wind Speed Around People Outdoors: Differences from Natural Wind in Thermal Comfort Evaluation

**Abstract** This chapter elucidates the critical differences between disturbed wind speed generated by human movement and ambient natural wind, and their distinct impacts on outdoor thermal comfort evaluation. The background necessitating this distinction is first introduced. It is demonstrated that disturbed wind speed, characterized by higher velocity and altered spectral properties compared to natural wind, plays a dominant role in convective heat loss for moving individuals. The factors influencing the characteristics of disturbed wind are explored. By accounting for this self-generated airflow, the accuracy of thermal comfort prediction for active populations in outdoor settings can be significantly improved. Furthermore, the implications for urban design and ventilation strategies are discussed. Finally, a summary of key findings and future research directions is presented.

## 1 Introduction

In modern urban settings, favoring walking over motorized transport for short journeys is increasingly promoted for its health advantages. Simultaneously, thermal comfort is receiving significant focus as a key determinant of living quality and urban habitability within spatial planning and architectural design [1]. Pleasant thermal conditions indoors and outdoors play a vital role in enhancing urban development across ecological, economic, social, and functional domains [2]. Moreover, sustained public concern has encouraged continued investigation, resulting in steady progress within the field.

Research on indoor thermal comfort over the last decades has predominantly addressed static or uniform thermal states. The predicted mean vote (PMV) model, for example, is a widely referenced tool mainly applied in thermally steady situations [3]. As noted by Nicol and Humphreys, the PMV index is most reliable under moderately stable environmental conditions [4]. Hence, its utility is reduced when assessing comfort in dynamically changing settings. Conversely, the adaptive thermal comfort (ATC) model has emerged as a complementary framework, gaining broader acceptance. This perspective treats individuals as proactive agents who engage with

 
Z. Fang et al., *Outdoor Thermal Comfort and Heat Stress Mitigation*, Environmental Science and Engineering, https://doi.org/10.1007/978-981-95-8899-2_2

and adjust to their milieu through physiological, psychological, and behavioral adaptations [5–8]. Metabolic rate represents a core physiological parameter in each of these models.

Thermal regulation in humans depends on energy interchange with the environment, where metabolic activity is indispensable. As a measure of energetic output, metabolic rate strongly influences thermal perception and comfort levels [9]. Common practice in thermal comfort studies involves estimating metabolic rate using "activity diaries", whereby a fixed value is assigned based on a person's activity category [10, 11]. Such constant values are incorporated into PMV calculations. However, during walking, metabolic rate changes continually. Relying on a uniform value is therefore often inaccurate for predictive thermal modeling, particularly as individual characteristics such as gender, age, and body mass index (BMI) are commonly neglected [12, 13]. Despite its importance, dynamic metabolic processes remain inadequately investigated.

To address these gaps, considerable research has aimed at enhancing metabolic rate assessment. For instance, Hasan et al. [14] deployed wearable devices to provide continuous metabolic rate feedback for building occupants. Havenith et al. [15] examined the role of individual factors in thermal comfort evaluation and reported that standardized metabolic rates from ISO 8996 are not sufficiently accurate. Interest has also grown in studying metabolic rate during movement to better understand comfort in transient scenarios. Colton and Larsen [16], for example, analyzed metabolic shifts in response to physical activity in workplaces, confirming that variations in metabolic rate affect thermal sensation and proposing improved estimation approaches.

Further studies have explored related aspects: Schnorr et al. [17] examined metabolic recovery following exercise, while Fabbri et al. [18] employed survey-based data from children to evaluate a method for estimating metabolic rate from instantaneous heart rate. Ji et al. [19] introduced an advanced technique using $CO_2$ concentration analysis within controlled chambers to study metabolic variations and thermal comfort during physical exertion. Although these efforts yield insights into metabolic activity during structured exercises, few studies concentrate on thermal comfort during walking—despite its prevalence as a primary mode of everyday transportation in modern societies.

Building on earlier work [20], the current research examines thermal comfort under conditions of natural walking, accounting for diverse external influences and the integrated role of bodily regulation. Numerous models have been suggested to evaluate outdoor thermal comfort [11, 21–27], including those targeting physiological control mechanisms [28]. To facilitate sustainable urban design, improved predictive tools for outdoor comfort are being developed [29, 30], along with optimization strategies [31, 32]. Given that walking mainly occurs outdoors—where changeable parameters such as solar radiation, ambient temperature, and wind interact complexly [33, 34]—the body's heat balance is particularly challenging to model. To mitigate the variability inherent in field studies, this project utilizes a naturally ventilated indoor environment without climate control, thereby transposing outdoor experimental requirements into a regulated setting. Under these conditions, most environmental parameters are stable, and air movement serves as the principal variable affecting walking subjects. This investigation provides a basis for further studies conducted outdoors.

This chapter addresses this gap through a controlled climate chamber study. The primary objectives are: (1) to quantitatively compare the magnitude and spectral characteristics of disturbed wind against natural wind; (2) to analyze the dynamic interplay between disturbed wind, physiological response, and subjective thermal perception during walking; and (3) to discuss the critical implications of these findings for improving the accuracy of outdoor thermal comfort evaluation and urban design strategies.

# 2 Methods

## 2.1 *Experimental Facility*

This experimental work was carried out in the indoor badminton facility at Guangzhou University; a schematic of the test arrangement is provided in Fig. 1. To monitor environmental conditions, three thermal comfort data loggers were placed at the mid-point and each end of the court. All measurements taken in the proximity of test subjects were made at a height of 1.1 m [35]. Core weather-related variables were acquired in accordance with ASHRAE Standard 55-2017 [36].

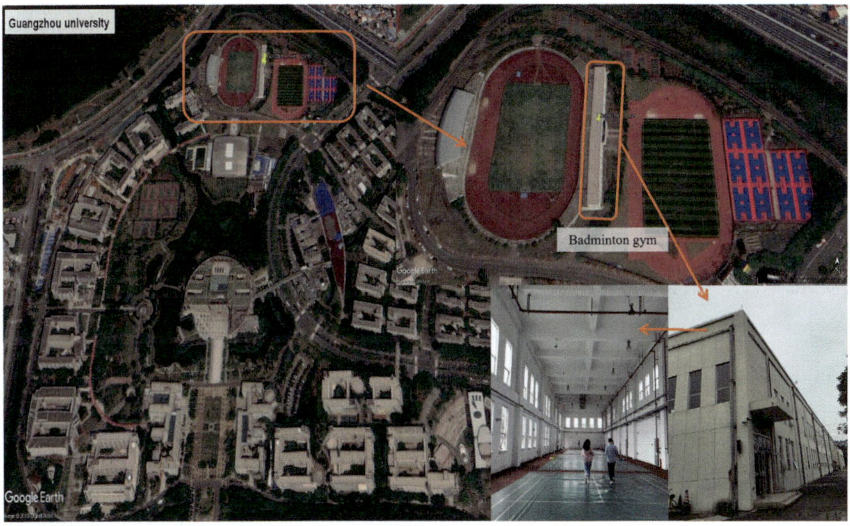


**Fig. 1** Testing environment

**Table 1** Detail information of test condition

| Classification | Air temperature $T_a$ (°C) | Relative humidity RH (%) | Black globe temperature $T_g$ (°C) | Air velocity V (m/s) |
|---|---|---|---|---|
| Max. value | 24.14 | 84.15 | 24.20 | 1.22 |
| Min. value | 20.72 | 74.8 | 20.81 | 0.00 |
| Average value | 22.10 | 79.13 | 22.19 | 0.16 |

## *2.2 Experimental Design*

### 2.2.1 Testing Conditions

The experimental sessions were held in Guangzhou during the spring of 2019, from March to April. Monitoring results showed that the average indoor air temperature was 22 ± 2 °C, while relative humidity levels remained around 80% ± 10%. More detailed information on testing conditions can be found in Table 1.

### 2.2.2 Testing Procedure

The research aimed to examine dynamic variations in auditory canal temperature, heart rate, metabolic rate, and body heat storage under different walking intensities, and to assess their influence on thermal comfort. Subjects were asked to walk at mean speeds of 1.2, 1.4, 1.6, and 2.0 m/s. The speeds of 1.2 and 1.4 m/s were defined as slow walking, and 2.0 m/s as fast walking. A metronome was used to help maintain the required pace. Throughout the tests, participants carried a wireless instrument measuring wind speed and temperature at navel height—approximately between 80 and 115 cm—accounting for variations among individuals.

The testing protocol included three phases: Phase 1 featured a 30 min sitting period for baseline preparation; Phase 2 required 20 min of walking; and Phase 3 involved a 10 min seated recovery. A diagram of the experimental procedure is provided in Fig. 2.

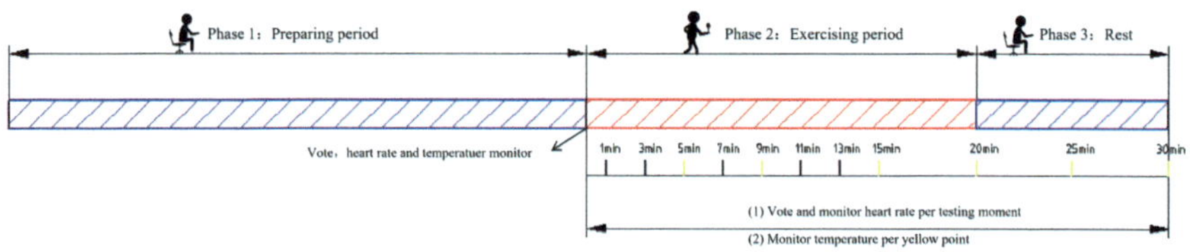


**Fig. 2** Experiment procedure

### 2.2.3 Testing Instrument

Indoor environmental variables, including air temperature ($T_a$), relative humidity (RH), globe temperature ($T_g$), and air velocity ($V_a$), were continuously monitored. Physiological parameters such as auditory canal temperature and heart rate were also recorded. Heart rate was monitored with a smart wristband, and auditory canal temperature was sensed using an infrared thermometer. It is noted in previous studies that rectal temperature is a more reliable indicator of core body temperature than auditory canal temperature [37, 38]. Greenleaf and Castle [37] observed differences of up to 1 °C between rectal and auditory canal measurements, and Modell et al. [38] expressed reservations about the clinical use of infrared tympanic thermometers. Nonetheless, owing to practical considerations during dynamic movement, the auditory canal was selected as the most feasible site. Equipment used is detailed in Table 2.

## 2.3 *Subjective Questionnaire and Physiological Measurement*

Thirty healthy university students—18 males and 12 females—participated in the study. Their average age was approximately 22.3 years and average BMI was 19.2, with male subjects displaying higher basal metabolic rates (BMR). Each participant completed every test condition across all walking speeds. At the beginning, the experimental process was explained to all subjects, and baseline physical characteristics—height, weight, BMI, and BMR—were measured using standardized instruments. Additional demographic data are summarized in Table 3. Clothing insulation

**Table 2** Measurement instruments of this experiment

| Instrument | Type | Parameter | Measuring range | Accuracy | Sampling rate (s) |
|---|---|---|---|---|---|
| Thermal comfort level recorder | SSDZY-1 | $T_a$ (°C) | − 20 to + 80 °C | ± 0.3 °C | 2 |
| | | RH (%) | 0.01–99.9% RH | ± 2% RH (10–90% RH) | 2 |
| | | $T_g$ (°C) | − 20–80 °C | ± 0.3 °C | 2 |
| Wireless universal wind speed and temperature recorder | WFWZY-1 | $V_a$ (m/s) | 0.05–5.00 m/s | 5% ± 0.05 m/s | 2 |
| Smart bracelet | 3XMSH05HM | HR (bpm) | 30–250 bpm | ± 3 bpm | 2 |
| Infrared thermometer | TB-300 | T (°C) | 32.0–42.2 °C | ± 0.2 °C | 1 |

**Table 3** Detail information of subjects

| Gender | Sample size | Age (years) | Height (cm) | Weight (kg) | BMI | BMR |
|---|---|---|---|---|---|---|
| Male | 18 | 22.5 ± 1.5 | 175.3 ± 5 | 66.0 ± 8 | 19.3 ± 1.5 | 1481.3 ± 149.2 |
| Female | 12 | 21.5 ± 2.5 | 161.0 ± 3 | 50.5 ± 1.5 | 19.0 ± 1.3 | 1197.0 ± 20.0 |
| Average | - | 22.3 | 172.0 | 58.3 | 19.2 | 1339.1 |

**Table 4** Subjective vote scale

| Thermal sensation | Blowing feeling | Thermal preference | Humidity preference | Acceptability |
|---|---|---|---|---|
| − 3 cold<br>− 2 cool<br>− 1 slightly cool<br>0 neutral<br>+ 1 slightly warm<br>+ 2 warm<br>+ 3 hot | − 3 much too strong<br>− 2 too strong<br>− 1 slightly strong<br>0 just right<br>+ 1 slightly week<br>+ 2 too week<br>+ 3 much too week | − 1 cooler<br>0 no change<br>+ 1 warmer | − 1 lower<br>0 no change<br>+ 1 higher | − 1 unacceptable<br>1 acceptable |

($I_{clo}$) was estimated based on the ASHRAE handbook [39]; all participants wore a uniform set of clothing consisting of a long-sleeved shirt, thin trousers, socks, and sneakers, with a total insulation value of 0.61 clo.

A two-part questionnaire was implemented [40]: the first section collected thermal sensation (TSV) and air movement sensation (MSV) votes during rest, and the second section captured these votes in addition to physiological data during walking. Subjective rating scales were consistent with ASHRAE Standard 55 [36] and ISO 7730 [41], as detailed in Table 4.

## 2.4 *Data Processing*

### 2.4.1 The Principle of Metabolic Rate Measurement

Usually, the metabolic rate of the human body during a period can be calculated by the amount of $O_2$ inhaled and the amount of $CO_2$ exhaled [42]. In addition, the human metabolic rate can be measured [3] with a surrogate method, e.g., heart rate. The surrogate method is simpler and more convenient than the breathing method with inhaled $O_2$ and exhaled $CO_2$ concentrations. Therefore, in this study, the heart rate was measured in the subject as a proxy for the metabolic rate. The relationship between the measured metabolic rate and the heart rate was calculated by the following Eq. (3):

$$M' = \frac{\mathrm{HR} - \mathrm{HR}_0}{\mathrm{RM}} + M_0, \tag{1}$$

Where M′ is the metabolic rate ($W/m^2$); $M_0$ is the metabolic rate in the inactive state ($W/m^2$); HR is the heart rate at the moment; $HR_0$ is the heart rate at rest, under thermally neutral conditions; and RM is the increase in the heart rate per unit of metabolic rate, which is stated by the following formula,

$$\mathrm{RM} = \frac{\mathrm{HR}_{\max} - \mathrm{HR}_0}{\mathrm{MWC} - \mathrm{M}_0}, \tag{2}$$

Where $HR_{max}$ is the maximum heart rate, $HR_{max}$ = 205–0.62 × A, and MWC is the maximum working capacity described in the following formulas:

$$\mathrm{MWC}_{\mathrm{male}} = (41.7 - 0.22 \times A) \times W^{0.666} \quad \left(\mathrm{W/m^2}\right) \tag{3}$$

$$\mathrm{MWC}_{\mathrm{female}} = (35.0 - 0.22 \times A) \times W^{0.666} \quad \left(\mathrm{W/m^2}\right) \tag{4}$$

Where A is the age in years, and W is the weight in kg.

### 2.4.2 The Principle of Heat Storage Measurement

According to Eqs. (5–8), the heat storage (S) of the human body can be calculated [43, 44] as

$$\mathrm{S} = \mathrm{M} - \mathrm{W} - \mathrm{R} - \mathrm{C} - \mathrm{E}. \tag{5}$$

The value of R (radiant heat transfer), C (convective heat transfer), and E (evaporating heat transfer) can be described in the following formulas:

$$R = f_{\mathrm{cl}}\, h_{\mathrm{r}}\, (t_{\mathrm{cl}} - t_{\mathrm{r}}) \tag{6}$$

$$f_{\mathrm{cl}} = \frac{1}{1 + 0.155 \times (h_{\mathrm{c}} + h_{\mathrm{r}}) \times I_{\mathrm{cl}}} \tag{6.1}$$

$$h_{\mathrm{r}} = f_{\mathrm{rad}}\, \varepsilon \left[ (t_{\mathrm{cl}} + 273.15)^2 + (t_{\mathrm{r}} + 273.15)^2 \right] \times [(t_{\mathrm{cl}} + 273.15) + (t_{\mathrm{r}} + 273.15)] \times 5.67 \times 10^{-8} \tag{6.2}$$

$$t_{\mathrm{cl}} = 35.7 - 0.032M - 0.18 I_{\mathrm{cl}} \left\{ -3.4 \times 10^{-8} f_{\mathrm{cl}} \left[ (t_{\mathrm{cl}} + 273.15)^4 - (t_{\mathrm{r}} + 273.15)^4 \right] + f_{\mathrm{cl}} h_{\mathrm{c}} (t_{\mathrm{cl}} - t_{\mathrm{a}}) \right\} \tag{6.3}$$

where $f_{cl}$ is the clothing area coefficient; $h_r$ is the radiation heat transfer coefficient (W/($m^2$ °C)); $t_{cl}$ is the clothing surface temperature (°C); $t_r$ is the mean radiant temperature (°C); $f_{rad}$ is the correction coefficient of effective surface area of human body; ε is the surface emissivity of human body; $I_{cl}$ is the effective thermal resistance of clothing ($m^2$ kPa/W); and $t_a$ is the air temperature (°C).

$$C = f_{cl}\, h_c\, (t_{cl} - t_a) \tag{7}$$

$$h_c = 8.6\, v^{0.6} \tag{7.1}$$

where $h_c$ is the convective heat transfer coefficient (W/(m$^2$ °C)).

$$E = E_{sk} + (C_{res} + E_{res}) \tag{8}$$

$$E_{sk} = \omega(p_{sk} - p_a) R_{ecl}\, h_e \tag{8.1}$$

$$R_{ecl} = \frac{1}{1 + 0.143 \times h_c \times I_{cl}} \tag{8.2}$$

$$\mathrm{LR} = \frac{h_e}{h_c} \tag{8.3}$$

$$t_{sk} = 33.876 - 0.641M \tag{8.4}$$

$$C_{res} + E_{res} = 0.0014M\,(34 - t_a) + 0.0173M\,(5.87 - p_a) \tag{8.5}$$

where ω is the moisture index of skin; $p_{sk}$ is the vapour pressure of water on the skin surface (kPa); $p_a$ is the vapor pressure of water in air (k$P_a$); $R_{ecl}$ is the wet permeability coefficient of clothing; $h_e$ is the coefficient of evaporating heat transfer (W/(m$^2$ kPa)); and tsk is the skin temperature.

### 2.4.3 Data Analysis

Measurements including mean thermal sensation vote (MTSV), mean air movement sensation vote (MMSV), auditory canal temperature, heart rate, and metabolic rate were plotted over time. Associations among variables were evaluated using linear regression.

All recorded data were compiled and processed within Excel 2019, including data input, organization, and averaging. Statistical analyses, including fitting of linear regression models and calculation of $R^2$ values, were performed using IBM SPSS Statistics 20. Figures and plots were created using Origin 9 and SPSS, based on the processed thermal comfort response data.

# 3 Results

## 3.1 Variation in the Thermal Environment Parameters

Table 5 compares multiple parameters of the thermal environment. From the data, air velocity showed notable variations among tests. Most other parameters exhibited no statistically significant differences across experimental conditions. It follows that variations in thermal comfort at various walking speeds are influenced not only by metabolic rate but also by the intensity of air movement caused by physical motion. This underscores the value of evaluating disturbed air velocities at different walking speeds in the present investigation.

Figure 3 displays the airflow variations associated with different walking speeds. The average natural wind speed was 0.13 m/s. As walking pace increased, air velocity rose above the natural wind level. Under slow walking, the induced airflow was consistent, averaging about 0.75 m/s. During fast walking, the disturbance produced at 2.0 m/s was more substantial, with a mean value of 1.11 m/s.

Figure 4 shows the spectral distributions for each air velocity condition. The power spectral density of natural wind was lowest across all frequencies and increased with walking speed, confirming that movement disturbs the nearby air. This increase contributed to higher turbulent energy—disturbance from faster walking introduced more kinetic energy.

## 3.2 Analysis of the Subjective Evaluations

Throughout the tests, subjects' Mean Thermal Sensation Vote (MTSV) and Mean Moisture Sensation Vote (MMSV) were collected. Figure 5 illustrates the percentage breakdown of votes under different walking conditions. The threshold of thermal perception during walking is clearly visible. Results show that over 50% of responses indicated neutral sensation (0) during slow walking. By contrast, during fast walking, thermal sensation ratings were consistently above neutral, with the combined percentages for + 1, + 2, and + 3 near 60%. Only a small number of subjects reported neutral feeling. During slow walking, the total for positive votes was only 30%. Hence, it was relevant to analyze thermal sensation differences across walking speeds.

Figure 6 presents the change of MTSV over time. Under stable temperature and humidity, the slope of MTSV increased progressively with walking speed. Faster speeds led to higher final steady-state values. Across all speeds, MTSV rose from 0 to 15 min and stabilized between 15 and 20 min. It returned to the initial resting level within 10 min after walking stopped.

Figure 7 describes variations in MMSV over time. At the beginning of exercise, MMSV started low but increased rapidly within the first minute. From 1 to 20 min, it decreased steadily and gently. The slope "k" for each speed reflects the overall downward trend in MMSV. During recovery, individuals walking at 2.0 m/

**Table 5** Comparison of thermal environment parameters

| Classification | | Air temperature $T_a$ (°C) | | | Relative humidity RH (%) | | | Black globe temperature $T_g$ (°C) | | | Air velocity V (m/s) | | |
|---|---|---|---|---|---|---|---|---|---|---|---|---|---|
| | | Max. | Min. | Average | Max. | Min. | Average | Max. | Min. | Average | Max. | Min. | Average |
| Stable state | | 24.14 | 20.72 | 22.10 | 84.15 | 74.80 | 79.13 | 24.20 | 20.81 | 22.19 | 1.22 | 0.00 | 0.16 |
| Walking state | 1.2 m/s | 24.03 | 20.72 | 22.30 | 86.30 | 74.80 | 80.47 | 24.10 | 20.78 | 22.49 | 0.91 | 0.34 | 0.79 |
| | 1.4 m/s | 23.88 | 21.24 | 22.16 | 81.26 | 77.10 | 79.69 | 23.91 | 21.54 | 22.34 | 0.99 | 0.49 | 0.81 |
| | 1.6 m/s | 24.09 | 21.54 | 22.67 | 87.60 | 76.10 | 77.98 | 24.20 | 21.83 | 22.79 | 1.13 | 0.36 | 0.87 |
| | 2.0 m/s | 24.14 | 21.61 | 22.24 | 84.15 | 73.68 | 76.24 | 24.21 | 20.81 | 22.03 | 1.40 | 0.41 | 1.11 |

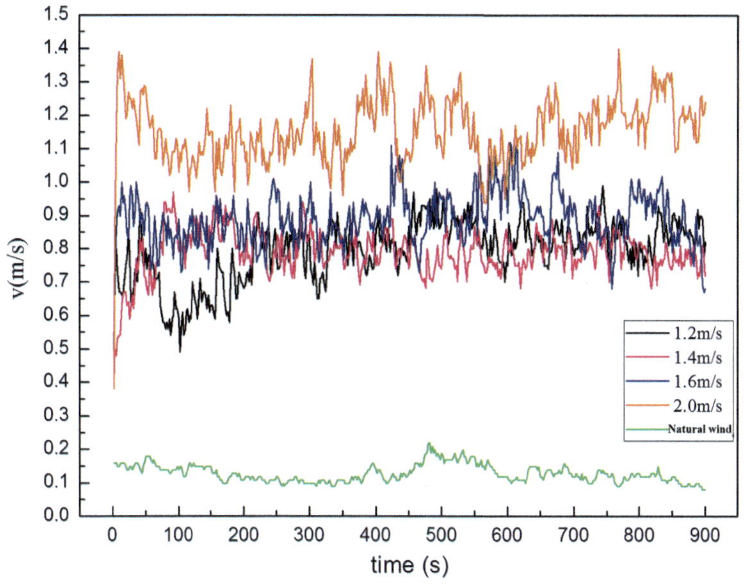


**Fig. 3** The disturbance air velocity

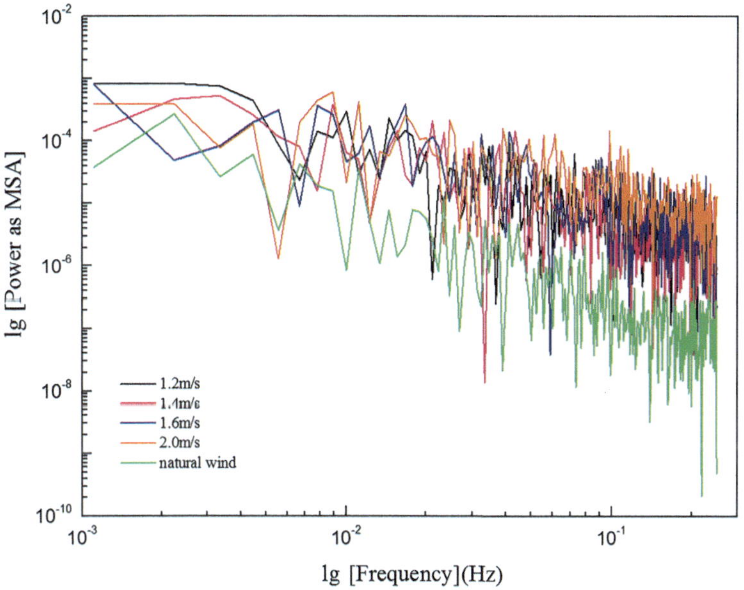


**Fig. 4** The spectral distribution characteristic

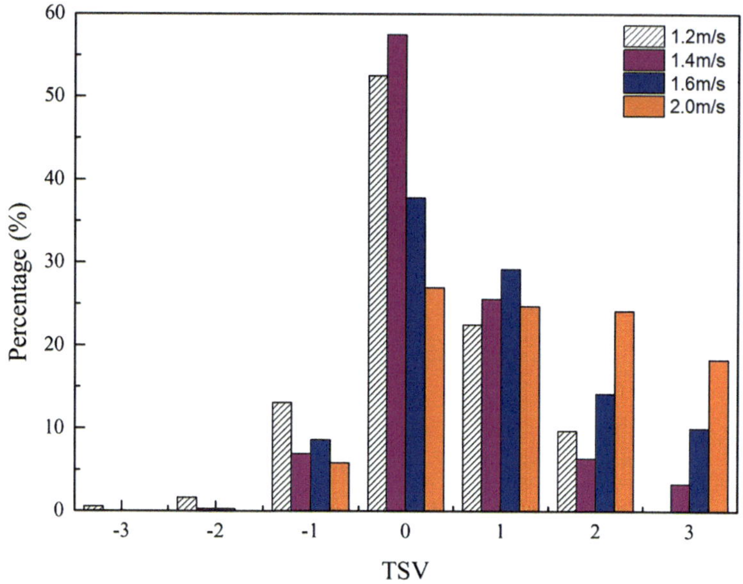


**Fig. 5** The percentage of different voting values

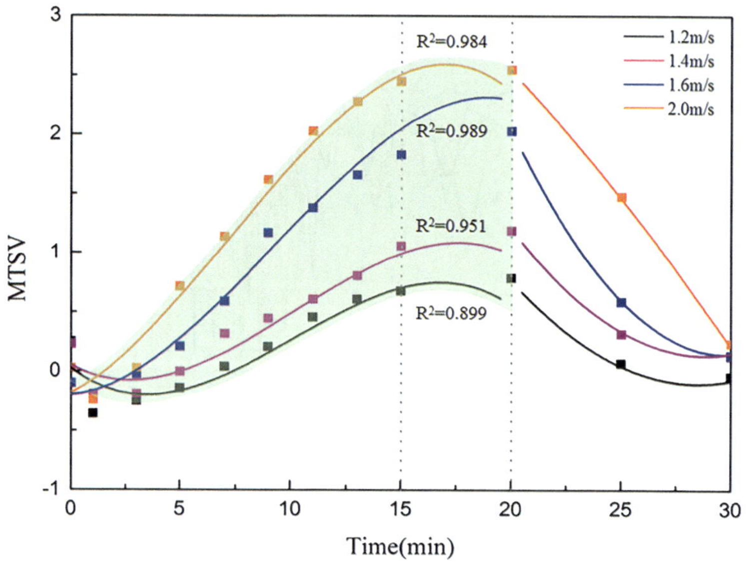


**Fig. 6** The variation of MTSV with time

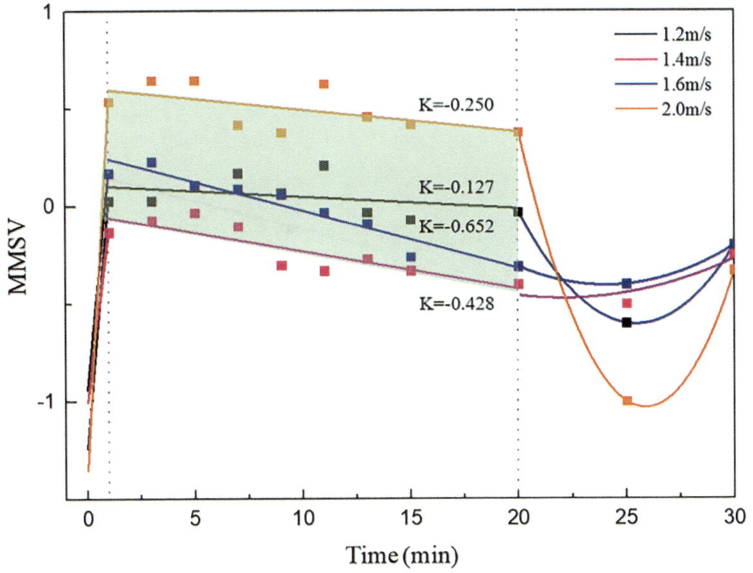


**Fig. 7** The variation of MMSV with time

s showed marked MMSV fluctuations, dropping quickly to a slightly cool feeling (− 1) before returning to around − 0.5. Other cases showed similar behavior, with MMSV decreasing slightly and eventually stabilizing near − 0.5.

A regression analysis was carried out between MTSV and walking speed to predict thermal sensation, as shown in Fig. 8. The $R^2$ value of approximately 0.98 indicates a strong linear dependence. The model shows that MTSV became zero at a walking speed of 1.03 m/s, indicating neutral thermal sensation. Within the speed range of 0.65–1.4 m/s, MTSV remained within ± 0.5, suggesting that thermal comfort was maintained. This model enables the prediction of thermal sensation at different walking speeds under similar conditions.

## 3.3 Variations of Physiological Indices

Previous studies have confirmed strong positive correlations between heart rate and exercise intensity, and between heart rate and metabolic rate. As exercise intensity increases, metabolic demand goes up, leading to elevated heart rates [42]. Therefore, heart rate can indicate both exercise intensity and metabolic activity. Similarly, auditory canal temperature is positively related to metabolism. Throughout the experiment, heart rate and auditory canal temperature were recorded continuously.

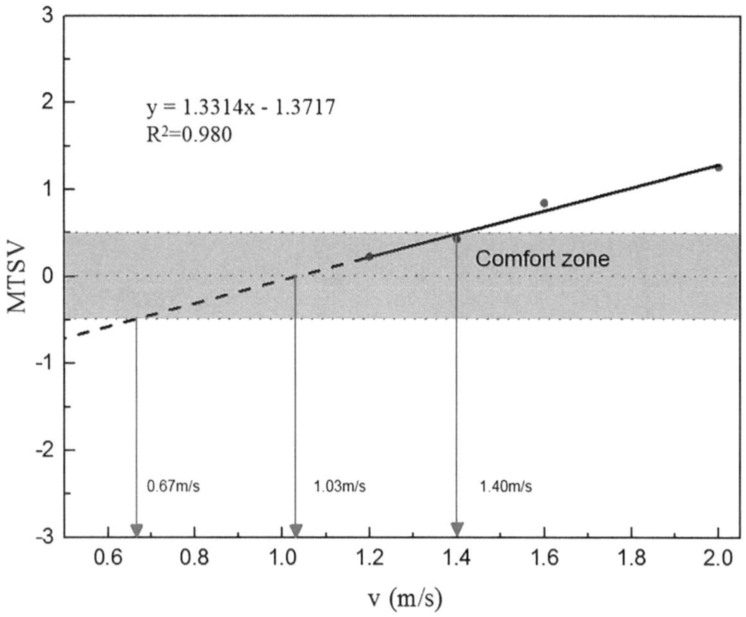


**Fig. 8** The relationship between v and MTSV in exercising period

Measured heart rates ranged from 60 to 110 bpm, and auditory canal temperatures varied from 36.2 to 37.0 °C. Typically, heart rates stayed below 100 bpm, and temperature fluctuations were within 1 °C [45, 46]. Relative changes in both variables were analyzed based on average values.

Figure 9 demonstrates average auditory canal temperature changes. In Fig. 9a, $\Delta T = T(time(i)) - T(time(i = 0))$; in Fig. 9b, $\Delta T = T(time(i)) - T(time(i - 1))$. Figure 9a reveals a declining trend in average auditory canal temperature for all walking speeds—faster walking led to a quicker decrease. Although many laboratory exercise studies under steady conditions show increased body temperature [47–49], this walking study differed due to enhanced airflow disturbance. When walking begins, body heat production does not reach the skin surface immediately, causing a delay. However, airflow disturbance starts right away, augmenting convective heat loss from the skin and initially reducing average body temperature. Then, as heat production propagates, temperature rises. Figure 9b indicates that temperature differences were kept within ± 0.075 °C for all speeds, demonstrating relative thermal stability. During rest, temperatures recovered quickly—within 10 min, ΔT (relative to initial values) was below 0.125 °C. Participants who walked faster experienced faster recovery of auditory canal temperature.

A regression model was developed for the maximum decrease in auditory canal temperature ($\Delta T_{max}$) and walking speed, as shown in Fig. 10. A strong linear relationship was observed ($R^2 \approx 0.84$), enabling the estimation of the average maximum temperature decrease at a given speed.

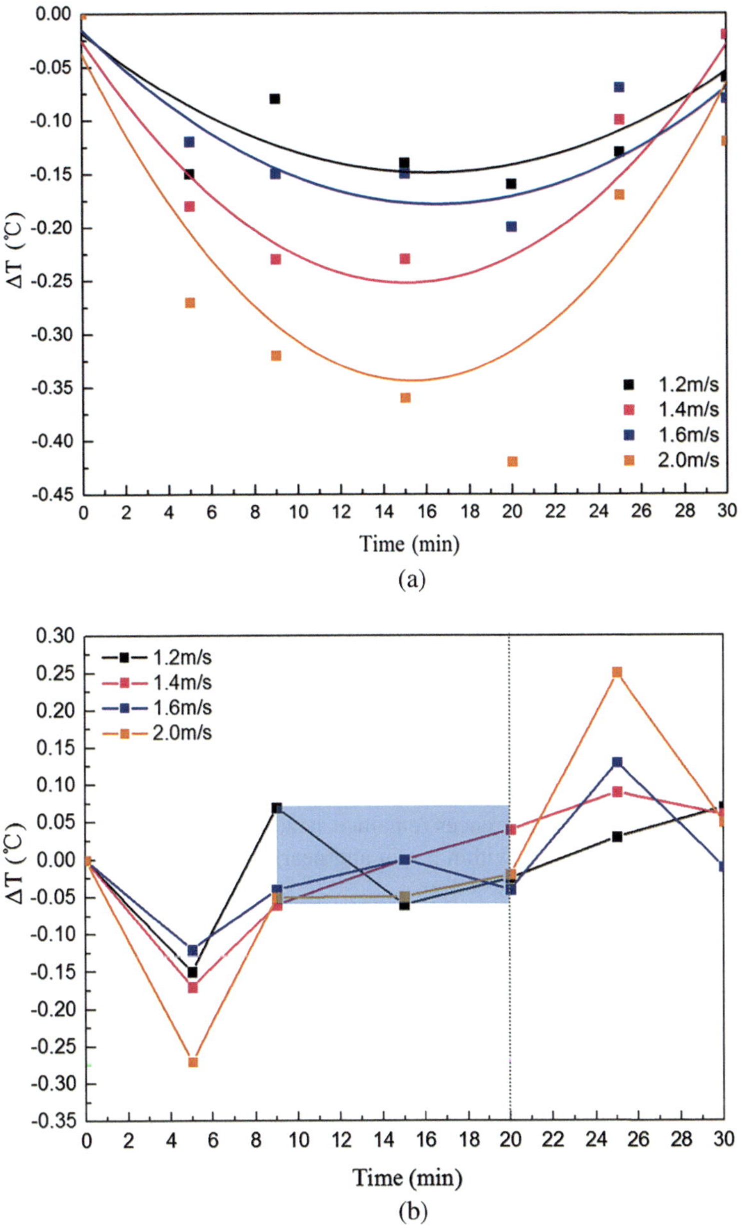


**Fig. 9** The average changes of body temperature: **a** compare with the initial value, **b** compare with the last value

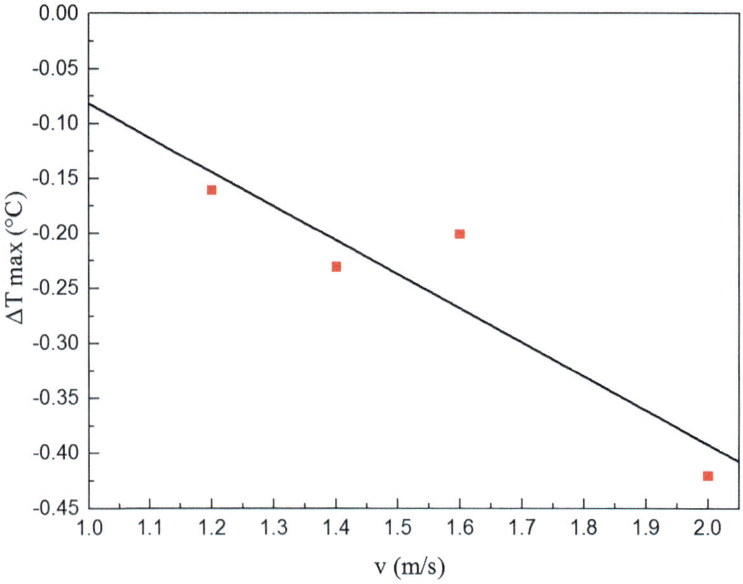


**Fig. 10** The relationship between $\Delta T_{max}$ and v

Figure 11 presents average heart rate variations. In Fig. 11a, ΔHeart rate = Heart rate(time(i)) – Heart rate(time(i = 0)); in Fig. 11b, ΔHeart rate = Heart rate(time(i)) – Heart rate(time(i – 1)). Figure 11a indicates a rapid heart rate increase within the first minute of exercise, followed by a slower rise. From Fig. 11b, it is clear that between 3 and 20 min, heart rate differences remained steady within ± 10 bpm. During rest, heart rates dropped rapidly within 5 min and nearly returned to pre-exercise levels after 10 min, with ΔHR below 7 bpm.

A regression equation was established for the maximum heart rate change ($\Delta HR_{max}$) and walking speed, as shown in Fig. 12. The high $R^2$ (0.92) confirms a strong linear relationship, assisting in predicting the average maximum heart rate change.

Using Eqs. (1–3), the average metabolic rate (M) was computed for each walking speed. Figure 13 demonstrates how the mean metabolic rate varied over time. ASHRAE Standard 55-2017 gives estimated metabolic rates of 2.6 met for walking at 1.2 m/s and 3.8 met at 1.8 m/s. By interpolation, the values for 1.4 m/s, 1.6 m/s, and 2.0 m/s are 3.0 met, 3.4 met, and 4.0 met, respectively. Figure 13 indicates that actual metabolic rates were below these estimates in the first 5 min of walking but aligned with them by the fifth minute.

Figure 14 shows changes in heat storage (S) over time. During slow walking, S remained positive throughout. During fast walking, however, S first became negative before turning positive after 1 min and then increasing.

Figures 9, 10, 11, 12 and 13 reveal that auditory canal temperature dropped rapidly in the first 5 min of walking, while metabolic rate increased. This suggests that the

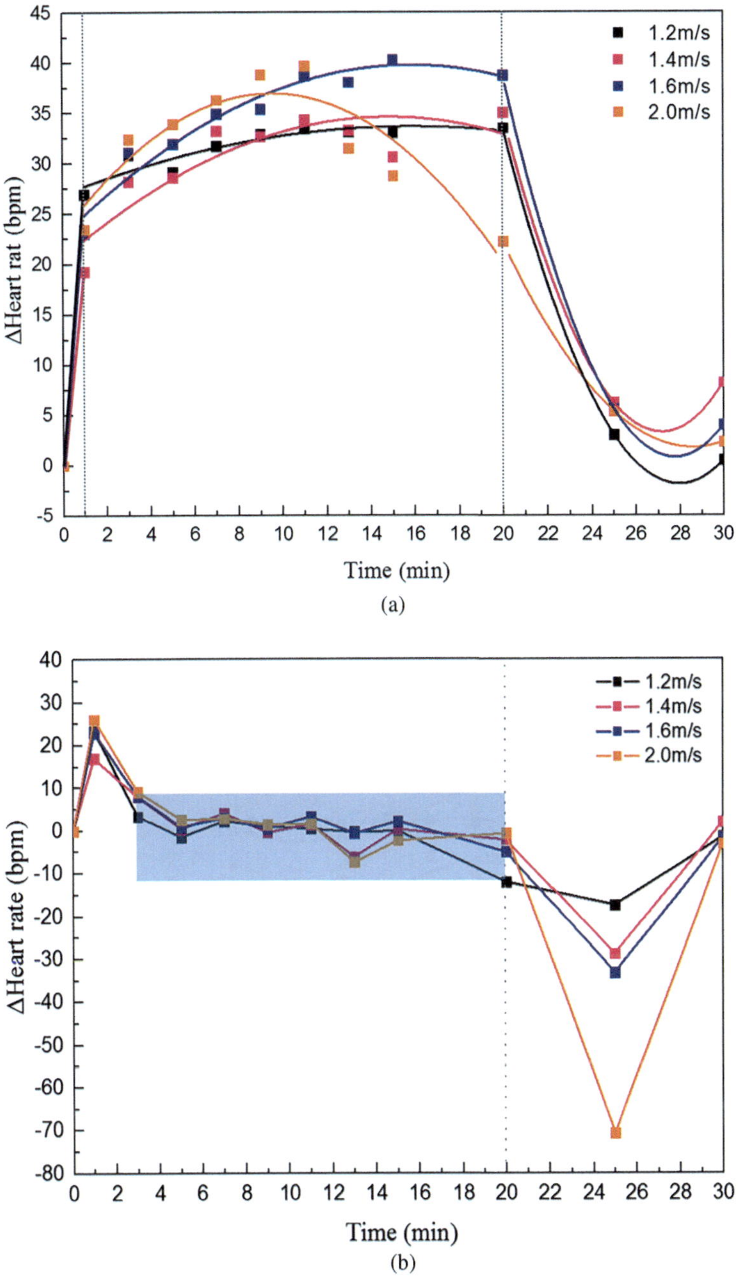


**Fig. 11** The average changes of heart rate: **a** compare with the initial value, **b** compare with the last value

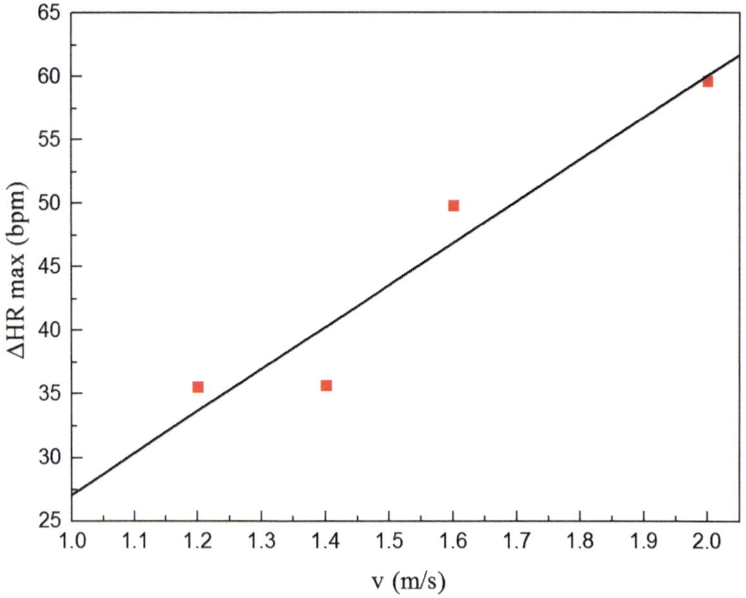


**Fig. 12** The relationship between $\Delta HR_{max}$ and v

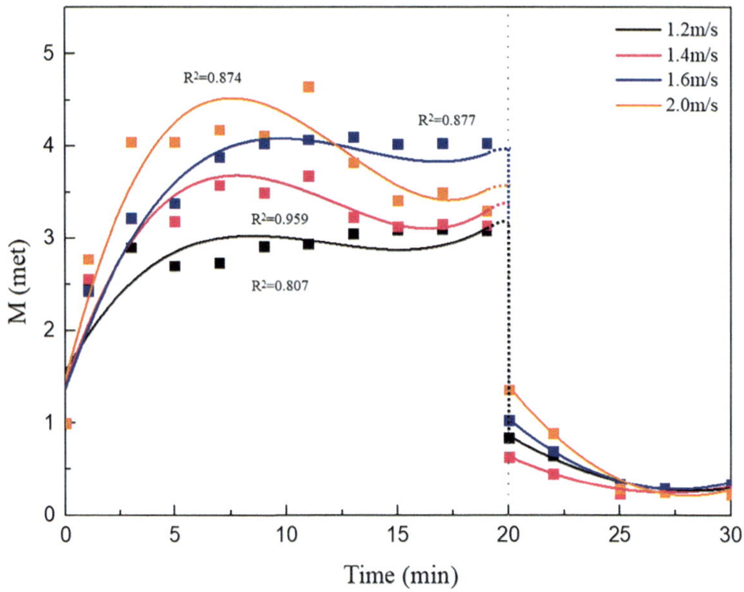


**Fig. 13** The variation of average metabolic rate with time

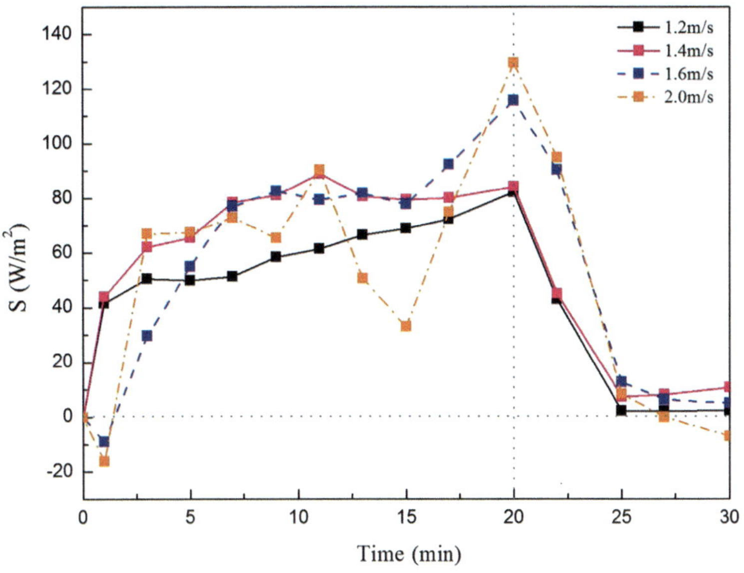


**Fig. 14** The variation of heat storage with time

body's heat production did not immediately compensate for the increased convective heat loss due to air movement.

From 5 to 20 min, except at 2.0 m/s, metabolic rates fluctuated slightly around a stable level—heat production offset skin heat loss, maintaining thermal balance through autonomic regulation. At 2.0 m/s, metabolic rate was steady from 5 to 11 min but then declined due to increased sweat evaporation caused by wind-induced airflow and heat loss. Once heat production equaled dissipation, the body automatically reduced metabolism. During rest, metabolic rate recovered to about 0.25 met, lower than the typical 1 met when sitting, suggesting that heart rate-based calculations slightly underestimated the true metabolic rate.

Combining heat storage, MTSV, and MMSV analysis, it was found that during slow walking, metabolic heat production exceeded dissipation. During fast walking, initial airflow disturbance enhanced skin heat loss, leading to negative S values and a brief slight cold sensation. Afterwards, the body increased heat production to balance dissipation, turning S positive. After 11 min, S decreased rapidly due to increased sweat evaporation, while metabolic rate remained stable, reducing heat storage. During this period, MMSV changed little, but MTSV rose steadily, indicating that airflow disturbance compensated for thermal sensation. By 15 min, sweat had largely evaporated, and S began to rise again.

A regression model was fitted to analyze the relationship between S and M, as shown in Fig. 15. For all speeds except 2.0 m/s, S increased with $|\Delta T|$. During rest, S decreased as $|\Delta T|$ diminished. At 2.0 m/s, the relationship between S and $\Delta T$ during exercise included two stages: initially similar to other speeds (represented by a quadratic polynomial), and when ΔT was below − 0.37, a linear relationship occurred.

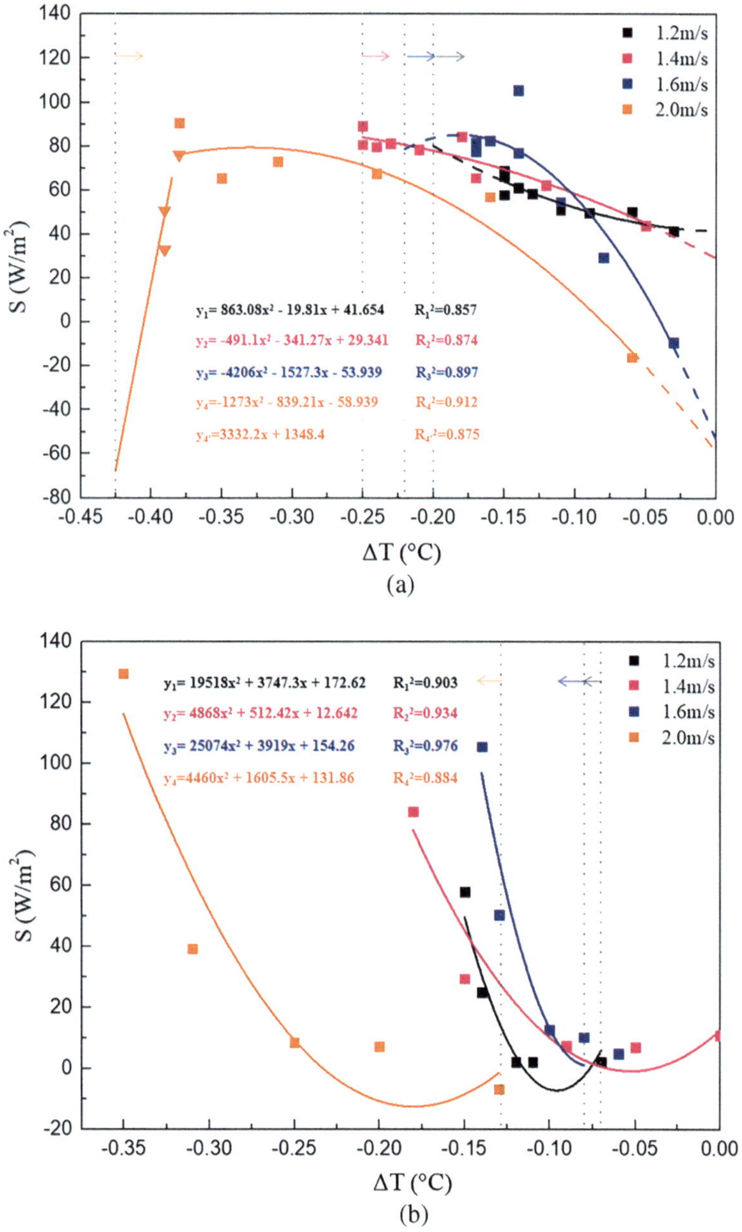


**Fig. 15** The relationship between ΔT and S: **a** exercising period, **b** rest period

Overall, a quadratic polynomial accurately captured the correlation between ΔT and S. Thus, changes in auditory canal temperature were not synchronized with heat storage variations—heat accumulation during exercise manifested as a temperature increase during rest.

The results also revealed a correlation between MTSV and S, as shown in Fig. 16. During exercise, a cubic polynomial accurately described this relationship. For slow walking, S stabilized when MTSV was between 0 and 0.8; for fast walking, stabilization occurred between 0.5 and 1.5. Outside these ranges, S increased rapidly with MTSV. At 2.0 m/s, limited data and high variability prevented trend fitting when MTSV was between 2 and 3. During rest, S decreased with MTSV, following a quadratic polynomial.

# 4 Discussion

## *4.1 Effect of Body Heat Balance Regulation Mechanism on Thermal Comfort*

In this work, key physiological markers linked to thermal regulation—namely, auditory canal temperature and heart rate—were evaluated and interpreted. The metabolic rate was estimated employing a heart rate-based approach [18]. These results were then contrasted with MTSV and MMSV to gauge thermal comfort at multiple walking speeds.

Findings revealed noticeable differences in the behavior of these variables within the thermal equilibrium system. Data examination showed that heart rate adapted quickly once movement began. In comparison, auditory canal temperature reacted more slowly, becoming measurable only after sustained activity. Earlier research supports the use of heart rate as a valid measure of exercise intensity [50], in agreement with these observations. Thus, it can be reasoned that heart rate variations relate more immediately to changes in metabolic rate than do auditory canal temperature shifts.

Additionally, the study uncovered a timing gap between subjective thermal comfort reports and the body's internal thermal regulation. Subjective perception commonly lagged behind physiological shifts. Even as bodily responses engaged quickly, conscious perception did not change at once. Similarly, after exercise ended, metabolic effects on thermal sensitivity lingered briefly. Miao et al. [51] documented rises in both local skin temperatures and subjective thermal sensation with prolonged exertion. Research by Cheung [52] into the tie between thermal sensation and physiological response indicated that controlled stress during multi-joint activities might ease discomfort, and that cognitive approaches could alleviate thermal stress and support sustained effort. Gagge [53] presented a two-node model distinguishing skin temperature from thermal sensation, introducing the ideas of a "fast" cold response and a "slower" warm response. Current evidence supports these concepts.

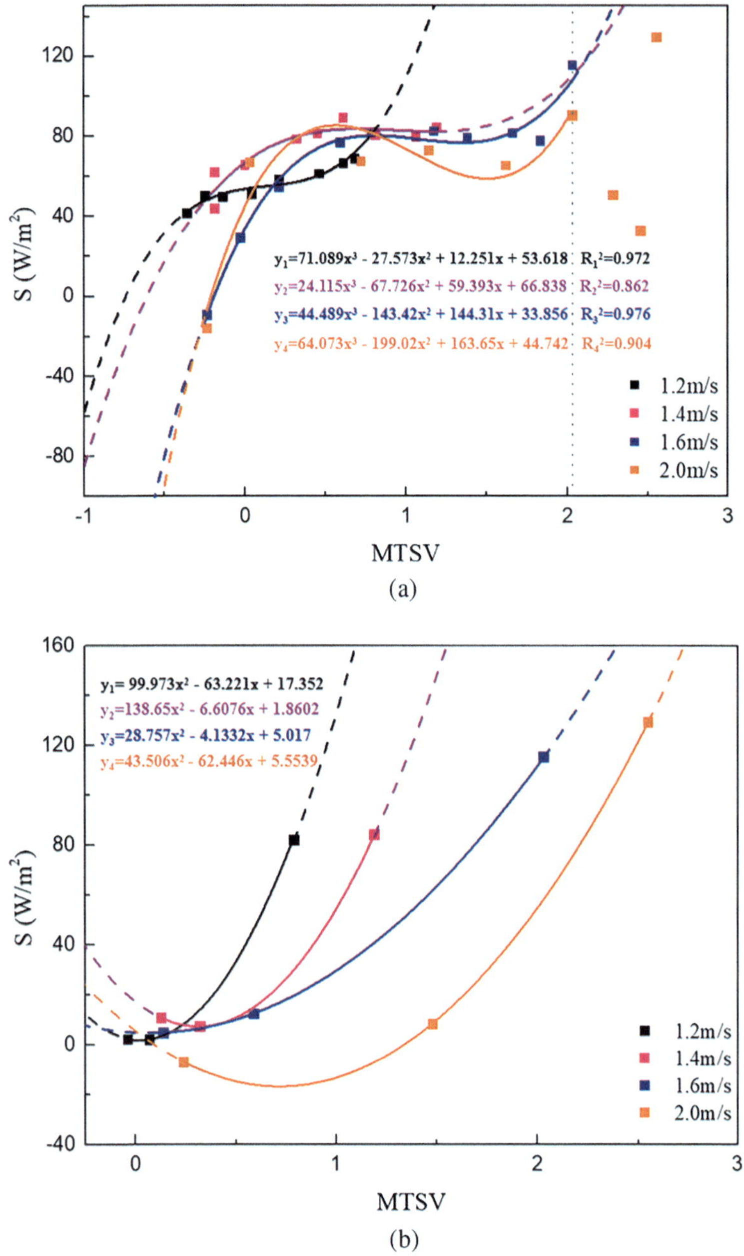


**Fig. 16** The relationship between MTSV and S: **a** exercising period, **b** rest period

Ji et al. [19] studied metabolic rate and thermal comfort in bicycle exercises. They noted that reaching metabolic stability took 5–6 min after exercise onset, and reverting to rest conditions required 7–9 min after finishing. In the present walking-based study, metabolic equilibrium was achieved within 3–5 min, and recovery concluded within 4–5 min post-walking. This indicates that walking permits faster metabolic adjustment and recovery than cycling. Thus, indoor environment standards should consider not only a person's activity before entering but also their movement while inside.

Regression analysis identified relationships among maximum auditory canal temperature variation, greatest heart rate change, and walking speed. Hence, walking speed enables estimation of extreme physiological responses. Examining ΔT and S showed a direct link underscoring the lag in thermal reaction. Similarly, relating MTSV to S demonstrated that changes in S influence thermal perception; even with stable S during activity, thermal feeling kept rising. This emphasizes the lasting impact of stored heat on sensory experience, where prior movement affects current perception. Therefore, interior climate design should account for the continuing influence of walking, beyond fixed external conditions.

## 4.2 *Adaptability of PMV*

MTSV was collected and PMV calculated in this investigation. Figure 17 shows differences between MTSV and PMV across test conditions. Some misalignment between TSV and PMV predictions was found. Figure 17 indicates that PMV over-estimated thermal sensation during walking at 1.2, 1.4, and 1.6 m/s. During seated preparation, however, PMV and TSV values corresponded more closely, supporting PMV's use in steady rather than dynamic settings [54].

The PMV model is designed to predict thermal feeling in stable conditions after acclimation [55, 56]. It combines six factors: activity level, clothing insulation, air temperature, mean radiant temperature, humidity, and air velocity. Yet the model uses steady-state inputs and often simplifies—e.g., averaged metabolic rates—instead of dynamic data. Since TSV reflects real-time sensation during movement like walking, mismatch with PMV is expected. Results show PMV often overestimated votes, except at 2.0 m/s, where closer alignment occurred, likely due to compensatory air movement.

## 4.3 *Effect of Disturbance Air Velocity on Thermal Comfort*

Analogous to auditory sensitivity to certain sound frequencies, the body is also tuned to specific airflow traits. Several reports [45, 46] have studied how airflow velocity and spectral traits affect thermal sensation in seated people. Luo et al. [57] found thermal comfort could be preserved under warm conditions (28 and 30 °C) within

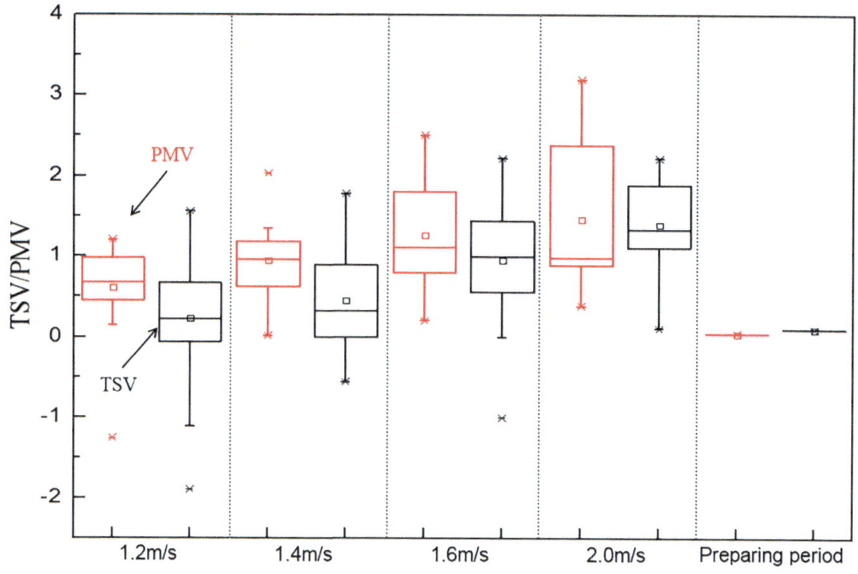


**Fig. 17** The comparison between TSV and PMV of five conditions

optimal airflow intervals of 0.4–0.6 m/s and 0.7–0.9 m/s. Xia et al. [58] observed that stronger airflow pulsation increased draft sensation risk and created a model predicting draft under isothermal conditions. Still, few studies have investigated airflow impacts on thermal comfort during walking.

Ouyang et al. [59] detailed spectral differences between natural and steady airflow, reporting that the β value (average negative slope of log power spectrum curves) for natural wind usually lies between 1.4 and 1.7. As Fig. 4 illustrates, natural wind herein had a β of 1.55, within this span. As walking speed increased, β decreased. In Fig. 8, MTSV stayed within comfort during slower walking, with β values of 1.15 and 1.22—closer to natural wind than faster walking. Zhou et al. [60] proved in climate chamber trials that seated subjects preferred simulated natural wind (with β near natural) over mechanical airflow. Our results support this.

Furthermore, outcomes showed that higher walking speeds increased airflow disruption and energy exchange with the body, boosting surface heat loss. Beyond certain speed and duration thresholds, airflow offset internal heat production. This compensation was clear at 2.0 m/s. Clearly, activity affects thermal comfort, and walking-induced airflow alters heat balance differently than in static cases. Thus, indoor thermal planning must fully include human motion states.

## 4.4 Limitation of this Study

Some limitations should be noted. Figure 11a shows a slight drop in average heart rate after 10 min at 2 m/s. Although heart rate declines during exertion are rare in literature, studies by Buchheit et al. [61] and Boullosa et al. [62] used higher-intensity running than 2 m/s walking, perhaps explaining the discrepancy. Also, tests by Parak et al. [63] were indoor without significant airflow disturbance from gait. We suggest the brief heart rate dip may stem from autonomic control: as sweat evaporation increases surface cooling, the body reaches thermal balance and triggers metabolic reduction, including cardiovascular adjustment, lowering heart rate [64]. Beyond instrument error, this deserves further study.

# 5 Conclusions

This chapter focused on measuring and analyzing disturbed wind speed generated by walking and its implications for thermal comfort evaluation. The main conclusions are:

Walking at different speeds produced distinct airflow disturbances, with induced wind speeds ranging from 0.75 m/s during slow walking to over 1.1 m/s at 2.0 m/s, far exceeding the background natural wind level of 0.13 m/s.

The spectral distribution of disturbed airflow showed steeper slopes than natural wind, confirming that walking-generated turbulence differs both in intensity and frequency characteristics.

These airflow disturbances significantly altered convective and evaporative heat transfer, modifying body heat storage and thermal sensation in ways not captured by steady-state models.

Compared with natural wind, disturbed wind caused by walking provided a compensatory cooling effect at moderate speeds but introduced discomfort risks at higher speeds due to increased evaporation and draft perception.

Overall, the findings demonstrate that disturbed wind speed induced by human movement must be distinguished from natural wind when evaluating outdoor thermal comfort. Incorporating these differences into predictive models is essential for accurately assessing comfort in dynamic, pedestrian-oriented urban environments.

**Acknowledgements** This work was supported by the National Natural Science Foundation of China (Project No. 51978180) and the Opening Funds of State Key Laboratory of Building Safety and Built Environment and National Engineering Research Center of Building Technology (Project No. BSBE2018-03). The authors thank all students who participated in the surveys.

## References


1. F. Golasi, E.L. Salata, M.C. Vollaro, Complying with the demand of standardization in outdoor thermal comfort: a first approach to the global outdoor comfort index (GOCI). Build. Environ. **130**, 104–119 (2018)
2. E. García-López, C. Heard, A study of the social acceptability of a proposal to improve the thermal comfort of a traditional dwelling. Appl. Therm. Eng. **75**(22), 1287–1295 (2015)
3. P.O. Fanger, *Thermal Comfort: Analysis and Application in Environmental Engineering* (Danish Technology Press, Copenhagen, 1970)
4. M.A. Humphreys, J.F. Nicol, The validity of ISO-PMV for predicting comfort votes in every-day thermal environments. Energy Build. **34**(6), 667–684 (2002)
5. G.S. Brager, R.J. de Dear, Thermal adaptation in the built environment: a literature review. Heat. Vent. Air Cond. **27**(1), 83–96 (2011)
6. R.J. de Dear, Developing an adaptive model of thermal comfort and preference. ASHRAE Trans. **104**(1), 73–81 (1998)
7. Z. Fang, S. Zhang, Y. Cheng, A.M.L. Fong, M.O. Oladokun, Z. Lin, H. Wu, Field study on adaptive thermal comfort in typical air conditioned classrooms. Build. Environ. **133**, 73–82 (2018)
8. Z. Fang, X. Feng, J. Liu, Z. Lin, C.M. Mak, J. Niu, K.T. Tse, X. Xu, Investigation into the differences among several outdoor thermal comfort indices against field survey in subtropics. Sustain. Cities Soc. **44**, 676–690 (2019)
9. F. Formenti, D. Constantinteodosiu, Y. Emmanuel, et al., Regulation of human metabolism by hypoxia-inducible factor. Proc. Natl. Acad. Sci. USA **107**, 12722–12727 (2010)
10. ISO, EN ISO 8996, *Ergonomics of the Thermal Environment—Determination of Metabolic Rate, International organization for Standardization*, Geneva, 2004
11. Z. Fang, Z. Lin, C.M. Mak, J. Niu, K.T. Tse, Investigation into sensitivities of factors in outdoor thermal comfort indices. Build. Environ. **128**, 129–142 (2018)
12. S. Haddad, P. Osmond, S. King, Metabolic rate estimation in the calculation of PMV for children. in *Cutting Edge International Conference of the Architectural Science Association*, 2013
13. G. Ma, X. Ding, Impact of metabolic rate on human thermal comfort. Constr. Energy Conserv. **7**, 37–38 (2009)
14. M.H. Hasan, F. Alsaleem, M. Rafaie, Sensitivity study for the PMV thermal comfort model and the use of wearable devices biometric data for metabolic rate estimation. Build. Environ. **110**, 173–183 (2016)
15. G. Havenith, I. Holmér, K. Parsons, Personal factors in thermal comfort assessment: Clothing properties and metabolic heat production. Energ. Build. **34**(6), 581–591 (2002)
16. T.F. Colton, E. Larsen, Exercise physiology: the response of metabolic rate to physical activity, in: *Proceedings of the 23rd Workshop/Conference of the Association for Biology Laboratory Education (ABLE)* **23**, 235–254 (2002)
17. J.D. Schnorr, G. Dörner, F. Stahl, et al., Human heat balance during post exercise recovery: separating metabolic and nonthermal effects. Am. J. Phys. **294**(2), 1586–1592 (2008)
18. K. Fabbri, *Indoor Thermal Comfort Perception: A Questionnaire Approach Focusing on Children* (Springer, New York City, 2015)
19. W. Ji, M. Luo, B. Cao, Y. Zhu, Y. Geng, B. Lin, A new method to study human metabolic rate changes and thermal comfort in physical exercise by $CO_2$ measurement in an airtight chamber. Energy Build. **177**, 402–412 (2018)
20. K. George, H.A. Flocas, M. Andreas, Dynamic modeling of human thermal comfort after the transition from an indoor to an outdoor hot environment. Int. J. Biometeorol. **59**, 205–216 (2015)
21. Z. Fang, X. Feng, X. Xu, X. Zhou, Z. Lin, Y. Ji, Investigation into outdoor thermal comfort conditions by different seasonal field surveys in China, Guangzhou. Int. J. Biometeorol. **7**, 1–12 (2019)

22. W. Liu, D. Yang, X. Sheng, P. Yang, Indoor clothing insulation and thermal history: a clothing model based on logistic function and running mean outdoor temperature. Build. Environ. **135**, 142–152 (2018)
23. W. Liu, W. Zhong, P. Wargocki, Performance, acute health symptoms and physiological responses during exposure to high air temperature and carbon dioxide concentration. Build. Environ. **114**, 96–105 (2017)
24. D. Yang, J. Xiong, W. Liu, Adjustments of the adaptive thermal comfort model based on the running mean outdoor temperature for Chinese people: a case study in Changsha China. Build. Environ. **114**, 357–365 (2017)
25. D. Lai, C. Chen, W. Liu, Y. Shi, C. Chen, An ordered probability model for predicting outdoor thermal comfort. Energ. Build. **168**, 261–271 (2018)
26. D. Lai, W. Liu, T. Gan, K. Liu, Q. Chen, A review of mitigating strategies to improve the thermal environment and thermal comfort in urban outdoor spaces. Sci. Total Environ. **661**(15), 337–353 (2019)
27. D. Lai, C. Chen, Comparison of the linear regression, multinomial logit, and ordered probability models for predicting the distribution of thermal sensation. Energy Build. **188**(2019b), 269–277 (2019b)
28. A. Zhang, Y. Sun, The integral regulation of the overall view and the steady state of the human body. J. Shaanxi Chinese Med. Coll. **26**(2), 16–17 (2003)
29. Y. Du, C.M. Mak, T. Huang, J. Niu, Towards an integrated method to assess effects of lift-up design on outdoor thermal comfort in Hong Kong. Build. Environ. **125**, 261–272 (2017)
30. D. Lai, C. Chen, W. Liu, et al., An ordered probability model for predicting outdoor thermal comfort. Energy Build. **168**, 261–271 (2018)
31. Y. Du, C.M. Mak, Y. Li, A multi-stage optimization of pedestrian level wind environment and thermal comfort with lift-up design in ideal urban canyons. Sustain. Cities Soc. **46**, 101424 (2019)
32. A. Chatzidimitriou, S. Yannas, Street canyon design and improvement potential for urban open spaces; the influence of canyon aspect ratio and orientation on microclimate and outdoor comfort. Sustain. Cities Soc. **33**, 85–101 (2017)
33. M. Xu, B. Hong, R. Jiang, L. An, T. Zhang, Outdoor thermal comfort of shaded spaces in an urban park in the cold region of China. Build. Environ. **155**, 408–420 (2019)
34. Y. Xie, T. Huang, J. Li, J. Liu, C.M. Mak, Z. Lin, Evaluation of a multi-nodal thermal regulation model for assessment of outdoor thermal comfort: sensitivity to wind speed and solar radiation. Build. Environ. **132**, 45–56 (2018)
35. ISO 7726, *Ergonomics of the thermal environment – Instruments for measuring physical quantities*, 1998
36. R.A. Engineers, ASHRAE Handbook: Fundamentals, 2017
37. J.E. Greenleaf, B.L. Castle, External auditory canal temperature as an estimate of core temperature. J. Appl. Physiol. **32**(2), 194–198 (1972)
38. J.G. Modell, C.R. Katholi, S.M. Kumaramangalam, E.C. Hudson, D. Graham, Unreliability of the infrared tympanic thermometer in clinical practice: a comparative study with oral mercury and oral electronic thermometers. South. Med. J. **91**(7), 649–654 (1998)
39. ANSI/ASHRAE Standard 55-2017, *Thermal Environmental Conditions for Human Occupancy American Society of Heating Refrigerating and air-Conditioning Engineers (ASHRAE)*, Atlanta, Georgia, 2017
40. ISO, International Standard 7730, *Ergonomics of the Thermal Environment – Analytical Determination and Interpretation of Thermal Comfort Using Calculation of the PMV and PPD Indices and Local Thermal Comfort Criteria*, International Standard Organization, Geneva, 2005
41. Y. Nishi, Measurement of thermal balance of man. Stud. Environ. Sci. **10**, 29–39 (1981)
42. W. Li, Y. Zhang, L. Zhang, et al., The research of the sport burden grade standard corresponding to the heart rate. Nat. Sci. J. Harbin Normal Univ. **24**(5), 99–102 (2008)
43. T. Sakoi, T. Mochida, Y. Kurazumi, K. Kuwabara, Y. Horiba, S. Sawada, Heat balance model for a human body in the form of wet bulb globe temperature indices. J. Therm. Biol. **71**, 1–9 (2018)

44. H. Wang, S. Hu, Experimental study on comfortable skin temperature under sweating condition. Heat. Ventilat. Air Condition. **5**, 110–113 (2014)
45. Z. Fang, H. Liu, B. Li, X. Du, A. Baldwin, Investigation of the effects of temperature for supplied air from a personal nozzle system on thermal comfort of air travelers. Build. Environ. **12**, 82–97 (2018)
46. W.L. Kenney, J. Wilmore, D. Costill, *Physiology of Sport and Exercise*, 6th edn. (Human Kinetics, Champaign, IL, 2015)
47. C. Byrne, C.L. Lim, The ingestible telemetric body core temperature sensor: a review of validity and exercise applications. Br. J. Sports Med. **41**(3), 126–133 (2007)
48. C.J. Ekblom, J.E. Greenleaf, L.H. Greenleaf, Temperature regulation during exercise dehydration in man. Acta Physiol. Scand. **79**(4), 475–483 (1970)
49. B. Saltin, L. Hermansen, Esophageal, rectal, and muscle temperature during exercise. J. Appl. Physiol. **21**(6), 1757–1762 (1966)
50. Y. Zhu, X. Ruan, Z. Xu, X. Tian, E. Zheng, Research for the important relationship between activity measurement and heart rate. Med. Equip. **06**, 5–7 (2003)
51. M. Miao, H. Lu, M. Cheng, Body surface temperature change and subjective thermal sensation assessment before and after exercise. J. Text. Res. **4**, 39–45 (2018)
52. S.S. Cheung, Interconnections between thermal perception and exercise capacity in the heat. Scand. J. Med. Sci. Sports **20**, 53–59 (2010)
53. A.P. Gagge, Rational temperature indices of man's thermal environment and their use with a 2-node model of his temperature regulation. Fed. Proc. **32**(5) (1973)
54. P.O. Fanger, J. Toftum, Extension of the PMV model to non-air-conditioned buildings in warm climates. Energy Build. **34**(6), 533–536 (2002)
55. B.W. Jones, Capabilities and limitations of thermal models for use in thermal comfort standards. Energy Build. **34**(6), 653–659 (2002)
56. R. Maiti, PMV model is insufficient to capture subjective thermal response from Indians. Int. J. Ind. Ergon. **44**(3), 349–361 (2014)
57. M. Luo, J. Yu, Y. Yang, O. Qin, Y. Zhu, Application of dynamic airflow on split air conditioning unit and its impacts on human thermal comfort: isothermal condition. Heat. Ventilat. Air Condition. **5** (2014)
58. Y. Xia, J. Niu, R. Zhao, J. Burnett, Effects of turbulent air on human thermal sensations in a warm isothermal environment. Indoor Air (Print) **10**(4) (2000)
59. Q. Ouyang, W. Dai, H. Li, Study on dynamic characteristics of natural and mechanical wind in built environment using spectral analysis. Build. Environ. **41**, 418–426 (2006)
60. X. Zhou, Q. Ouyang, G. Lin, et al., Impact of dynamic airflow on human thermal response. Indoor Air **16**(5), 348–355 (2006)
61. M. Buchheit, R. Solano, G.P. Millet, Heart-rate deflection point and the second heart-rate variability threshold during running exercise in trained boys. Pediatr. Exerc. Sci. **19**(2), 192–204 (2007)
62. D.A. Boullosa, E.S. Barros, S. Del Rosso, F.Y. Nakamura, A.SLeicht., Reliability of heart rate measures during walking before and after running maximal efforts. Int. J. Sports Med. **35**(12), 999–1005 (2014)
63. J. Parak, I. Korhonen, Evaluation of wearable consumer heart rate monitors based on photopletysmography, in *36th Annual International Conference of the IEEE Engineering in Medicine and Biology Society* (2014, pp. 3670–3673)
64. W.I. Cranston, Temperature regulation. Surgery **24**(12), 446–451 (1941)

# Ground Temperature and Ground View Factor in Outdoor Thermal Environment Evaluations

**Abstract** Mean radiant temperature (MRT) is strongly influenced by ground radiation, which depends on both the ground view factor (Fg-p) and ground temperature (Tg). However, their roles are often oversimplified and rarely examined in detail. This chapter systematically evaluates the combined effects of Fg-p and Tg on MRT through field microclimate measurements in Shanghai and Monte Carlo ray-tracing simulations. The analysis shows that variations in human height (70–230 cm) exert negligible influence on MRT when the distance from the radiant ground boundary exceeds 40 m, and that differences in ground area become insignificant when this distance is no less than 70 m. Under uniform heating conditions, reductions in Tg are directly reflected in decreased MRT, while in non-uniform thermal environments, the extent of cooler surfaces substantially affects MRT estimation. Using averaged non-uniform Tg values may lead to overprediction of MRT by more than 2 °C. These findings highlight the importance of accurately incorporating both ground temperature and ground view factor into outdoor thermal comfort evaluations to avoid systematic errors and to improve the reliability of predictive models.

## 1 Introduction

Against the backdrop of global climate change and rapid urbanization, profound changes in exterior thermal conditions have considerably influenced daily human life, health, and overall quality of living [1–3]. When outdoor thermal environments are pleasant, individuals are more inclined to engage in leisure, recreational, and occupational activities outside [4, 5]. Involvement in outdoor pursuits not only improves physical and mental health but also aids in decreasing energy usage in buildings and mitigating carbon emissions [6, 7]. Nonetheless, extremely hot conditions frequently deter outdoor engagements, particularly within urban areas in the summer [8, 9].

The exterior thermal setting is mainly shaped by air temperature, wind velocity, relative humidity, and thermal radiation [10–12]. In summer, thermal radiation often emerges as the principal element affecting human thermal comfort outdoors, typically quantified as the mean radiant temperature (MRT, °C) [13, 14]. MRT is described as



Z. Fang et al., *Outdoor Thermal Comfort and Heat Stress Mitigation*, Environmental Science and Engineering, https://doi.org/10.1007/978-981-95-8899-2_3

the uniform temperature of a hypothetical black enclosure that would result in the same net radiant heat transfer with an individual as the real non-uniform surroundings [15]. Estimating MRT in open-air contexts is notably intricate and affected by numerous parameters [16–21].

An essential aspect influencing MRT is radiation from the ground ($R_g$, W/m$^2$) [22, 23], which is especially relevant since the ground makes up a substantial portion of the radiant surroundings around a person. $R_g$ encompasses both reflected and emitted radiant energy, affected by incoming solar radiation and characteristics of the ground surface like absorptivity, emissivity, and reflectivity [24–27]. The impact of $R_g$ on MRT relies on the view factor from the ground to the person ($F_{g-p}$) and its associated radiant intensity.

As depicted in Fig. 1, the view factor $F_{i-j}$ indicates the proportion of radiation exiting surface i that is received by surface j [28], computed via Eq. (1). This factor can be established through experimental approaches such as photographic methods [29, 30] or numerical modeling [31–34]. With progress in technology, various computational techniques for radiative heat transfer have been devised, including Monte Carlo ray tracing [35, 36], the discrete transfer method [37, 38], the discrete ordinates method [39, 40], and the finite volume method [41, 42]. Of these, the Monte Carlo ray-tracing approach is generally considered the most precise and is commonly used to determine view factors, especially for setups including large to small surfaces [33, 43–46].

$$F_{i-j} = \frac{1}{A_i} \int_{A_i} \int_{A_j} \frac{\cos\theta_i \cos\theta_j}{\pi R^2} dA_j dA_i \tag{1}$$

According to Eq. (1), a human body in a standing position can receive radiation from the ground across a long distance (as shown in Fig. 1, R). In accordance with the reversibility of the radiant ray, $F_{g-p}$ equals the fraction of the rays emitted by the

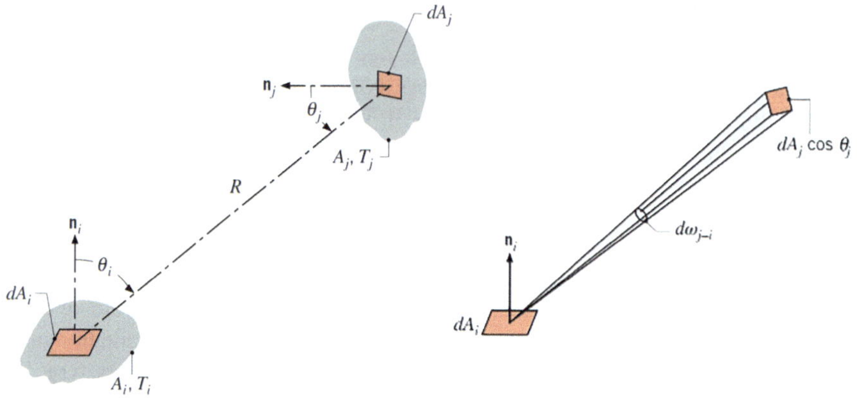


**Fig. 1** View factor between elemental surfaces of the areas $dA_i$ and $dA_j$ (Bergman et al. 2011)

human body and absorbed by the ground, as shown in Fig. 2. Because the human body resembles a cylinder or elliptical column with a diffuse and gray surface, if the ground surface is sufficiently large, we can assume that approximately half the rays fall to the ground. Therefore, $F_{g-p}$ in a large area is nearly equivalent to 0.5. Numerous numerical software, such as RayMan [47, 48] and ENVI-met [19, 49, 50], have adopted $F_{g-p} = 0.5$ to calculate the radiation ($R_{g-p}$, W/m$^2$) absorbed by the human body from the ground. However, the treatment of $F_{g-p}$ as 0.5 is not appropriate for non-uniform $R_g$. The ground typically consists of several areas with different amounts of radiation. To calculate the radiation from this ground to the human body, the $F_{g-p}$ of various ground areas must be considered, and it changes depending on such factors as human height, ground area size, and ground area position relative to the human body. However, the effect of the $F_{g-p}$ difference caused by these factors on the MRT has rarely been quantified.

Based on Eq. (1), an upright individual can receive terrestrial emitted and reflected energy across extended distances (represented as R in Fig. 1). Owing to the principle of radiative reciprocity, $F_{g-p}$ is identical to the proportion of radiation released by the human form that the ground absorbs, as illustrated in Fig. 2. Since the human body is similar to a cylinder or elliptical shape with a diffuse gray surface, and assuming a vast ground area, it can be roughly assumed that half of the emitted radiation reaches the ground. Hence, $F_{g-p}$ across an extensive region is almost equal to 0.5. Several numerical instruments, such as RayMan [47, 48] and ENVI-met [19, 49, 50], have employed $F_{g-p} = 0.5$ to approximate the radiation ($R_{g-p}$, W/m$^2$) that the human body absorbs from the ground. However, assuming $F_{g-p}$ as 0.5 is not appropriate under non-uniform $R_g$ circumstances. The ground generally comprises several zones with divergent radiative attributes. To precisely calculate radiation transfer to a person, the $F_{g-p}$ for different ground sections must be accounted for, as it fluctuates with human stature, ground segment dimensions, and relative location. Yet, the influence of such $F_{g-p}$ differences on MRT remains inadequately measured.

**Fig. 2** Rays from the human body onto the ground

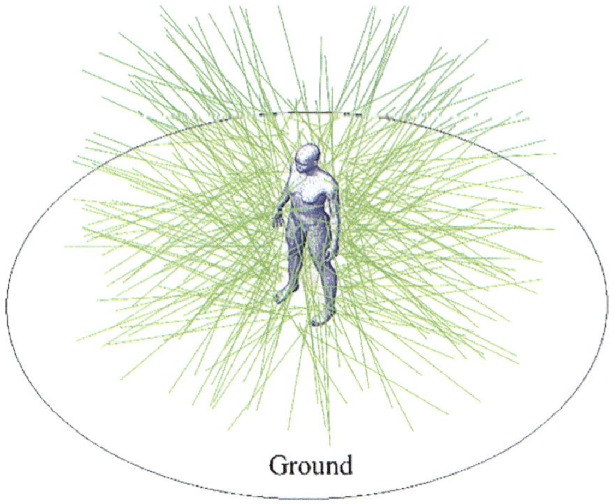

The majority of $R_g$ is emitted by the ground because the reflectivity of most ground surfaces is so small (or their emissivity is so great) that their reflected radiation is negligible. According to the Stefan–Boltzmann law, the emitted radiation is related to the ground temperature ($T_g$, °C). Therefore, decreasing $T_g$ is an effective strategy to reduce $R_g$. Several studies suggest that "cold" ground materials can greatly improve the thermal environment of cities in hot climates [51–55]. Moreover, increasing ground surface albedo [56–60] and its evaporative flux resulting from planting grass or watering [61, 62] can effectively suppress surface heat sensation and promote pavement cooling. In addition to ground heat dissipation, improving the thermal boundary conditions of the ground, such as including shade coverage, is a very efficient method for reducing $T_g$ [63–68]. In a shaded environment, $T_g$ is normally the same as or lower than the air temperature, and it is significantly lower than the temperature of the ground under direct solar radiation [67, 69]. For example, the temperature of shaded ground is 19 °C cooler than that of the ground under direct solar radiation [67]. Although some field studies have explored the influence of $R_g$ on the MRT [58, 70, 71], the effect of $T_g$ on the MRT has rarely been quantified.

The dominant component of $R_g$ stems from radiation emitted by the ground, owing to the quite minimal reflectivity—or conversely, the high emissivity—of most terrestrial surfaces, which renders reflected radiation negligible. In accordance with the Stefan–Boltzmann law, the emitted radiation is a function of ground temperature ($T_g$, , °C). Consequently, reducing $T_g$ presents a viable strategy for decreasing $R_g$. Multiple studies demonstrate that employing "cool" pavement materials can markedly improve thermal conditions in urban areas experiencing hot weather [51, 52, 54, 55, 57]. Additionally, elevating the albedo of ground surfaces [56–60] and boosting evaporative flux through vegetation or irrigation [61, 62] can efficiently mitigate surface overheating and support pavement cooling. Beyond promoting ground heat dissipation, improving the thermal boundary conditions—such as by incorporating shade—also serves as a highly effective measure to lower $T_g$ [63–68]. In shaded environments, $T_g$ generally matches or falls below air temperature and is considerably lower than that of unshaded ground under solar exposure [66, 67]. For example, shaded ground may be up to 19 °C cooler than sunlit surfaces [66]. Although a number of field studies have explored the influence of $R_g$ on MRT [58, 70, 71], the relationship between $T_g$ and MRT has rarely been quantified.

As depicted in Fig. 3, non-uniform $T_g$ often arises from material heterogeneity and environmental diversity [72], a condition particularly evident in partially shaded terrain. To escape strong solar radiation in summer, individuals frequently choose shaded locations, for instance under trees (Fig. 3c) or beneath umbrellas. Because shaded ground tends to be cooler than areas in direct sunlight, the radiation emitted from these patches varies. Moreover, the ground-to-person view factor ($F_{g-p}$) also differs. Therefore, it is necessary to compute $R_{g-p}$ using Eq. (2), in which $F_{gl-p}$ and $F_{gh-p}$ indicate the view factors from the lower and higher temperature ground areas to the human body, respectively, while $R_{gl}$ and $R_{gh}$ refer to radiation (W/m$^2$) from the cooler and warmer zones, respectively.

$$R_{g-p} = F_{gl-p}R_{gl} + F_{gh-p}R_{gh} \tag{2}$$

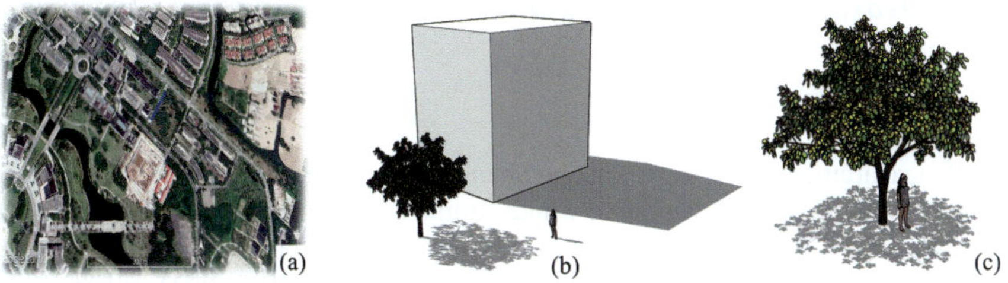


**Fig. 3** Non-uniform ground temperature scenarios

Radiation from terrain exhibiting non-uniform temperature distributions is frequently approximated using a mean temperature value [19, 47–50]. This mean ground temperature ($T_{ga}$, °C) is obtained from Eq. (3), in which n indicates the quantity of distinct ground zones with different temperatures, and $T_{gi}$ (i = 1, 2, ..., n) refers to the temperature of each ground segment (°C). Owing to the differences between $T_{ga}$ and the true $T_{gi}$, inaccuracies arise in the radiation estimates for individual areas. For example, taking the average of temperatures in shaded and sunlit zones overestimates the temperature in shaded regions and underestimates it in sunlit areas. Therefore, the simplification of employing averaged non-uniform ground temperatures introduces inaccuracies in $R_{g-p}$ and, as a result, distorts MRT values.

$$T_{ga} = \frac{T_{g1} + T_{g2} + \cdots + T_{gn}}{n} \tag{3}$$

In light of this, the present study seeks to measure the effect of $R_g$—incorporating both $F_{g-p}$ and $T_g$—on MRT through microclimate monitoring and Monte Carlo ray-tracing simulations. The outcomes are expected to facilitate more precise theoretical and numerical assessments of MRT in open-air environments, especially those with non-uniform ground temperatures, thus supporting better planning of outdoor thermal conditions.

## 2 Methodology

Microclimate data collection was conducted to acquire $T_g$, $R_g$, MRT, and air temperature; these records were used to assess the influence of $F_{g-p}$ and $T_g$ on MRT. The variation of $F_{g-p}$ over different ground zones and for individuals of various heights was simulated via the Monte Carlo ray-tracing method. Following this, the effects of human height and ground surface area on $F_{g-p}$ and MRT were investigated, focusing particularly on how the accuracy of $F_{g-p}$ affects MRT precision. Finally, both uniform and non-uniform changes in $T_g$ and their impacts on MRT were analyzed.

## 2.1 Measurement and Calculation

### 2.1.1 Measurement

Field measurements were performed in Shanghai, China (Fig. 4), a city characterized by hot summers and cold winters, where summer average peak temperatures often surpass 30 °C. Data acquisition occurred at two locations within Tongji University's Jiading campus (31°17′8″ N, 121°12′26″ E; indicated as P1 and P2 in Fig. 4) on October 2, 2021. The time intervals for measurements were 11:30:00 to 12:57:30 at P1 and 13:16:45 to 14:34:00 at P2. Air temperature varied from 31 to 32 °C and was recorded at 15-s intervals with a HOBO U12 Temperature Data Logger. A six-direction integrated radiometer—equipped with six longwave radiometers (pyrgeometers) and six shortwave radiometers (pyranometers) (Fig. 5, see Table 1 for specs)—was used to measure radiation. This instrument simultaneously captures radiative fluxes from upward, downward, and the four main horizontal directions [63, 73]. Sensors were installed at a height of 1.1 m on a sunlit concrete pavement to mimic radiation exposure on a person in an open summer environment. Using a 15-s sampling interval, 351 and 310 data sets were obtained from the two sites, respectively. These radiation measurements were subsequently applied to determine $R_g$, $T_g$, and MRT, as explained in Sects. 2.1.2 and 2.1.3.

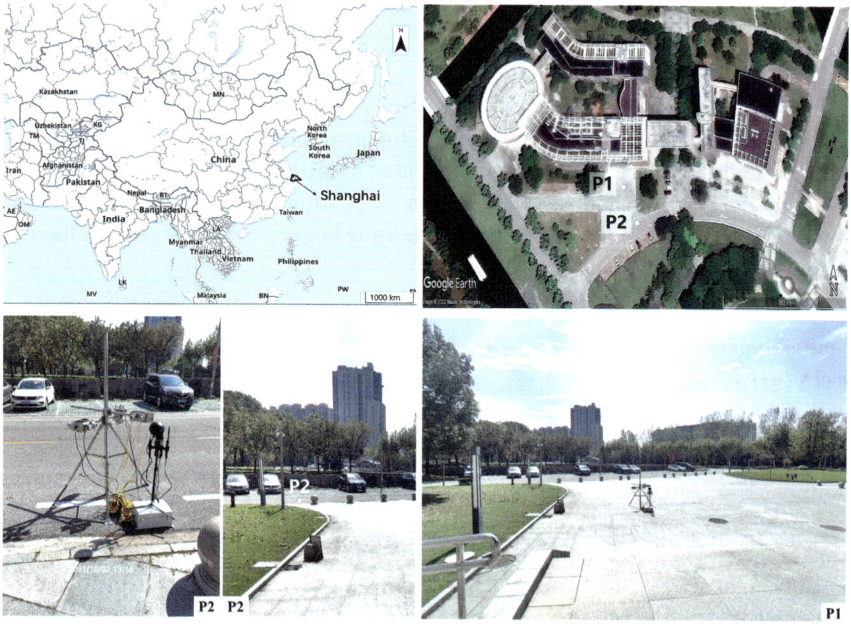


Fig. 4 Field measurement locations

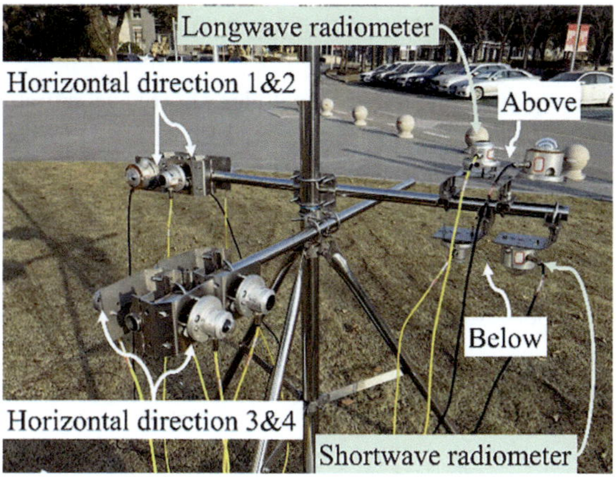


**Fig. 5** Six-direction radiation integral instrument

**Table 1** Instrument specifications

| Instrument | Spectral area (μm) | Sensitivity (μV/W/m$^2$) | Response time (s) | Field of view | Amount | Recording interval (s) |
|---|---|---|---|---|---|---|
| CGR3 Pyrgeometer | 4.5–42 | 5–15 | < 18 | 150° | 6 | 15 |
| CMP3 Pyranometer | 0.3–2.8 | 10–32 | 20 | 180° | 4 | |
| CMP3 Pyranometer | 0.285–3 | 10–15 | 20 | 180° | 2 | |

### 2.1.2 Calculation of Mean Radiant Temperature

The MRT can be calculated using Eq. (4) [74], where $S_{\text{str}}$ is the radiation absorbed by a human body (W/m$^2$), $K\downarrow$ and $L\downarrow$ are the shortwave and longwave radiation from above, respectively (W/m$^2$), $K\uparrow$ and $L\uparrow$ are the shortwave and longwave radiation from below, respectively (W/m$^2$), $K_{\text{hor}-i}$ and $L_{\text{hor}-i}$ are the shortwave and longwave radiation from the horizontal plane, including the four cardinal directions, respectively (W/m$^2$), $a_k$ and $a_l$ are the human surface absorptivity for shortwave and longwave radiation, respectively ($a_k = 0.7$, $a_l = 0.97$), $\sigma$ is the Stefan-Boltzmann constant ($\sigma = 5.67 \times 10–8$ W/(m$^2$ K$^4$)), and $\varepsilon_p$ is the emissivity of the human body ($\varepsilon_p = 0.97$).

$$\text{MRT} = \sqrt{\frac{s_\pi}{\varepsilon_p \cdot \sigma}} = \sqrt{\frac{0.06(\alpha_k K\downarrow + \alpha_L L\downarrow) + 0.06(\alpha_k K\uparrow + \alpha_L L\uparrow) + 0.22\left(\alpha_k \sum_1^i K_{hor-i} + \alpha_2 \sum_1^i L_{hor-i}\right)}{\varepsilon_p \cdot \sigma}} \tag{4}$$

### 2.1.3 Calculation of Ground Radiation and Ground Temperature

The $R_g$ from the ground includes shortwave radiation ($K_g$, W/m$^2$) and longwave radiation ($L_g$, W/m$^2$), as shown in Eq. (5). It is partly absorbed by the human body, as shown in Eq. (6), where $L_g$ includes the emitted radiation ($L_{ge}$, W/m$^2$) and the reflected radiation ($(1-\alpha_g)L_{gi}$, W/m$^2$) of the ground, as shown in Eq. (7) and Fig. 6. According to the Stefan-Boltzmann law, $L_{ge}$ can be calculated by Eq. (8), where $\varepsilon_g$ is the ground emissivity, which is assumed to be equal to $\alpha_g$ in the measurement. According to Eq. (9), $T_g$ can be obtained if $L_g$, $L_{gi}$, and $\varepsilon_g$ or $\alpha_g$ are known.

$$R_g = K_g + L_g \tag{5}$$

$$R_{g-p} = K_{g-p} + I_{g-p} = \alpha_k \times F_{g-p} \times K_g + \alpha_1 \times F_{g-p} \times I_g \tag{6}$$

$$L_y = L_{ge} + \left(1 - \alpha_y\right)L_{gi} \tag{7}$$

$$I_e = \varepsilon_e \sigma (T_e + 273.15)^4 \tag{8}$$

$$T_g = \sqrt{\frac{L_{gc}}{\varepsilon_g \sigma}} - 273.15 = \sqrt{\frac{L_{gi} - \left(1 - \alpha_g\right)L_{gi}}{\varepsilon_g \sigma}} - 273.15 \tag{9}$$

The measurement radiation from above can be approximated as the radiation incident onto the ground, i.e. $K_{up} = K_{gi}$ (total shortwave radiation incident onto the ground, W/m$^2$) and $L_{up} = L_{gi}$. Similarly, the radiation from below can be approximated as the radiation from the ground, i.e. $K_{down} = K_g$ (total shortwave radiation from the ground, W/m$^2$) and $L_{down} = L_g$. Therefore, $R_g$, $R_{g-p}$, and $T_g$ in the measurement can be calculated according to the Eqs. (5)–(9) with $\varepsilon_g = \alpha_g = 0.96$.

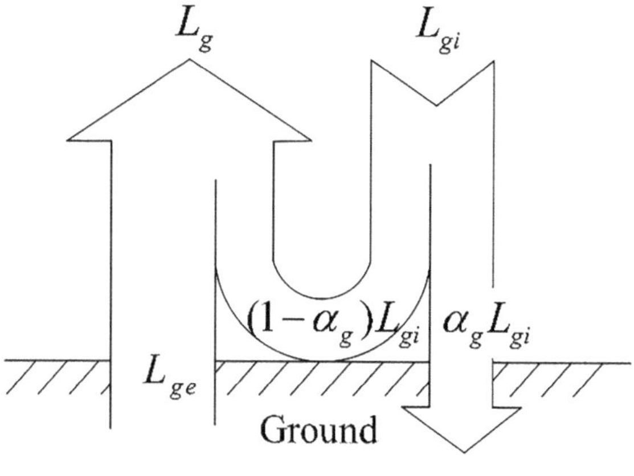


**Fig. 6** Ground longwave radiation diagram ($L_{gi}$: longwave radiation incident onto the ground, W/m$^2$; $\alpha_g$: ground absorptivity; $L_{ge}$: radiation emitted from the ground, W/m$^2$; $L_g$: total longwave radiation from the ground, W/m$^2$)

## 2.2 Simulation for Ground View Factor

### 2.2.1 Model and Calculation

Seventeen human models (Fig. 7) of different heights in an upright posture were generated with MakeHuman software (http://www.makehumancommunity.org/). Heights spanned from 70 to 230 cm, covering the majority of the world's standing population. The 70 cm height corresponds to a young child starting to stand, and 230 cm is close to the typical upper height limit. Most adults are between 150 and 190 cm tall [75]. These models were integrated into TAITherm software (ThermoAnalytics).

In addition, 31 flat rectangular ground models of various sizes $2L \times 2L$ (Fig. 8) were developed in TAITherm, with side length L extending from 1 to 100 m (refer to Appendix A). Each human model was placed at the center of a ground model, resulting in 527 scenarios (17 × 31) incorporating different statures and ground dimensions. $F_{g-p}$ for each case was calculated using Monte Carlo ray-tracing within TAITherm.

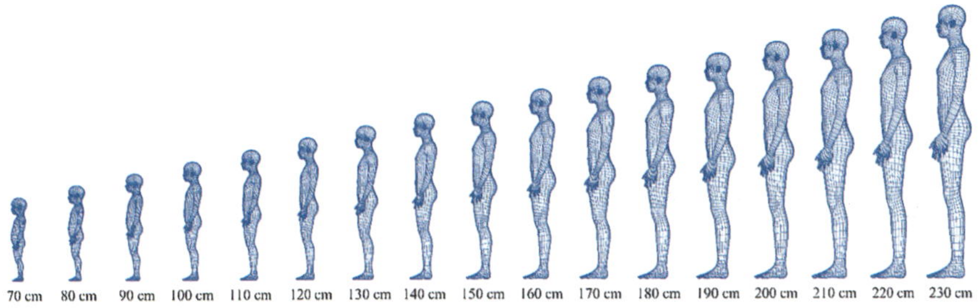


**Fig. 7** Standard weight with various height human models in standing posture

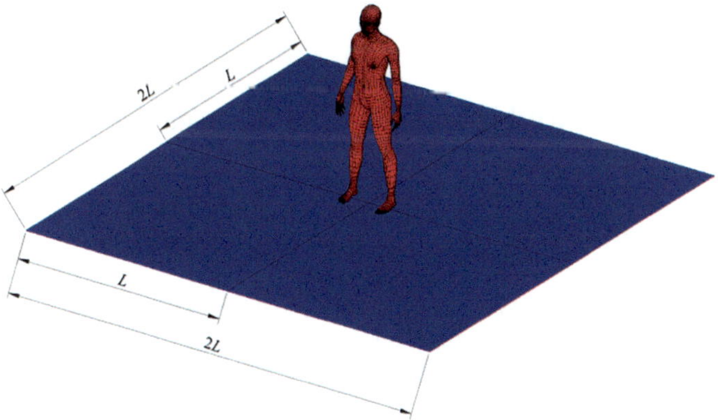


**Fig. 8** Human body-ground numerical model

**Table 2** Ground view factors of L = 1 m for 230 cm tall body with varying ray numbers from each mesh

| Case | Number of rays launched from each mesh | $F_{g-p}$ |
|---|---|---|
| 1 | 513 | 0.141 |
| 2 | 3201 | 0.142 |
| 3 | 18,436 | 0.142 |

### 2.2.2 Mesh and Ray Number

Consistent with Monte Carlo ray-tracing theory [35, 36], increasing the number of meshes and rays enhances precision but also computational load. An independence test was performed to identify appropriate parameters. With the ground modeled as a radiative boundary, only the human model's mesh and ray settings needed refinement. The number of rays emitted per mesh (513, 3201, or 18,436) was tested on a model containing 14,398 meshes. The tallest model (230 cm) on a ground with L = 1 m was selected for testing due to its highest ray requirements. As indicated in Table 2, the $F_{g-p}$ value for case 1 was 0.001 less than those for cases 2 and 3, with negligible difference between cases 2 and 3 even though case 3 had six times more rays. Therefore, the settings from case 2 were adopted: 14,398 meshes per human model and 3201 rays per mesh.

### 2.2.3 Simulation Validation

Due to the intricacies of outdoor radiation exchange, validation of simulations was based on straightforward theoretical cases (cases A and B in Table 3). As shown in Figs. 9 and 10, all surfaces were blackbodies maintained at 100 °C. Per the Stefan-Boltzmann law, the emissive power and total energy from surface i were 1099.3 W/m$^2$ and 11.0 W, respectively. The radiation absorbed by surface j was simulated as 2.2 W in both cases, giving $F_{i-j} \approx 0.2$, consistent with analytical results in Table 3, thereby verifying the simulation's reliability.

## 2.3 *Effect of Ground View Factor and Ground Temperature on the MRT*

### 2.3.1 Ground View Factor

When the surroundings around a person are categorized into ground and other surfaces, MRT can be calculated using Eq. (12), where $R_{o-p}$ denotes radiation from non-ground surfaces (W/m$^2$). Hence, $R_{o-p} = S_{\text{str}} - R_{g-p}$, as specified in Eq. (13). Under the assumption that Ro-p remains unchanged during measurement, the effect of $F_{g-p}$ precision on MRT was assessed for deviations between 0.001 and 0.20. This evaluation further addressed how human height and ground area influence $F_{g-p}$ and, in turn, MRT.

**Table 3** View factor calculations for simple cases (Bergman et al., 2011)

| Geometry | Relation | Equation No |
|---|---|---|
| Case A: Aligned Parallel Rectangles 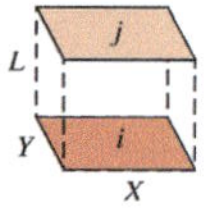 | $\overline{X} = X/L, \overline{Y} = Y/L$ <br> $F_{i-j} = \frac{1}{\pi W}\left(\begin{array}{c} W\tan^{-1}\frac{1}{W} + H\tan^{-1}\frac{1}{H} - \left(H^2+W^2\right)^{1/2}\tan^{-1}\frac{1}{(H^2+W^2)^{1/2}} \\ +\frac{1}{4}\ln\left\{\frac{(1+W^2)(1+H^2)}{1+W^2+H^2}\left[\frac{W^2(1+H^2+H^2)}{(1+W^2)(W^2+H^2)}\right]^{W^2}\times\left[\frac{H^2(1+H^2+W^2)}{(1+H^2)(H^2+W^2)}\right]^{H^2}\right\}\end{array}\right)$ | (10) |
| Case B: Perpendicular Rectangles with a Common Edge 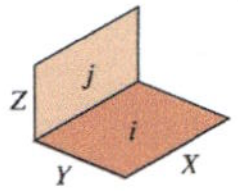 | $H = Z/X, W = Y/X$ <br> $F_{i-j} = \frac{1}{\pi W}\left(\begin{array}{c} W\tan^{-1}\frac{1}{W} + H\tan^{-1}\frac{1}{H} - \left(H^2+W^2\right)^{1/2}\tan^{-1}\frac{1}{(H^2+W^2)^{1/2}} \\ +\frac{1}{4}\ln\left\{\frac{(1+W^2)(1+H^2)}{1+W^2+H^2}\left[\frac{W^2(1+H^2+H^2)}{(1+W^2)(W^2+H^2)}\right]^{W^2}\times\left[\frac{H^2(1+H^2+W^2)}{(1+H^2)(H^2+W^2)}\right]^{H^2}\right\}\end{array}\right)$ | (11) |

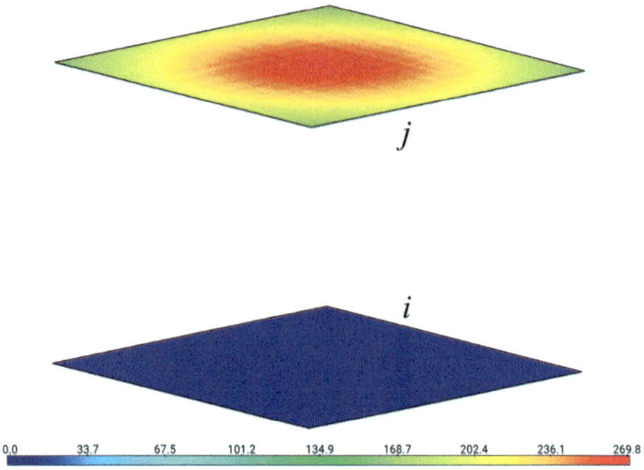


**Fig. 9** Incident radiation (W/m²) from face i to face j in Case A with X = Y = L = 100 mm

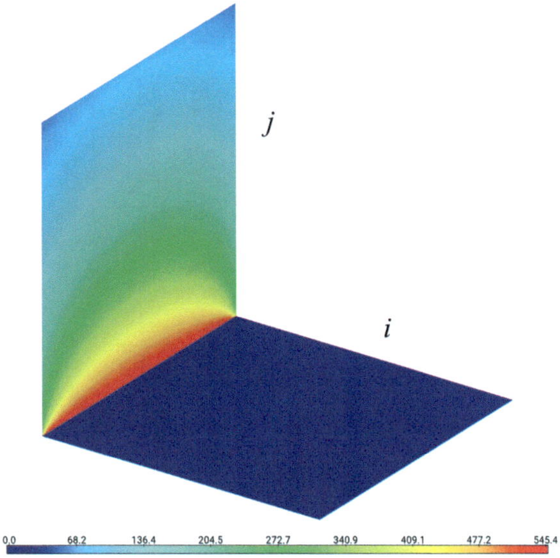


**Fig. 10** Incident radiation (W/m²) from face i to face j in Case B with X = Y = Z = 100 mm

$$\mathrm{MRT} = 4\sqrt{\frac{S_{str}}{\varepsilon_p \cdot \sigma}} = \sqrt{\frac{R_{q-p} + R_{o-p}}{\varepsilon_p \cdot \sigma}} \tag{12}$$

$$\begin{aligned} R_{o-p} &= S_{str} - R_{q-p} \\ &= \left[ 0.06(a_k K \downarrow + a_l L \downarrow) + 0.06(a_k K \uparrow + a_l L \uparrow) + 0.22\left( a_k \sum_{1}^{4} K_{no-i} + a_l \sum_{1}^{4} L_{no-i} \right) \right] \\ &- \alpha_k \times F_{g-p} \times K \uparrow + \alpha_l \times F_{g-p} \times L \uparrow \end{aligned} \tag{13}$$

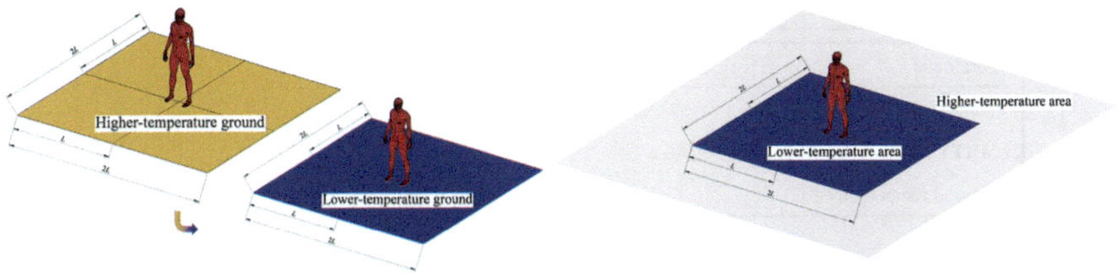


(a) Uniform temperature change (b) Non-uniform temperature change

**Fig. 11** Diagram of uniform and non-uniform changes in ground temperature

### 2.3.2 Ground Temperature

Several assumptions were made to analyze the effect of uniform and non-uniform $T_g$ variations on MRT across different heights, based on collected radiation data: First, $R_{g-p}$ was considered constant. Second, $T_g$ was assumed either fully (uniform change) or partially (non-uniform change) equal to the air temperature at the time of measurement, as shown in Fig. 11. Because the measured $T_g$ was higher than air temperature, the upper and lower temperatures in Fig. 11 represent ground and air temperatures, respectively. Third, ground size was taken as sufficiently large for $F_{g-p}$ to be 0.5. Fourth, ground radiation was treated as isotropic with $\varepsilon = 0.96$. Lastly, persons of different heights were positioned at the center of a cooler ground area with L from 1 to 10 m (Fig. 11b). Based on these assumptions, variations in $R_{g-p}$ were calculated according to Sect. 2.1.3, and related changes in MRT were obtained using Eq. (12).

# 3 Results and Discussion

## *3.1 Effect of Ground View Factor on the MRT*

### 3.1.1 Precision of Ground View Factor

The precision of $F_{g-p}$ plays a crucial role in determining MRT. Figure 12 demonstrates that an $F_{g-p}$ deviation no greater than 0.005 leads to an MRT variation of less than 0.5 °C. If the $F_{g-p}$ deviation increases to 0.01, the resulting MRT difference stays under 1 °C. Nevertheless, when the $F_{g-p}$ deviation attains 0.1, the MRT discrepancy rises to 7.7 °C—a substantial variation that considerably influences outdoor thermal comfort. These findings imply that an $F_{g-p}$ precision level of 0.01 is sufficient for MRT calculation, which corresponds with earlier suggestions regarding view factor accuracy for assessing MRT outdoors [73].

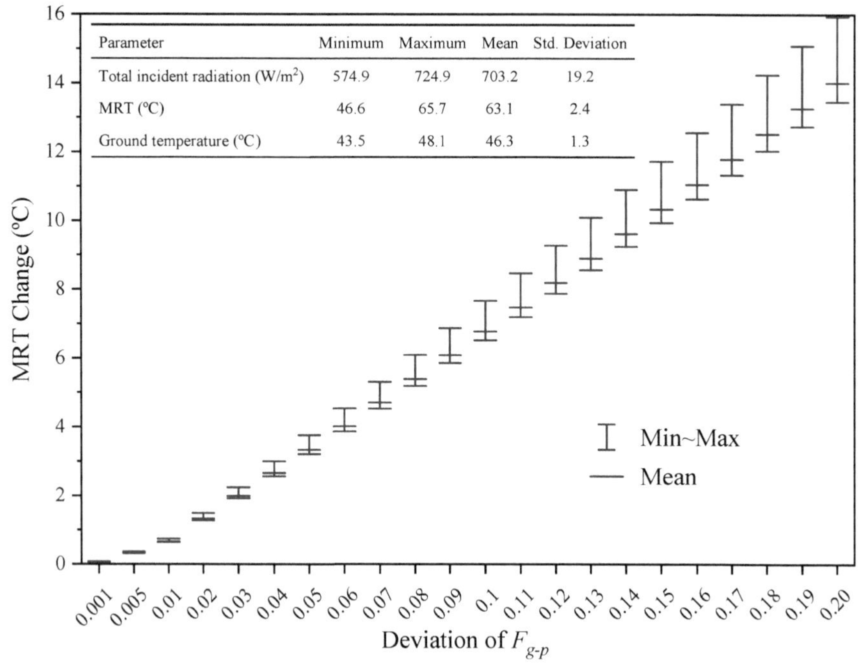


**Fig. 12** The MRT changes caused by $F_{g-p}$ deviations

### 3.1.2 Human Height and Ground Area

Discrepancies in $F_{g-p}$ may stem from variations in human stature and ground surface dimensions. The $F_{g-p}$ distributions corresponding to different human heights are provided in Fig. 13 and Appendix A. As Fig. 13b reveals, $F_{g-p}$ values differ markedly in ground regions close to the person, with shorter body heights correlating with larger $F_{g-p}$. Consequently, more pronounced height differences result in greater $F_{g-p}$ variations in such zones. For example, at L = 1 m, the $F_{g-p}$ difference between 70 and 230 cm tall individuals is 0.183, potentially leading to an MRT difference around 14 °C, as shown in Fig. 12. However, as the ground area enlarges, the $F_{g-p}$ difference decreases. At L = 40 m, the $F_{g-p}$ difference between 70 and 230 cm tall persons is merely 0.004. Referring to Sect. 3.1.1, an $F_{g-p}$ deviation below 0.01 exerts a negligible impact on MRT. Hence, for L > 40 m, MRT variations caused by height differences from 70 to 230 cm can be ignored. Likewise, for L > 11 m, MRT differences for heights between 150 and 190 cm also become insignificant. Moreover, if an average $F_{g-p}$ value (consult Appendix A) for various heights is applied in MRT estimation, the ground area over which $F_{g-p}$ differences are negligible is markedly smaller: L = 11 m for individuals between 70 and 230 cm tall, and L = 5 m for those between 150 and 190 cm tall. Should a lower MRT precision be acceptable, the requisite L can be even smaller.

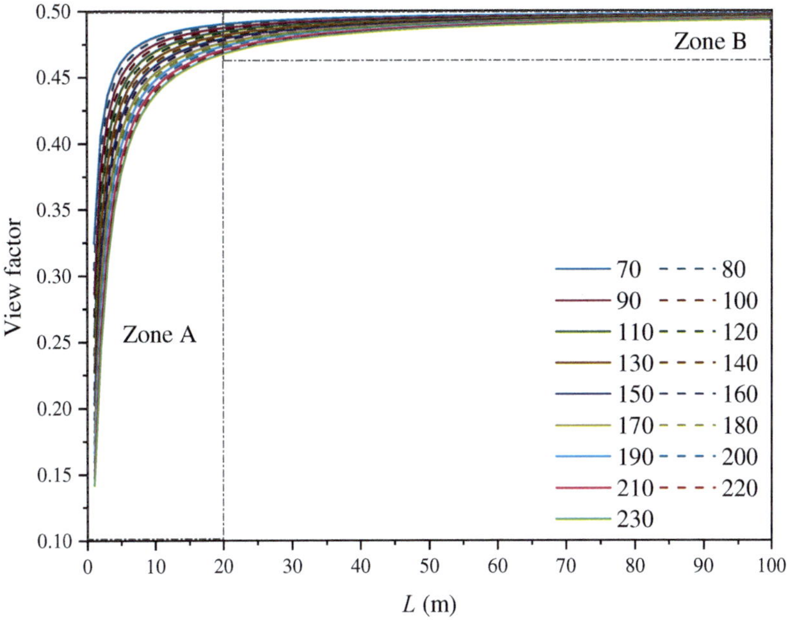


(a) View factor of $L$ equal to 1 to 100 m

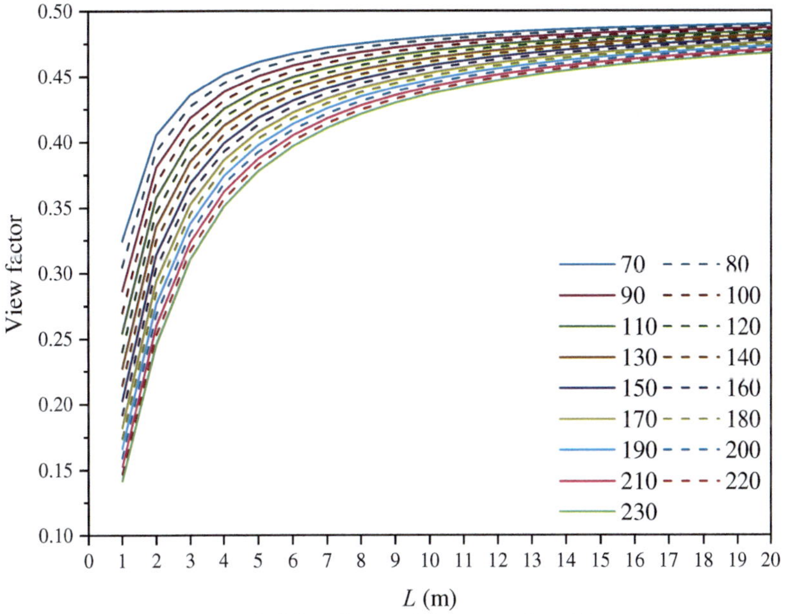


(b) View factor of $L$ equal to 1 to 10 m

**Fig. 13** View factor of the ground to the human body for different heights

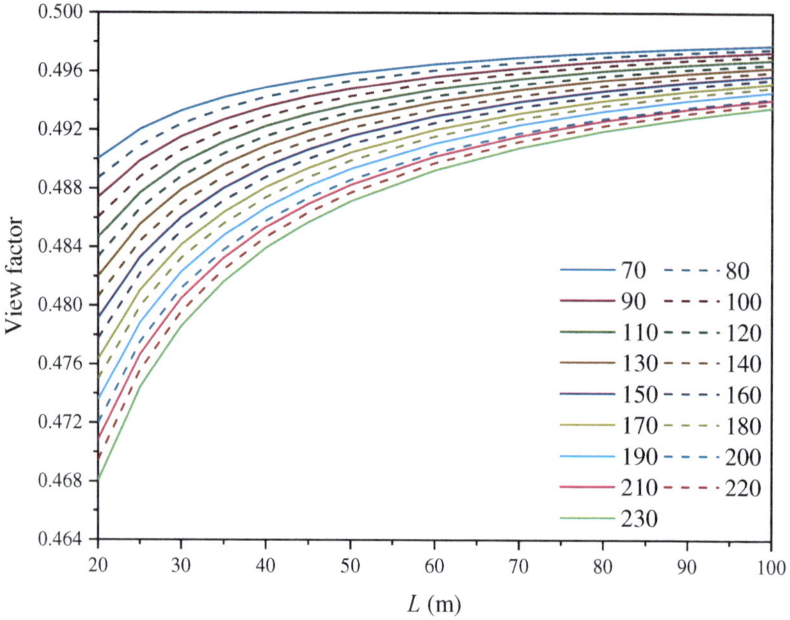


(c) View factor of $L$ equal to 20 to 100m

**Fig. 13** (continued)

According to Fig. 13, the $L$ not only affects the $F_{g-p}$ difference among various heights, but also impacts the $F_{g-p}$ variation at the same height. Moreover, the $F_{g-p}$ for different heights are very close to 0.5 with an increase in $L$. According to Sect. 3.1.1, although the environment enclosing the human body includes a vast ground area where $F_{g-p}$ is 0.5, a smaller ground area is sufficient to meet the precision requirement of the MRT calculation. A smaller ground area helps reduce modeling time and computational resources in numerical simulation. Thus, a smaller ground area should be employed. As shown in

Table 4, if the actual $F_{g-p}$ of the radiant ground is 0.5, the $F_{g-p}$ deviations for $F_{g-p} \approx 0.49$, $F_{g-p} \approx 0.48$, and $F_{g-p} \approx 0.47$ are 0.01, 0.02, and 0.03, respectively. The size of the ground area is different for various human heights and different $F_{g-p}$ deviations. With an $F_{g-p}$ deviation of 0.01 ($F_{g-p} \approx 0.49$), the $L$ of the ground area that must be considered for different heights is between 20 to 70 m. This indicates that a ground size of $L=$ 70 m is sufficient for human heights less than 230 cm under an MRT deviation of 1 °C. Similarly, the ground size that must be considered for various heights under different $F_{g-p}$ deviations can be obtained, as in .

Table 4, which shows that the ground size can be significantly reduced because of the greater $F_{g-p}$ deviations.

As illustrated in Fig. 13, the variable L affects not only the spread of $F_{g-p}$ across different heights but also the $F_{g-p}$ magnitude at a specific height. Furthermore, as L grows, $F_{g-p}$ values for all heights approach 0.5. In line with Sect. 3.1.1,

**Table 4** L (m) of radiant ground area under different $F_{g-p}$

| Height (cm) | **70** | **80** | **90** | **100** | **110** | **120** | **130** | **140** | **150** |
|---|---|---|---|---|---|---|---|---|---|
| $F_{g-p} \approx 0.49$ | 20 | 25 | 30 | 30 | 35 | 35 | 40 | 40 | 45 |
| $F_{g-p} \approx 0.48$ | 10 | 12 | 13 | 14 | 16 | 17 | 18 | 20 | 25 |
| $F_{g-p} \approx 0.47$ | 7 | 8 | 9 | 10 | 11 | 12 | 12 | 13 | 14 |
| Height (cm) | **160** | **170** | **180** | **190** | **200** | **210** | **220** | **230** | |
| $F_{g-p} \approx 0.49$ | 45 | 50 | 60 | 60 | 60 | 60 | 70 | 70 | |
| $F_{g-p} \approx 0.48$ | 25 | 25 | 30 | 30 | 30 | 30 | 35 | 35 | |
| $F_{g-p} \approx 0.47$ | 15 | 16 | 17 | 18 | 19 | 20 | 25 | 25 | |

although an extensive ground area with $F_{g-p} = 0.5$ may enclose the human body, a more limited ground extent frequently suffices to fulfill MRT accuracy criteria. Employing a smaller ground area assists in conserving modeling time and computational resources in numerical simulations. Thus, utilizing a reduced ground area is advisable. Table 4 summarizes that when the true $F_{g-p}$ of the radiant ground is 0.5, the deviations for $F_{g-p} \approx 0.49$, $F_{g-p} \approx 0.48$, and $F_{g-p} \approx 0.47$ are 0.01, 0.02, and 0.03, respectively. The requisite ground dimension depends on both human height and tolerable $F_{g-p}$ deviation. For an $F_{g-p}$ deviation of 0.01 (i.e., $F_{g-p} \approx 0.49$), the necessary L for different heights spans 20 m to 70 m. This indicates that a ground size of L = 70 m is suitable for persons up to 230 cm tall when an MRT deviation of 1 °C is acceptable. Correspondingly, the minimal ground sizes essential for diverse heights under varying $F_{g-p}$ deviations can be established, as presented in Table 4, showing that ground dimensions may be significantly decreased when larger $F_{g-p}$ deviations are allowed.

## 3.2 *Effect of Uniform Ground Temperature on MRT*

### 3.2.1 Ground Radiation

The shortwave and longwave radiation from the six directions (above, below, and four horizontal cardinal directions) were obtained through radiation measurement, as shown in Figs. 14 and 15, respectively. As measured in other investigations [36, 76, 77], the shortwave radiation fluctuates owing to intermittent blocking of the sun by clouds at the measurement time of 11:30 to 12:05; this fluctuation causes MRT fluctuations, as shown in Fig. 16. As can be seen in Fig. 14, although the shortwave radiation incident onto the ground ($K \downarrow$) was the largest among the six directions, the shortwave radiation from the ground ($K \uparrow$) was the smallest because of the small shortwave reflectivity of the ground. Therefore, the shortwave radiation absorbed by the human body from the ground was small.

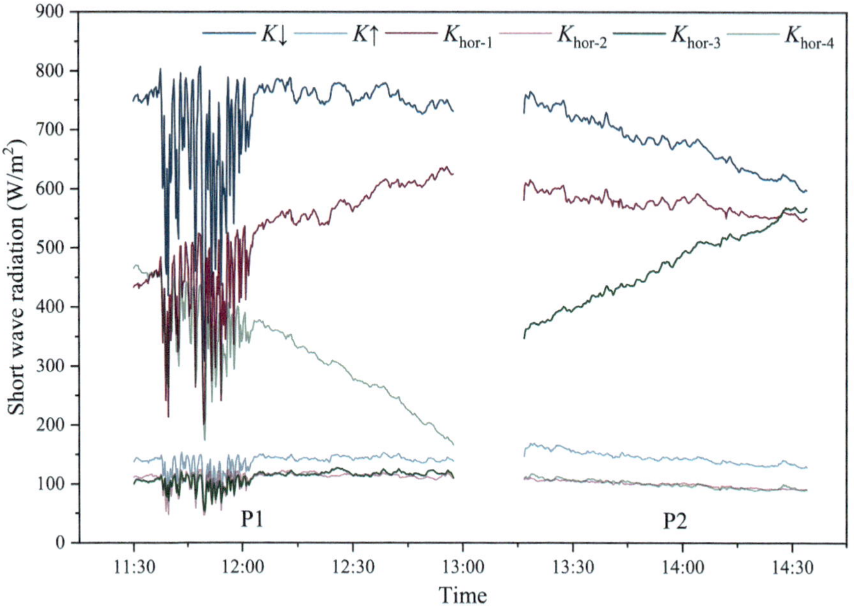


**Fig. 14** Shortwave radiation in the measurement

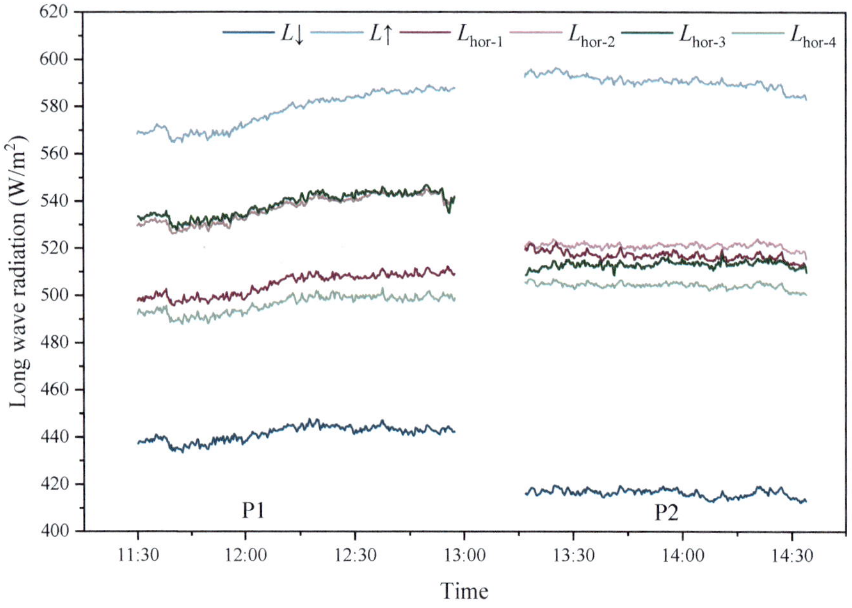


**Fig. 15** Longwave radiation in the measurement

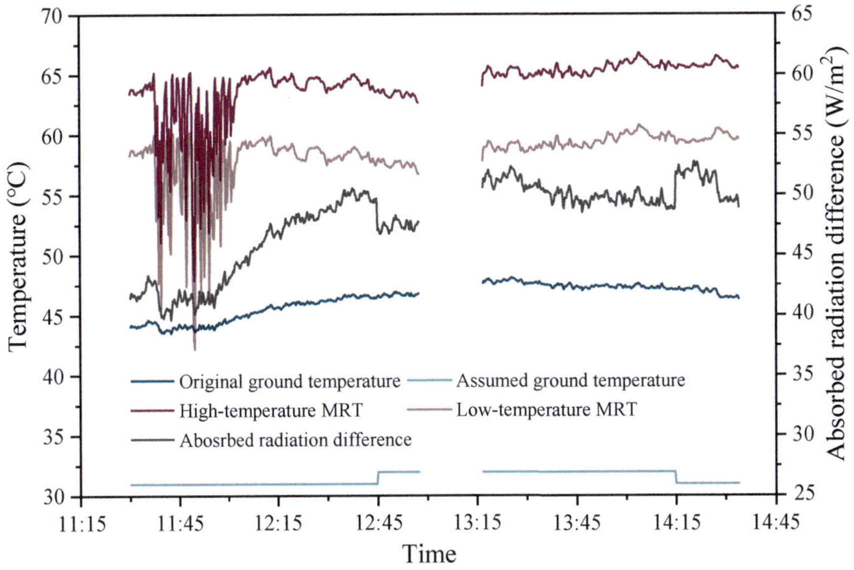


**Fig. 16** Ground temperature and its effect on MRT

However, although $L\downarrow$ was the smallest longwave radiation among the six directions, $L\uparrow$ was the largest, as shown in Fig. 15. The reason is that the $L_{ge}$ in $L\uparrow$ was large owing to the high $T_g$ caused by the hot outdoor environment, especially the strong direct solar radiation. According to the measurement, the proportion of $L_{ge}$ in $L_g$ was approximately 97%, and its share in $R_g$ was 76%–89%. Therefore, $R_g$ is significantly affected by the $L_{ge}$ related to $T_g$.

### 3.2.2 Ground Temperature

According to Eq. (9), the $T_g$ in the measurement was 43.5–48.1 °C, as shown in Fig. 16. Under the assumption that $T_g$ in the measurement changes to match the air temperature at the same time, $T_g$ decreases by 12.5–16.4 °C. Because of the decrease in $T_g$, the $R_g$ absorbed by the human body was reduced by 39.3–52.4 W/m$^2$, which led to an MRT reduction of 4.9 °C–6.3 °C, as shown in Fig. 16. According to a previous study [13], this MRT change may cause the outdoor thermal comfort sensation to change by half a thermal grade.

## 3.3 *Effect of Non-uniform Ground Temperature on MRT*

### 3.3.1 Non-uniform Temperature

As shown in Fig. 17, the size of the lower-temperature region notably affects MRT. For areas where L ≤ 10 m, each 1 m increase in L leads to an average MRT change of about 1 °C across various human heights. Nevertheless, the rate at which the MRT difference grows declines with increasing L, owing to a reduced gradient in $F_{g-p}$ change within the cooler zone, illustrated in Fig. 13. Variations in MRT are also evident among people of different heights on surfaces incorporating a lower-temperature segment, with these contrasts being particularly evident under conditions of substantial height disparity and limited cooler areas, as indicated in Fig. 18. For example, the greatest MRT difference between individuals measuring 70 cm and 230 cm tall is nearly 0.6 °C when L = 10 m, but rises to roughly 2.3 °C when L = 1 m. This effect stems from more pronounced differentiations in $F_{g-p}$ inside the lower-temperature region.

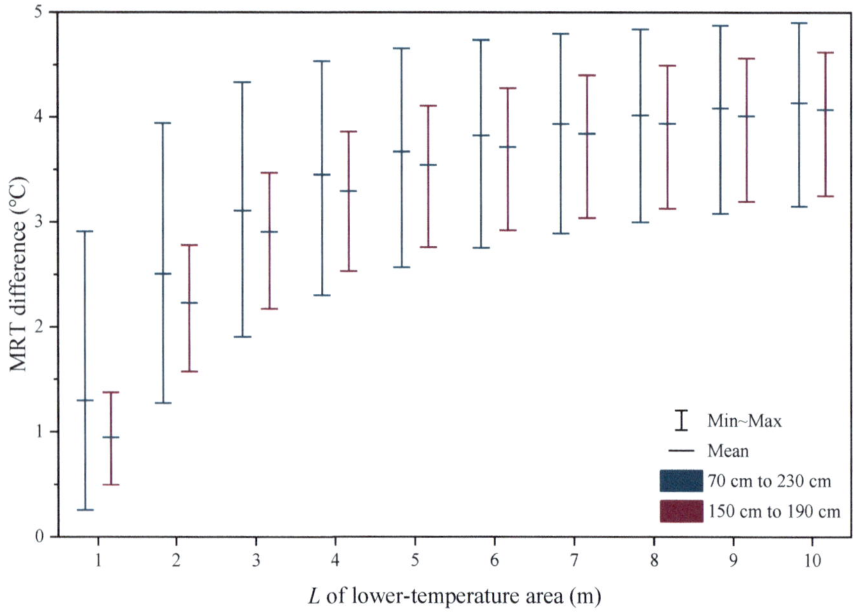


**Fig. 17** MRT difference caused by the lower-temperature area (the difference is equal to the results of measurement minus non-uniform temperature change)

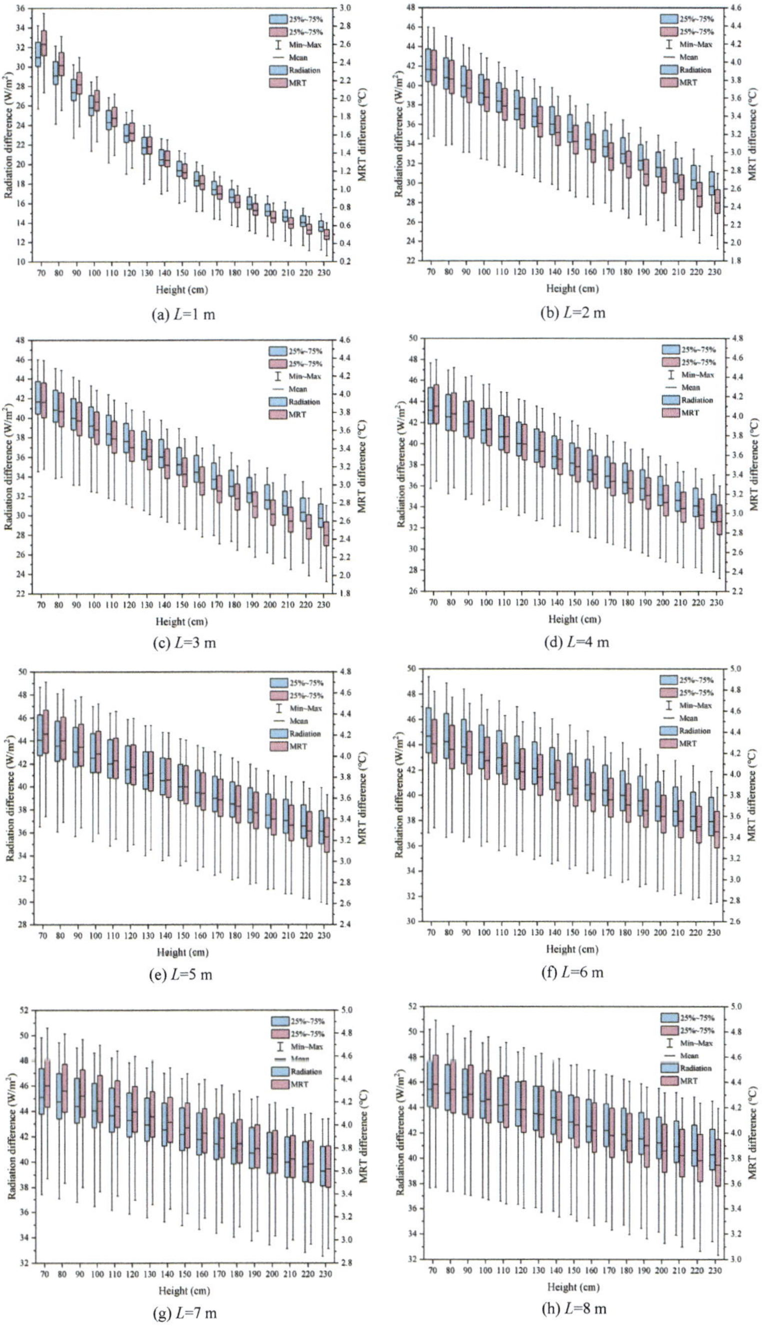


**Fig. 18** Difference of radiation absorbed by the human surface and the MRT for various human heights (the difference are equal to the results of measurement minus non-uniform temperature change)

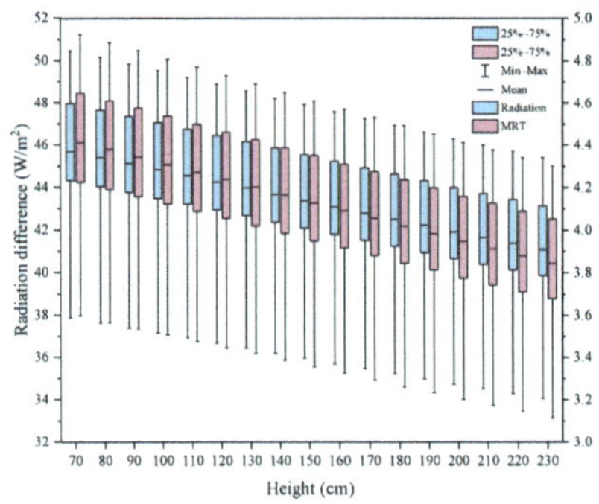


(i) $L$=9 m

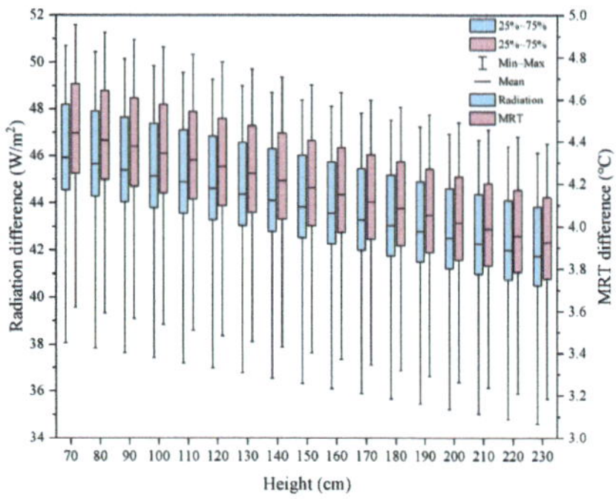


(j) $L$=10 m

**Fig. 18** (continued)

### 3.3.2 Average Temperature

Comparisons were conducted between radiation and MRT values derived from an averaged ground temperature (i.e., mean of non-uniform temperatures) and those from the true non-uniform temperature terrain, presented in Fig. 19. Underestimation of MRT occurs in some scenarios where the lower-temperature zone has L below 2 m, as shown in Fig. 19a, b. These encompass instances with human height $\geq$ 110 cm at L = 1 m, and those with height $\geq$ 220 cm at L = 2 m, where the maximum underestimation reaches about 1.5 °C. In the majority of cases, however, employing averaged non-uniform temperatures leads to increasingly significant MRT overestimation as the lower-temperature area enlarges. For instance, an overestimation of nearly 2.9 °C occurs due to a ground radiation deviation of approximately 23 W/m$^2$ when a person 70 cm tall stands on ground with a lower-temperature area of L = 10 m. For heights ranging from 150 to 190 cm, the mean maximum MRT difference is around 2.5 °C, arising from a radiation deviation close to 20 W/m$^2$, which similarly manifests for a lower-temperature zone where L = 10 m.

The underestimation or overestimation of the MRT depends on the deviation of ground radiation absorbed by the human body ($\Delta R_{g-p}$, W/m$^2$) caused by averaging the non-uniform temperature. The $\Delta R_{g-p}$ is related to the deviation in radiation emitted from the ground and absorbed by the human body ($\Delta R_{ge-p}$, W/m$^2$). The $\Delta R_{ge-p}$ can be calculated by Eqs. (14)–(17), where the $F_{g-p,i}$ ($i = 1, 2, \ldots$, n) are the view factors from the ground areas with different temperatures to the human body, $R_{ge-p}(nu)$ is the emitted radiation from the ground areas with different temperatures (non-uniform temperature ground) absorbed by the human body (W/m$^2$), and $R_{ge-p}(av)$ is the emitted radiation from the average-temperature ground absorbed by the human body (W/m$^2$). For the case $i = 2$, $\Delta R_{ge-p}$ can be calculated by Eq. (18), where the difference between values of $a_l \varepsilon_g \sigma (T_{ga}^4 - T_{g1}^4)$ and $a_l \varepsilon_g \sigma (T_{g2}^4 - T_{ga}^4)$ is small in most outdoor cases. Then, $\Delta R_{ge-p}$ is mainly determined by $F_{g-p,1}$ and $F_{g-p,2}$. For a case where $F_{g-p,1}$ is clearly larger than $F_{g-p,2}$ and $T_{g1}$ is lower than

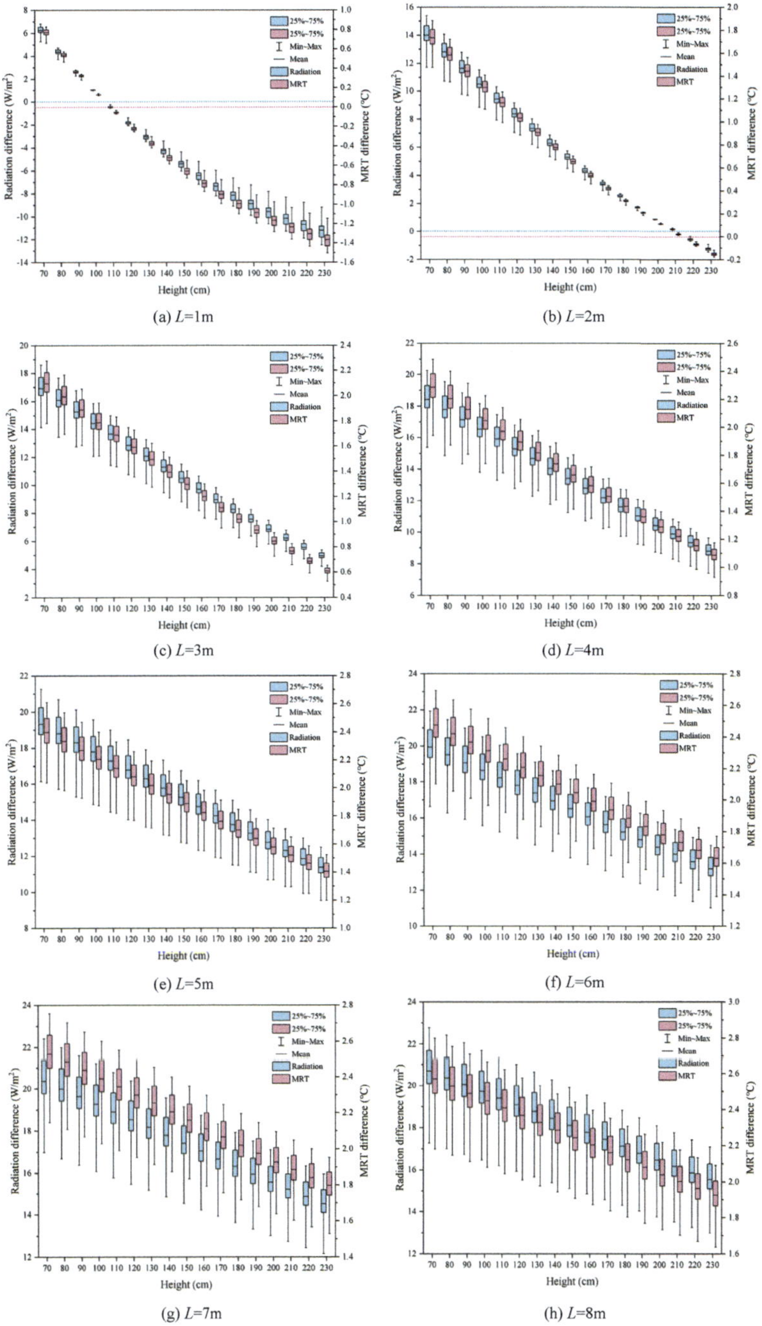


**Fig. 19** Difference of radiation and MRT between the average temperature ground (averaging the non-uniform temperatures) and the non-uniform temperature ground (the values are equal to the results of average temperature ground minus non-uniform temperature ground)

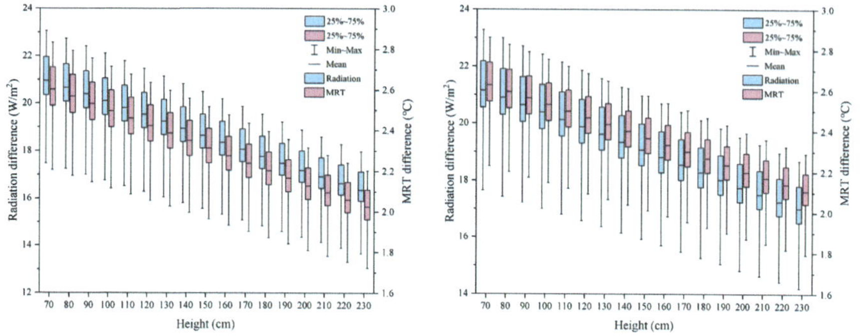


(i) $L$=9m

(j) $L$=10m

**Fig. 19** (continued)

$T_{g2}$, the radiation and MRT derived by averaging $T_{g1}$ and $T_{g2}$ overestimate the real value, as in most cases shown in Fig. 19.

$$F_{g-p} = F_{g-p,1} + F_{g-p,2} + \cdots + F_{g-p,n} \tag{14}$$

$$R_{gs-p}(\lambda) = \alpha_1 \left( F_{g-p,1}\varepsilon_g \sigma T_{g1}^4 + F_{g-p,2}\varepsilon_g \sigma T_{g2}^4 + \cdots + F_{g-p,n}\varepsilon_g \sigma T_{gn}^4 \right) \tag{15}$$

$$R_{ge-p}(\alpha v) = \alpha_l R_{g-p}\varepsilon_g \sigma T_{ga}^4 \tag{16}$$

$$\Delta R_{ge-p} = R_{ge-p}(av) - R_{ge-p}(nu) \tag{17}$$

$$\begin{aligned} \Delta R_{ge-p} &= \alpha_l \varepsilon_g \sigma \left( F_{g-p,1} T_{g0}^4 + F_{g-p,2} T_{g\alpha}^4 - F_{g-p,1} T_{g1}^4 - F_{g-p,2} T_{g2}^4 \right) \\ &= F_{g-p,1}\alpha_l \varepsilon_g \sigma \left( T_{g1}^4 - T_{g0}^4 \right) - F_{g-p,2}\alpha_l \varepsilon_g \sigma \left( T_{g2}^4 - T_{g\alpha}^4 \right) \end{aligned} \tag{18}$$

## 3.4 Limitations and Future Study

The impact of $F_{g-p}$ and $T_g$ on MRT was assessed using microclimate data collected in Shanghai, China. Consequently, the numerical results obtained are particular to this measurement context and could vary under different geographical and climatic conditions. Additional measurements at varied locations, during multiple seasons and times of day, are advised in subsequent studies. Moreover, the conclusions were drawn based on a set of specific assumptions. In actual outdoor environments, MRT is affected not solely by ground radiation, but also by radiative contributions from surrounding elements such as structures, vegetation, and the sky [45, 78, 79]. A decline in $R_g$ resulting from lower temperature may occur alongside shifts in radiation from other directions. For example, when a building casts shade on the ground, shortwave radiation from that direction decreases relative to unshaded scenarios. Similarly, shade from a tree canopy at high solar altitude reduces both shortwave and longwave radiation from above, partially due to transpiration processes [63, 77, 80–82]. To isolate the influence of $T_g$ on MRT, radiation from other directions was treated as unchanged despite fluctuations in $T_g$.

Apart from the assumptions mentioned, the ground was conceptualized as an idealized configuration—uniformly concrete, extensive, and level. In reality, ground surfaces can include grass, water, or other materials with distinct emissivities [24, 83]. These surfaces emit differing levels of radiation even at the same $T_g$. Additionally, the ground area around a person may be restricted due to obstacles such as buildings, leading $F_{g-p}$ to differ with ground extent. All these variables affect $R_{g-p}$ and modify the relationship between $T_g$ and MRT. Still, the analytical approach presented in this study for evaluating $T_g$'s effect on MRT can be applied to other ground types and layouts.

The role of $T_g$ in influencing MRT has been clearly established, particularly in hot climatic regions. Several studies report substantial reductions in $T_g$ in shaded environments [63, 64]. Furthermore, even under strong solar radiation, shaded grass temperatures beneath trees are commonly below air temperature [53]. Thus, integrating trees and grass in open spaces represents a beneficial approach for enhancing thermal comfort and justifies more extensive research.

Nevertheless, because shading cannot be implemented in all situations, studying how to cool ground surfaces under direct solar radiation remains important. Recently, advanced photonic materials have been engineered that remain cooler than ambient air under solar exposure. These materials display very high reflectivity within the solar energy spectrum (0.3–4 μm), along with high absorptivity and emissivity in the mid-infrared band [64–67, 84]. While such materials may reduce energy consumption in building envelopes, they could elevate heat exposure for people if reflected radiation is absorbed by the human body. To lessen $R_{g-p}$, both longwave ground emission ($L_{ge}$) and reflected radiation must be minimized. Therefore, surfaces characterized by low reflectivity and high thermal conductivity are likely to perform more effectively.

## 4 Conclusion

This chapter investigated the influence of ground view factor and ground temperature on outdoor mean radiant temperature using Monte Carlo ray-tracing simulations and field measurements. The key conclusions are:

(1) In Shanghai's summer conditions, the impact of $F_{g-p}$ on MRT stays under 1 °C provided that the $F_{g-p}$ deviation remains below 0.01, suggesting a negligible effect on outdoor thermal comfort. Hence, differences in $F_{g-p}$ attributable to human height have limited influence on MRT for persons between 70 and 230 cm

tall when the distance (□) from the radiant ground boundary exceeds 40 m. This threshold decreases to 11 m for individuals with heights from 150 to 190 cm. Additionally, the influence of $F_{g-p}$ variation due to the size of the radiant ground region can be ignored for humans up to 230 cm tall when □ is no less than 70 m across large ground surfaces.

(2) $T_g$ plays a significant role in shaping MRT, especially for ground areas close to the human body. Under uniform ground temperature conditions, a $T_g$ decrease of 16.4 °C can cause up to a 6.3 °C reduction in MRT during summer in Shanghai. In scenarios with non-uniform $T_g$, an increase of 1 m in the dimension □ of a cooler region—where temperature is 12.5 °C to 16.4 °C below adjacent zones—corresponds to an MRT shift of around 1 °C for people with heights between 70 and 230 cm. Also, applying an averaged non-uniform $T_g$ value could overestimate MRT by roughly 2.5 °C for individuals 150–190 cm tall, and by 2.9 °C for those 70 cm tall.

In summary, the results highlight that ground-related parameters cannot be neglected in thermal comfort modeling. Incorporating detailed ground temperature and ground view factor improves the precision of MRT estimation and provides practical guidance for climate-sensitive urban design.

**Acknowledgements** This study was supported by the National Natural Science Foundation of China [grant number 52178086].

## Appendix A

View factors of different human heights in various radiant ground areas

| L (m) | Human height (cm) | | | | | | | | | | | | | | | | | |
|---|---|---|---|---|---|---|---|---|---|---|---|---|---|---|---|---|---|---|
| | 70 | 80 | 90 | 100 | 110 | 120 | 130 | 140 | 150 | 160 | 170 | 180 | 190 | 200 | 210 | 220 | 230 | Average |
| 1 | 0.324 | 0.305 | 0.286 | 0.270 | 0.254 | 0.240 | 0.227 | 0.214 | 0.203 | 0.192 | 0.182 | 0.174 | 0.166 | 0.159 | 0.153 | 0.147 | 0.142 | 0.214 |
| 2 | 0.406 | 0.393 | 0.381 | 0.369 | 0.358 | 0.347 | 0.336 | 0.325 | 0.314 | 0.304 | 0.294 | 0.285 | 0.277 | 0.268 | 0.261 | 0.253 | 0.246 | 0.319 |
| 3 | 0.436 | 0.427 | 0.419 | 0.410 | 0.402 | 0.394 | 0.385 | 0.377 | 0.369 | 0.361 | 0.353 | 0.345 | 0.338 | 0.331 | 0.324 | 0.317 | 0.311 | 0.370 |
| 4 | 0.452 | 0.445 | 0.438 | 0.432 | 0.425 | 0.419 | 0.413 | 0.406 | 0.399 | 0.393 | 0.386 | 0.380 | 0.374 | 0.368 | 0.363 | 0.357 | 0.351 | 0.400 |
| 5 | 0.461 | 0.456 | 0.450 | 0.445 | 0.440 | 0.435 | 0.429 | 0.424 | 0.419 | 0.413 | 0.408 | 0.403 | 0.398 | 0.392 | 0.388 | 0.383 | 0.378 | 0.419 |
| 6 | 0.468 | 0.463 | 0.459 | 0.454 | 0.450 | 0.445 | 0.441 | 0.436 | 0.432 | 0.427 | 0.423 | 0.418 | 0.414 | 0.409 | 0.405 | 0.401 | 0.397 | 0.432 |
| 7 | 0.472 | 0.468 | 0.464 | 0.461 | 0.457 | 0.453 | 0.449 | 0.445 | 0.441 | 0.437 | 0.433 | 0.430 | 0.426 | 0.422 | 0.418 | 0.415 | 0.411 | 0.441 |
| 8 | 0.476 | 0.472 | 0.469 | 0.466 | 0.462 | 0.459 | 0.455 | 0.452 | 0.448 | 0.445 | 0.442 | 0.438 | 0.435 | 0.431 | 0.428 | 0.425 | 0.422 | 0.448 |
| 9 | 0.478 | 0.475 | 0.472 | 0.469 | 0.466 | 0.463 | 0.460 | 0.457 | 0.454 | 0.451 | 0.448 | 0.445 | 0.442 | 0.439 | 0.436 | 0.433 | 0.430 | 0.454 |
| 10 | 0.480 | 0.478 | 0.475 | 0.472 | 0.470 | 0.467 | 0.464 | 0.461 | 0.459 | 0.456 | 0.453 | 0.450 | 0.448 | 0.445 | 0.442 | 0.440 | 0.437 | 0.459 |
| 11 | 0.482 | 0.480 | 0.477 | 0.475 | 0.472 | 0.470 | 0.467 | 0.465 | 0.462 | 0.460 | 0.457 | 0.455 | 0.452 | 0.450 | 0.447 | 0.445 | 0.442 | 0.462 |
| 12 | 0.484 | 0.481 | 0.479 | 0.477 | 0.475 | 0.472 | 0.470 | 0.468 | 0.465 | 0.463 | 0.461 | 0.458 | 0.456 | 0.454 | 0.452 | 0.449 | 0.447 | 0.465 |
| 13 | 0.485 | 0.483 | 0.481 | 0.479 | 0.477 | 0.475 | 0.472 | 0.470 | 0.468 | 0.466 | 0.464 | 0.462 | 0.460 | 0.457 | 0.455 | 0.453 | 0.451 | 0.468 |
| 14 | 0.486 | 0.484 | 0.482 | 0.480 | 0.478 | 0.476 | 0.474 | 0.472 | 0.470 | 0.468 | 0.466 | 0.464 | 0.462 | 0.460 | 0.458 | 0.457 | 0.455 | 0.470 |
| 15 | 0.487 | 0.485 | 0.483 | 0.481 | 0.480 | 0.478 | 0.476 | 0.474 | 0.472 | 0.470 | 0.469 | 0.467 | 0.465 | 0.463 | 0.461 | 0.459 | 0.458 | 0.472 |
| 16 | 0.488 | 0.486 | 0.484 | 0.483 | 0.481 | 0.479 | 0.478 | 0.476 | 0.474 | 0.472 | 0.470 | 0.469 | 0.467 | 0.465 | 0.464 | 0.462 | 0.460 | 0.474 |
| 17 | 0.488 | 0.487 | 0.485 | 0.484 | 0.482 | 0.480 | 0.479 | 0.477 | 0.476 | 0.474 | 0.472 | 0.471 | 0.469 | 0.467 | 0.466 | 0.464 | 0.463 | 0.475 |
| 18 | 0.489 | 0.488 | 0.486 | 0.485 | 0.483 | 0.482 | 0.480 | 0.478 | 0.477 | 0.475 | 0.474 | 0.472 | 0.471 | 0.469 | 0.468 | 0.466 | 0.465 | 0.477 |
| 19 | 0.490 | 0.488 | 0.487 | 0.485 | 0.484 | 0.482 | 0.481 | 0.480 | 0.478 | 0.477 | 0.475 | 0.474 | 0.472 | 0.471 | 0.469 | 0.468 | 0.466 | 0.478 |
| 20 | 0.490 | 0.489 | 0.487 | 0.486 | 0.485 | 0.483 | 0.482 | 0.481 | 0.479 | 0.478 | 0.476 | 0.475 | 0.474 | 0.472 | 0.471 | 0.469 | 0.468 | 0.479 |
| 25 | 0.492 | 0.491 | 0.490 | 0.489 | 0.488 | 0.487 | 0.486 | 0.484 | 0.483 | 0.482 | 0.481 | 0.480 | 0.479 | 0.478 | 0.477 | 0.476 | 0.474 | 0.483 |

(continued)

(continued)

| L (m) | Human height (cm) | | | | | | | | | | | | | | | | |
|---|---|---|---|---|---|---|---|---|---|---|---|---|---|---|---|---|---|
| | 70 | 80 | 90 | 100 | 110 | 120 | 130 | 140 | 150 | 160 | 170 | 180 | 190 | 200 | 210 | 220 | 230 | Average |
| 30 | 0.493 | 0.492 | 0.492 | 0.491 | 0.490 | 0.489 | 0.488 | 0.487 | 0.486 | 0.485 | 0.484 | 0.483 | 0.482 | 0.481 | 0.480 | 0.480 | 0.479 | 0.486 |
| 35 | 0.494 | 0.493 | 0.493 | 0.492 | 0.491 | 0.490 | 0.490 | 0.489 | 0.488 | 0.487 | 0.486 | 0.486 | 0.485 | 0.484 | 0.483 | 0.482 | 0.482 | 0.488 |
| 40 | 0.495 | 0.494 | 0.494 | 0.493 | 0.492 | 0.492 | 0.491 | 0.490 | 0.490 | 0.489 | 0.488 | 0.487 | 0.487 | 0.486 | 0.485 | 0.485 | 0.484 | 0.489 |
| 45 | 0.495 | 0.495 | 0.494 | 0.494 | 0.493 | 0.492 | 0.492 | 0.491 | 0.491 | 0.490 | 0.489 | 0.489 | 0.488 | 0.487 | 0.487 | 0.486 | 0.486 | 0.491 |
| 50 | 0.496 | 0.495 | 0.495 | 0.494 | 0.494 | 0.493 | 0.493 | 0.492 | 0.492 | 0.491 | 0.490 | 0.490 | 0.489 | 0.489 | 0.488 | 0.488 | 0.487 | 0.492 |
| 60 | 0.497 | 0.496 | 0.496 | 0.495 | 0.495 | 0.494 | 0.494 | 0.493 | 0.493 | 0.493 | 0.492 | 0.492 | 0.491 | 0.490 | 0.490 | 0.490 | 0.489 | 0.493 |
| 70 | 0.497 | 0.497 | 0.496 | 0.496 | 0.496 | 0.495 | 0.495 | 0.494 | 0.494 | 0.494 | 0.493 | 0.493 | 0.492 | 0.492 | 0.492 | 0.491 | 0.491 | 0.494 |
| 80 | 0.497 | 0.497 | 0.497 | 0.496 | 0.496 | 0.496 | 0.495 | 0.495 | 0.495 | 0.494 | 0.494 | 0.494 | 0.493 | 0.493 | 0.493 | 0.492 | 0.492 | 0.495 |
| 90 | 0.498 | 0.497 | 0.497 | 0.497 | 0.496 | 0.496 | 0.496 | 0.496 | 0.495 | 0.495 | 0.495 | 0.494 | 0.494 | 0.494 | 0.493 | 0.493 | 0.493 | 0.495 |
| 100 | 0.498 | 0.498 | 0.497 | 0.497 | 0.497 | 0.497 | 0.496 | 0.496 | 0.496 | 0.495 | 0.495 | 0.495 | 0.495 | 0.494 | 0.494 | 0.494 | 0.494 | 0.496 |

# References


1. M.A. Eusuf, M.A. Mohit, M.S. Eusuf, M. Ibrahim, Impact of outdoor environment to the quality of life. Proc. Soc. Behav. Sci. **153**, 639–654 (2014)
2. D. Lai, B. Chen, K. Liu, Quantification of the influence of thermal comfort and life patterns on outdoor space activities, in *Building simulation*, vol. 13 (Springer, 2020), pp. 113–125
3. J.K. Vanos, E. Kosaka, A. Iida, M. Yokohari, A. Middel, I. Scott-Fleming et al., Planning for spectator thermal comfort and health in the face of extreme heat: the Tokyo 2020 Olympic marathons. Sci. Total Environ. **657**, 904–917 (2019)
4. P.K. Cheung, C.Y. Jim, Improved assessment of outdoor thermal comfort: 1-hour acceptable temperature range. Build. Environ. **151**, 303–317 (2019)
5. J. Huang, C. Zhou, Y. Zhuo, L. Xu, Y. Jiang, Outdoor thermal environments and activities in open space: an experiment study in humid subtropical climates. Build. Environ. **103**, 238–249 (2016)
6. L.E. McCurdy, K.E. Winterbottom, S.S. Mehta, J.R. Roberts, Using nature and outdoor activity to improve children's health. Curr. Probl. Pediatr. Adolesc. Health Care **40**, 102–117 (2010)
7. T. Sugiyama, C.W. Thompson, Outdoor environments, activity and the well-being of older people: conceptualising environmental support. Environ Plan A **39**, 1943–1960 (2007)
8. Z. Fang, X. Feng, X. Xu, X. Zhou, Z. Lin, Y. Ji, Investigation into outdoor thermal comfort conditions by different seasonal field surveys in China, Guangzhou. Int. J. Biometeorol. **63**, 1357–1368 (2019)
9. T.-P. Lin, Thermal perception, adaptation and attendance in a public square in hot and humid regions. Build. Environ. **44**, 2017–2026 (2009)
10. Z. Fang, Z. Lin, C.M. Mak, J. Niu, K.-T. Tse, Investigation into sensitivities of factors in outdoor thermal comfort indices. Build. Environ. **128**, 129–142 (2018)
11. O. Potchter, P. Cohen, T.-P. Lin, A. Matzarakis, Outdoor human thermal perception in various climates: a comprehensive review of approaches, methods and quantification. Sci. Total Environ. **631**, 390–406 (2018)
12. T. Xi, Q. Li, A. Mochida, Q. Meng, Study on the outdoor thermal environment and thermal comfort around campus clusters in subtropical urban areas. Build. Environ. **52**, 162–170 (2012)
13. X. Feng, Z. Zheng, Y. Yang, Z. Fang, Quantitative seasonal outdoor thermal sensitivity in Guangzhou, China. Urban Climate **39**, 100938 (2021)
14. Z. Li, X. Feng, X. Fan, J. Sun, Z. Fang, Effect of direct solar projected area factor on outdoor thermal comfort evaluation: a case study in Shanghai, China. Urban Climate **41**, 101033 (2022)
15. ASHRAE, *2017 ASHRAE Handbook: Fundamentals* (ASHRAE, 2017).
16. P.J. Crank, A. Middel, M. Wagner, D. Hoots, M. Smith, A. Brazel, Validation of seasonal mean radiant temperature simulations in hot arid urban climates. Sci. Total Environ. **749**, 141392 (2020)
17. J. Du, C. Sun, L. Liu, X. Chen, J. Liu, Comparison and modification of measurement and simulation techniques for estimating Tmrt in summer and winter in a severely cold region. Build. Environ. **199**, 107918 (2021)
18. J. Du, C. Sun, Q. Xiao, X. Chen, J. Liu, Field assessment of winter outdoor 3-D radiant environment and its impact on thermal comfort in a severely cold region. Sci. Total Environ. **709**, 136175 (2020)
19. C.V. Gál, N. Kántor, Modeling mean radiant temperature in outdoor spaces, a comparative numerical simulation and validation study. Urban Climate **32**, 100571 (2020)
20. N. Kántor, J. Unger, The most problematic variable in the course of human-biometeorological comfort assessment—the mean radiant temperature. Central Eur. J. Geosciences **3**, 90–100 (2011)
21. H. Lee, H. Mayer, D. Schindler, Importance of 3-D radiant flux densities for outdoor human thermal comfort on clear-sky summer days in Freiburg, Southwest Germany. Meteorologische Zeitschrift **23**, 315–330 (2014)
22. F. Lindberg, S. Onomura, C. Grimmond, Influence of ground surface characteristics on the mean radiant temperature in urban areas. Int. J. Biometeorol. **60**, 1439–1452 (2016)

23. S. Manavvi, E. Rajasekar, Estimating outdoor mean radiant temperature in a humid subtropical climate. Build. Environ. **171**, 106658 (2020)
24. M. Hendel, S. Parison, A. Grados, L. Royon, Which pavement structures are best suited to limiting the UHI effect? A laboratory-scale study of Parisian pavement structures. Build. Environ. **144**, 216–229 (2018)
25. Y. Qin, J.E. Hiller, Understanding pavement-surface energy balance and its implications on cool pavement development. Energy Build. **85**, 389–399 (2014)
26. M. Santamouris, Using cool pavements as a mitigation strategy to fight urban heat island—a review of the actual developments. Renew. Sustain. Energy Rev. **26**, 224–240 (2013)
27. C. Tong, X. Li, L. Duanmu, H. Wang, Prediction of the temperature profiles for shallow ground in cold region and cold winter hot summer region of China. Energy Build. **242**, 110946 (2021)
28. T.L. Bergman, F.P. Incropera, D.P. DeWitt, A.S. Lavine, *Fundamentals of Heat and Mass Transfer* (John Wiley & Sons, 2011)
29. P.O. Fanger, *Thermal Comfort: Analysis and Applications in Environmental Engineering* (Danish Technical Press, 1970)
30. C. Underwood, E. Ward, The solar radiation area of man. Ergonomics **9**, 155–168 (1966)
31. D. Aviv, J. Gros, H. Alsaad, E. Teitelbaum, C. Voelker, J. Pantelic et al., A data-driven ray tracing simulation for mean radiant temperature and spatial variations in the indoor radiant field with experimental validation. Energy Build. **254**, 111585 (2022)
32. D. Aviv, M. Hou, E. Teitelbaum, H. Guo, F. Meggers, Simulating invisible light: adapting lighting and geometry models for radiant heat transfer, in *Proceedings of Symposia Simulation Architectural Urban Design* (2020), pp. 311–318
33. E. Krayenhoff, A. Christen, A. Martilli, T. Oke, A multi-layer radiation model for urban neighbourhoods with trees. Bound.-Layer Meteorol. **151**, 139–178 (2014)
34. C.Y. Park, D.K. Lee, E.S. Krayenhoff, H.K. Heo, S. Ahn, T. Asawa et al., A multilayer mean radiant temperature model for pedestrians in a street canyon with trees. Build. Environ. **141**, 298–309 (2018)
35. J.R. Howell, M. Perlmutter, *Monte Carlo Solution of Thermal Transfer Through Radiant Media Between Gray Walls* (1964)
36. J.R. Mahan, *The Monte Carlo Ray-Trace Method in Radiation Heat Transfer and Applied Optics* (John Wiley & Sons, 2019)
37. P. Coelho, M. Carvalho, *A Conservative Formulation of the Discrete Transfer Method* (1997)
38. F. Lockwood, N. Shah, A new radiation solution method for incorporation in general combustion prediction procedures, in *Symposium (International) on Combustion*, vol. 18 (Elsevier, 1981), pp. 1405–1414
39. S. Chandrasekhar, *Radiative Transfer: Courier Corporation* (2013)
40. W. Fiveland, Three-dimensional radiative heat-transfer solutions by the discrete-ordinates method. J. Thermophys. Heat Transfer **2**, 309–316 (1988)
41. E. Chui, P.M. Hughes, G. Raithby, Implementation of the finite volume method for calculating radiative transfer in a pulverized fuel flame. Combust. Sci. Technol. **92**, 225–242 (1993)
42. E. Chui, G. Raithby, P. Hughes, Prediction of radiative transfer in cylindrical enclosures with the finite volume method. J. Thermophys. Heat Transfer **6**, 605–611 (1992)
43. A. Frank, W. Heidemann, K. Spindler, Modeling of the surface-to-surface radiation exchange using a Monte Carlo method. J. Phys. Conf. Ser. **745**, 032143 (IOP Publishing, 2016)
44. J.R. Howell, M.P. Mengüç, K. Daun, R. Siegel, *Thermal Radiation Heat Transfer* (CRC press, 2020)
45. C.Y. Park, D.K. Lee, E.S. Krayenhoff, H.K. Heo, J.H. Hyun, K. Oh et al., Variations in pedestrian mean radiant temperature based on the spacing and size of street trees. Sustain. Cities Soc. **48**, 101521 (2019)
46. Z.-H. Wang, Monte Carlo simulations of radiative heat exchange in a street canyon with trees. Sol. Energy **110**, 704–713 (2014)
47. A. Matzarakis, F. Rutz, H. Mayer, Modelling radiation fluxes in simple and complex environments—application of the RayMan model. Int. J. Biometeorol. **51**, 323–334 (2007)

48. A. Matzarakis, F. Rutz, H. Mayer, Modelling radiation fluxes in simple and complex environments: basics of the RayMan model. Int. J. Biometeorol. **54**, 131–139 (2010)
49. M. Bruse, H. Fleer, Simulating surface–plant–air interactions inside urban environments with a three dimensional numerical model. Environ. Model. Softw. **13**, 373–384 (1998)
50. S. Huttner, *Further Development and Application of the 3D Microclimate Simulation ENVI-met*. Universitätsbibliothek Mainz (2012)
51. L. Doulos, M. Santamouris, I. Livada, Passive cooling of outdoor urban spaces. The role of materials. Solar Energy **77**, 231–249 (2004)
52. C. Georgakis, S. Zoras, M. Santamouris, Studying the effect of "cool" coatings in street urban canyons and its potential as a heat island mitigation technique. Sustain. Cities Soc. **13**, 20–31 (2014)
53. C.Y. Park, E.J. Yoon, D.K. Lee, J.H. Thorne, Integrating four radiant heat load mitigation strategies is an efficient intervention to improve human health in urban environments. Sci. Total Environ. **698**, 134259 (2020)
54. M. Santamouris, G.Y. Yun, Recent development and research priorities on cool and super cool materials to mitigate urban heat island. Renew. Energy (2020)
55. H. Wu, B. Sun, Z. Li, J. Yu, Characterizing thermal behaviors of various pavement materials and their thermal impacts on ambient environment. J. Clean. Prod. **172**, 1358–1367 (2018)
56. D. Lai, W. Liu, T. Gan, K. Liu, Q. Chen, A review of mitigating strategies to improve the thermal environment and thermal comfort in urban outdoor spaces. Sci. Total Environ. **661**, 337–353 (2019)
57. Y. Qin, P. Wei, J. Wang, T. Kanghao, *Experimental Study on the Thermal Characteristics of Urban Mockups With Different Paved Streets* (2021)
58. M. Taleghani, The impact of increasing urban surface albedo on outdoor summer thermal comfort within a university campus. Urban Climate **24**, 175–184 (2018)
59. S. Tsoka, T. Theodosiou, K. Tsikaloudaki, F. Flourentzou, Modeling the performance of cool pavements and the effect of their aging on outdoor surface and air temperatures. Sustain. Cities Soc. **42**, 276–288 (2018)
60. N. Xie, H. Li, A. Abdelhady, J. Harvey, Laboratorial investigation on optical and thermal properties of cool pavement nano-coatings for urban heat island mitigation. Build. Environ. **147**, 231–240 (2019)
61. M. Amani-Beni, B. Zhang, J. Xu, Impact of urban park's tree, grass and waterbody on microclimate in hot summer days: a case study of Olympic Park in Beijing, China. Urban Forestry & Urban Greening **32**, 1–6 (2018)
62. M. Hendel, L. Royon, The effect of pavement-watering on subsurface pavement temperatures. Urban Climate **14**, 650–654 (2015)
63. A. Middel, E.S. Krayenhoff, Micrometeorological determinants of pedestrian thermal exposure during record-breaking heat in Tempe, Arizona: introducing the MaRTy observational platform. Sci. Total Environ. **687**, 137–151 (2019)
64. M. Napoli, L. Massetti, G. Brandani, M. Petralli, S. Orlandini, Modeling tree shade effect on urban ground surface temperature. J. Environ. Qual. **45**, 146–156 (2016)
65. N. Nasrollahi, Y. Namazi, M. Taleghani, The effect of urban shading and canyon geometry on outdoor thermal comfort in hot climates: a case study of Ahvaz, Iran. Sustain. Cities Soc. **65**, 102638 (2021)
66. M.A. Rahman, V. Dervishi, A. Moser-Reischl, F. Ludwig, H. Pretzsch, T. Rötzer, et al., Comparative analysis of shade and underlying surfaces on cooling effect, in *Urban Forestry & Urban Greening* (2021a), p. 127223
67. A. Speak, L. Montagnani, C. Wellstein, S. Zerbe, The influence of tree traits on urban ground surface shade cooling. Landsc. Urban Plan. **197**, 103748 (2020)
68. Y. Wang, H. Akbari, The effects of street tree planting on Urban Heat Island mitigation in Montreal. Sustain. Cities Soc. **27**, 122–128 (2016)
69. M.A. Rahman, V. Dervishi, A. Moser-Reischl, F. Ludwig, H. Pretzsch, T. Rötzer, et al., Comparative analysis of shade and underlying surfaces on cooling effect. Urban Forestry Urban Greening **63**, 127223 (2021b)

70. L. Chen, Y. Zhang, J. Han, X. Li, An investigation of the influence of ground surface properties and shading on outdoor thermal comfort in a high-altitude residential area. Frontiers Archit. Res. **10**, 432–446 (2021)
71. Z. Fang, X. Feng, J. Liu, Z. Lin, C.M. Mak, J. Niu et al., Investigation into the differences among several outdoor thermal comfort indices against field survey in subtropics. Sustain. Cities Soc. **44**, 676–690 (2019)
72. Y. Back, P.M. Bach, A. Jasper-Tönnies, W. Rauch, M. Kleidorfer, A rapid fine-scale approach to modelling urban bioclimatic conditions. Sci. Total Environ. **756**, 143732 (2021)
73. S. Thorsson, F. Lindberg, I. Eliasson, B. Holmer, Different methods for estimating the mean radiant temperature in an outdoor urban setting. Int. J. Climatol. J. R. Meteorol. Soc. **27**, 1983–1993 (2007)
74. V.D. Ingeneure, *VDI 3787, Part II: Environmental Meteorology: Methods for the Human-biometeorological Evaluation of Climate and Air Quality for the Urban and Regional Planning at Regional Level*. Part I: Climate. VDI/DIN-Handbuch Reinhaltung der Luft, vol. 1 (1998)
75. M. Roser, C. Appel, H. Ritchie, *Human Height. Our World in Data* (2013)
76. N. Kántor, A. Kovács, Á. Takács, Small-scale human-biometeorological impacts of shading by a large tree. Open Geosciences **8**, 231–245 (2016)
77. S. Zheng, J.-M. Guldmann, Z. Liu, L. Zhao, Influence of trees on the outdoor thermal environment in subtropical areas: an experimental study in Guangzhou, China. Sustain. Cities Soc. **42**, 482–497 (2018)
78. I. Azcarate, J.Á. Acero, L. Garmendia, E. Rojí, Tree layout methodology for shading pedestrian zones: thermal comfort study in Bilbao (Northern Iberian Peninsula). Sustain. Cities Soc. **72**, 102996 (2021)
79. A. Chatzidimitriou, S. Yannas, Street canyon design and improvement potential for urban open spaces; the influence of canyon aspect ratio and orientation on microclimate and outdoor comfort. Sustain. Cities Soc. **33**, 85–101 (2017)
80. S. Green, Radiation balance, transpiration and photosynthesis of an isolated tree. Agric. For. Meteorol. **64**, 201–221 (1993)
81. N. Kántor, C.V. Gál, Á. Gulyás, J. Unger, The impact of façade orientation and woody vegetation on summertime heat stress patterns in a central European square: comparison of radiation measurements and simulations. Adv. Meteorol. (2018)
82. H. Sinoquet, X. Le Roux, B. Adam, T. Ameglio, F.-A. Daudet, RATP: a model for simulating the spatial distribution of radiation absorption, transpiration and photosynthesis within canopies: application to an isolated tree crown. Plant Cell Environ. **24**, 395–406 (2001)
83. S.-A. Tan, T.-F. Fwa, Influence of pavement materials on the thermal environment of outdoor spaces. Build. Environ. **27**, 289–295 (1992)
84. X. Yin, R. Yang, G. Tan, S. Fan, Terrestrial radiative cooling: Using the cold universe as a renewable and sustainable energy source. Science **370**, 786–791 (2020)

# Techniques for Radiant Temperature Measurement: Validation and Comparative Analysis

**Abstract** Evaluating outdoor mean radiant temperature (MRT) is essential for understanding and designing thermal conditions in open spaces. The six-direction integral method (SDIM) provides the most accurate measurement but is financially and technically demanding. This chapter introduces a modified two-hemisphere method (MTHM) as a more practical alternative. Field measurements in Shanghai were used to compare MRT values derived from the two-hemisphere method (THM) and the SDIM. Results showed that THM deviates by less than 3 °C in shaded areas and up to 5 °C under direct sunlight, mainly due to its omission of anisotropic sky diffuse radiation. The proposed MTHM incorporates this component, thereby improving both the accuracy and stability of MRT estimates. Validation confirmed that MTHM substantially reduces deviations from SDIM, making it a reliable tool for outdoor thermal comfort evaluation. This methodological advancement supports more accessible MRT assessment and strengthens the evidence base for climate-sensitive urban design.

## 1 Introduction

The quality of the outdoor thermal environment has a deep effect on daily living, health conditions, work performance, and energy use in buildings [1–6] in the context of climate change and urban growth. To begin, extreme outdoor heat or cold jeopardizes human safety and wellness, especially for people working outdoors [7]. Every year, extreme heat leads to a significant number of fatalities worldwide, causing many countries to provide public cooling facilities. Additionally, the thermal atmosphere outside is the most critical element affecting thermal comfort in open areas [8–12]. A comfortable external thermal setting promotes outdoor work and recreational activity, which in turn supports better mental and physical health [13, 14]. Lastly, the energy demand of buildings is highly dependent on the outdoor thermal climate. The closer the thermal environment around a structure is to neutral—that is, neither too cold nor too hot—the less energy is required for operating HVAC (Heating, ventilation, and air conditioning) systems [4, 5].



Z. Fang et al., *Outdoor Thermal Comfort and Heat Stress Mitigation*, Environmental Science and Engineering, https://doi.org/10.1007/978-981-95-8899-2_4

The outdoor thermal environment is governed by factors such as air temperature, relative humidity, wind velocity, and mean radiant temperature (MRT) [10, 15, 16]. MRT—characterized as the uniform temperature of a theoretical black enclosure resulting in identical radiant heat exchange with an individual as the real surroundings [17]—is the most influential parameter for outdoor thermal comfort [3, 8, 9, 18]. A variation of 12 °C in MRT during summer alters a person's thermal sensation by one full level [9]. Furthermore, microclimate management strategies can substantially modify MRT [19–22]. During summer, shading significantly reduces MRT, thereby alleviating outdoor heat stress. Conversely, in winter, direct solar radiation markedly increases MRT, reducing the impact of cold stress. Nevertheless, the intricate nature of outdoor radiation makes MRT challenging to quantify [23]. Multiple measurement approaches, including the globe thermometer method (GHM), the two-hemisphere method (THM), and the six-direction integral method (SDIM), have been devised to evaluate outdoor MRT [24–27].

Based on principles of thermal equilibrium, the GHM derives MRT using environmental variables such as air temperature (°C), wind speed (m/s), and globe temperature (°C). Globe temperature is measured using a black-globe thermometer, which houses a temperature sensor—such as a mercury thermometer bulb, thermocouple, or resistance probe—within a blackened sphere [28]. MRT can subsequently be computed using Eq. (1) or (2), where ε represents the emissivity of the globe and D denotes its diameter (m). Although the GHM offers operational convenience, it is prone to inaccuracies in outdoor settings due to the instrument's delayed response time and discrepancies in absorptivity and form between the globe and the human body [18, 29–33]. As a result, a number of studies have introduced improved versions of the GHM [24, 27]. In practice, the measurement mechanism of the GHM constrains its use in outdoor environments where shortwave and longwave radiation vary dynamically. Therefore, radiation-integration techniques such as THM and SDIM are better suited for outdoor MRT assessment.

Natural convection:

$$\mathrm{MRT} = \left[ \left(T_g + 273\right)^4 + \frac{0.25 \times 10^8}{\varepsilon_g} \left( \frac{|T_g - T_a|}{D} \right)^{1/4} \times \left(T_g - T_a\right) \right]^{1/4} - 273 \tag{1}$$

Forced convection:

$$\mathrm{MRT} = \left[ \left(T_g + 273\right)^4 + \frac{1.1 \times 10^8 \times \mathrm{v}_a^{0.6}}{\varepsilon_g \times D^{0.4}} \left(T_g - T_a\right) \right]^{1/4} - 273 \tag{2}$$

The integral radiation theory-based computational method is formulated in Eq. (3), where $S_{str}$ signifies the total radiative energy absorbed per unit area of the human body (W/m$^2$), $\sigma$ corresponds to the Stefan-Boltzmann constant ($\sigma = 5.67 \times 10^{-8}$ W/(m$^2 \cdot K^4$)), $\varepsilon_p$ indicates the emissivity of human tissue ($\varepsilon_p = 0.97$), $F_i$ denotes the view factor from surrounding surfaces to the human body, $a_k$ and $a_l$ refer to the absorptivity of human surface to shortwave and longwave radiation, respectively ($a_k = 0.7$, $a_l = 0.97$), and $K_i$ and $L_i$ characterize the shortwave and longwave radiation incident from the environment onto the human surface, respectively (W/m$^2$) [34]. Although this methodology is applicable to outdoor MRT estimation, accurately quantifying radiation contributions from all surrounding surfaces poses challenges due to variations in radiative properties and geometric view factors among different surfaces. As a result, simplified procedures such as the SDIM and THM become indispensable for practical outdoor MRT quantification.

$$\mathrm{MRT} = \sqrt[4]{\frac{S_{\mathrm{str}}}{\varepsilon_{\mathrm{p}} \cdot \sigma}} = \sqrt[4]{\frac{\sum_{\mathrm{i}=1}^{\mathrm{n}} f_I \cdot a_K \cdot K_I}{\varepsilon_{\mathrm{p}} \cdot \sigma} + \frac{\sum_{\mathrm{i}=1}^{\mathrm{n}} f_I \cdot a_L \cdot L_I}{\varepsilon_{\mathrm{p}} \cdot \sigma}} \tag{3}$$

As shown in Fig. 1, the six-direction integral method (SDIM) represents the human body as a cubic form with distinct view factors: 0.06 for radiation from above and below, and 0.22 for each of the four horizontal cardinal directions. These coefficients correspond to the proportional radiative area on the human surface, which is divided into six planes of a cube. However, the approach does not account for variations in view factors that occur with different body orientations. Although the actual view factors for the front and rear surfaces exceed those of the lateral sides, their mean value (0.22) is employed. Therefore, the SDIM requires six pyranometers and six pyrgeometers to measure incident shortwave and longwave radiation from above, below, and the four horizontal directions. Multiple studies have applied the SDIM to evaluate outdoor mean radiant temperature or validate MRT simulation results, as it is currently acknowledged as the most accurate method for outdoor MRT measurement [18, 26]. Nonetheless, the requirement for twelve radiation instruments makes this technique both expensive and practically cumbersome.

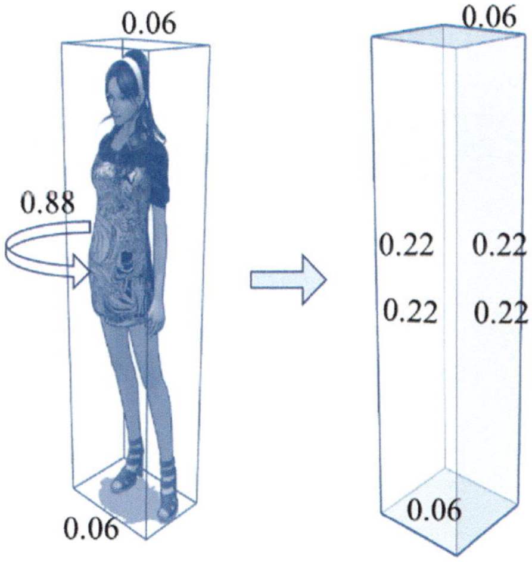


**Fig. 1** SDIM simplification for the human body shape (the numbers on the figure are the view factors for each surface)

The two-hemisphere method (THM) provides a more cost-efficient and operationally simpler solution compared to the SDIM, requiring just two pyranometers and two pyrgeometers. As shown in Fig. 2, this technique divides the outdoor radiation environment into upper and lower hemispheres, allowing concurrent measurement of diffuse shortwave and longwave radiation [35, 36]. Additionally, direct solar radiation ($I^*$, W/m$^2$) impacting human subjects in the THM is calculated using the direct solar projected area factor ($f_p$), defined as the proportion between the direct solar radiation projected area ($A_P$) and the body's effective radiative area ($A_{eff}$), as specified in Eq. (4), where $A_D$ represents the nude skin surface area, $f_{cl}$ is the projected area factor, and $f_{eff}$ signifies the effective radiative area factor [37]. Prominent outdoor microclimate simulation software [36, 38], including RayMan [39, 40] and ENVI-met [41, 42], have implemented the THM methodology. However, multiple investigations have documented MRT discrepancies in the THM approach, with extreme variations reaching 15 °C [25, 43, 44]. Although linear regression was used to enhance the THM in Lee's research [25], the root causes of these deviations remain insufficiently examined. Furthermore, this adapted THM version might produce less reliable outcomes in different locations since the regression equations are based on particular measurement data and may not generalize well to other environments.

$$f_{\mathrm{p}}=\frac{a_{\mathrm{p}}}{a_{\mathrm{eff}}}=\frac{a_{\mathrm{p}}}{a_D \cdot f_{\mathrm{ci}} \cdot f_{\mathrm{eff}}} \tag{4}$$

Accordingly, this study aims to determine the origins of MRT deviation in the THM and to introduce a corresponding refinement strategy. These efforts will

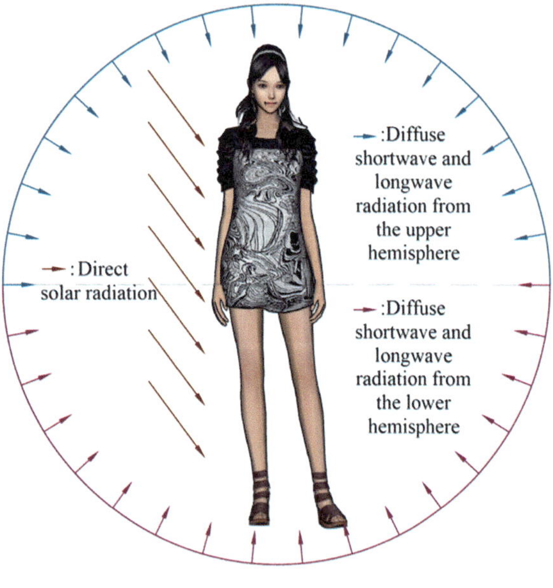


**Fig. 2** THM simplification for the radiant outdoor environment

enhance the accuracy of outdoor MRT measurement, thereby supporting more effective planning of outdoor thermal environments.

## 2 Methodology

The research steps, procedures, and objectives of each research step in this study are shown in Fig. 3. First, radiation measurements were performed to acquire the MRT values of the THM and the SDIM, which were estimated using traditional methods. Second, the MRT difference between the THM and the SDIM was quantified, and its underlying causes were explored. Third, a modified method (MTHM) based on the THM and considering anisotropic sky diffuse radiation was proposed. Finally, the MTHM performance was validated by comparing the MRT difference between the MTHM and the SDIM.

### 2.1 Measurement

The radiation measurement was conducted in Shanghai, China, which has a hot-summer and cold-winter building climate. In Shanghai summer, the mean maximal horizontal global solar irradiance is approximately 679.2 W/m$^2$, and the mean $T_a$ is more than 24 °C. In winter, the mean maximal horizontal global solar irradiance is approximately 423.1 W/m$^2$, and the mean $T_a$ is around 6.6 °C [42, 45]. The measurements were carried out on sunny days during summer and winter at 4 locations (shown in Fig. 4) within the Jiading campus of Tongji University (31°17′8" N, 121°12′26" E), as shown in Table 1, where the P1 was shaded by a tree canopy, and other locations were exposed to direct sunlight. A HOBO U12 Temperature Data Logger recorded $T_a$ every 30 s, which ranged between 31.0 °C and 34.3 °C in the summer and 8.7 °C and 9.0 °C in the winter. As shown in Fig. 5, the shortwave and longwave radiation from above, below, and horizontal cardinal directions were recorded every 15 s by six pyranometers and six pyrgeometers (specifications listed in Table 2) at a height of approximately 1.1 m above the ground.

### 2.2 SDIM MRT Calculation

The shortwave ($K_{\text{abs}}$, W/m$^2$) and longwave ($L_{\text{abs}}$, W/m$^2$) radiation absorbed by the human body in the SDIM were determined through Eqs. (5) and (6), respectively [34]. $K\downarrow$ and $L\downarrow$ denote the downward shortwave and longwave radiation(W/m2), $K\uparrow$ and $L\uparrow$ a represent upward radiation from the ground (W/m2), and $K_{\text{hor}-i}$ and $L_{\text{hor}-i}$ are the shortwave and longwave radiation from four horizontal cardinal

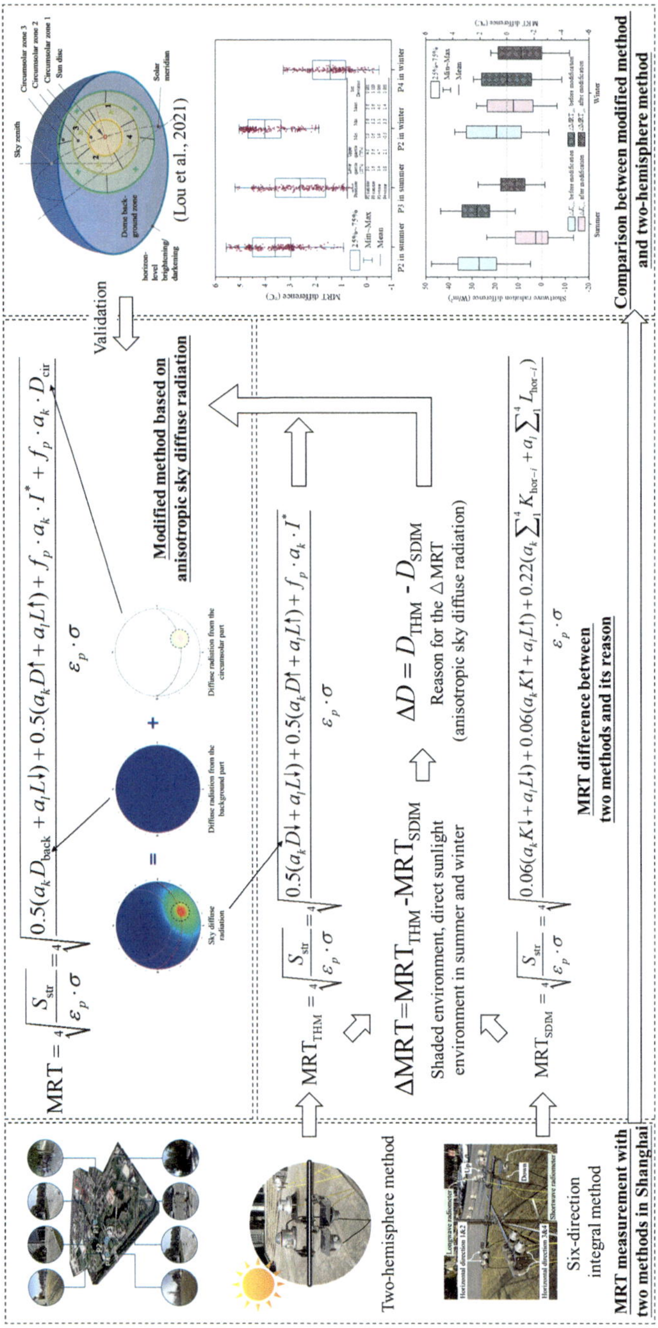


**Fig. 3** An infographic showing the research steps, procedures, and objectives of each research step

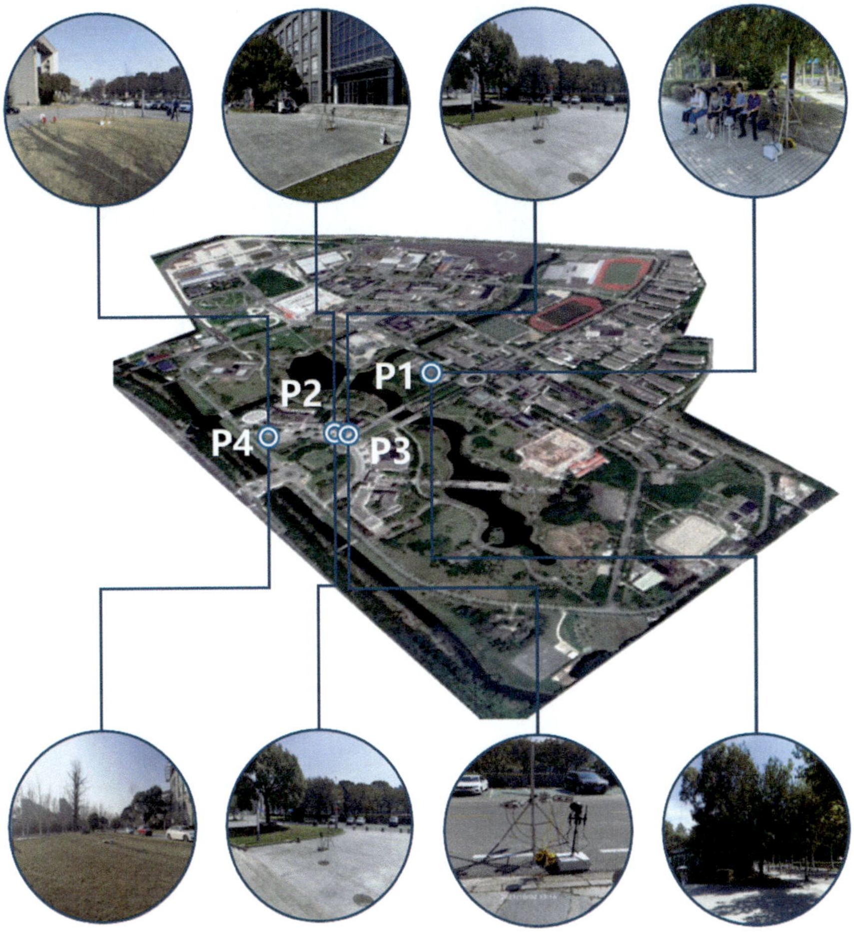


**Fig. 4** Measurement locations

**Table 1** Measurement location and date

| Season | Location | Local date and time | Shade situation | Solar zenith ($\theta$, degree) | Air temperature (°C) |
|---|---|---|---|---|---|
| Summer | P1 | 2021/09/27 13:30:00–14:07:00 | Under the shade | 43 to 46 | 32.6 to 33.7 |
| | P2 | 2021/10/02 11:30:00–12:57:30 | In direct sunlight | 34 to 38 | 31.0 to 33.8 |
| | P3 | 2021/10/02 13:15:00–14:30:00 | | 41 to 53 | 33.2 to 34.3 |
| Winter | P2 | 2021/12/29 12:40:00–13:40:45 | | 56 to 62 | 8.7 to 9.6 |
| | P4 | 2021/12/29 14:00:00–14:59:00 | | 52 to 70 | 8.9 to 9.6 |

**Fig. 5** Six-direction radiation measurement instrument

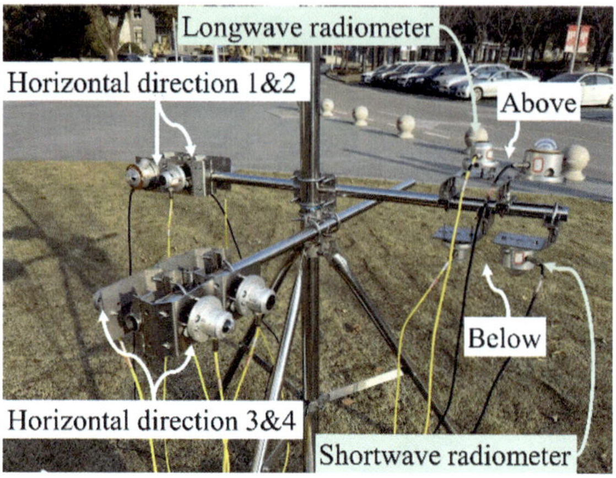


**Table 2** Instrument specifications

| Instrument | Spectral range | Sensitivity | Response time | Field of view | Amount | Recording interval |
|---|---|---|---|---|---|---|
| Kipp & Zonen CGR3 Pyrgeometer | 4.5 to 42 μm | 5 to 15 μV/W/m$^2$ | < 18 s | 150° | 6 | 15 s |
| Kipp & Zonen CMP3 Pyranometer | 0.3 to 2.8 μm | 10 to 32 μV/W/m$^2$ | 20 s | 180° | 4 | |
| LSI DPA154 Pyranometer | 0.29 to 3 μm | 10 to 15 μV/W/m$^2$ | 20 s | 180° | 2 | |

directions, respectively (W/m$^2$). Finally, the MRT value for the THM ($MRT_{SDIM}$) was obtained through Eq. (7).

$$K_{abs} = 0.06a_K(K\downarrow + K\uparrow) + 0.22a_K\sum_{i=1}^{4} K_{hor-i} \tag{5}$$

$$I_{abs} = 0.06a_I(L\downarrow + L\uparrow) + 0.22a_I\sum_{i=1}^{4} L_{hor-i} \tag{6}$$

$$MRT_{SDIM} = 4\sqrt{\frac{S_{str}}{E_p \cdot \sigma}} = 4\sqrt{\frac{K_{abs} + L_{abs}}{E_p \cdot \sigma}} \tag{7}$$

As the SDIM is currently considered the most accurate approach for MRT measurement, its results were treated as reference values in comparison with the THM. To evaluate the discrepancy between the THM and SDIM, the radiation recorded by the SDIM was separated into upper and lower hemispheres according to the THM framework, as depicted in Fig. 6. The absorbed shortwave and longwave

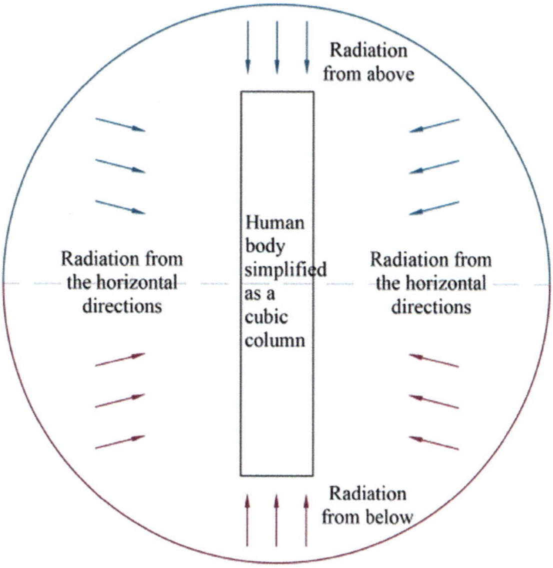


**Fig. 6** Radiation from dual hemispheres acting upon a human body modeled as a cubic column within the SDIM framework

radiation by the human body from the upper ($K_{abs,UH}$, $L_{abs,UH}$, W/m$^2$) and lower ($K_{abs,LH}$, $L_{abs,LH}$, W/m$^2$) hemispheres were then derived using Eqs. (8) to (11).

$$K_{abs,UH} = 0.06a_K K\downarrow + 0.11a_K \sum_{i=1}^{4} K_{hor-i} \tag{8}$$

$$L_{abs,UH} = 0.06a_L L\downarrow + 0.11a_L \sum_{1}^{4} L_{hor-i} \tag{9}$$

$$K_{abs,LH} = 0.06a_K K\uparrow + 0.11a_K \sum_{i=1}^{4} K_{hor-i} \tag{10}$$

$$L_{abs,LH} = 0.06a_L L\uparrow + 0.11a_L \sum_{i=1}^{4} L_{hor-i} \tag{11}$$

## 2.3 THM MRT Calculation

$K_{abs}$ and $L_{abs}$ in the THM were computed through Eqs. (12) and (13), and the $MRT_{TMH}$ was obtained using Eq. (14), where $D\downarrow$ and $D\uparrow$ denote the diffuse shortwave radiation from the upper and lower hemispheres, respectively (W/m$^2$). In most measurement scenarios relevant to this study, $D\uparrow$ can be considered approximately equal to $K\uparrow$ in the majority of circumstances where our measurement is applicable. $I^*$ approaches zero, while $D\downarrow$ is nearly equivalent to $K\downarrow$. $D\downarrow$ and $I^*$ were derived from $K\downarrow$ in the direct sunlight environment using a fraction of direct and diffuse solar radiation collected from the SOLCAST (https://www.solcast.com/). SOLCAST can provide the data, including Global Horizontal Irradiance (GHI, W/m$^2$), Diffuse Horizontal Irradiance (DHI, W/m2), and Direct Normal Irradiance (DNI, W/m$^2$) with a minimum time series of 5 min, and its accuracy has been validated [46, 47]. Then, $D\downarrow$ and $I^*$ can be calculated using Eqs. (15) and (16), respectively. The DHI / GHI and DNI / GHI can be assumed to be constant within 5 min because they fluctuate very little during that period, as shown in Fig. 7. $f_p$ in Eqs. (12) and (14) were derived from our previous study with an interval of 15 s for a standing posture human body [8], as shown in Fig. 7. Then, $K_{abs,UH}$, $L_{abs,UH}$, $K_{abs,LH}$, and $L_{abs,LH}$ in the THM can be can calculated utilizing Eqs. (17) to (20).

$$K_{abs} = 0.5a_{\kappa}(D\downarrow + D\uparrow) + f_p \cdot a_{\kappa} \cdot I^* \quad (12)$$

$$L_{abs} = 0.5a_l(L\downarrow + L\uparrow) \quad (13)$$

$$MRT_{THM} = \sqrt[4]{\frac{S_{str}}{\varepsilon_p \cdot \sigma}} = \sqrt[4]{\frac{0.5(a_L L\downarrow + a_K D\downarrow) + 0.5(a_L L\uparrow + a_K D\uparrow) + f_p \cdot a_K \cdot I^*}{\varepsilon_p \cdot \sigma}} \quad (14)$$

$$Dl = Kl \times \frac{DHI}{GHI} \quad (15)$$

$$I^* = K\downarrow \times \frac{DNI}{GHI} \quad (16)$$

$$K_{abs,UH} = \left(0.5a_{\kappa} \times \frac{DHI}{GHI} + f_p \cdot a_{\kappa} \times \frac{DNI}{GHI}\right) \cdot K\downarrow \quad (17)$$

$$L_{abs,UH} = 0.5a_l L\downarrow \quad (18)$$

$$K_{abs,LH} = 0.5a_{\kappa} K^{\dagger} \quad (19)$$

$$L_{abs,LH} = 0.5a_l L^{\dagger} \quad (20)$$

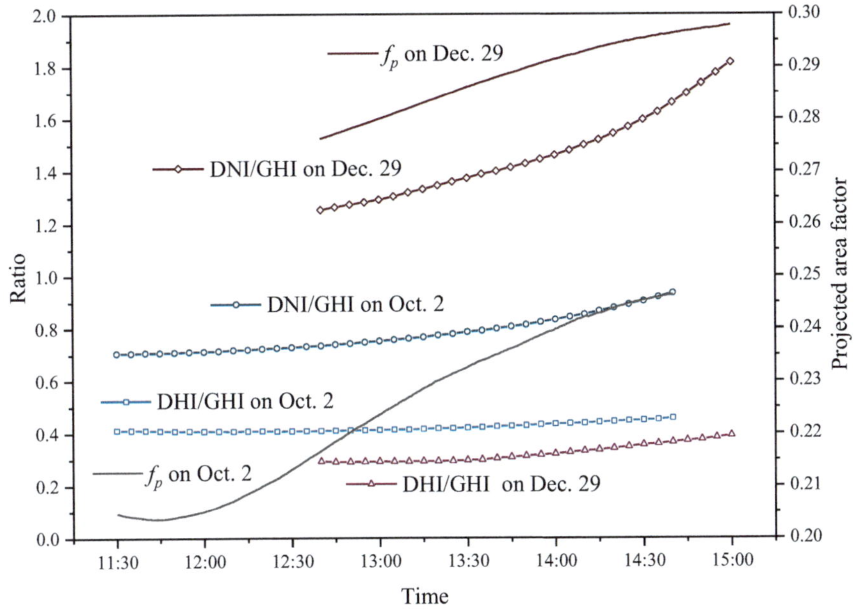


**Fig. 7** Ratios of DHI/GHI and DNI/GHI, as well as the projected area factor for a human body in a standing position

## *2.4 Modification Method (MTHM)*

Based on the analysis of MRT differences between the THM and SDIM in Sect. 4, the deviation in THM MRT ($MRT_{TMH}$) results from overestimated sky diffuse radiation due to unaccounted anisotropic sky conditions. Multiple studies have confirmed that sky diffuse radiation exhibits strong anisotropy, with radiation intensity near the solar disk being substantially higher than in other sky regions [48–52]. Accordingly, we propose a modified approach to adjust the sky diffuse radiation incident on the human body for improved MRT accuracy within the THM framework, as expressed in Eq. (21), where $D_{back}$ is the diffuse solar radiation from the sky background part (W/m$^2$), $D_{cir}$ diffuse solar radiation from the circumsolar region (W/m$^2$). The parameter $f_p$ is used to compute $D_{cir}$ incident on the human surface. As shown in Fig. 8, the sum of $D_{back}$ and $D_{cir}$ equals the total anisotropic sky diffuse radiation. It should be noted that $D_{back}$ is isotropic and originates from the entire sky dome, including the circumsolar area.

$$\mathrm{MRT} = \sqrt{\frac{S_{str}}{\varepsilon_p \cdot \sigma}}$$

$$= \sqrt{\frac{0.5(a_l L\uparrow + a_k D_{\text{back}}) + 0.5(a_l L\downarrow + a_k D\uparrow) + f_p \cdot a_k \cdot I^* + f_p \cdot a_k \cdot D_{\text{cir}}}{\varepsilon_p \cdot \sigma}} \tag{21}$$

As shown in Fig. 9, $D_{\text{cir}}$ and $D_{\text{back}}$ were derived using a simplified model based on Lou's study [53]. Per Eq. (22), $D_{\text{cir}}$ equals the product of $F_C$ and the total horizontal sky diffuse radiation ($E_{HD}$, W/m$^2$), and $F_C$ is defined as the ratio between the difference of circumsolar diffuse radiation ($E_C$, W/m$^2$) and background diffuse radiation ($E_B$, W/m$^2$) to $E_{HD}$. Noting the distinction between $E_C$ and $D_{\text{cir}}$, where $E_C$ is the radiation with the background radiation at the circumsolar part but $D_{\text{cir}}$ is the radiation without the background radiation. Moreover, $E_B$ is approximately equal to $D_{\text{back}}$. . In Lou's model, the circumsolar parts include zone 1 ($0° < \chi < 12°$), zone 2 ($12° < \chi < 24°$), and zone 3 ($24° < \chi < 36°$), and the background part consists of the entire sky. Then, the solid angles for zone 1, 2, 3, and the background part are 0.137 sr, 0.102 sr, 0.164 sr, and $2\pi$ sr, respectively. Therefore, the $F_C$ is equal to the sum of $F_{C1}$, $F_{C2}$, and $F_{C3}$ ($F_{Ci}$ is the ratio of ($E_{Ci}$-$E_B$) to $E_{HD}$, $i = 1, 2, 3$), which can be determined using Eqs. (23) to (25) [53], where $\alpha$ is the solar altitude angle (degree), and $K_t$ is the clearness index, which is defined as the ratio of global solar irradiance measured at ground level and its counterpart estimated at the top of the atmosphere [54]. $K_t$ was set to be 0.8 as a result of the clear, low-turbidity sky present for the direct sunlight measurement [55]. Moreover, $E_{HD}$ is roughly equivalent to $D\downarrow$ obtained using Eq. (15). Thus, $D_{\text{cir}}$ is readily accessible. Similarly, the $D_{\text{back}}$ was calculated using Eq. (26), where $F_B$ is the ratio of $E_B$ to $E_{HD}$ and can be obtained by Eq. (27) [53].

$$D_{\text{cir}} = f_C E_{HD} = \frac{E_C - E_B}{E_{HD}} E_{HD} \tag{22}$$

$$T_{C1} = 0.36K_T^2(1 + \exp(-6.29 + 7.9K_T + 0.92\sin\alpha))^{-1} \tag{23}$$

$$f_{C2} = 0.68K^2(1 + \exp(-4.66 + 5.51K_r + 1.15\sin\alpha))^{-1} \tag{24}$$

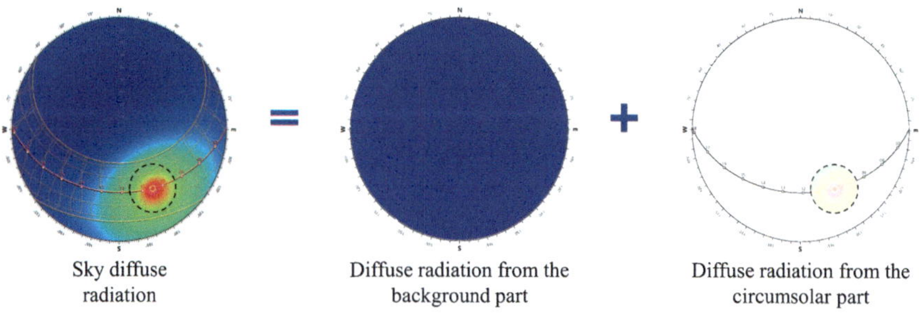


**Fig. 8** Diffuse radiation from sky background and circumsolar parts

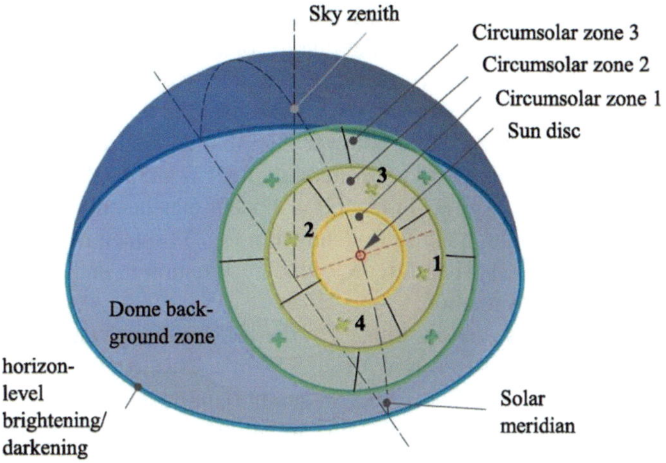


**Fig. 9** Circumsolar and background regions of the sky [53]. The ranges of zone 1, zone 2, and zone 3 are $0° < \chi < 12°$, $12° < \chi < 24°$, and $24° < \chi < 36°$, respectively. The $\chi$ represents the angular separation between the sun and a sky element, measured with the center of the celestial hemisphere as the vertex

$$f_{C3} = \left(149.6K_T^3 + 0.91\sin^2\alpha_S\right)(1 + \exp(2.43 + 4.81K_T + 0.84\sin\alpha))^{-1} \quad (25)$$

$$D_{\text{back}} = f_B E_{HD} = \frac{E_B}{E_{HD}} E_{HD} \quad (26)$$

$$f_B = \left(1.036 - 0.74K_r + 0.31K_r^2\right) + (1 + \exp(1.46 - 0.96K_r + 9.73\sin\alpha))^{-1} \quad (27)$$

# 3 Results

## 3.1 Shaded Environment

As shown in Fig. 10, $K_{abs}$ was underestimated by values ranging from 8.5 W/m$^2$ to 11.8 W/m$^2$ between the lower and upper quartiles (25% to 75%) in the box plot, with a maximum underestimation reaching 12.9 W/m$^2$. This discrepancy occurs because the THM underestimated $K_{abs,UH}$, as seen in Fig. 11, whereas the SDIM provides a more accurate determination. The radiation received by a sensor from a surface is proportional to both the view factor and the surface's radiant emission. In shaded conditions, the view factors from horizontal environmental surfaces to horizontal radiation sensors exceed those to vertical sensors. Conversely, view factors from vertical surfaces to vertical sensors are larger than those to horizontal sensors. As illustrated in Fig. 12, $K_{hor}$ is substantial due to solar radiation scattered by the sky, distant façades, and the ground. However, $K\uparrow$ is relatively low because tree canopy shading reduces radiation incident on the ground near the downward-facing sensor. Consequently, $K_{abs,LH}$ in the THM is lower than that derived from Eq. (10) in the SDIM, resulting in an average underestimation of 8.9 W/m2 by the THM, as shown in Fig. 11. Additionally, $K_{abs,UH}$ in the THM closely matches $K_{abs,UH}$ in the SDIM for the majority of measurement times (shown in Fig. 11) since $K\downarrow$ is comparable to $K_{hor}$ (shown in Fig. 12). Therefore, $K_{abs}$ underestimation in the THM is mostly a result of $K_{abs,LH}$ underestimation.

Similarly, $L_{abs}$ was also underestimated by the THM, although the underestimation is less than $K_{abs}$ and ranges from 4 W/m$^2$ to 5 W/m$^2$. . Due to comparable longwave radiation contributions from above, below, and horizontal directions (Fig. 12), both $L_{abs,UH}$ and $L_{abs,LH}$ in the THM and SDIM are similar. As a result, the MRT in the THM was underestimated by 0.8 °C to 2.8 °C, primarily due to the underestimation of $K_{abs}$ Based on our earlier research [9], this level of MRT underestimation in the THM is generally acceptable for evaluating outdoor thermal comfort in most situations.

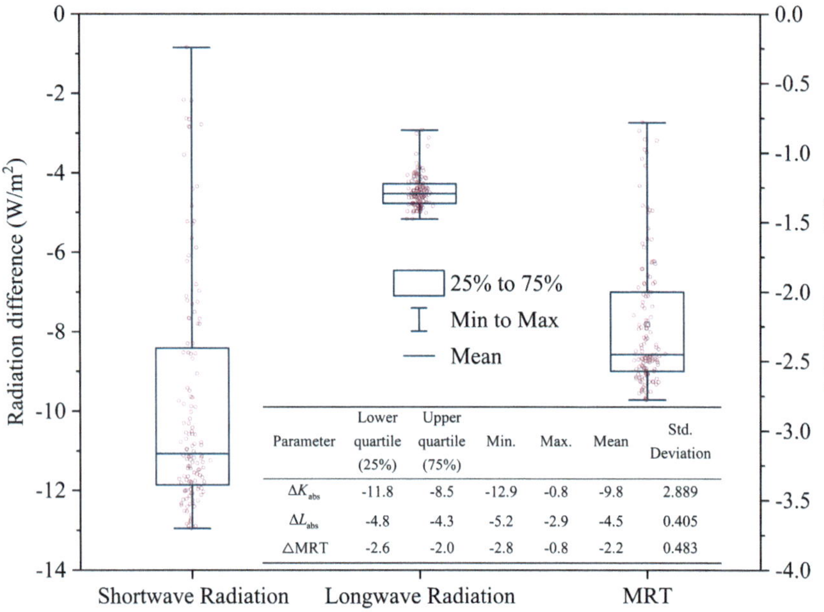


| Parameter | Lower quartile (25%) | Upper quartile (75%) | Min. | Max. | Mean | Std. Deviation |
|---|---|---|---|---|---|---|
| $\Delta K_{abs}$ | -11.8 | -8.5 | -12.9 | -0.8 | -9.8 | 2.889 |
| $\Delta L_{abs}$ | -4.8 | -4.3 | -5.2 | -2.9 | -4.5 | 0.405 |
| ΔMRT | -2.6 | -2.0 | -2.8 | -0.8 | -2.2 | 0.483 |

**Fig. 10** Difference between the HTM minus the SDIM in the absorbed radiation and MRT at P1

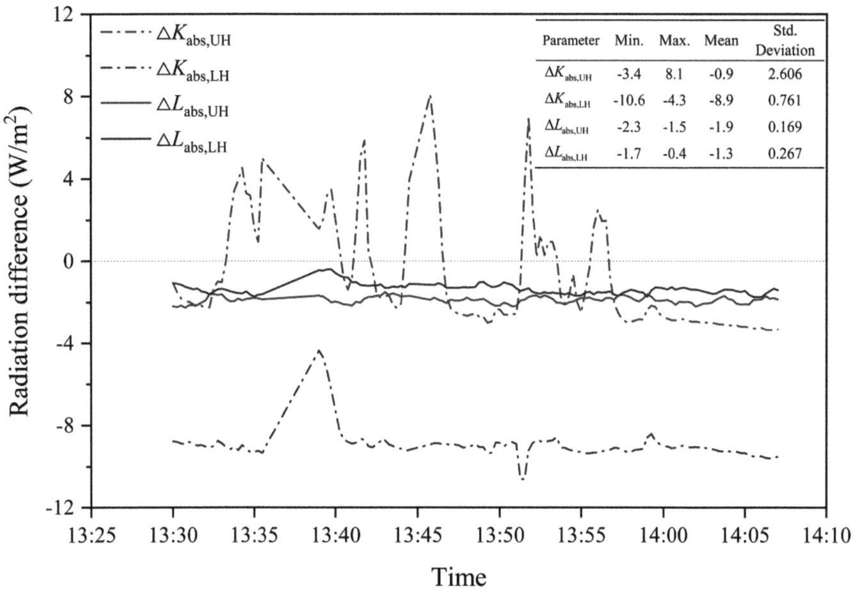


**Fig. 11** Difference between the HTM minus the SDIM in the absorbed radiation from the upper and lower hemispheres at P1

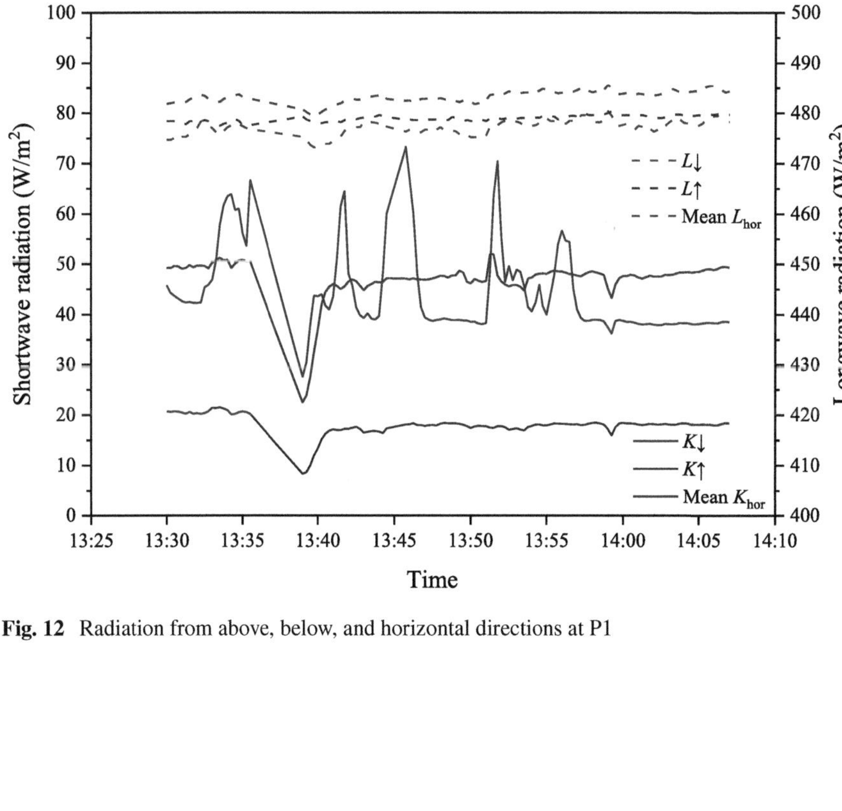


**Fig. 12** Radiation from above, below, and horizontal directions at P1

## 3.2 *Direct Sunlight Environment*

### 3.2.1 Summer Environment

In direct sunlight environment during the summer, $K_{abs}$ was significantly overestimated in the THM for the majority of the measurement time, as depicted in Fig. 13. The maximal $K_{abs}$ differences in the P2 and P3 are 47.6 W/m$^2$ and 45.0 W/m$^2$, respectively, which significantly greater than their value (12.9 W/m$^2$) in the shaded environment. As shown in Fig. 14, $K\downarrow$ is obviously greater than $K\uparrow$ and mean $K_{hor}$ due to the direct and diffuse solar radiation from above. Thus, $K_{abs,UH}$ in the THM is great, with a large view factor of 0.5, and it is significantly greater than it is in the SDIM, as shown in Fig. 15. $K_{abs,UH}$ overestimation in the THM is 71.6 W/m$^2$ on average, with a maximum of 86.6 W/m$^2$ in the measurement. In contrast, $K_{abs,LH}$ is underestimated by the THM for the same reasons as it is in the shaded environment. Noting that the shortwave radiation or $K_{abs}$ is highly variable at certain times [56–58], such as the measurement time of 11:30 to 12:05 shown in Fig. 14 and Fig. 15, as a result of clouds intermittently obscuring the sun. As in the shaded environment, the $L_{abs}$ was underestimated by the THM, as shown in Fig. 13. However, the reason for this underestimation differs from that of the shaded environment, which is the underestimation of the total $L_{abs}$ coming from the fact that $L_{abs,UH}$ underestimation is greater than $L_{abs,LH}$ overestimation, as shown in Fig. 15. $L_{abs,UH}$ underestimation results from the little $L\downarrow$ due to the minimal longwave radiation from the sky. $L_{abs,LH}$ overestimation is caused by the large $L\uparrow$, which is due to the great radiation emitted from the ground due to the high temperature resulting from direct sunlight [59, 60]. $L_{abs,UH}$ underestimation and $L_{abs,LH}$ overestimation are comparable, resulting in a total $L_{abs}$ deviation variation in the THM that is less than 11 W/m$^2$ (shown in Fig. 13). In the TMH, the MRT was overestimated by up to 5 °C due to the more significant $K_{abs}$ overestimation.

### 3.2.2 Winter Environment

As observed in summer under direct sunlight, $K_{abs}$ is also predominantly overestimated by the THM during winter (Fig. 16), though the overestimation is reduced and mainly ranges between 5.4 W/m2 and 34.8 W/m$^2$. The explanation is that $K\downarrow$ in winter becomes smaller than in summer ( as shown in Fig. 17), causing $K_{abs,UH}$ to decrease as well (shown in Fig. 18). In contrast, $L_{abs}$ is underestimated by the THM in winter (shown in Fig. 16) as it is in summer, but $L_{abs}$ underestimation grows greater, reaching a maximum of 21.7 W/m$^2$. As shown in Fig. 18, the winter $L_{abs,LH}$ is comparable to that in summer. Nevertheless, $\Delta L_{abs,LH}$ decreases due to reduced ground-emitted radiation resulting from lower ground temperatures in winter. Consequently, the total $L_{abs}$ deviation in winter exceeds that in summer.

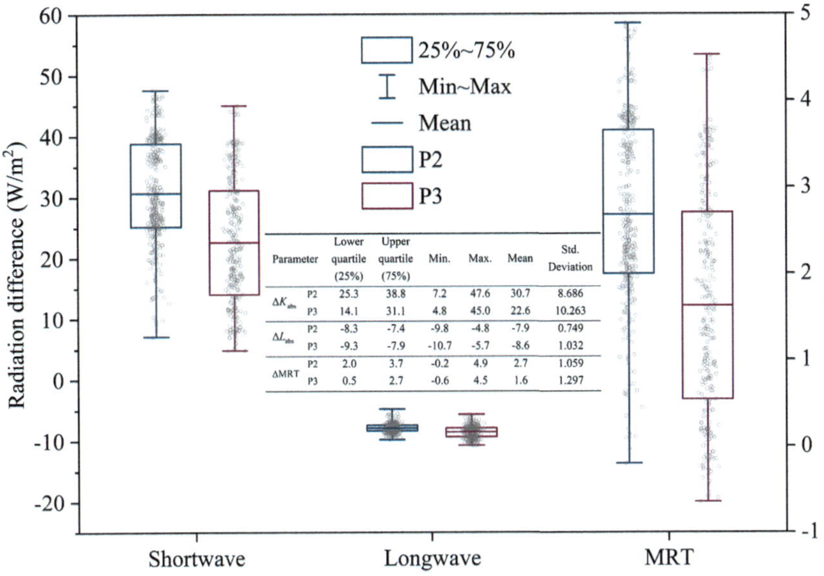


| Parameter | | Lower quartile (25%) | Upper quartile (75%) | Min. | Max. | Mean | Std. Deviation |
|---|---|---|---|---|---|---|---|
| $\Delta K_{abs}$ | P2 | 25.3 | 38.8 | 7.2 | 47.6 | 30.7 | 8.686 |
| | P3 | 14.1 | 31.1 | 4.8 | 45.0 | 22.6 | 10.263 |
| $\Delta L_{abs}$ | P2 | -8.3 | -7.4 | -9.8 | -4.8 | -7.9 | 0.749 |
| | P3 | -9.3 | -7.9 | -10.7 | -5.7 | -8.6 | 1.032 |
| ΔMRT | P2 | 2.0 | 3.7 | -0.2 | 4.9 | 2.7 | 1.059 |
| | P3 | 0.5 | 2.7 | -0.6 | 4.5 | 1.6 | 1.297 |

**Fig. 13** Between the HTM minus the SDIM in the absorbed radiation and MRT at P2 and P3 during the summer

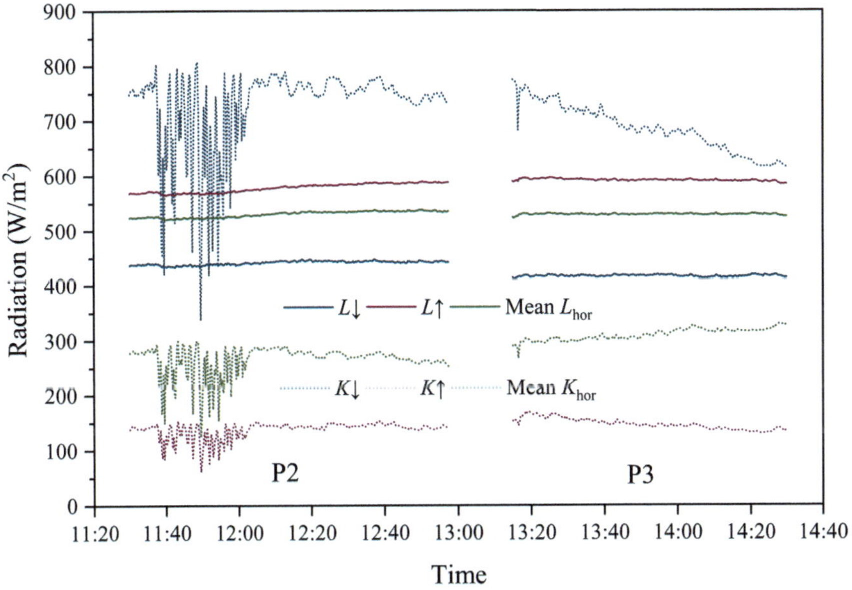


**Fig. 14** Radiation from above, below, and horizontal directions at P2 and P3 during the summer

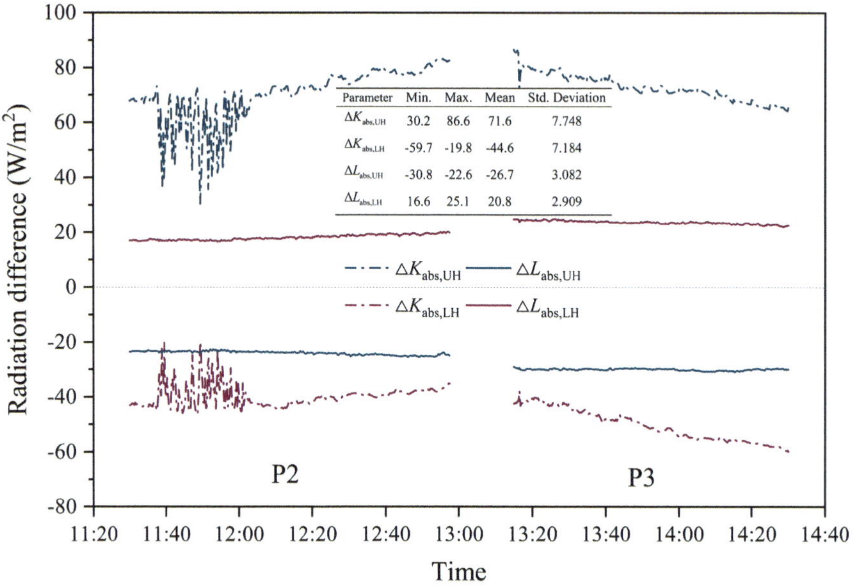


**Fig. 15** Difference between the HTM minus the SDIM in the absorbed radiation from the upper and lower hemispheres at P2 and P3 during the summer

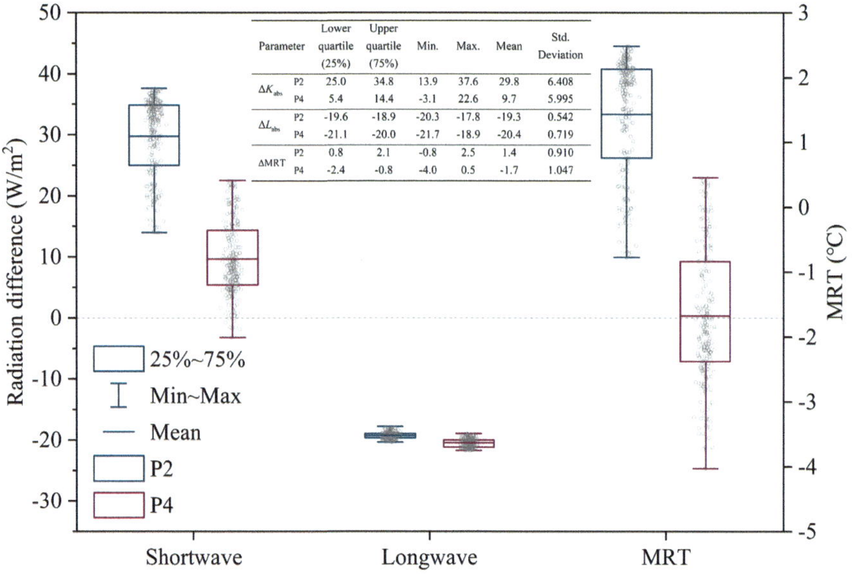


**Fig. 16** Difference between the HTM minus the SDIM in the absorbed radiation and MRT at P2 and P4 during the winter

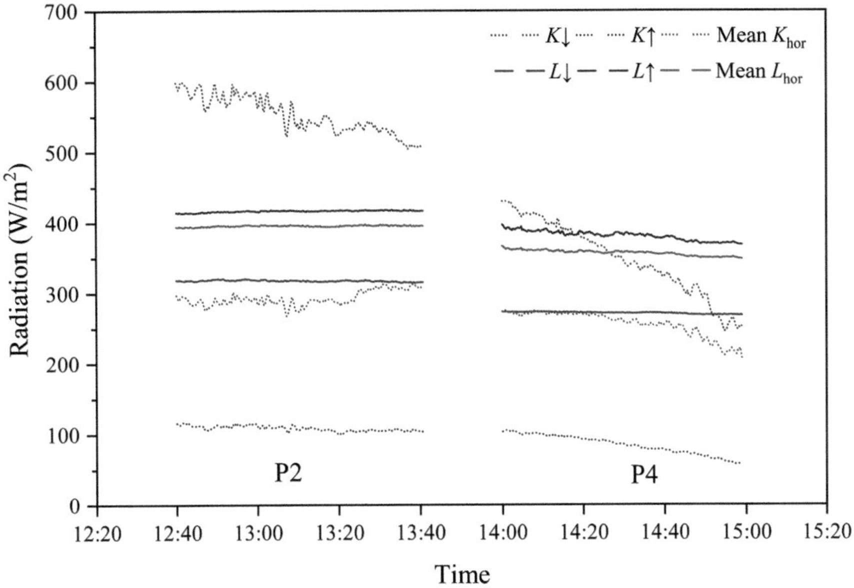


**Fig. 17** Radiation from above, below, and horizontal directions at P2 and P4 during the winter

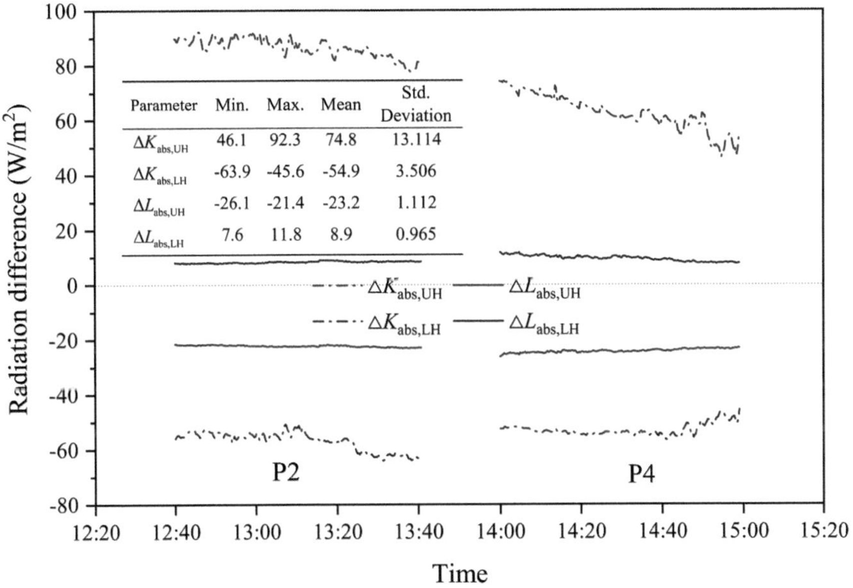


| Parameter | Min. | Max. | Mean | Std. Deviation |
|---|---|---|---|---|
| $\Delta K_{abs,UH}$ | 46.1 | 92.3 | 74.8 | 13.114 |
| $\Delta K_{abs,LH}$ | -63.9 | -45.6 | -54.9 | 3.506 |
| $\Delta L_{abs,UH}$ | -26.1 | -21.4 | -23.2 | 1.112 |
| $\Delta L_{abs,LH}$ | 7.6 | 11.8 | 8.9 | 0.965 |

**Fig. 18** Difference between the HTM minus the SDIM in the absorbed radiation from the upper and lower hemispheres at P2 and P4 during the winter

Owing to both the smaller overestimation of $K_{abs}$ and the greater underestimation of $L_{abs}$ within the THM, the mean MRT deviation in winter ranges from 1.4 °C to 1.7 °C, which is less pronounced than the summer deviation.

# 4 Discussion

According to the results shown in Sect. 3, the deviation in $MRT_{TMH}$ under partially shaded conditions is smaller than that observed in direct sunlight environments. In shaded settings, the MRT deviation during measurement remains below 3 °C due to the relatively uniform radiation distribution (as illustrated in Fig. 11), which is considered negligible for most outdoor thermal comfort evaluations [9]. This implies that MRT in shaded regions can be accurately measured using the THM, a method that is both more economical and more practical than the SDIM. However, in sun-exposed environments, significant non-uniformity in radiation can cause substantially larger deviations in $MRT_{TMH}$, as demonstrated in Figs. 14 and 17. The maximum deviation reaches nearly 5 °C, which may lead to a half-grade shift in thermal sensation [9]. Similar results, including even greater MRT deviations under varying measurement conditions, have been reported in the study by Lee et al. [25].

## 4.1 *Effect of Shortwave Radiation Deviation on the MRT*

In direct sunlight environments, $K_{abs}$ was largely overestimated by the THM during most measurement periods, while $L_{abs}$ was consistently underestimated in both summer and winter, as shown in Figs. 13 and 16. Therefore, the slight MRT deviation observed in the THM may be attributed to the overestimation of $K_{abs}$ being nearly balanced by the underestimation of $L_{abs}$, rather than indicating high accuracy in measuring either radiation component. In reality, the opposing effects of $K_{abs}$ overestimation and $L_{abs}$ underestimation on the MRT are canceled out by each other, since the MRT is determined by the sum of $K_{abs}$ and $L_{abs}$ underestimation on MRT partially cancel each other, since MRT is determined by the total of shortwave and longwave radiation. Although this mutual compensation reduces the apparent MRT deviation, the reliability of the MRT estimation remains limited. Hence, greater measurement precision for both shortwave and longwave radiation should be pursued.

As shown in Figs. 15 and 18, $K_{abs,UH}$ in the THM was significantly overestimated during most measurement times, especially when $K\downarrow$ was high (Figs. 14 and 17). To assess the influence of $K_{abs}$ deviation in the THM on MRT, we calculated MRT using $K_{abs}$ from the THM and $L_{abs}$ from the SDIM, as illustrated in Fig. 19. The results demonstrate that MRT differences caused by $K_{abs}$ deviation are greater than those resulting from combined $K_{abs}$ and $L_{abs}$ inaccuracies. T The MRT differences range from –0.5 °C to 5.6 °C, with most values being positive due to systematic

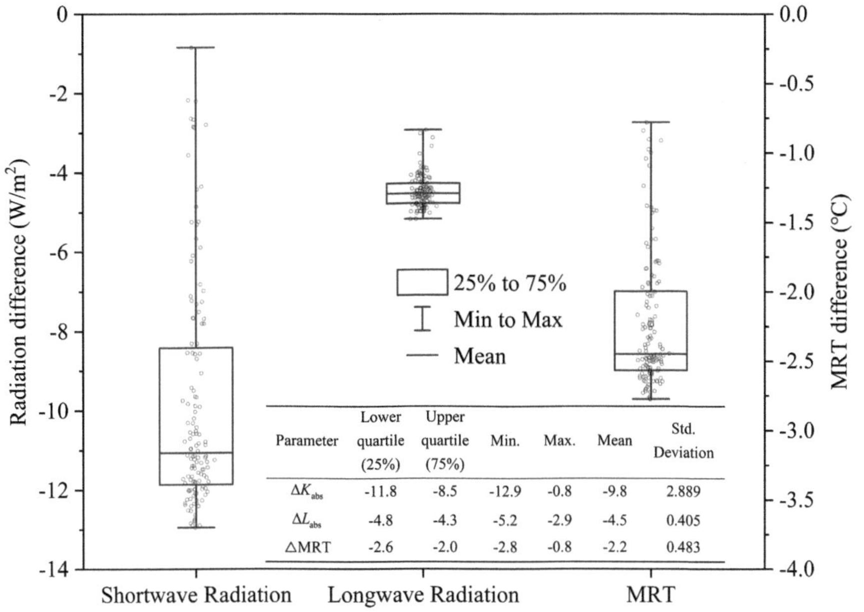


| Parameter | Lower quartile (25%) | Upper quartile (75%) | Min. | Max. | Mean | Std. Deviation |
|---|---|---|---|---|---|---|
| $\Delta K_{abs}$ | -11.8 | -8.5 | -12.9 | -0.8 | -9.8 | 2.889 |
| $\Delta L_{abs}$ | -4.8 | -4.3 | -5.2 | -2.9 | -4.5 | 0.405 |
| ΔMRT | -2.6 | -2.0 | -2.8 | -0.8 | -2.2 | 0.483 |

**Fig. 19** Effect of the $K_{abs}$ deviation in the THM on the MRT (the MRT difference is the result of $MRT_{TMH}$ minus $MRT_{SDIM}$)

overestimation of $K_{abs}$ by the THM. Additionally, MRT differences are smaller in winter compared to summer, where the maximum deviation reaches 5.6 °C.

## 4.2 Deviation from Sky Radiation

As illustrated in Fig. 20, it has been established that sky diffuse radiation exhibits strong anisotropy [50–52, 61, 62], with the circumsolar region being markedly more intense than other sky areas (or the background region). The results indicate that disregarding anisotropic sky diffuse radiation could lead to an MRT discrepancy of up to 3 °C [63]. In the measurements, the overestimation of $K_{abs}$ within the THM, resulting from the overestimation of $K_{abs,UH}$, is also a consequence of exaggerated accounting for anisotropic sky diffuse radiation.

Radiation from the circumsolar and background regions is predominantly measured using the upward-facing shortwave radiation sensor (pyranometer) in the THM, and it is applied to determine $K_{abs,UH}$ with a view factor of 0.5, as expressed in Eq. (17). In reality, $D_{cir}$ exhibits directionality similar to $I^*$, , which changes with the sun's position. Therefore, the radiation ($D_{cir,abs}$, W/m$^2$) absorbed by a human body from $D_{cir}$ in the THM can be estimated using Eq. (28). However, the correct value of $D_{cir,abs}$ should be computed using Eq. (29), where $f_p$ may not be equal to $0.5 \cos\theta$ in

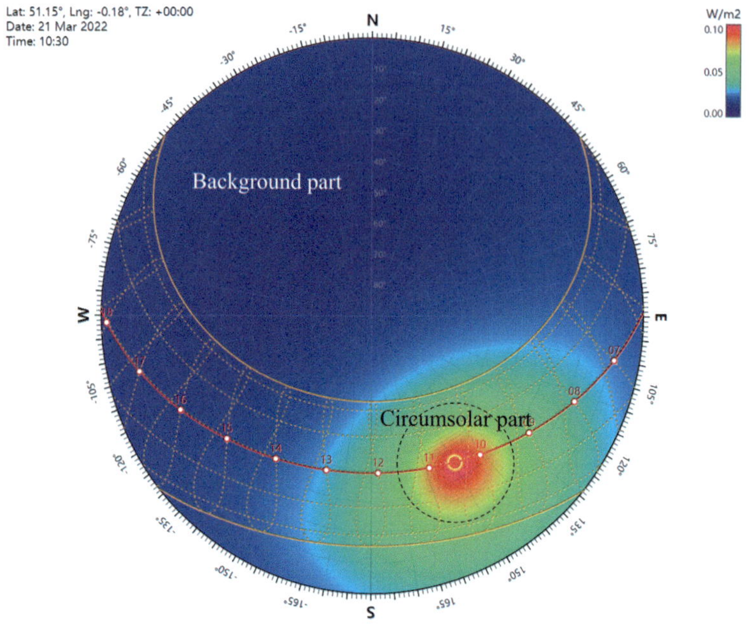


**Fig. 20** Anisotropic sky diffuse radiation

Eq. (28). Consequently, $D_{\text{cir,abs}}$ is overestimated by the THM when the "$0.5\cos\theta$" is greater than $f_p$, and vice versa.

$$D_{\text{cir,abs}} = \alpha_\kappa 0.5\cos\theta D_{\text{cir}} \tag{28}$$

$$D_{\text{cir,abs}} = \alpha_{\kappa,f} D_{\text{cir}} \tag{29}$$

As indicated in Fig. 21, the difference remains above zero and diminishes progressively during the measurement conducted on October 2, which corresponds to the overestimation of $K_{\text{abs,UH}}$ within the THM (shown in Fig. 15). Likewise, the pattern of the difference observed on December 29 also reflects the variation in $K_{\text{abs,UH}}$ illustrated in Fig. 18. However, the $K_{\text{abs}}$ difference is inconsistent with the difference of $0.5\cos\theta$ minus $f_p$. The principal reason is that the SDIM may underestimate $D_{cir}$ due to the discrepancy between the direction of horizontal sensors and the direction of $D_{cir}$. Actually, $D_{cir}$ can originate from a variety of azimuths, but there are only four measurement directions for horizontal sensors in the SDIM if the sensors are not rotated [33]. Over time, $K_{\text{abs,UH}}$ in the TMH will be underestimated due to the decline of $0.5\cos\theta$ and the increase of $f_p$ [8], which can be found in the paper by Lee et al. [25].

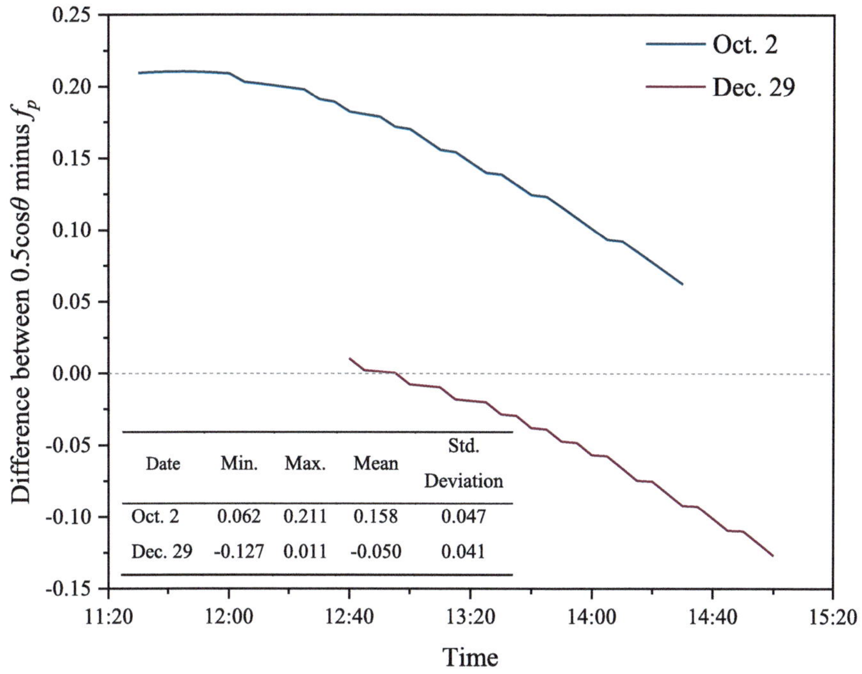


**Fig. 21** Difference between 0.5 cos $\theta$ minus $f_p$

## 4.3 *MTHM Performance*

Based on the analysis in Sect. 4.1, we introduce the MTHM to refine the calculation of sky diffuse radiation within the THM framework, as described in Sect. 2.4. As shown in Fig. 22, the $K_{\text{abs}}$ deviation significantly decreases in the summer after the MTHM is applied. Thus, the maximal MRT deviation drops from around 5 °C to less than 3 °C, and the mean MRT deviation (25% to 75% in the box plot) decreases from 2.2 °C to 0.5 °C. Similarly, the $K_{\text{abs}}$ deviation decreases in winter following adjustment by the MTHM. However, it does not appear that the MRT deviation has diminished. The reason is that the $K_{\text{abs,LH}}$ underestimation has not changed, even though the $K_{\text{abs,UH}}$ overestimation has. Thus, the total $K_{\text{abs}}$ deviation may increase in the MTHM compared to the THM. However, the MTHM is more robust than the THM since $K_{\text{abs,UH}}$ is more accurate. Noting that $K_{\text{abs}}$ overestimation persists in both summer and winter shows that the $K_{\text{abs,UH}}$ overestimation remains. The principal reason is the $D_{cir}$ underestimation by the SDIM. Fortunately, the $K_{\text{abs}}$ overestimation has been greatly decreased. Overall, the MTHM substantially improves MRT accuracy compared to the THM during summer. Although the MTHM may not enhance MRT accuracy under certain winter conditions, such as those in our measurements, it does improve the robustness of MRT estimates. Additionally, most MRT deviations in both the THM and MTHM remain relatively small during winter and are generally acceptable

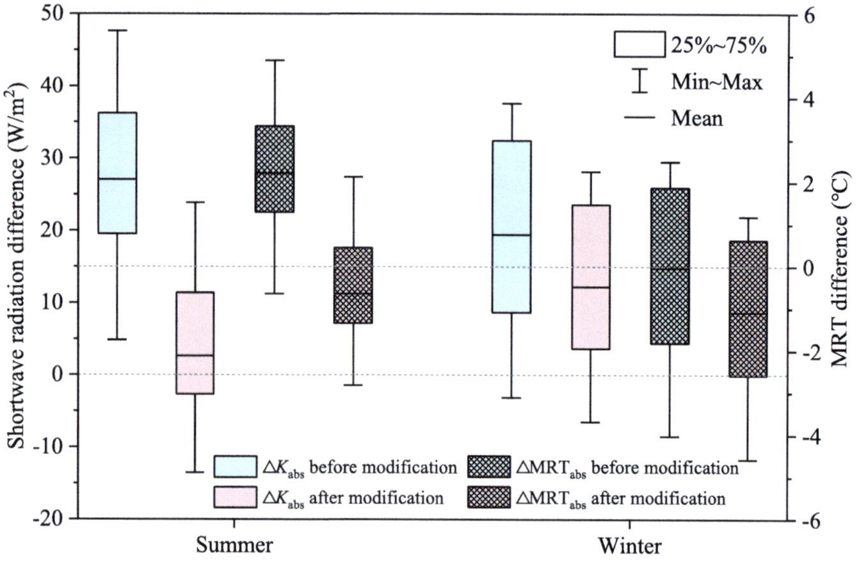


**Fig. 22** Shortwave radiation and MRT difference between the THM minus the MTHM

for outdoor thermal comfort evaluation. Therefore, the performance of the MTHM is superior to that of the THM, making it well-suited for outdoor MRT measurements used in assessing thermal comfort.

## 5 Conclusions

This chapter evaluated different methods for measuring outdoor mean radiant temperature (MRT), focusing on the comparison between the traditional two-hemisphere method (THM) and the six-direction integral method (SDIM), and introduced a modified two-hemisphere method (MTHM). The analysis showed that THM performs reasonably well under shaded conditions, with deviations from SDIM generally below 3 °C, but its accuracy declines under direct solar exposure, particularly in summer, due to limitations in representing anisotropic sky diffuse radiation. By incorporating anisotropic diffuse components, the MTHM substantially improves accuracy and robustness, yielding results that align more closely with SDIM while retaining practical advantages in field applications.

Although the modified approach demonstrates clear benefits, further validation across longer time spans, diverse climatic contexts, and seasonal variations is required. Refinements in pyrheliometer configuration and diffuse radiation modeling could also enhance performance. Overall, the MTHM offers a balanced solution between accuracy and feasibility, providing a reliable tool for outdoor MRT assessment and supporting more robust evaluations of outdoor thermal comfort.

**Acknowledgements** This study was supported by the National Natural Science Foundation of China [grant number 52178086].

## References

1. D. Lai, B. Chen, K. Liu, Quantification of the influence of thermal comfort and life patterns on outdoor space activities, in *Building simulation* (Springer, 2020), pp. 113–-125
2. M.A. Eusuf, M.A. Mohit, M.S. Eusuf, M. Ibrahim, Impact of outdoor environment to the quality of life. Proc. Soc. Behav. Sci. **153**, 639–654 (2014)
3. H. Guo, D. Aviv, M. Loyola, E. Teitelbaum, N. Houchois, F. Meggers, On the understanding of the mean radiant temperature within both the indoor and outdoor environment, a critical review. Renew. Sustain. Energy Rev. **117**, 109207 (2020)
4. J. He, A. Hoyano, T. Asawa, A numerical simulation tool for predicting the impact of outdoor thermal environment on building energy performance. Appl. Energy **86**(9), 1596–1605 (2009)
5. F.S. de La Flor, S.A. Domınguez, Modelling microclimate in urban environments and assessing its influence on the performance of surrounding buildings. Energy Build. **36**(5), 403–413 (2004)
6. S. Thorsson, J. Rocklöv, J. Konarska, F. Lindberg, B. Holmer, B. Dousset, D. Rayner, Mean radiant temperature–a predictor of heat related mortality. Urban Climate **10**, 332–345 (2014)
7. M. Nikolopoulou, *Thermal Comfort in Urban Spaces, Urban Microclimate Modelling for Comfort and Energy* Studies (Springer, 2021), pp. 55–77
8. Z. Li, X. Feng, X. Fan, J. Sun, Z. Fang, Effect of direct solar projected area factor on outdoor thermal comfort evaluation: a case study in Shanghai, China. Urban Climate **41**, 101033 (2022)
9. X. Feng, Z. Zheng, Y. Yang, Z. Fang, Quantitative seasonal outdoor thermal sensitivity in Guangzhou, China. Urban Climate **39**, 100938 (2021)
10. D. Lai, W. Liu, T. Gan, K. Liu, Q. Chen, A review of mitigating strategies to improve the thermal environment and thermal comfort in urban outdoor spaces. Sci. Total Environ. **661**, 337–353 (2019)
11. Z. Fang, X. Feng, J. Liu, Z. Lin, C.M. Mak, J. Niu, K.-T. Tse, X. Xu, Investigation into the differences among several outdoor thermal comfort indices against field survey in subtropics. Sustain. Cities Soc. **44**, 676–690 (2019)
12. Z. Fang, X. Feng, X. Xu, X. Zhou, Z. Lin, Y. Ji, Investigation into outdoor thermal comfort conditions by different seasonal field surveys in China, Guangzhou. Int. J. Biometeorol. **63**(10), 1357–1368 (2019)
13. J. Huang, C. Zhou, Y. Zhuo, L. Xu, Y. Jiang, Outdoor thermal environments and activities in open space: an experiment study in humid subtropical climates. Build. Environ. **103**, 238–249 (2016)
14. P.K. Cheung, C.Y. Jim, Improved assessment of outdoor thermal comfort: 1-hour acceptable temperature range. Build. Environ. **151**, 303–317 (2019)
15. J. Li, N. Liu, The perception, optimization strategies and prospects of outdoor thermal comfort in China: a review. Build. Environ. **170**, 106614 (2020)
16. M. Taleghani, Outdoor thermal comfort by different heat mitigation strategies—a review. Renew. Sustain. Energy Rev. **81**, 2011–2018 (2018)
17. ASHRAE, *2017 ASHRAE Handbook: Fundamentals*, ASHRAE2017
18. N. Kántor, J. Unger, The most problematic variable in the course of human-biometeorological comfort assessment—the mean radiant temperature. Central Eur. J. Geosciences **3**(1), 90–100 (2011)
19. M.A. Rahman, V. Dervishi, A. Moser-Reischl, F. Ludwig, H. Pretzsch, T. Rötzer, S. Pauleit, Comparative analysis of shade and underlying surfaces on cooling effect. Urban Forestry & Urban Greening 127223 (2021)

20. M. Napoli, L. Massetti, G. Brandani, M. Petralli, S. Orlandini, Modeling tree shade effect on urban ground surface temperature. J. Environ. Qual. **45**(1), 146–156 (2016)
21. M. Aminipouri, A.J. Knudby, E.S. Krayenhoff, K. Zickfeld, A. Middel, Modelling the impact of increased street tree cover on mean radiant temperature across Vancouver's local climate zones. Urban Forestry & Urban Greening **39**, 9–17 (2019)
22. C.Y. Park, D.K. Lee, E.S. Krayenhoff, H.K. Heo, J.H. Hyun, K. Oh, T.Y. Park, Variations in pedestrian mean radiant temperature based on the spacing and size of street trees. Sustain. Cities Soc. **48**, 101521 (2019)
23. P.Y. Fan, K.P. Chun, A. Mijic, D.N.-Y. Mah, Q. He, B. Choi, C.K.C. Lam, O. Yetemen, Spatially-heterogeneous impacts of surface characteristics on urban thermal environment, a case of the Guangdong-Hong Kong-Macau Greater Bay Area. Urban Climate **41**, 101034 (2022)
24. J.K. Vanos, K. Rykaczewski, A. Middel, D.J. Vecellio, R.D. Brown, T.J. Gillespie, Improved methods for estimating mean radiant temperature in hot and sunny outdoor settings. Int. J. Biometeorol. **65**(6), 967–983 (2021)
25. H. Lee, S. Jo, S. Park, A simple technique for the traditional method to estimate mean radiant temperature. Int. J. Biometeorol. **66**(3), 521–533 (2022)
26. E. Krüger, F. Minella, A. Matzarakis, Comparison of different methods of estimating the mean radiant temperature in outdoor thermal comfort studies. Int. J. Biometeorol. **58**(8), 1727–1737 (2014)
27. S. Thorsson, F. Lindberg, I. Eliasson, B. Holmer, Different methods for estimating the mean radiant temperature in an outdoor urban setting. Int. J. Climatol. J. R. Meteorol. Soc. **27**(14), 1983–1993 (2007)
28. E. ISO, Ergonomics of the thermal environment—instruments for measuring physical quantities. Geneva Int Stand Organ (1998)
29. P. Tang, Q. Li, Evaluation of the observation methods of outdoor mean radiant temperature in a subtropical city. Build. Environ. **207**, 108462 (2022)
30. W. Ouyang, Z. Liu, K. Lau, Y. Shi, E. Ng, Comparing different recalibrated methods for estimating mean radiant temperature in outdoor environment. Build. Environ. 109004 (2022)
31. Y.-C. Chen, T.-P. Lin, A. Matzarakis, Comparison of mean radiant temperature from field experiment and modelling: a case study in Freiburg, Germany. Theoretical Appl. Climatol. **118**(3), 535–551 (2014)
32. N. Kántor, A. Kovács, T.-P. Lin, Looking for simple correction functions between the mean radiant temperature from the "standard black globe" and the "six-directional" techniques in Taiwan. Theoret. Appl. Climatol. **121**(1), 99–111 (2015)
33. R.D. Brown, Correcting the error in measuring radiation received by a person: introducing cylindrical radiometers. Sensors **19**(23), 5085 (2019)
34. V.D. Ingeneure, VDI 3787, Part II: Environmental Meteorology: Methods for the Human-biometeorological Evaluation of Climate and Air Quality for the Urban and Regional Planning at Regional Level, Part I: Climate. VDI/DIN-Handbuch Reinhaltung der Luft 1 (1998)
35. D. Robinson, A. Stone, A simplified radiosity algorithm for general urban radiation exchange. Build. Serv. Eng. Res. Technol. **26**(4), 271–284 (2005)
36. C.V. Gál, N. Kántor, Modeling mean radiant temperature in outdoor spaces, a comparative numerical simulation and validation study. Urban Climate **32**, 100571 (2020)
37. P.O. Fanger, *Thermal Comfort: Analysis and Applications in Environmental Engineering* (Danish Technical Press, 1970)
38. C.K.C. Lam, H. Lee, S.-R. Yang, S. Park, A review on the significance and perspective of the numerical simulations of outdoor thermal environment. Sustain. Cities Soc. **71**, 102971 (2021)
39. A. Matzarakis, F. Rutz, H. Mayer, Modelling radiation fluxes in simple and complex environments—application of the RayMan model. Int. J. Biometeorol. **51**(4), 323–334 (2007)
40. A. Matzarakis, F. Rutz, H. Mayer, Modelling radiation fluxes in simple and complex environments: basics of the RayMan model. Int. J. Biometeorol. **54**(2), 131–139 (2010)
41. M. Bruse, H. Fleer, Simulating surface–plant–air interactions inside urban environments with a three dimensional numerical model. Environ. Model. Softw. **13**(3–4), 373–384 (1998)

42. S. Huttner, Further development and application of the 3D microclimate simulation ENVI-met, Mainz, Univ, Diss (2012)
43. H. Staiger, A. Matzarakis, Accuracy of mean radiant temperature derived from active and passive radiometry. Atmosphere **11**(8), 805 (2020)
44. S. Park, S.E. Tuller, Comparison of human radiation exchange models in outdoor areas. Theoret. Appl. Climatol. **105**(3), 357–370 (2011)
45. Mohurd, *Standard for Weather Data of Building Energy Efficiency* (2015)
46. J.M. Bright, Solcast: validation of a satellite-derived solar irradiance dataset. Sol. Energy **189**, 435–449 (2019)
47. J.M. Bright, S. Killinger, N.A. Engerer, Data article: distributed PV power data for three cities in Australia. J. Renew. Sustain. Energy **11**(3), 035504 (2019)
48. W. Yao, Z. Li, X. Wang, Q. Zhao, Z. Zhang, L. Lin, A simplified model exploration research of new anisotropic diffuse radiation model. Energy Convers. Manage. **126**, 724–735 (2016)
49. Q.-H. Tao, Z. Li, J. Zheng, X. Chen, Model of solar diffuse radiation transmission through circular perforated louvers and experimental verification. Energy Build. **142**, 49–55 (2017)
50. Z. Li, H. Xing, G. Augenbroe, Criterion based selection of sky diffuse radiation models. Sustain. Cities Soc. **50**, 101692 (2019)
51. R. Perez, R. Seals, J. Michalsky, All-weather model for sky luminance distribution—preliminary configuration and validation. Sol. Energy **50**(3), 235–245 (1993)
52. R. Perez, P. Ineichen, R. Seals, J. Michalsky, R. Stewart, Modeling daylight availability and irradiance components from direct and global irradiance. Sol. Energy **44**(5), 271–289 (1990)
53. S. Lou, D.H. Li, I.Y. Lun, Y. Huang, W. Chen, D. Xia, Y. Yang, Radiance of the circumsolar and background parts of skydome for buildings under random obstructions. Energy Build. **236**, 110796 (2021)
54. B.Y. Liu, R.C. Jordan, The interrelationship and characteristic distribution of direct, diffuse and total solar radiation. Sol. Energy **4**(3), 1–19 (1960)
55. L. Yang, X. Gao, Z. Li, D. Jia, Quantitative effects of air pollution on regional daily global and diffuse solar radiation under clear sky conditions. Energy Rep. **8**, 1935–1948 (2022)
56. N. Kántor, A. Kovács, Á. Takács, Small-scale human-biometeorological impacts of shading by a large tree. Open Geosciences **8**(1), 231–245 (2016)
57. J.R. Mahan, *The Monte Carlo Ray-Trace Method in Radiation Heat Transfer and Applied Optics* (John Wiley & Sons, 2019)
58. S. Zheng, J.-M. Guldmann, Z. Liu, L. Zhao, Influence of trees on the outdoor thermal environment in subtropical areas: an experimental study in Guangzhou, China. Sustain. Cities Soc. **42**, 482–497 (2018)
59. M.A. Rahman, V. Dervishi, A. Moser-Reischl, F. Ludwig, H. Pretzsch, T. Rötzer, S. Pauleit, Comparative analysis of shade and underlying surfaces on cooling effect. Urban Forestry & Urban Greening **63**, 127223 (2021)
60. A. Speak, L. Montagnani, C. Wellstein, S. Zerbe, The influence of tree traits on urban ground surface shade cooling. Landsc. Urban Plan. **197**, 103748 (2020)
61. R. Perez, R. Stewart, C. Arbogast, R. Seals, J. Scott, An anisotropic hourly diffuse radiation model for sloping surfaces: description, performance validation, site dependency evaluation. Sol. Energy **36**(6), 481–497 (1986)
62. I. García, M. de Blas, B. Hernández, C. Sáenz, J.L. Torres, Diffuse irradiance on tilted planes in urban environments: evaluation of models modified with sky and circumsolar view factors. Renew. Energy **180**, 1194–1209 (2021)
63. N. Wallenberg, F. Lindberg, B. Holmer, S. Thorsson, The influence of anisotropic diffuse shortwave radiation on mean radiant temperature in outdoor urban environments. Urban Climate **31**, 100589 (2020)

# Human Factors in Outdoor Thermal Comfort

# Role of Human Factors in Thermal Comfort: Metabolic Rate and Activity Levels

**Abstract** Metabolic rate is a fundamental variable in thermal comfort research, as it determines the rate of internal heat production and directly affects human thermal sensation. This chapter reviews recent progress in the quantification of metabolic rate and its integration into outdoor thermal comfort evaluation. The literature indicates that while thermal indices such as PMV rely on fixed metabolic rate assumptions, real-world activity levels vary dynamically and contribute significantly to discrepancies in model predictions. Advances in interdisciplinary research, including physiological monitoring, wearable sensing, and metabolic modeling, have improved the ability to capture activity-related variations in comfort studies. Nevertheless, challenges remain in terms of accuracy, practicality, and the integration of metabolic rate with other human and environmental factors. Representing metabolic rate as a static parameter or considering it in isolation often leads to misinterpretation of field results. This chapter highlights the necessity of treating metabolic rate and activity levels as dynamic, context-dependent variables, and discusses how their systematic inclusion can improve the accuracy of thermal comfort indices and inform human-centered urban design.

## 1 Introduction

As living standards improve, the importance of thermal comfort has become increasingly evident [1]. People now prefer engaging in outdoor physical activities rather than staying indoors during leisure time, shifting research attention from indoor to semi-open and outdoor environments [1]. However, due to the urban heat island effect, outdoor discomfort in summer is common—especially during activities with high energy expenditure [2, 3]. Therefore, efforts are needed to enhance both built and natural thermal environments and to optimize human thermal perception [4–6].

Thermal comfort is shaped by the interaction of environmental and human factors [7, 8]. Because of geographical variation, significant differences exist not only in climate [9] but also in thermal comfort levels indoors [10] and outdoors [11, 12]. This has led to the development of various evaluation standards and predictive models.



Z. Fang et al., *Outdoor Thermal Comfort and Heat Stress Mitigation*, Environmental Science and Engineering, https://doi.org/10.1007/978-981-95-8899-2_5

Using different climatic categorizations, this work consolidates key factors, assessment scales, and frameworks of thermal comfort, aiming to identify central variables and practical challenges in accurate evaluation.

Metabolic rate (M), a fundamental element of the heat balance equation, reflects internal heat production and is a primary determinant of thermal sensation [13]. It fluctuates dynamically and is often approximated through standardized activity logs that assume fixed values by activity type. This approach introduces errors, as metabolic response is also influenced by body condition and mental state [14]. Although exercise intensity strongly affects M, thermoregulatory processes also respond to ambient conditions [15]. Thus, even at constant activity levels, environment significantly alters metabolic rate. This review analyzes studies across regions and activity intensities, highlighting methods for assessing M.

Thermal indices combine environmental factors into composite indicators of physiological load [16]. They serve as criteria for comfort evaluation and for environmental design. Epstein and colleagues [17] cataloged 40 indices for thermal assessment. Widely used measures such as PMV [13] and SET* [18] remain central in comfort studies [19, 20], while PET [21] and UTCI [22] were later developed for outdoor application [23, 24]. Despite their wide use, accuracy remains limited, as metabolic rate—a key input—cannot be measured precisely. Prior studies proposed improvements, but variability in predictions is still largely due to imprecise M estimates [20, 23]. Hence, this work reviews studies on refining thermal indices, evaluates how errors in M estimation affect predictions, and suggests directions for further development.

The purpose of this chapter is to synthesize knowledge, highlight challenges in improving comfort models across diverse conditions and activity intensities, and guide future research. Accurate M estimation is essential for informing architectural and urban planning. By stressing thermal safety and using strategies such as passive ventilation, greenery, and climate-sensitive design, this work aims to support comfort improvements within a human-oriented, energy-efficient framework. The specific objectives are: (1) summarize influencing factors of thermal comfort, assess current models, and identify key variables and challenges in accurate evaluation; (2) consolidate recent research from varied environments and activities, focusing on metabolic rate quantification methods; and 3) analyze how errors in M estimation affect thermal indices and propose improvement strategies for future development.

## 2 Research Method

This investigation executed the identification and selection of relevant literature using a structured methodology compliant with appropriate guidelines [25]. The process for gathering and screening publications is shown in Fig. 1. The core database employed was Web of Science, with further searches conducted in Google Scholar and Scopus. The search strategy incorporated: "thermal comfort" or "thermal sensation" in combination with "metabolic rate" or "metabolism" or "activity" or "exercise" or "dynamic activity/exercise". To examine methodological approaches concerning metabolic rate across diverse environments, terms designating specific locales—including "indoor (space)", "outdoor (space)", "semi-open (space)", and "semi-open/transitional (space)"—were successively merged into the main search string. This methodology assisted in constructing the conceptual foundation of the review.

In the period before 2009, across building science and energy application disciplines, thermal comfort received relatively little emphasis compared to dominant themes such as energy efficiency, where the yearly publication count featuring the keywords "thermal comfort and metabolic rate" did not surpass nine. Accordingly, the search scope was restricted to articles released within the preceding 15 years. From the initial database query, 336 documents were retrieved, as indicated in Fig. 2. Academic work focusing on metabolic rate in thermal comfort contexts has grown considerably each year from 2009 to 2022. According to publications spanning multiple spatial categories, research pertaining to indoor settings has persistently formed the majority of recent scholarly output, although studies on outdoor spaces have also exhibited an upward trend throughout the past five years. In addition, despite increased scholarly attention to semi-open spaces since 2017, a relatively small volume of data was accumulated during the five years prior.

This assessment included exclusively peer-reviewed conference and journal articles. Several authoritative standards and reference manuals—among them the ASHRAE standard, ISO 8996, ISO 7730, and the Compendium of Physical Activities—were also referenced to obtain essential information regarding the conceptualization, calculation, and appraisal of metabolic rate. After preliminary screening, 275 articles received in-depth evaluation, and their associated reference lists as well as recommended readings were scrutinized carefully. A final set of 116 publications directly relevant to the research focus was selected for comprehensive examination.

During the analysis of thermal conditions and comfort perception, keywords representing distinct climatic regions—for instance, "tropical climate", "arid climate", and "subtropical climate"—were added incrementally to isolate suitable articles (Chap. 3). The program VOSviewer, which facilitates the construction and graphical representation of bibliometric networks, was used to organize and summarize techniques for measuring metabolic rate in Chap. 4 [26]. By introducing supplementary terms like "measurement" or "methods" or "measuring methods" and "device" or "instrument", 37 pertinent publications were isolated from the total 116 for keyword co-occurrence examination. Within the segment analyzing the variance between thermal indices incorporating metabolic rate and actual thermal sensation (Chaps. 6 and 7), established indices such as "PMV", "SET*", "UTCI", and "PET", alongside the full name and abbreviation "TSV" (thermal sensation vote), were integrated into the fundamental keywords to refine the screening process. The full organizational scheme of the review is provided in Fig. 3.

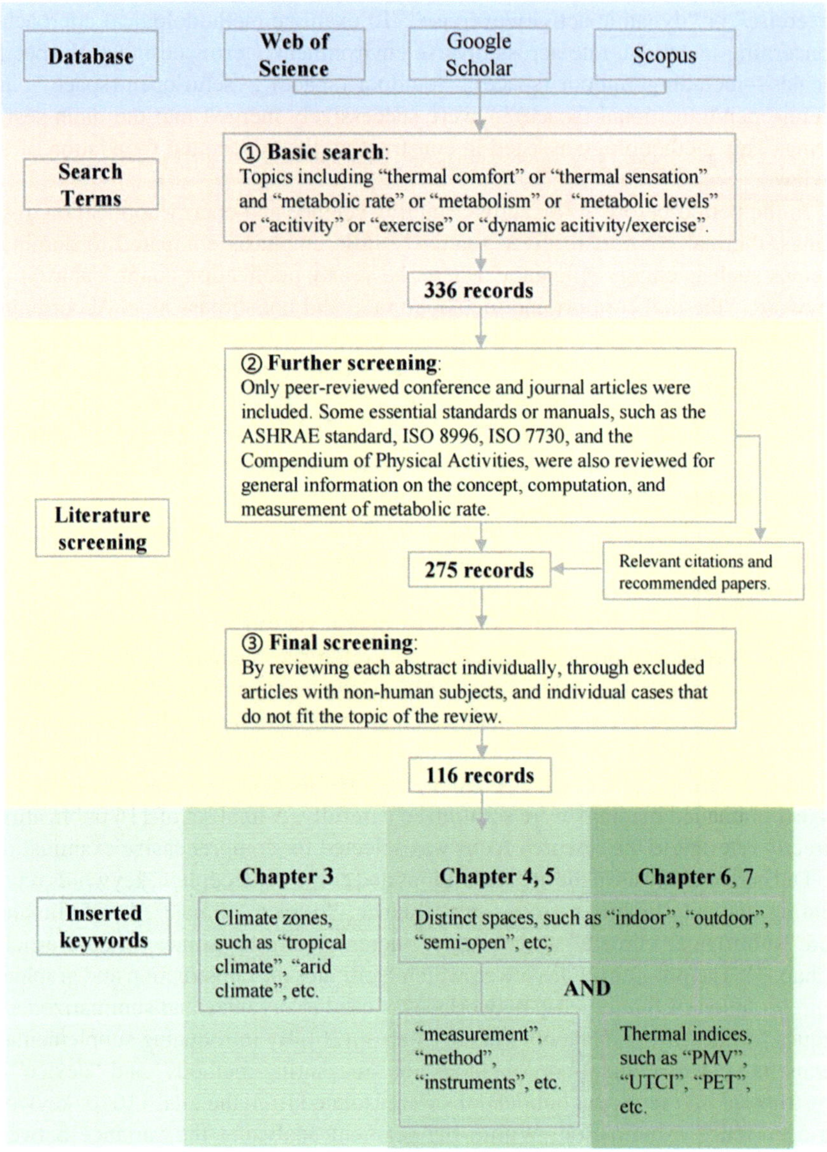


**Fig. 1** Literature search and screening process

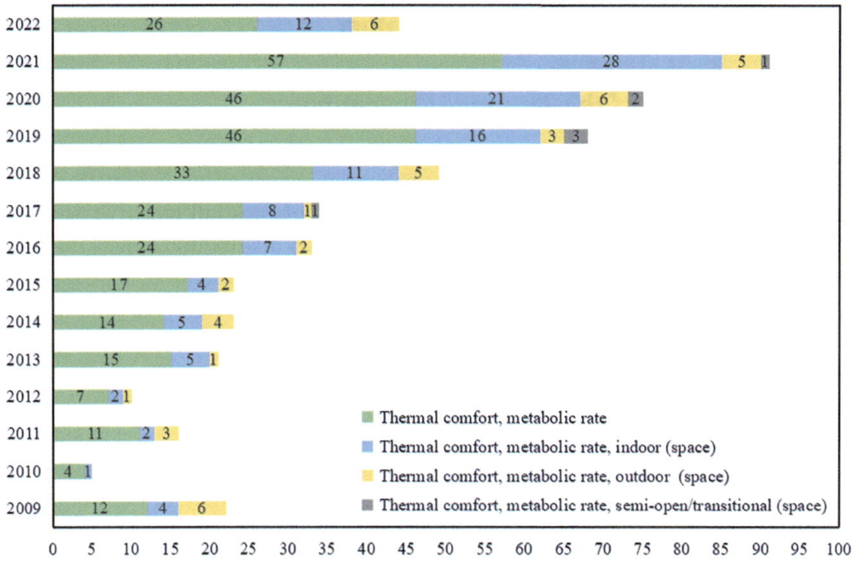


**Fig. 2** The number of published articles related to keywords in different years

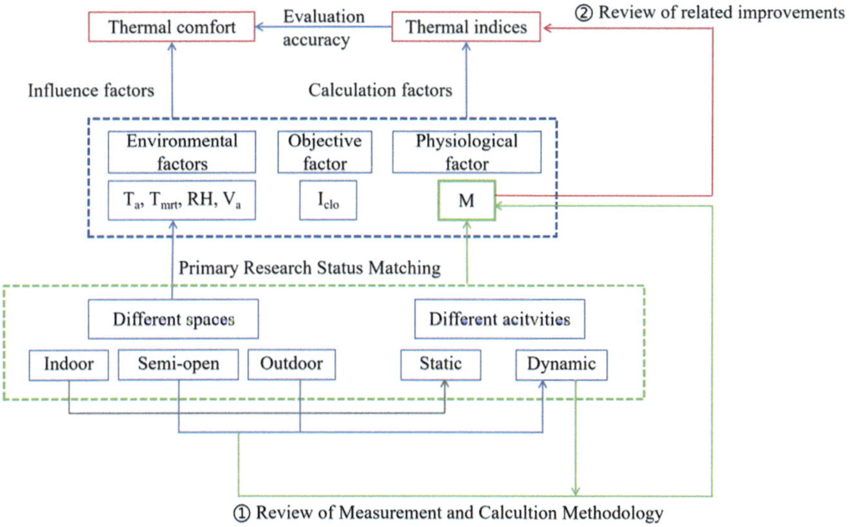


**Fig. 3** Review framework

# 3 Thermal Environment and Comfort

## *3.1 Parameters of the Thermal Environment*

The planet's diverse geographical regions present wide variation in climatic conditions. According to Köppen's classification, global climates fall into five principal zones that align with vegetation and human population distribution (Fig. 4). Differences in air temperature ($t_a$), relative humidity (RH), precipitation, and seasonal averages (Table 3) create distinct human thermal comfort requirements. This section evaluates comfort research in built and natural environments using the Köppen classification to link thermal conditions with human comfort across contexts.

Thermal comfort is influenced by six elements: $t_a$, RH, air velocity ($V_a$), mean radiant temperature ($t_r$), and human factors—clothing insulation ($I_{clo}$) and metabolic rate (M). Among environmental variables, ta is most commonly used, affecting thermal balance through heat exchange [28]. RH alters evaporation from skin, changing body temperature and perception [29]. Va influences convective and evaporative cooling, while radiation strongly affects comfort in outdoor settings [13]. Clothing insulation modifies heat transfer; standards such as ISO 7730 [7] and ASHRAE 55 [8] provide reference values.

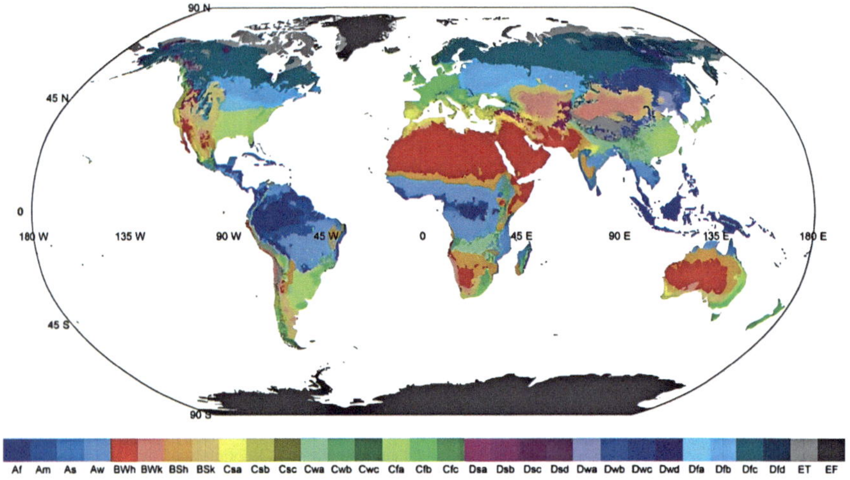


| First letter | Second letter | | Third letter |
|---|---|---|---|
| A: Tropical | f: Fully humid | T: Tundra | h: Hot arid |
| B: Dry | m: Monsoon | F: Frost | k: Cold arid |
| C: Mild temperate | s: Dry summer | | a: Hot summer |
| D: Snow | w: Dry winter | | b: Warm summer |
| E: Polar | W: Desert | | c: Cool summer |
| | S: Steppe | | d: Cold summer |

**Fig. 4** Köppen climate classification (For the specific legend, please refer to [9, 27])

Metabolic rate is a key determinant of comfort, shaping internal heat production and dissipation across environments [30]. It interacts with ta, RH, Va, and tr, influencing equilibrium and sensation [31]. Variability in M produces diverse adaptive responses [32, 33]. Higher M increases heat load, making individuals less tolerant in warm settings [34]. Lower M reduces heat production, requiring greater insulation and raising cold sensitivity [35]. High humidity limits evaporation, so intense activity under humid conditions elevates heat storage and discomfort [36]. $V_a$ and $t_r$ further regulate heat loss and generation [37].

In sum, metabolic rate is integral to thermoregulation, acting with other factors to shape thermal balance and comfort. Recognizing its variability and dynamic nature is essential for accurate evaluation and for creating thermally comfortable environments.

## *3.2 Evaluation of Thermal Comfort*

### 3.2.1 Evaluated Scale (TSV)

Since the beginning of the twentieth century, numerous researchers have sought to refine and advance the assessment frameworks for thermal comfort [31, 38–41]. Table 1 presents currently applied thermal sensation vote (TSV) evaluation scales, along with their developers and suitable application contexts. The 7-point scale represents the most widely adopted approach for evaluating both indoor and outdoor environments. The 9-point and 7-point thermal sensation scales exhibit substantial overlap, though the 9-point version differs primarily through the inclusion of "very cold" and "very hot" categories to accommodate more extreme thermal conditions, similar to those referenced in automotive studies [42]. These scaling systems are typically applied in thermal sensation voting for severe indoor settings and outdoor evaluations where broader variations in thermal perception may occur.

### 3.2.2 Thermal Comfort Models

Table 2 offers a comparison of the conceptual foundations and variables included in four widely utilized thermal indices. Each index integrates key environmental factors—such as air temperature, relative humidity, wind speed, and mean radiant temperature—into its computational approach. Although human-related elements receive differing levels of emphasis among indices, metabolic rate is invariably integrated. This underscores that the correctness of the metabolic rate value input during thermal index calculation directly influences the results, emphasizing the necessity of precise metabolic rate estimation to ensure the indices' practical relevance. Notably, the UTCI model assesses thermal environments through temperature, radiation, wind, and humidity, yet depends on a constant metabolic rate (135 $W/m^2$) and clothing insulation determined by air temperature [74]. Despite being considered

**Table 1** Evaluated scale of thermal sensation

| Category | Scale | Source | Content | Application range | | |
|---|---|---|---|---|---|---|
| Categorical scale | 5-point scale | Houghten [38] | Too cold–comfortably cool–very comfortable–comfortably warm–too warm | Suitable for arid areas, subcold zone | Indoor | Outdoor [43, 44] |
| | 7-point scale | Bedford [39] | Much too cool–too cool–comfortably cool–comfortable–comfortably warm–too warm–much too warm | Commonly used | Indoor [45–56] | Outdoor [23, 57–65] |
| | | Gagge et al. [31] | Cold–cool–slightly cool–comfortable–slightly warm–warm–hot | | | |
| | | ASHRAE standard [8], ISO-10551 [66] | Cold–cool–slightly cool–neutral–slightly warm–warm–hot | | | |
| | 9-point scale | Originate from Matzarakis and Mayer [40] | Very cold–cold–cool–slightly cool–neutral–slightly warm–warm–hot–very hot | Mostly used in tropics and subtropics | Indoor (usually the extreme conditions) [67–69] | Outdoor [62, 70–73] |
| | | ASHRAE standard [8], ISO-10551 [66] | | | | |
| Numerical rating scales | 10-point scale | Sharifi et al. [41] | Numbers from 1 to 10 represent the thermal sensation from the coldest to the hottest, and the corresponding value of thermoneutral is about 6.9. The quantitative correspondence with the traditional 7-point scale is as follows: Cold (-) –cool (8.3)–slightly cool (7.6)-neutral (6.9)–slightly warm (6.3)–warm (5.5)–hot (2.3) | Suitable for long-term evaluation | Indoor/outdoor | |

among the most physiologically informed indices due to its sophisticated modeling of human responses, further developments should aim to integrate guidance on apparel selection that corresponds to differing activity levels (and associated metabolic rates).

**Table 2** Comparison of the four thermal indices

| Thermal index | Theoretical Basis | Thermal sensation Categories | Factors included | | | | | | | | |
|---|---|---|---|---|---|---|---|---|---|---|---|
| | | | Environmental factor | | | | Human factor | | | | |
| | | | $t_a$ | RH | $V_a$ | $t_r$ | $t_s$ | $\omega$ | $t_{core}$ | $I_{clo}$ | M |
| PMV [13] | Thermal comfort equation | 7-point scale: cold-cool-slightly cool-comfortable-slightly warm-warm-hot | ○ | ○ | ○ | ○ | Δ | × | × | ○ | ○ |
| SET* (°C) [75] | Gagge's two-node model | 8-point scale: very cold (10 to14.5)-cool (14.5 to 17.5)-slightly cool (17.5 to 22.2)-comfortable (22 2 to 25.6)-slightly warm (25.6 to 30)-warm (30 to 34.5)-hot (34.5 to 37.5)-very hot (above 37.5) | ○ | ○ | ○ | ○ | Δ | Δ | Δ | ○ | ○ |

(continued)

**Table 2** (continued)

| Thermal index | Theoretical Basis | Thermal sensation Categories | Factors included | | | | | | | | |
|---|---|---|---|---|---|---|---|---|---|---|---|
| | | | Environmental factor | | | | Human factor | | | | |
| | | | $t_a$ | RH | $V_a$ | $t_r$ | $t_s$ | ω | $t_{core}$ | $I_{clo}$ | M |
| PET* (°C) [76] | Munich Energy Balance Model for Individuals (MEMI) | 9-point scale: very cold-(4) cold-(8) cool-(13) slightly cool-(18) comfortable-(23)slightly warm-(29)warm-(35)hot-(41)very hot | ○ | ○ | ○ | ○ | Δ | × | Δ | ○ | ○ |
| UTCI (°C) [77] | UTCI-Fiala multi-node model | 10-point scale: extreme cold stress (below −40)-very strong cold stress (−40 to −27)-strong cold stress (−27 to −13)-moderate cold stress (−13 to 0)-slight cold stress (0 to 9)-no thermal stress (9 to 26)-moderate heat stress (26 to 32)-strong heat stress (32 to 38)-very strong heat stress (38 to 46)-extreme heat stress (46) | ○ | ○ | ○ | ○ | Δ | Δ | Δ | Δ | $\Delta M = 135 W/m^2$ |

*Note* i. $t_s$ is the average skin temperature, ω is the skin moisture degree, $t_{core}$ is the core temperature, and M is the metabolic rate

ii. ○ represents factors used as input parameters, × represents factors not considered, and Δ represents factors considered but not be used as the input parameters.

**Table 3** Factors affecting thermal comfort in different climate zones

| World climate zoning | | Affecting factors in thermal comfort | | | | | | | |
|---|---|---|---|---|---|---|---|---|---|
| | | Category | | Environmental factor | | | | Human factor | |
| | | | | $t_a$ | RH | $V_a$ | $t_r$ | $I_{clo}$ | M |
| A | Tropical climate | Indoor | Southeast Asia [46] | √ | | | √ | | |
| | | | Kumasi, Ghana [47] | √ | √ | | √ | √ | √ |
| | | | Mexico [48] | √ | √ | √ | √ | | |
| | | | Hainan, China [49] | √ | √ | √ | √ | √ | √ |
| | | Outdoor | Dhaka [57] | √ | √ | √ | √ | √ | √ |
| | | | Singapore [79] | √ | √ | √ | √ | √ | √ |
| | | | Kuala Lumpur [59] | √ | √ | √ | √ | √ | √ |
| | | | Chandigarh, India [80] | √ | √ | √ | √ | √ | √ |
| B | Arid climate | Indoor | Turpan, China [60] | √ | | √ | √ | √ | √ |
| | | | Beheira, Egypt [51] | √ | √ | √ | √ | | |
| | | | Adrar, North Africa [52] | √ | √ | √ | √ | | |
| | | Outdoor | Mendoza Metropolitan Area, Argentina [44] | √ | √ | √ | | | |
| | | | Egypt [81] | √ | | √ | √ | √ | √ |
| | | | Riyadh, Saudi Arabia [82] | √ | √ | √ | √ | √ | √ |
| | | | Turpan, Chin [60] | √ | | √ | √ | √ | √ |
| | | | Haryana, India [61] | √ | √ | √ | √ | √ | √ |
| C | Subtropical climate | Indoor | Bogotá, Colombia [45] | √ | √ | √ | √ | √ | √ |
| | | | Florianópolis, Brazil [53] | √ | √ | √ | | | |
| | | | Chitwan, Nepal [54] | √ | √ | √ | √ | | |
| | | | Islamabad, Pakistan [55] | √ | | √ | √ | | |

(continued)

**Table 3** (continued)

| World climate zoning | | Affecting factors in thermal comfort | | | | | | | |
|---|---|---|---|---|---|---|---|---|---|
| | | Category | | Environmental factor | | | | Human factor | |
| | | | | $t_a$ | RH | $V_a$ | $t_r$ | $I_{clo}$ | M |
| | | | Guangzhou, China [56] | √ | √ | √ | √ | √ | √ |
| | | Outdoor | Hongkong, China [58] | √ | √ | √ | √ | √ | √ |
| | | | Chandigarh, India [80] | √ | √ | √ | √ | √ | √ |
| | | | Guangzhou, China [23] | √ | √ | √ | √ | √ | √ |
| | | | Hongkong, China [73] | √ | √ | √ | √ | √ | √ |
| | | | Curitiba, Brazil [63] | √ | √ | √ | √ | √ | √ |
| D | Continental climate | Indoor | Bayannur, China [50] | √ | | √ | √ | √ | √ |
| | | | Beijing, China [83] | √ | √ | √ | √ | √ | √ |
| | | | Seoul, Korea [43] | √ | √ | √ | √ | | |
| | | Outdoor | Toronto, Canada [84] | √ | √ | √ | √ | √ | √ |
| | | | Moscow, Russia [85] | √ | √ | √ | √ | √ | √ |
| | | | Beijing, China [86] | √ | √ | √ | √ | √ | √ |
| | | | Seoul, Korea [64] | √ | √ | √ | √ | √ | √ |
| | | | Tianjin, China [65] | √ | √ | √ | √ | √ | √ |
| E | Polar climate | Indoor | No record | | | | | | |
| | | Outdoor | | | | | | | |

## 3.3 *The Role of Metabolic Rate as the Main Human Factor Affecting Thermal Comfort*

Using the Köppen climate classification as an organizing structure, Table 3 enables the gathering and integration of relevant literature to facilitate a broad review of notable thermal comfort research worldwide. This part focuses specifically on how metabolic rate is addressed across different studies and its resulting impact on thermal comfort findings. Research conducted in both interior and exterior settings was examined to appraise the effect of human factors. Indoor-based studies have typically placed limited emphasis on human variables; common techniques entail approximating clothing insulation through questionnaires and translating recent activities into MET values using conversion tables for PMV estimation (classified as incorporating $I_{clo}$ and M), or assigning predetermined values for clothing insulation and metabolic rate (classified as excluding $I_{clo}$ and M) [46, 48]. Moreover, the Griffiths approach is commonly employed to calculate the comfort temperature ($t_c$) via the equation $t_c = t_a + (0 - TSV)/G$ when relative humidity is not considered (with G denoting the Griffiths constant, typically valued at 0.5/°C) [78].

Even though thermal comfort models account for $I_{clo}$ and M, Table 3 shows that real indoor conditions are usually stable. Indoor activities are mainly sedentary—such as sitting, typing, and standing [45, 47, 49, 50, 56, 60, 83]—and correspond to low metabolic rates. Hence, M is often not dynamically integrated into models; many studies adopt fixed values and focus on environmental variables [48, 51–55, 83]. In contrast, outdoor comfort evaluation must address fluctuating conditions, varied activities, and higher metabolic demands, making assessment more complex [87]. Although comfort models are applied outdoors [88–90] and M is a key input, most studies estimate it from questionnaire-based averages rather than real-time monitoring [87]. Achieving accurate real-time evaluation of thermal environments therefore remains a major challenge [91].

# 4 Metabolic Rate

## *4.1 Analysis of the Factors Affecting the Metabolic Rate*

Metabolic rate is a core element in classical thermal equilibrium models [13] and strongly influences thermal comfort. Although M may stabilize within 3–5 min during specific activities [92], limited heat dissipation can cause production to exceed loss, leading to heat storage (S) shifting toward the hot end of balance equations. Rising activity thus increases heat retention and discomfort.

Resting Metabolic Rate (RMR) reflects energy expenditure during sedentary states and depends on body mass, height, age, sex, fat percentage, and muscle mass [93, 94]. Larger body size increases M, while aging reduces it. Muscle tissue shows higher metabolic turnover than fat [95]. Beyond physiology, M is also affected by activity, emotional state, the specific dynamic action of food (SDA), and ambient conditions [96]. This review summarizes these factors and highlights research gaps (Fig. 5).

Physical activity markedly raises M. Even mild exertion increases oxygen intake, energy use, and heat generation, with workload rising 10–20 times above rest [97]. Hence, metabolic indicators are widely used for assessing energy expenditure.

Mental state also alters metabolism. Calm cognitive tasks cause minor increases, but stress or arousal elevate heat output via muscle tension and hormone secretion.

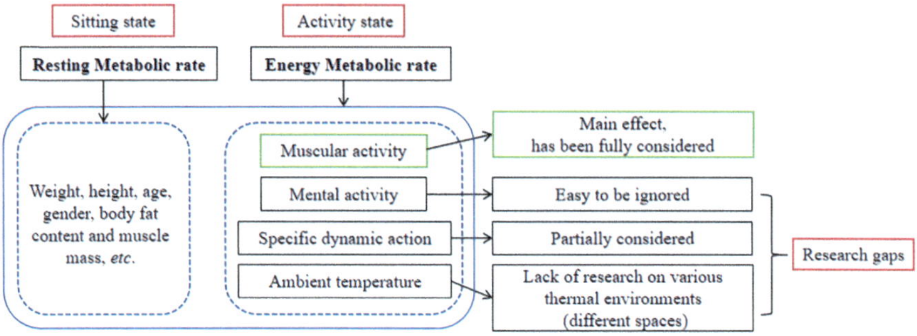


**Fig. 5** Review framework of factors affecting metabolic rate

Symptoms include faster heartbeat or warmth. Studies by Brouwer and Hogervorst linked emotional states to heart rate and metabolic changes [98–101], yet thermal comfort research often neglects these effects.

SDA, the thermic effect of feeding, raises internal heat after meals. Mason et al. found peaks 3–4 h post-consumption [102], with rates returning to baseline in 4–6 h [103]. Pre-experiment diet control (e.g., excluding alcohol or stimulants) is often inconsistent [104, 105], underscoring the need for standardized protocols.

Exercise also promotes white-to-brown adipose tissue (BAT) conversion, boosting energy consumption [106–108]. BAT generates heat through adaptive non-shivering thermogenesis [109]. Cold exposure strongly activates BAT, raising expenditure [110–112], whereas thermally neutral environments suppress it [113]. Reviews confirm BAT's key role in cold adaptation [114, 115]. Ambient thermal conditions also directly influence M: Johnson et al. found energy expenditure inversely related to temperature between 15 and 28 °C [116], and Saxton reported ~ 12% rises in M per 1 °C rectal temperature increase [117]. Thermal comfort studies indoors [92, 118], outdoors [119], and semi-open [23] settings further reveal environmental variation. Yet common estimation techniques often ignore these context-specific effects.

In sum, errors in M estimation stem from multiple interacting factors rather than a single cause. Addressing these complexities is crucial for improving measurement accuracy and refining thermal comfort evaluation.

## *4.2 Measurement Method of Metabolic Rate*

A review of applicable literature was carried out prior to assessing different metabolic rate determination methods. Figure 6 illustrates a keyword co-occurrence analysis produced using VOSviewer software during the literature retrieval phase. Approaches to measuring metabolic rate fall into three principal categories. The first strategy approximates metabolic rate by comparing activity levels with standardized reference values, where "ISO" emerges as a prominent keyword. The second strategy involves either direct or indirect calculation of metabolic rate through experimental assessment of exhaled carbon dioxide concentration (or inhaled oxygen concentration). An alternative methodology uses wearable technology to record heart rate, which is then converted to metabolic rate through established correlation formulae, with relevant keywords encompassing "vest" and "heart rate."

From this analysis, Table 4 consolidates the metabolic rate measurement techniques identified in the literature survey, detailing corresponding instruments, advantages, constraints, primary application contexts, and related computational formulae. These techniques are grouped into direct and indirect categories. Direct methodologies acquire metabolic rate through single reference checks or immediate energy expenditure computation, whereas indirect approaches require supplementary empirical formula conversions or additional reference consultations to establish metabolic rate.

As shown in Table 4, each metabolic measurement method has distinct advantages and limitations. The most common approach uses activity comparison tables, requiring only activity identification without instruments or complex calculations. However, ISO and ASHRAE data mainly derive from Passmore and Durnin's 1967 review of European and North American participants, with values largely unchanged since then [128]. Method 1 is practical but inaccurate, as ethnic and geographical factors significantly influence metabolic rate [129], a finding supported by many studies [130–132].

Methods 2 and 3 measure oxygen consumption and carbon dioxide production, directly quantifying metabolism. Classified as Level 4 "Expertise" in ISO 8996, gas

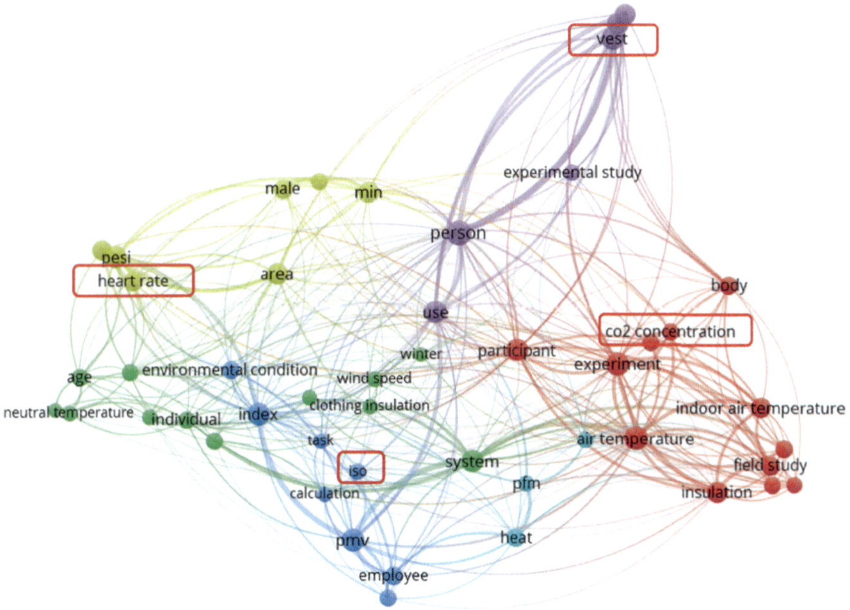


**Fig. 6** Key words related to metabolic rate measurement methods

**Table 4** Summary of metabolic rate measurement methods

| Methods | | Advantages | Disadvantages | Devices | Application space | Calculation equations | Source of uncertainties |
|---|---|---|---|---|---|---|---|
| Direct methods | 1. "Activity diaries"- refer to standards such as ISO 7730, ISO 8996, CPA, ASHRAE standard 55, ASHRAE handbook-Fundamental | Convenience | i May not find the corresponding activity state;<br>ii Ignore individual differences;<br>iii Low accuracy | - | Indoor, Semi-open, Outdoor | Annex B—Metabolic rates of different types of space [7]; Annex A to D- Evaluation of metabolic rate at level 1/2/3/4 [14]; CPA homepage: https://sites.google.com/site/compendiumofphysicalactivities/ [120]; ASHRAE standard 55- table 5-1 Metabolic rate for typical tasks [8]; ASHRAE handbook-Fundamental Chapter 9-Table 4 [121] | Missing matching activities; personal errors |
| | 2. Calculate the amount of $O_2$ inhaled and $CO_2$ exhaled | i Real-time records;<br>ii High accuracy; | i High cost;<br>ii With bulky devices;<br>iii Bring extra thermal resistance | Breathing mask | Indoor, Semi-open (Seldom), Outdoor (Seldom) | $M = \frac{5.88 \times (0.23RQ + 0.77) V_{O_2}}{A_d}$.<br>Where RQ is the respiratory quotient, $V_{O2}$ is the Oxygen consumption rate (L/min), $A_d$ is the Dubois surface area ($m^2$) [14] | Personal errors; systematic errors; random errors |
| | 3. Indirect measurement of carbon dioxide production in closed rooms | i Convenience;<br>ii Low cost;<br>iii Do not need to wear additional equipment | Only appropriate for airtight laboratories, i.e., not appropriate for many real-world situations | $CO_2$ sensors | Indoor | $M = \frac{\Delta C_{CO_2} \times V \times TEO \times 1000}{0.85 \times \Delta T \times A_d}$<br>Where $\Delta C_{CO2}$ is the concentration changes of $CO_2$ during a period (ppm), V is the volume of the experimental chamber (L), TEO is the thermal equivalence of Oxygen, the value is 20.36 kJ/L, $\Delta$ T is the time interval of calculation (s) [14] | Personal errors; systematic errors; random errors |

(continued)

**Table 4** (continued)

| Methods | | Advantages | Disadvantages | Devices | Application space | Calculation equations | Source of uncertainties |
|---|---|---|---|---|---|---|---|
| | 4. Generalized exponential model for prediction of walking metabolic rate | i Convenience;<br>ii Low cost; | i Only suitable for walking activities<br>ii Limited speed range from 0.4 to 1.9 m/s | - | Indoor, Semi-open, Outdoor | $V_{O_{2total}} = V_{O_{2rest}} + 3.85 + \frac{5.97V^2}{Ht}$<br>Where $V_{O_{2total}}$ and $V_{O_{2rest}}$ represent the rates of oxygen uptake during the walking period and the rest period (mL/kg min), V represents the walking speeds (m/s), Ht represents the height of individual (m) [122] | Personal errors; systematic errors; random errors |
| Indirect methods | 5. Measure heart rate and matched to metabolic rate by correlation | i Convenience;<br>ii Low cost;<br>iii Real-time records | i Relatively low accuracy;<br>ii Poor contact sometimes occurs | Wearable detectors, such as wearable sports watches, chest-worn heart rate belt, finger oximeters, etc | Indoor, Semi-open, Outdoor | $M = M_0 + \frac{(HR - HR_0)}{RM}$<br>Where $M_0$ is the metabolic rate in the inactive state ($W/m^2$), HR is the heart rate at the moment, $HR_0$ is the heart rate in beats per minute at rest, and RM is the increase in heart rate per unit of the metabolic rate [14] | Personal errors; systematic errors; random errors |

(continued)

**Table 4** (continued)

| Methods | Advantages | Disadvantages | Devices | Application space | Calculation equations | Source of uncertainties |
|---|---|---|---|---|---|---|
| 6. Based on measurements of acceleration | i Convenience; ii Low cost; iii Continuous monitoring | i Too many empirical formulas for energy expenditure conversion with uncertain accuracy; ii Limited activity assessment | Wearable accelerometer (usually triaxial) | Indoor, Semi-open, Outdoor | Some typical algorithm examples: Freedson et al. [123, 124]; Hendelman et al. [125]; Crouter et al. [126]; Yngve et al. [127] etc | Personal errors; systematic errors; random errors |
| 7. Image recognition technology combined with method 1 | i Convenience; ii High accuracy; iii Time-saving | iii Underdeveloped; iv Only recognizes pre-set activity levels | Cameras capable of capturing body movements, such as the Kinect camera | Indoor, Semi-open, Outdoor | – | Missing matching activities; personal errors; systematic errors; random errors |

analysis is regarded as the accuracy gold standard [14]. Method 4, the Ludlow and Weyand HWS model, predicts walking metabolism from velocity and body mass [122, 133], but applies only to level walking at limited speeds.

Method 5 combines heart rate data from wearables with ISO 8996 empirical formulas [14]. Rated Level 3, it offers moderate accuracy—better than Method 1 but below Methods 2–4—and is widely adopted for its portability and real-time capability. Method 6 estimates energy expenditure (EE) from accelerometer data using regression models [134–137], though predictions often vary by device and lack a universal standard [138, 139]. Method 7, still emerging, applies image recognition to detect activities and estimate metabolic rate via Method 1.

Methods 2–7 share common uncertainty sources: random and personal errors, instrument calibration issues, and poor maintenance. In Method 2, respiratory airflow fluctuations introduce error despite high-precision analyzers, though averaging data over minutes improves reliability [140]. Method 3 avoids this by assessing $CO_2$ concentration changes in enclosed spaces [141]. Method 5 accuracy depends heavily on heart rate monitor precision, with improved devices producing estimates closer to true values. Standardized protocols can reduce many uncertainties.

By contrast, Method 1 has intrinsic errors, as unlisted activities must be approximated by similar ones. Methods 4, 6, and 7, due to major limitations or early development, are seldom applied in current thermal comfort research.

### 4.2.1 Methods Often Employed in Indoor Environments

A review of over 200 studies on metabolic rate estimation confirms that the "activity diary" remains the dominant tool in thermal comfort (TC) research across diverse environments. However, significant inconsistencies persist among estimation strategies in different spatial contexts.

Indoor TC research is relatively advanced due to stable environments and predictable activity patterns. While many studies still use "activity diaries," alternative approaches have been explored. For example, Teli et al. argued that existing TC models, largely based on adult parameters, are unsuitable for children [142]. Children show higher resting metabolic rates per body weight, which decline with age. They tested four adaptation methods, following Havenith's suggestion that children's metabolic rate should be ~ 10% lower than adults [143]. Yet, earlier studies showed that weight, surface area, and metabolic rate are not strictly linear [144], and growth alters surface-to-weight ratios [143]. Thus, applying adult values risks large errors, though these initiatives provide useful frameworks. Almeida et al. also showed that body surface adjustments improve accuracy for children [145]. Ito and Nishi developed a portable accelerometer-based device for real-time metabolic assessment [146].

Metabolic rate can also be calculated from $O_2$ consumption and $CO_2$ production [147]. Staudenmayer et al. applied the K4b2 device to compute M-values across activity intensities [148]. Zhai et al. used COSMED K10 during office tasks, showing ISO and ASHRAE standards overestimate sitting/standing by 10–20%, while ISO underestimates walking by 5–9% and ASHRAE by > 20% [131]. Nomoto et al. improved accuracy by combining indirect calorimetry with 3D surface measurements, finding daily student metabolic rates lower than ISO 8996 and ASHRAE values [132]. These studies confirm the superior reliability of Method 2. Its drawback is the need for uncomfortable, costly oxygen masks, restricting application to labs. Portable versions have improved feasibility for outdoor studies [149–151], but heavy devices may still affect participants' performance.

Alternative mask-free methods infer M from surrounding $CO_2$ fluctuations. Ruiz et al. proposed combining $CO_2$ analysis with indoor air quality (IAQ) assessment [152]; Ji et al. applied this in a sealed-room study [141], and Fabbri validated its outcomes [153]. However, this method is limited to enclosed spaces and unsuitable for outdoor TC research.

### 4.2.2 Methods that Suitable for Outdoor Environments and Others

Numerous investigators have applied new technologies to improve heart rate–based metabolic estimation, a well-established approach for tracking M fluctuations [13, 154]. Among heart rate evaluation methods, photoplethysmography (PPG), electrocardiography (ECG), and oscilloscopy are most common, forming the basis of many portable devices [141]. ECG, first introduced in 1903 [155], remains widely used; chest-worn belts provide high precision [156] but require direct skin contact, which may cause discomfort and add thermal insulation. PPG, enhanced by advances in photonics, now supports popular fitness wearables [157–159] and pulse oximeters [160]. Their lightweight, wireless design enables real-time data collection in indoor [92], transitional [23], and outdoor [161] settings. Hasan et al. used an accelerometer-equipped wristband (Fitbit) to infer M from activity-based energy expenditure [162], while Revel et al. designed a cost-effective monitoring device for real-time use [163]. Based on ISO 8996 principles [14], this method links heart rate changes to muscular effort, with studies noting a 7% variation in M estimates and a 0.05 PMV discrepancy, demonstrating its potential for continuous evaluation.

Advances in computing have also enabled intelligent image-based systems for M estimation. Deep learning classifiers can detect facial features and body movements with high accuracy [164, 165]. Many researchers integrate image recognition with the "activity diary" approach, using action detection to infer M. Building on Choi et al. [166], Park et al. developed a DNN-based framework to identify office postures and estimate M [167]. Na et al. later introduced HRnet and a Kinect-based vision model [168], followed by METnet, which computes M from multiple routine activities using Kinect imagery [169]. Image-assisted metabolic assessment is rapidly advancing and is expected to become a central tool for future thermal comfort prediction.

### 4.2.3 Methods Commonly Used in Measuring Exercise Metabolic Rates

Individuals commonly partake in dynamic activities such as walking, running, and cycling. Accurate quantification of metabolic rate during these activities presents greater difficulty than under static circumstances. Surveys of literature indicate that most indoor thermal comfort research has concentrated on stationary positions (e.g., seated, standing), with relatively scant attention paid to dynamic thermal conditions. Thus, this section reviews studies undertaken in semi-open and outdoor settings and consolidates frequently employed methods for approximating exertion metabolic rate in thermal comfort science. Table 5 itemizes computational approaches extracted from metabolic calculation sections of relevant publications.

As shown in Table 5, many studies still estimate metabolic rates by matching activity types to standard protocols. This approach is simple and provides reference values but overlooks individual variation in age, body composition, and fitness, leading to errors. The Compendium of Physical Activities (CPA) offers a broader database than ISO standards and is increasingly adopted for its wider range of MET values [180]. Wearable devices are also growing in use, as they account for personal variation and enable real-time M estimation via links with heart rate, oxygen uptake, and other metrics. This improves accuracy over reference tables but requires costly electronics. Moreover, heart rate monitors add insulation, potentially affecting results [181], and indirect estimation increases procedural complexity.

Beyond thermal comfort, walking-related metabolic prediction models have been widely studied in sports science, physiology, and military training. Pandolf et al. (1970s) proposed an EE model for slow walking (<0.7 m/s), incorporating weight, load, speed, slope, and terrain for occupational health [182, 183], but it was limited to low activity. Epstein et al. later created a model for running (2.2–3.2 m/s) [184], while Drain et al. found Pandolf's model underestimated modern military load carriage [185]. Building on earlier findings [133], Ludlow and Weyand [122] developed a predictive model using height, weight, and speed, validated through large datasets and machine learning. It forecasts M for walking at 0.4–1.9 m/s. Moore et al. further derived a step-frequency–based regression equation with < 0.01 met error at normal speeds (0.89–1.56 m/s) [186]. These models are simple and accurate within tested ranges but cannot be generalized to non-walking activities or diverse terrains.

In summary, current methods for exercise metabolic rate—whether wearable technology, standard tables, or walking models from other fields—remain limited. Developing portable, cost-effective, and personalized monitoring instruments remains a key goal. For now, wearable devices continue to be the benchmark for dynamic thermal comfort studies in semi-open and outdoor settings.

**Table 5** Summary of the methods for calculating exercise metabolic rates

| Source | Location | Activity Type | Metabolic rate calculation Methodology |
|---|---|---|---|
| Lu and Li (2020) [170] | Transition space (Guangzhou, China) | Sitting, 3 km/h walking, and 6 km/h walking | Refer to [14] Table C.1 (Establish the relationship between Age, weight, HR, and M for calculation) |
| Zhang et al. [23], Fang et al. [20] | Transition space (Guangzhou, China) | 1.2 m/s, 1.4 m/s, 1.6 m/s and 2.0 m/s walking | Measure HR through a sports bracelet, and establish the relationship between metabolic rate and heartrate through the formula |
| Fang et al. (2018) [119]; Fang et al. [171] | Outdoor space (Guangzhou, China) | Random questionnaire | Refer to [14] Table B.3 (Activity dairy) |
| Niu et al. (2020) [172] | Outdoor space (Xi' in China) | Meditating, sitting, standing, walking (4.5–5.1 km/h), Rope jumping (slow pace) | Refer to the 2011 Compendium of Physical Activities [120] |
| Niu et al. (2022) [173] | Outdoor space (Chongqing, China) | Sitting, reading, singing, Taichi, walking, conditioning exercise, running | Refer to the 2011 Compendium of Physical Activities [120] |
| Nikolopoulou and Lykoudis (2006, review paper) [174] | Outdoor space (European countries) | Random questionnaire | Refer to ISO 7730 Table B.1 |
| Mahmoud (2011) [175] | Outdoor space (Cairo, Egypt) | Random questionnaire | Record activity level (No mention of which specification is referenced) |
| Kenny et al. (2009) [176] | Outdoor space (Canada) | Walking, running, and cycling | Refer to the 2000 Compendium of Physical Activities [177] |

(continued)

**Table 5** (continued)

| Source | Location | Activity Type | Metabolic rate calculation Methodology |
|---|---|---|---|
| Vanos et al. (2012) [178] | Outdoor space (Toronto, Canada) | Walking, running, and cycling | $M = VO_{2act}/3.5$ met, refer to the list of activities and the corresponding metabolic rate ($M_{act}$) and activity speed ($v_a$) lists proposed by Strath et al |
| Monteiro (2013) [179] | Outdoor space (Sao Paulo, Brazil) | Random questionnaire | Record activity level (No mention of which specification is referenced) |

# 5 Effects of Metabolic Rate on Thermal Comfort in Different Thermal Environments

## 5.1 *The Effects of Metabolic Rate Indoors*

Indoor thermal conditions are more controllable, and regulation of metabolic activity is closely tied to comfort. Environmental changes trigger passive metabolic adjustments, enabling the body to maintain dynamic balance [187]. Threlkeld showed that moderate activity allows people to remain within thermal zones without impairing heat balance, though this varies with activity level and clothing [188]. Yang et al. applied the heat balance equation to identify acceptable ranges, finding a linear relationship among temperature, humidity, wind, and metabolic change; two anthropometric and four environmental factors influenced comfort [189]. Luo et al. confirmed that human heat output depends strongly on ambient temperature and clothing insulation [115]. In stable environments such as air-conditioned rooms, comfort is achieved by aligning M with surrounding conditions. Zhai et al. demonstrated in a gym that expectations and ventilation requirements vary with activity intensity, and airflow can be adjusted to meet thermal needs [37].

## 5.2 *The Effects of Metabolic Rate in Outdoor Space*

Outdoor activities generally involve higher and more variable metabolic rates. Fang et al. reported values from 1 to 3.8 met, with ~ 70% near 2 met [119]. Yang et al. showed that a 0.2 met rise reduced acceptable indoor temperature by 2.3 °C, narrowing comfort ranges [189]. Fang et al. further found that increasing M from

1.0–2.0 to 2.6 met raised UTCI by 3.9 °C and PET by 2.5–7.8 °C, highlighting index sensitivity to M [119]. Thus, accurate outdoor comfort assessment requires careful integration of metabolic variability.

Unlike indoors, outdoor conditions shift abruptly, with solar exposure altering surface temperatures by up to 5.2 °C [190], and wind speed strongly affecting convective and evaporative heat loss. Uğursal and Culp showed that high M can be tolerated in warm conditions with appropriate airflow [191]. Indoors, M changes correlate well with sensation votes [92], but this link weakens outdoors due to rapid shifts in solar radiation and wind. While temperature and radiation dominate outdoor comfort [174], long-term exposure fosters adaptation. Nikolopoulou and Lykoudis [174] observed that low-M activities occurred at higher seasonal temperatures, reflecting adaptation. Tang et al. found altered thermal votes among students after nine days of outdoor training despite similar metabolic states [192], and Fang et al. confirmed this process in construction workers in Southern China [161]. Hence, outdoors, M interacts with environmental conditions and adaptation, making it inadequate to model its effect as a fixed value or in isolation.

## *5.3 The Effects of Metabolic Rate in Semi-open Space*

Transitional spaces, compared with fully exposed outdoors, often shield from direct sunlight—an important cause of summer discomfort—while retaining some indoor stability. Yet factors such as ground radiation, wind speed, and direction remain variable, giving these areas partial outdoor unpredictability. Thus, the role of metabolic rate in comfort must be evaluated case by case. As transitional spaces usually experience milder conditions, M's influence often interacts with environmental traits. Zhang et al. showed that incorporating dynamic M changes, especially from walking speed, improved thermal index predictions, raising thermal sensation accuracy above 80% [23]. Fang et al. reported similar improvements with upgraded SET* models [20], suggesting M should be integrated into indices for transitional as well as outdoor spaces.

Lu et al. examined tourists and residents on tropical islands and confirmed physiological adaptation in semi-open settings [193]. Thermal satisfaction was closely linked to wind and temperature. As in outdoor contexts, adaptation moderates M's impact on comfort. Overall, dynamic shifts in M strongly affect thermal index performance in transitional spaces, highlighting the need to include M in comfort evaluations.

### 5.4 *Quantitative Analysis of the Impact of Metabolic Rate Changes on Thermal Comfort*

Building on the examination in Sects. 5.1–5.3, this portion aggregates patterns of metabolic rate variation caused by external/internal influences and their outcomes for thermal comfort—mainly subjective thermal sensation votes—as recorded in the assessed literature. particulars are included in Table 6:

Table 7 summarizes quantitative analyses of metabolic rate effects on thermal comfort. Activity-induced M influences comfort through shifts in temperature intervals (air, skin, indices), preferred temperature ranges, and thermal sensation votes. Rising M increases heat generation, creating a warmer perception in the same environment. In cold settings, however, higher M helps preserve core temperature but also elevates heat loss demand, sometimes intensifying cold sensation [196]. Thus, thermal sensation reflects both environment and M level, as shown in Fang et al.'s fitted equation for MTSV against environmental variables and M [171].

Neutral temperature ($n_T$) is the point of thermal neutrality, typically lower at higher M and higher at reduced M—a pattern confirmed in Table 7. Preferred temperature (Tpref), defined as the crossover of "prefer warmer" and "prefer cooler" responses [197], also shifts with M: higher M aligns with preference for cooler conditions. Outdoor studies of nT and Tpref are limited due to greater complexity than indoors. In subtropical summers, outdoor sensations often exceed neutral values, with preferences leaning cooler. Consequently, many outdoor investigations rely directly on MTSV for evaluating thermal comfort.

Contemporary research emphasizes the significance of metabolic rate in human thermal comfort investigations, since differences in metabolic rate can directly alter thermal perception even with constant environmental parameters. Hence, precise and efficient monitoring of metabolic rate is crucial for progressing thermal comfort science. Owing to the variety in background conditions and metabolic rates among existing studies, it is difficult to systematize or standardize the measurement of metabolic rate's impact on thermal comfort. Nonetheless, these results establish a groundwork for additional research and supply useful guidance for directing future inquiry.

## 6 Effects of Metabolic Rate on Thermal Indices

### 6.1 *The Role of Metabolic Rate in Thermal Index Model Calculations*

Section 3.2.2 summarized the factors used in thermal comfort models. As a key human variable, metabolic rate is included in all thermal indices. To assess its specific role, this section examines how M functions within the mathematical structures

**Table 6** Influencing factors of metabolic rate change

| Reference | Method | Environmental parameters | Research objectives | Results | Effects of metabolic rate | Scenarios |
|---|---|---|---|---|---|---|
| Luo et al. [115] | Breathing mask (Method 2) ($M = 1.1$met) | Temperatures from neutral (26 °C) to cold (24/21/18/16 °C) and hot (31/28 °C); RH = 50%; $V_a$ = 0.1 m/s | Explore the influencing factors of human metabolic heat production | Changes in metabolic rate can be expressed as a second-order polynomial equation with independent variances of TSV or PMV: $\Delta M = 0.013TSV^2 - 0.0164TSV + 0.01(R^2 = 0.97)$; $\Delta M = 0.0067PMV^2 - 0.0126PMV + 0.0115(R^2 = 0.97)$ | When the interior environment changes slightly, the human body actively adjusts its metabolic rate by engaging the thermal balance mechanism, resulting in a dynamic thermal comfort state | Indoor |
| Yang et al. [189] | $\Delta M = M_{max} - 58.2$, $M_{max} = f(t_a, t_{rt}, RH, I_{clo}, V_a, M_{max})$ | $t_a$ = 19–33 °C; RH = 20%-90%; V = 0.25–2.5 m/s | Explore the allowable fluctuation range of human metabolic rate and thermal environment parameters under thermal comfort conditions | Changes in metabolic rate are jointly affected by $T_a$ and RH: $\Delta M = 183.104 - 5.25Ta - 56.609RH + 13.627$ | When the interior environment changes slightly, the human body actively adjusts its metabolic rate by engaging the thermal balance mechanism, resulting in a dynamic thermal comfort state | Indoor |
| Huang et al. [194] | Random questionnaire (Method 1) | Annual mean $t_a$ = 16–18 °C, RH = 70–80% | Exploring the equivalent metabolic rate used to predict MTSV in Chongqing, China | Winter: $M = -0.021t_a + 0.80$ ($R^2 = 0.48$); Autumn: $M = -0.0045t_a^2 + 0.17t_a - 1.01$ ($R^2 = 0.91$); Summer: $M = -0.04t_a + 1.44$ ($R^2 = 0.83$) | The corrected metabolic rate gives a more precise input for heat index computations that evaluate thermal settings | Indoor |

(continued)

**Table 6** (continued)

| Reference | Method | Environmental parameters | Research objectives | Results | Effects of metabolic rate | Scenarios |
|---|---|---|---|---|---|---|
| Uğursal and Culp [191] | Chest-worn heart rate belt (Method 4) | $t_a$ = 23.9 °C, 28.3 °C; RH = 45% ± 5%; supplied air up to 0.8 m/s | Explore the effects of temperature, metabolic rate and dynamic local airflow on thermal comfort | The thermal sensation votes of the simultaneous airflow with 2.13 met were similar to the no localized airflow condition with 1.55 Met | Simultaneous airflow compensated for the increase in metabolic rate | Indoor |
| Tang et al. [192] | Finger oximeters (Method 4) | $t_a$ = 27.6–34 °C; RH = 64.7%-91.1%; $V_a$ = 0.05–3.77 m/s | Explore the thermal safety and physiological adaptability of outdoor physical training for college students | Metabolic rate after training is positively correlated with PET: $M = 4.46PET + 7.24$ ($R^2 = 0.91$); $MTSV = 0.09PET - 0.41$ ($R^2 = 0.87$) (When PET = 41.5 °C, there is no difference in MTSV before and after training.) | PET is positively correlated with metabolic rate after physical training, and the thermal sensation after exercise is significantly different from that before training due to the occurrence of physiological thermal adaptation | Outdoor |
| Fang et al. [161] | Finger oximeters (Method 4) | $t_a$ = 26.55–47.19 °C; $t_r$ = 27.21–63.93 °C; RH = 45.8%-93.7%; $V_a$ = 0.05–3.77 m/s | Explore the summer thermal response of construction workers in South China | Heart rate (HR) has a good quadratic fitting relationship with $t_a$ and $t_r$, and has a positive linear relationship with thermal sensation: $HR = -0.02t_a^2 + 1.75t_a + 57.71$ ($R^2 = 0.73$); $HR = -0.01t_r^2 + 1.09t_r + 66.2$ ($R^2 = 0.69$); $HR = 5.99MTSV + 73.965$ | HR determines the value of M, and metabolic rate is affected by $t_a$ and $t_r$, which then affects MTSV | Outdoor |

(continued)

**Table 6** (continued)

| Reference | Method | Environmental parameters | Research objectives | Results | Effects of metabolic rate | Scenarios |
|---|---|---|---|---|---|---|
| Fang et al. [195] | Finger oximeters (Method 4) | $t_a$ = 27.7–33.1 °C; RH = 60.0–84.7%; | Explore the human thermal physiological response to wearing personal protective equipment | Metabolic rate during light work while seated is significantly affected by wet-bulb globe temperature (WBGT):<br>Before putting on personal protective equipment:<br>HR = 3.22WBGT-20.88 ($R^2$ = 0.73);<br>when wearing:<br>HR = 2.74WBGT + 5.58 ($R^2$ = 0.83);<br>after taking off:<br>HR = 5.05WBGT- 69.06 ($R^2$ = 0.72) | The change pattern of MTSV over time is synchronized with M | Semi-open |
| Fiala et al. [187] | From the literature, exercise intense = 0.8–10 met | $t_a$ = 5–50 °C, exercise intense = 0.8–10 met | Explore regulatory responses that address the weak link of core temperature prediction in mathematical models of active systems | The extra metabolism due to shivering, Sh (W), was obtained from the measured total heat output of the subjects, $M_{exp}$ (W), by eliminating the extra heat production caused by the exercise, H (W), and the basal heat production predicted by the passive system, Mbas (W):<br>$Sh = M_{exp} - H - M_{bas}$ | Shivering caused by low temperature brings additional metabolic heat production, which is a regulatory process to maintain body temperature | unspecified |

**Table 7** Quantitative effects of changes in metabolic rate on thermal comfort

| Reference | Changes in metabolic rate | Results in thermal comfort evaluation | Scenarios |
|---|---|---|---|
| Goto et al. [198] | The activity level transitions from sedentary to an increased workload of 20% | MTSV increased by 0.3 units | Indoor |
| McNall et al. [35] | Three metabolic rate conditions, including 1.7 met, 2.2 met, and 2.8 met | $n_T$ are 22 °C, 19 °C and 16 °C respectively | Indoor |
| Bai et al. [104] | Activity status is divided four levels (each lasting 20 min), and the metabolic rates are 1.2 met, 2.4 met, 3.0 met, and 3.7 met respectively | When $t_a = 25$ °C: $n_T$ are 25.1°C, 24.4 °C, 24.1 °C and 23.9 °C;<br>When $t_a = 30$°C: $n_T$ are 24.6, 24.3, 23.8, and 22.8 °C | Indoor |
| Fang et. al [199] | Sitting and two different exercise intensities were continued until the thermal sensation was stable (M was 1 met, 2.4 met and 4.4 met respectively) | When $t_a = 28$°C, MTSV are 0.67, 1.79 and 3.3;<br>When $t_a = 34$°C, MTSVare 2.95, 3.63 and 3.85 | Indoor |
| Wang and Hu [200] | Two activity levels lasted for 20 min each, with metabolic rates of 2 met and 5 met respectively, and ambient temperature of 26 °C | Wehn sitting, $n_{ST} = 32.8$ °C;<br>When M = 2 met, $n_{ST} =$ 32.3 °C;<br>When M = 5 met, $n_{ST} =$ 30.7 °C | Indoor |
| Zhai et al. [201] | Carry out activities with metabolic rates of 3 met, 3.5 met and 4.5 met respectively, lasting 30 min, with an initial ring temperature of 26 °C | Without mechanical ventilation: $T_{pref}$ are 23.3, 22.8, and 22.1 °C;<br>With mechanical ventilation: $T_{pref}$ are24.9, 24.1, and 23.6 °C | Indoor |
| Gao et al. [202] | When sitting, standing, walking at 1.2 km/h and walking at 2.4 km/h, the metabolic rates are 1 met, 1.1 met, 1.9 met and 2.5 met respectively | $T_{pref}$ are 25.85 °C, 25 °C, 24.1 °C and 23.2 °C | Indoor |
| Yang et al. [189] | Activity levels are 1 met and 1.2 mct | For every 0.2 met increase in metabolic rate, the acceptable temperature decreases by 2.3 °C | Indoor |
| Fang et al. [171] | Random questionnaire survey, metabolic rate range from 1 met to 2.6 met | $MTSV = 0.17t_a + 0.02RH - 0.373V_a + 0.014t_{mrt} + 0.161I_{clo} + 0.141M - 4.741$ ($R^2 = 0.56$) | Outdoor |
| Fang et al. [20] | Walk continuously for 20 min at four different walking speeds, metabolic rate range from 2.6 met to 4.0 met | $MTSV = 0.602\ M - 1.057V_s + 1.612$ ($R^2 = 0.76$) | Outdoor |

(continued)

**Table 7** (continued)

| Reference | Changes in metabolic rate | Results in thermal comfort evaluation | Scenarios |
|---|---|---|---|
| Fang et al. [119] | Random questionnaire survey, metabolic rate range from 1 met to 2.6 met | $\frac{\Delta MTSV}{\Delta M} = 0.00576\ln(TCI) - 0.0136$, where TCI represents thermal comfort indices, including PET and UTCI | Outdoor |
| Zhang et al. [23] | From sitting to continuing to exercise at a certain activity level for 20 min, M ranged from 0 to 4 met | $\Delta MTSV = 0.493\Delta M - 1.045$ | Semi-open |

Where $n_T$ refers to thermal neutral temperatures (°C), $n_{ST}$ refers to Neutral skin temperature (°C), $T_{pref}$ refers to preferred temperature (°C)

of index calculations (Table 8). Results from PMV, SET*, and PET are directly determined by the value of M. For UTCI, the reference air temperature—the model's final output—depends on the interaction of M with environmental parameters in a passive system. Overall, M alters index outputs and thereby affects the reliability of thermal environment evaluation.

## *6.2 Thermal Index Deviation from Actual Thermal Sensation—Related to Metabolic Rate*

Fanger's PMV model was designed for uniform, steady environments near neutrality and does not accurately represent body thermal state. Skin temperature and sweat rate are inferred solely from activity level, independent of climate [187, 204]. Studies confirm PMV is reliable only within a narrow metabolic range [189, 205, 206]; errors in M lead to distorted predictions. Luo et al. identified metabolic rate as a major source of PMV inaccuracy [118], and Chamra et al. showed that M and clothing insulation dominate PMV uncertainty [207]. Havenith et al. further noted that ISO 8996 methods often lack accuracy [208], limiting PMV's precision to classify buildings within 0.3 units. Since PMV was derived from votes of 1300 adults in climate chambers [13], it excludes other age groups. Ter Mors et al. found adult-based M estimates misrepresent children's comfort in classrooms [209], while Almeida et al. showed standardized methods overestimate children's sensations; adjusting M for body surface improved PMV-MTSV consistency [145].

Thermal index predictions also decline at high M, typical in outdoor or semi-open conditions. Fletcher reported PMV is unreliable at elevated activity [180]. Zhang et al. showed PMV fails to forecast dynamic sensations during walking indoors [92]. Fang et al. demonstrated that fluctuating M alters MTSV even when PET and UTCI remain constant [119]. Similarly, Zhang et al. found UTCI and PET matched rest conditions

**Table 8** The role of metabolic rate in thermal index calculations

| Thermal index | Theoretical Basis | Mathematical model | The role of M in thermal index calculations |
|---|---|---|---|
| PMV | Thermal comfort equation | $PMV = \left(0.303e^{-0.036M} + 0.028\right)\{M - W - 3.05[5.733 - 0.007(M - W) - P_a] - 0.42(M - W - 58.15) - 0.0173M(5.87 - P_a) - 0.0014(34 - t_a) - 3.96 \times 10^{-8} f_{cl}\left[(t_{cl} + 273)^4 - (t_r + 273)^4\right] - f_{cl}h_c(t_{cl} - t_a)\}$ | PMV is proposed based on the heat balance equation of indoor steady-state environment ($M - W = C + R + E_{sk} + C_{res} + S$). Professor Fanger defined human body heat load (L) as the difference between the body's heat production and heat dissipation, where M is the total heat production of the human body. PMV is a function of M and L established based on the thermal comfort equation (when $S = 0$ in the thermal balance equation). Therefore, the accurate calculation of M plays a decisive role in the accuracy of PMV |
| SET* (°C) | Gagge's two-node model | Core layer:<br>$M + M_{shiv} - W = C_{res} + E_{res} + (K + \rho_{bl}m_{bl}c_{bl})(T_{cr} - T_{sk}) + \frac{m_{cr}}{A}c_{cr}\frac{dT_{cr}}{dt}$<br>Skin layer: $(K + \rho_{bl}m_{bl}c_{bl})(T_{cr} - T_{sk}) = Q_{sk} + \frac{m_{sk}}{A}c_{sk}\frac{dT_{sk}}{dt}$<br>SET* computational formula: $Q_{sk} = h_c(T_{sk} - SET^*) + \omega h_c(P_{sk} - 0.5P_{SET^*})$ | Through the calculation of the two-node model, $Q_{sk}$, $T_{sk}$ and $\omega$ in the real environment can be obtained. Then an equation is established based on the same $Q_{sk}$ in the equivalent environment and the real environment to obtain the equivalent air temperature, that is SET*. Accurately obtaining parameters in the real environment becomes the key to accurately calculating SET*. Among them, M plays a decisive role in the calculation of the heat dissipation process of the core layer |
| PET* (°C) | MEMI | Heat flow from the core layer to the skin layer: $F_{cs} = \rho_{bl}m_{bl}c_{bl}(T_{cr} - T_{sk})$<br>Heat flow from the skin layer to the outer surface of the garment: $F_{scl} = \frac{T_{sk} - T_{cl}}{I_{cl}}$<br>Physiological equivalent temperature ($PET = t_a$):<br>$M - W - 3.05[5.733 - 0.007(M - W) - P_a] - 0.42(M - W - 58.15) - 0.0173M(5.87 - P_a) - 0.0014(34 - t_a) = 3.96 \times 10^{-8} f_{cl}\left[(t_{cl} + 273)^4 - (t_r + 273)^4\right] - f_{cl}h_c(t_{cl} - t_a)$ | The MEMI model is established based on the human body heat balance equation and some parameters of the two-node model. Compared with the first two models, its physiological sweat rate and heat flow calculation accuracy on each surface of the body has been improved. Therefore, M also plays a decisive role in the calculation of the heat dissipation process of the core layer |

(continued)

**Table 8** (continued)

| Thermal index | Theoretical Basis | Mathematical model | The role of M in thermal index calculations |
|---|---|---|---|
| UTCI (°C) | UTCI-Fiala multi-node model | UTCI is defined as the $t_a$ of the reference condition causing the same model response as the actual condition:<br>$UTCI(t_a, t_r, V_a, P_a) = t_a + Offset(t_a, t_r, V_a, P_a)$<br>Using fixed metabolic heat production value: $M = 2.3$ met ($135W/m^{-2}$) [203] | The UTCI-Fiala model subdivides body elements into spatial parts and individual tissue nodes, developed through regression statistics. Its dynamic system can predict the thermoregulatory responses of the central nervous system, such as vasoconstriction and dilation, shivering thermogenesis, and sweat excretion, etc. Therefore, although M is set to a fixed value in the model due to the selected activity level, the effect of metabolism on the passive system in different environments is dynamic (thermal stress response), with a direct impact on the air temperature of the reference environment |

Where $P_a$ is the partial pressure of water vapor (mmHg), $f_{cl}$ is the garment area coefficient, $t_{cl}$ is the average temperature of the clothing surface (°C), $M_{shiv}$ is the increased metabolic rate from shivering ($W/m^2$), $C_{res}$ and $E_{res}$ are respiratory sensible heat and respiratory latent heat dissipation ($W/m^2$), c is the specific heat capacity (J/(kg·°C)), ρ is the density(kg/L), m is flow rate ($L/(m^2 \cdot s)$), Subscript bl is blood, cr is core, sk is skin, $h_c$ is a comprehensive convective commutation coefficient considering the thermal resistance of clothing in a standard environment ($W/(m^2\ °C)$).

in transitional spaces but failed during walking [23]. Fang et al. suggested SET* for such spaces, yet its accuracy varied across walking phases [20].

Gender-based metabolic differences also influence index accuracy. Hema Latha observed women's M rises 13.4% during the luteal phase [210], while Molliet and Mady found women require higher comfort temperatures than men under the same activity [211]. Thus, using identical indices (e.g., PMV) for both genders introduces bias.

Table 9 summarizes the impact of M variability on thermal index calculations and prediction accuracy in existing studies.

The sources of prediction bias associated with metabolic rate in thermal indices are compiled in Fig. 7. In summary, accurate computation of metabolic rate is crucial for improving the predictive accuracy of thermal indices in comfort research. To expand the applicability and value of thermal indices, continued investigation into more precise methods for estimating individualized metabolic rates is essential.

## 6.3 Improvement of Thermal Index Models Related to Metabolic Rate

Fletcher analyzed metabolic rate data, including from the Compendium of Physical Activities (CPA), and showed that metabolic weighting reduces PMV bias at lower M ($\leq$2 met) [180]. Teli et al. applied a similar weighting to refine children's 1 met value by comparing physiological and activity energy ratios between children and adults [142]. Almeida et al. further showed that adjusting M by body surface area improves PMV–MTS agreement, estimating corrections of + 30% for ages 9–10 and + 210% for ages 4–5 [145]. These adjustments enhance thermal index reliability.

Fang et al. developed improved indices using the bin method, refining models for 1–2 met and 2.6 met conditions with $R^2 > 0.7$ outdoors [119]. Zhang et al. modified PET and UTCI by incorporating $\Delta M$ (instantaneous vs. initial M) to improve semi-open comfort predictions [23], while Fang et al. adjusted SET* using metabolic phases (ascent and steady) [20]. These corrections are illustrated in Fig. 8.

Other approaches emphasize personalization. Lee and Ham created a data-driven model using daily office M as predictors and confirmed activity-based M improves forecasting [214]. Anand et al. found ignoring differences in age, BMI, ethnicity, or diet significantly impairs prediction accuracy [215]. Na et al. used ASHRAE 55 activity standards to train a deep-learning METnet model, accurately estimating M-values from Kinect motion images [169].

Figure 9 outlines existing strategies for refining thermal index models. The literature review indicates that no current device or method offers broad applicability, high accuracy, ease of use, and rapid computation for metabolic rate determination. Additional research is needed to develop portable metabolic rate monitors suitable for diverse daily environments through extensive experimental testing. Advancements in thermal comfort research will be promoted by establishing a framework of

**Table 9** Impact of metabolic rate changes on the prediction accuracy of thermal index models

| Thermal index | Reference | Environmental parameters | Determine method of M | Sources of error | Deviation sources of thermal index model |
|---|---|---|---|---|---|
| PMV | Ter Mors et al. [209] | $\theta_{e,ref} = 15–25\ °C$ | Refer to ISO7730 standard, and correct the metabolic rate of Children by multiplying $\frac{1.7}{1.14}$. [212] | Insufficient consideration of individual characteristics (in this case the age and body surface area) | PMV underestimates the thermal sensation by 0.5–1.5 units |
| | Almeida et al. [145] | $t_a = 10.3–14.3\ °C$ | refer to ISO7730 standard, and 1) $M = 1.2 + \Delta M$ [10]; 2)$M = 1.2 \times \frac{A_{Du\ adult}}{A_{Du\ children}}$; ; 3) $M = 1.2\Delta M \times \frac{A_{Du\ adult}}{A_{Du\ children}}$ | Insufficient consideration of individual characteristics (in this case the age and body surface area) | PMV underestimates MTSV, and the mean absolute error (MAE) between PMV and MTSV is $MAE_{0)} = 0.81$, $MAE_{1)} = 0.63$; $MAE_{2)} = 0.61$, $MAE_{3)} = 0.71$ |
| | Fletcher [180] | $t_a = 10–25\ °C$; RH = 44–70%; $V_a = 0.15$ m/s | Use a weighting method that considers activities composed of multiple metabolic rates: $M = \frac{1}{T}\sum_{i=1}^{n} M_i t_i$, $PMV = \frac{1}{T}\sum_{i=1}^{n} PMV_i t_i$, and then select different source to obtain M values of different activities | High metabolic level (M >2 met); Different source to obtain M values | PMV calculated from different metabolic rate sources has biased prediction accuracy of actual thermal sensation. When $M \leq 2$, the deviations between the improved PMV value and MTSV corrected by ISO, CPA, $CPA_{rh}$, and $CPA_{sa}$ respectively fall within ± 0.5unit, while when M > 2 met, the bias exceeds ± 0.5unit |

(continued)

**Table 9** (continued)

| Thermal index | Reference | Environmental parameters | Determine method of M | Sources of error | Deviation sources of thermal index model |
|---|---|---|---|---|---|
| SET* (°C) | Fang et al. [20] | Indoors: $t_a = 22 \pm 2$ °C; RH $= 80 \pm 10\%$;<br>Semi-open space: $t_a = 34 \pm 2.5$ °C; RH $= 85 \pm 10\%$ | Use wearable sports watch to record heart rate data in real time | Ignore the dynamic changes during exercise | Different stages (changing or stabilizing) of the metabolic rate during exercise vary in the predictability of thermal sensation |
| | Ji et al. [213] | $t_a = t_r = 26$ °C; RH $= 50\%$ | Enumerating calculations from 1met-4met | The parameters calculated by two-node model, skin temperature, skin wettedness and skin heat lose in SET* standard environment were not strictly equal to those in actual environment | The clothing insulation ($I_s$) and the air velocity ($v_s$) of the standard environment were not of the fixed values but the functions of the actual metabolic rate:<br>$I_s = \frac{1.33}{M+0.74} - 0.095$<br>$v_s = \sqrt[0.53]{\frac{5.66(M-0.85)^{0.39}}{8.6}}, ifM < 1.072, v_s = 0.15$ |
| PET* (°C) | Zhang et al. [23] | $t_a = 33.9 \pm 1.1$ °C; RH $= 85.9 \pm 3.8\%$ | Use finger clip to record heart rate in real time | Ignore the dynamic changes during exercise | The deviation between PET and actual thermal sensation can be expressed by the following formula:<br>MTSV $= -0.235$PET $+ 0.626\Delta$M $- 18.858$ |
| | Fang et al. [119] | $t_{mrt} = 28.4$ °C; RH $= 70$–$83\%$ | Questionnaire survey | Different activity levels (1met-3.8met), large operating temperature ($t_{op}$) range ($t_{op} =$ 15–40 °C) | The deviation between PET and actual thermal sensation can be expressed by the following formula:<br>When M = 2.6met,<br>MTSV $= 4.97\ln(\text{PET}) - 14.247(R^2 = 0.86)$<br>When M = 1-2met,<br>MTSV $= 4.11\ln(\text{PET}) - 12.65(R^2 = 0.82)$ |

(continued)

**Table 9** (continued)

| Thermal index | Reference | Environmental parameters | Determine method of M | Sources of error | Deviation sources of thermal index model |
|---|---|---|---|---|---|
| UTCI (°C) | Zhang et al. [23] | $t_a = 33.9 \pm 1.1$ °C; RH $= 85.9 \pm 3.8\%$ | Use finger clip to record heart rate in real time | Ignore the dynamic changes during exercise | The deviation between UTCI and actual thermal sensation can be expressed by the following formula: $MTSV = -0.235UTCI + 1.225\Delta M + 7.745$ |
| | Fang et al. [119] | $t_{mrt} = 28.4$ °C; RH $= 70$–$83\%$ | Questionnaire survey | Different activity levels (1met-3.8met), large operating temperature ($t_{op}$) range ($t_{op} =$ 15–40 °C) | The deviation between UTCI and actual thermal sensation can be expressed by the following formula: When M = 2.6met, $MTSV = 4.382\ln(UTCI) - 13.046(R^2 = 0.74)$ When M = 1-2met, $MTSV = 4.834\ln(UTCI) - 15.184(R^2 = 0.83)$ |

Where $\theta_{e,ref}$ represents 4 day weighted running mean of daily min/max mean, including current day; $A_{Du}$ is DuBois body surface area ($m^2$); T is the total duration of the activity phase; $M_i$ is the metabolic rate of the active phase (met), and $t_i$ is the duration of the active phase; $CPA_{rh}$ is CPA M values relative to the reference human, $CPA_{rh} = \frac{CPA \times 81.34}{104.76}$ (met); $CPA_{sa}$ is the M value that align CPA values with ISO 8996 values, $CPA_{sa} = (CPA \times 0. .6236) + 0.259$ (met).

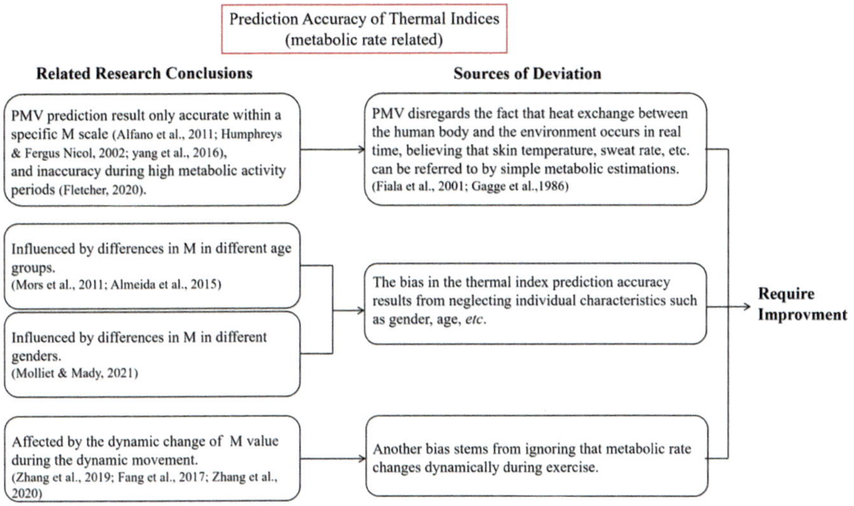


**Fig. 7** Summary of sources of metabolic rate-related thermal index prediction deviation

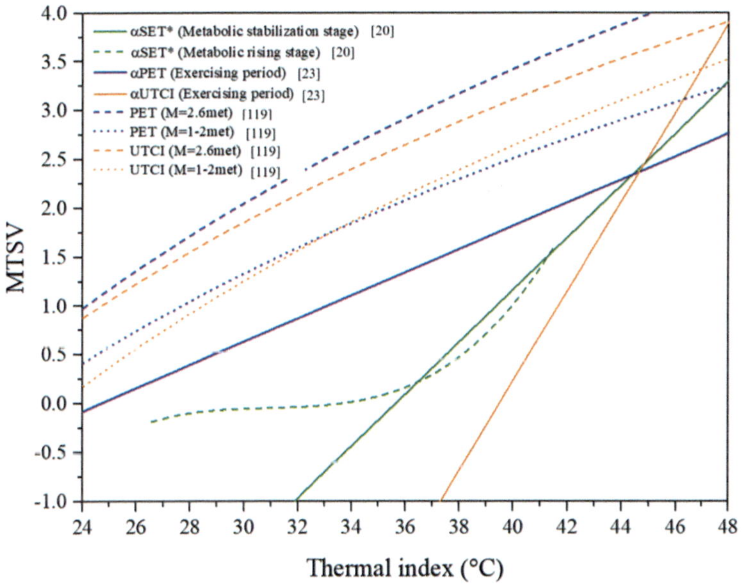


**Fig. 8** Summary of existing thermal index improving models via bin methods

personalized metabolic rate estimation methods adaptable to various conditions, age groups, and activity types.

## 7 Conclusion

This chapter examined the role of metabolic rate in thermal comfort research, with particular emphasis on its dynamic variability and interaction with environmental conditions. Metabolic rate emerged as a decisive factor in shaping thermal regulation, while inaccuracies in its estimation—especially during dynamic activities—remain a central challenge. Although activity diaries and standardized reference values are still widely applied, these approaches often fail to capture real-time fluctuations or individual differences.

Wearable devices currently represent the most widely adopted solution for monitoring metabolic rate in semi-open and outdoor environments. However, issues of portability, response time, and cost continue to limit their broader application. Future progress depends on the development of lightweight, affordable, and accurate personalized sensors capable of supporting continuous monitoring.

The findings further underline that the influence of metabolic rate differs across settings. In indoor environments, it closely tracks environmental changes, whereas in outdoor and transitional spaces it is moderated by radiation, wind, and physiological adaptation. As a result, treating metabolic rate as constant or isolated can lead to misleading conclusions. Improving the predictive accuracy of thermal indices therefore requires more precise, real-time assessment of metabolic activity, together with validation across larger and more diverse datasets.

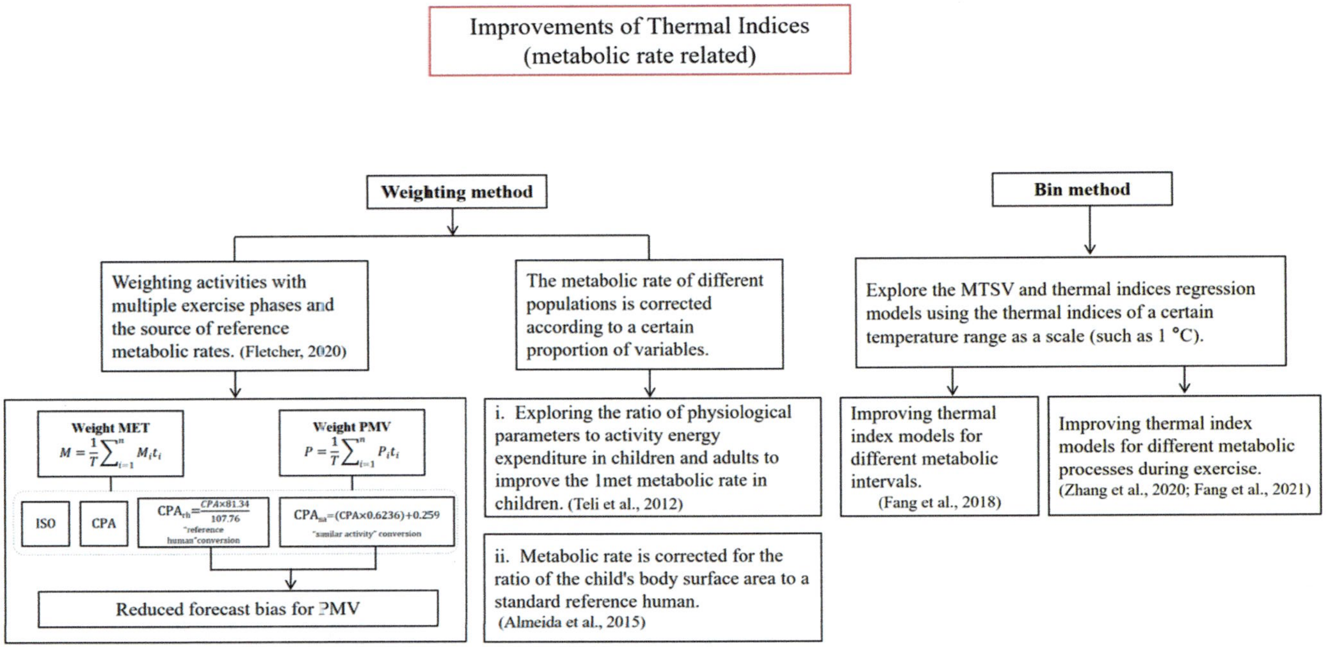


**Fig. 9** Existing improving methods of thermal indices

**Acknowledgements** This study was supported by the National Natural Science Foundation of China [Project No. 52278097], and a Theme-based Research Scheme Grant from the Research Grants Council of the Hong Kong Special Administrative Region, China (Project No. T22-504/21-R).

## References

1. I. Golasi, F. Salata, E. de Lieto Vollaro, M. Coppi, Complying with the demand of standardization in outdoor thermal comfort: a first approach to the Global Outdoor Comfort Index (GOCI). Build. Environ. **130**, 104–119 (2018)
2. G. Kalogeropoulos, A. Dimoudi, P. Toumboulidis, S. Zoras, Urban heat Island and thermal comfort assessment in a medium-sized Mediterranean City. Atmosphere **13**(7), 1102 (2022)
3. A. Ghaffarianhoseini, U. Berardi, A. Ghaffarianhoseini, K. Al-Obaidi, Analyzing the thermal comfort conditions of outdoor spaces in a university campus in Kuala Lumpur, Malaysia. Sci. Total Environ. **666**, 1327–1345 (2019)
4. K.-M. Wai, C. Yuan, A. Lai, K. Peter, Relationship between pedestrian-level outdoor thermal comfort and building morphology in a high-density city. Sci. Total Environ. **708**, 134516 (2020)
5. D. Lai, W. Liu, T. Gan, K. Liu, Q. Chen, A review of mitigating strategies to improve the thermal environment and thermal comfort in urban outdoor spaces. Sci. Total Environ. **661**, 337–353 (2019)
6. J. Huang, D. Shi, Z. Fang, Y. Gao, C. Zhuang, J. Zhai, Impact of short-term thermal experience on thermal sensation: a case study of Chongqing, China. Build. Environ. **179**, 106921 (2020)
7. ISO, ISO 7730, *Ergonomics of the Thermal Environment—Analytical Determination and Interpretation of Thermal Comfort Using Calculation of the PMV and PPD Indices and Local Thermal Comfort Criteria* (2005)
8. ASHRAE, ASHRAE Standard 55, *Thermal Environmental Conditions for Human Occupancy, ANSI/ASHRAE Standard 55–2021* (ASHRAE, Atlanta, 2021)
9. H.E. Beck, N.E. Zimmermann, T.R. McVicar, N. Vergopolan, A. Berg, E.F. Wood, Present and future Köppen-Geiger climate classification maps at 1-km resolution. Sci. Data **5**(1), 1–12 (2018)
10. B. Li, C. Du, R. Yao, W. Yu, V. Costanzo, Indoor thermal environments in Chinese residential buildings responding to the diversity of climates. Appl. Therm. Eng. **129**, 693–708 (2018)
11. S.Q.S. Hirashima, A. Katzschner, D.G. Ferreira, E.S. Assis, L. Katzschner, Thermal comfort comparison and evaluation in different climates. Urban Climate **23,** 219–230 (2018)
12. M.H. Elnabawi, N. Hamza, S. Dudek, Thermal perception of outdoor urban spaces in the hot arid region of Cairo, Egypt. Sustain. Cities Soc. **22**, 136–145 (2016)
13. P.O. Fanger, *Thermal comfort. Analysis and applications in environmental engineering, Thermal comfort. Analysis and applications in environmental engineering* (1970)
14. ISO, ISO 8996: 2021, *Ergonomics of the Thermal Environment—Determination of Metabolic Rate*. ISO2021
15. P.R. Höppe, Heat balance modelling. Experientia **49**, 741–746 (1993)
16. K. Parsons, *Human Thermal Environments*, 2nd edn (Taylor & Francis, London, UK, 2010) [Google Scholar]
17. Y. Epstein, D.S. Moran, Thermal comfort and the heat stress indices. Ind. Health **44**(3), 388–398 (2006)
18. A. Gagge, Rational temperature indices of man's thermal environment and their use with a 2-node model of his temperature regulation, in *Federation Proceedings* (1973), pp. 1572–1582
19. P. Wu, Y. Zhang, Z. Fang, Y. Gao, Comparison of thermal comfort in different kinds of building spaces: field study in Guangzhou, China. Indoor Built Environ. **31**(1), 186–202 (2022)

20. Z. Fang, Y. Zhang, Y. Yang, X. Zhou, Z. Zheng, Experimental investigation of standard effective temperature (SET*) adapted for human walking in an indoor and transitional thermal environment. Sci. Total Environ. **793**, 148421 (2021)
21. P. Höppe, The physiological equivalent temperature—a universal index for the biometeorological assessment of the thermal environment. Int. J. Biometeorol. **43**, 71–75 (1999)
22. P. Bröde, D. Fiala, K. Blazejczyk, Y. Epstein, I. Holmér, G. Jendritzky, B. Kampmann, M. Richards, H. Rintamäki, A. Shitzer, *Calculating UTCI Equivalent Temperature, Environmental Ergonomics XIII*, University of Wollongong, Wollongong (2009), pp. 49–53
23. Y. Zhang, J. Liu, Z. Zheng, Z. Fang, X. Zhang, Y. Gao, Y. Xie, Analysis of thermal comfort during movement in a semi-open transition space. Energy Build. **225**, 110312 (2020)
24. Z. Zheng, Y. Zhang, Y. Mao, Y. Yang, C. Fu, Z. Fang, Analysis of SET* and PMV to evaluate thermal comfort in prefab construction site offices: case study in South China. Case Stud. Thermal Eng. **26**, 101137 (2021)
25. H. Snyder, Literature review as a research methodology: an overview and guidelines. J. Bus. Res. **104**, 333–339 (2019)
26. N.J. Van Eck, L. Waltman, *Visualizing Bibliometric Networks, Measuring Scholarly Impact: Methods and Practice* (Springer, 2014), pp. 285–320
27. D. Chen, H.W. Chen, Using the Köppen classification to quantify climate variation and change: an example for 1901–2010. Environ. Dev. **6**, 69–79 (2013)
28. P.O. Fanger, Calculation of thermal comfort: introduction of a basic comfort equation, ASHRAE Trans, Part II **73**, III4. 1-III4. 20 (1967)
29. Y. Nishi, Measurement of thermal balance of man, in *Studies in Environmental Science* (Elsevier, 1981), pp. 29–39
30. D. Fiala, K.J. Lomas, M. Stohrer, A computer model of human thermoregulation for a wide range of environmental conditions: the passive system. J. Appl. Physiol. **87**(5), 1957–1972 (1999)
31. A.P. Gagge, J. Stolwijk, J. Hardy, Comfort and thermal sensations and associated physiological responses at various ambient temperatures. Environ. Res. **1**(1), 1–20 (1967)
32. A.P. Gagge, J. Stolwijk, B. Saltin, Comfort and thermal sensations and associated physiological responses during exercise at various ambient temperatures. Environ. Res. **2**(3), 209–229 (1969)
33. R. Gonzalez, Exercise physiology and sensory responses. Stud. Environ. Sci. **10**, 123–144 (1981)
34. G. Havenith, D. Fiala, Thermal indices and thermophysiological modeling for heat stress. Compr. Physiol. **6**(1), 255–302 (2011)
35. P. McNall, J. Jaax, F. Rohles, R. Nevins, W. Springer, Thermal comfort (thermally neutral) conditions for three levels of activity. ASHRAE Trans. **73**(1), 1.3 (1967)
36. O. Jay, G.P. Kenny, Heat exposure in the Canadian workplace. Am. J. Ind. Med. **53**(8), 842–853 (2010)
37. Y. Zhai, C. Elsworth, E. Arens, H. Zhang, Y. Zhang, L. Zhao, Using air movement for comfort during moderate exercise. Build. Environ. **94**, 344–352 (2015)
38. F.C. Houghten, Determination of the comfort zone. ASHVE Trans. 361–384 (1923)
39. T. Bedford, The warmth factor in comfort at work, in *A Physiological Study of Heating and Ventilation*, vol 76 (1936)
40. A. Matzarakis, H. Mayer, Another kind of environmental stress: thermal stress, WHO Newsl. **18**, 7–10 (1996)
41. S. Sharifi, W. Saman, A. Alemu, J. Boland, A proposed long-term thermal comfort scale. Build. Res. Inf. **49**(6), 661–678 (2021)
42. Y.D. Guan, M.H. Hosni, B.W. Jones, T.P. Gielda, Investigation of human thermal comfort under highly transient conditions for automotive applications-Part 1: Experimental design and human subject testing implementation. ASHRAE Trans. **109**, 885 (2003)
43. J. Han, S.-B. Kwon, C. Chun, Indoor environment and passengers' comfort in subway stations in Seoul. Build. Environ. **104**, 221–231 (2016)

44. M.A. Ruiz, E.N. Correa, Adaptive model for outdoor thermal comfort assessment in an Oasis city of arid climate. Build. Environ. **85**, 40–51 (2015)
45. A. García, F. Olivieri, E. Larrumbide, P. Ávila, Thermal comfort assessment in naturally ventilated offices located in a cold tropical climate, Bogotá. Build. Environ. **158**, 237–247 (2019)
46. R.-L. Hwang, W.-A. Chen, Identifying relative importance of solar design determinants on office building façade for cooling loads and thermal comfort in hot-humid climates. Build. Environ. **226**, 109684 (2022)
47. H. Mohammadpourkarbasi, I. Jackson, D. Nukpezah, I. Appeaning Addo, R. Assasie Oppong, Evaluation of thermal comfort in library buildings in the tropical climate of Kumasi, Ghana. Energy Build. **268**, 112210 (2022)
48. S. Lu, B. Pang, Y. Qi, K. Fang, Field study of thermal comfort in non-air-conditioned buildings in a tropical island climate. Appl. Ergon. **66**, 89–97 (2018)
49. S. Safarova, J. van Hoof, L. Law, K.K. Zander, S.T. Garnett, Thermal comfort in a tropical savanna climate: the case of home occupants in Darwin, Australia. Energy Build. **266**, 112074 (2022)
50. Z. Yang, W. Zhang, M. Qin, H. Liu, Comparative study of indoor thermal environment and human thermal comfort in residential buildings among cities, towns, and rural areas in arid regions of China. Energy Build. **273**, 112373 (2022)
51. I.I. El-Darwish, R.A. El-Gendy, Post occupancy evaluation of thermal comfort in higher educational buildings in a hot arid climate. Alex. Eng. J. **57**(4), 3167–3177 (2018)
52. A. Bassoud, H. Khelafi, A.M. Mokhtari, A. Bada, Evaluation of summer thermal comfort in arid desert areas. Case study: old adobe building in Adrar (South of Algeria). Build. Environ. **205**, 108140 (2021)
53. C.C. de Oliveira, R.F. Rupp, E. Ghisi, Influence of environmental variables on thermal comfort and air quality perception in office buildings in the humid subtropical climate zone of Brazil. Energy Build. **243**, 110982 (2021)
54. D.K. Shahi, H.B. Rijal, G. Kayo, M. Shukuya, Study on wintry comfort temperature and thermal improvement of houses in cold, temperate, and subtropical regions of Nepal. Build. Environ. **191**, 107569 (2021)
55. A. Asif, M. Zeeshan, S.R. Khan, N.F. Sohail, Investigating the gender differences in indoor thermal comfort perception for summer and winter seasons and comparison of comfort temperature prediction methods. J. Therm. Biol. **110**, 103357 (2022)
56. T. Tang, Y. Zhu, X. Zhou, Z. Guo, Y. Mao, H. Jiang, Z. Fang, Z. Zheng, X. Chen, Investigation of the effects of face masks on thermal comfort in Guangzhou, China. Build. Environ. **214**, 108932 (2022)
57. T. Sharmin, K. Steemers, M. Humphreys, Outdoor thermal comfort and summer PET range: a field study in tropical city Dhaka. Energy Build. **198**, 149–159 (2019)
58. K.K.-L. Lau, S.C. Chung, C. Ren, Outdoor thermal comfort in different urban settings of sub-tropical high-density cities: an approach of adopting local climate zone (LCZ) classification. Build. Environ. **154**, 227–238 (2019)
59. N. Aghamohammadi, C.S. Fong, M.H. Mohd Idrus, L. Ramakreshnan, U. Haque, Outdoor thermal comfort and somatic symptoms among students in a tropical city. Sustain. Cities Soc. **72**, 103015 (2021)
60. S. Chang, W. He, H. Yan, L. Yang, C. Song, Influences of vernacular building spaces on human thermal comfort in China's arid climate areas. Energy Build. **244**, 110978 (2021)
61. P. Kumar, A. Sharma, Assessing the outdoor thermal comfort conditions of exercising people in the semi-arid region of India. Sustain. Cities Soc. **76**, 103366 (2022)
62. S. Manavvi, E. Rajasekar, Evaluating outdoor thermal comfort in urban open spaces in a humid subtropical climate: Chandigarh, India. Build. Environ. **209**, 108659 (2022)
63. E. Krüger, Impact of site-specific morphology on outdoor thermal perception: a case-study in a subtropical location. Urban Climate **21**, 123–135 (2017)
64. M.-A. Jeong, S. Park, G.-S. Song, Comparison of human thermal responses between the urban forest area and the central building district in Seoul, Korea. Urban Forestry Urban Greening **15**, 133–148 (2016)

65. D. Lai, D. Guo, Y. Hou, C. Lin, Q. Chen, Studies of outdoor thermal comfort in northern China. Build. Environ. **77**, 110–118 (2014)
66. BSI, BS EN ISO 10551: 2019, *Ergonomics of the Thermal Environment. Assessment of the Influence of the Thermal Environment Using Subjective Judgment Scales* (British Standards Institution London, 2019)
67. T. Goto, J. Toftum, R. De Dear, P.O. Fanger, Thermal sensation and comfort with transient metabolic rates. Indoor Air **1**, 1038–1043 (2002)
68. C. Zuo, L. Luo, W. Liu, Effects of increased humidity on physiological responses, thermal comfort, perceived air quality, and Sick Building Syndrome symptoms at elevated indoor temperatures for subjects in a hot-humid climate. Indoor Air **31**(2), 524–540 (2021)
69. K. Velt, H. Daanen, Thermal sensation and thermal comfort in changing environments. J. Build. Eng. **10**, 42–46 (2017)
70. T. Tang, Y. Zhang, Z. Zheng, X. Zhou, Z. Fang, W. Liu, Detailed thermal indicators analysis based on outdoor thermal comfort indices in construction sites in South China. Build. Environ. **205**, 108191 (2021)
71. Y. Zhang, H. Chen, Q. Meng, Thermal comfort in buildings with split air-conditioners in hot-humid area of China. Build. Environ. **64**, 213–224 (2013)
72. Z. Fang, X. Feng, J. Liu, Z. Lin, C.M. Mak, J. Niu, K.-T. Tse, X. Xu, Investigation into the differences among several outdoor thermal comfort indices against field survey in subtropics. Sustain. Cities Soc. **44**, 676–690 (2019)
73. P.K. Cheung, C.Y. Jim, Subjective outdoor thermal comfort and urban green space usage in humid-subtropical Hong Kong. Energy Build. **173**, 150–162 (2018)
74. P. Bröde, D. Fiala, K. Błażejczyk, I. Holmér, G. Jendritzky, B. Kampmann, B. Tinz, G. Havenith, Deriving the operational procedure for the Universal Thermal Climate Index (UTCI). Int. J. Biometeorol. **56**, 481–494 (2012)
75. D.A. McIntyre, *Indoor climate* (Elsevier, 1980)
76. A. Matzarakis, H. Mayer, M.G. Iziomon, Applications of a universal thermal index: physiological equivalent temperature. Int. J. Biometeorol. **43**(2), 76–84 (1999)
77. K. Błażejczyk, P. Broede, D. Fiala, G. Havenith, I. Holmér, G. Jendritzky, B. Kampmann, A. Kunert, Principles of the New Universal Thermal Climate Index (UTCI) and its application to bioclimatic research in European scale. Miscellanea Geographica **14**(1), 91–102 (2010)
78. I. Griffths, Thermal comfort studies in buildings with passive solar features, field studies, UK: Report of the Commission of the European Community, ENS35 90 (1990)
79. J.A. Acero, L.A. Ruefenacht, E.J.Y. Koh, Y.S. Tan, L.K. Norford, Measuring and comparing thermal comfort in outdoor and semi-outdoor spaces in tropical Singapore. Urban Climate **42**, 101122 (2022)
80. M.S.E. Rajasekar, Evaluating outdoor thermal comfort in urban open spaces in a humid subtropical climate: Chandigarh, India. Build. Environ. **209**, 108659 (2022)
81. R. Abd Elraouf, A. Elmokadem, N. Megahed, O.A. Eleinen, S. Eltarabily, The impact of urban geometry on outdoor thermal comfort in a hot-humid climate. Build. Environ. 109632 (2022)
82. A. Alkhoudiri, I. Navarro, J.M. Fort, S. Alumran, Parametric comparative analysis of outdoor thermal comfort in a desert climate: a case study of single-family houses in Riyadh. Urban Climate **46**, 101300 (2022)
83. C. Xu, S. Li, X. Zhang, Application of the CPMV index to evaluating indoor thermal comfort in winter: case study on an office building in Beijing. Build. Environ. **162**, 106295 (2019)
84. M. Taleghani, U. Berardi, The effect of pavement characteristics on pedestrians' thermal comfort in Toronto. Urban Climate **24**, 449–459 (2018)
85. P.I. Konstantinov, M. Varentsov, E. Malinina, Modeling of thermal comfort conditions inside the urban boundary layer during Moscow's 2010 summer heat wave (case-study). Urban Climate **10**, 563–572 (2014)
86. L. An, B. Hong, X. Cui, Y. Geng, X. Ma, Outdoor thermal comfort during winter in China's cold regions: a comparative study. Sci. Total Environ. **768**, 144464 (2021)

87. L. Chen, E. Ng, Outdoor thermal comfort and outdoor activities: a review of research in the past decade. Cities **29**(2), 118–125 (2012)
88. Q. Zhao, Z. Lian, D. Lai, Thermal comfort models and their developments: a review. Energy Built Environ. **2**(1), 21–33 (2021)
89. F. Binarti, M.D. Koerniawan, S. Triyadi, S.S. Utami, A. Matzarakis, A review of outdoor thermal comfort indices and neutral ranges for hot-humid regions. Urban Climate **31**, 100531 (2020)
90. S. Shooshtarian, C.K.C. Lam, I. Kenawy, Outdoor thermal comfort assessment: a review on thermal comfort research in Australia. Build. Environ. **177**, 106917 (2020)
91. M.H. Elnabawi, N. Hamza, Behavioural perspectives of outdoor thermal comfort in urban areas: a critical review. Atmosphere **11**(1), 51 (2019)
92. Y. Zhang, X. Zhou, Z. Zheng, M.O. Oladokun, Z. Fang, Experimental investigation into the effects of different metabolic rates of body movement on thermal comfort. Build. Environ. **168**, 106489 (2020)
93. R.E. Hasson, C.A. Howe, B.L. Jones, P.S. Freedson, Accuracy of four resting metabolic rate prediction equations: effects of sex, body mass index, age, and race/ethnicity. J. Sci. Med. Sport **14**(4), 344–351 (2011)
94. K.R. Westerterp, Control of energy expenditure in humans. Eur. J. Clin. Nutr. **71**(3), 340–344 (2017)
95. A.M. Mengeste, A.C. Rustan, J. Lund, Skeletal muscle energy metabolism in obesity. Obesity **29**(10), 1582–1595 (2021)
96. D. Miller, Factors affecting energy expenditure. Proc. Nutr. Soc. **41**(2), 193–202 (1982)
97. J. Huang, H. Zhang, *Human and Thermal Environment*. Science Press
98. A.-M. Brouwer, M.A. Hogervorst, A new paradigm to induce mental stress: the Sing-a-Song Stress Test (SSST). Front. Neurosci. **8**, 224 (2014)
99. A.-M. Brouwer, M. Hogervorst, B. Reuderink, Y. van der Werf, J. van Erp, Physiological signals distinguish between reading emotional and non-emotional sections in a novel. Brain-Comput. Interfaces **2**(2–3), 76–89 (2015)
100. M.A. Hogervorst, A.-M. Brouwer, W.K. Vos, Physiological correlates of stress in individuals about to undergo eye laser surgery, in *2013 Humaine Association Conference on Affective Computing and Intelligent Interaction* (IEEE, 2013), pp. 473–478
101. A.-M. Brouwer, E. Van Dam, J.B. Van Erp, D.P. Spangler, J.R. Brooks, Improving real-life estimates of emotion based on heart rate: a perspective on taking metabolic heart rate into account. Front. Hum. Neurosci. **12**, 284 (2018)
102. E.H. Mason, E. Hill, D. Charlton, Abnormal specific dynamic action of protein, glucose, and fat associated with undernutrition. J. Clin. Investig. **4**(3), 353–387 (1927)
103. S.M. Secor, Specific dynamic action: a review of the postprandial metabolic response. J. Comp. Physiol. B. **179**, 1–56 (2009)
104. Y. Bai, Y. Zhang, Z. Wei, Y. Wang, Experimental study on human comfort responses after simulated summer commutes with double transients of temperature and metabolic rate. Build. Environ. **221**, 109253 (2022)
105. Y. Mao, Y. Zhu, Z. Guo, Z. Zheng, Z. Fang, X. Chen, Experimental investigation of the effects of personal protective equipment on thermal comfort in hot environments. Build. Environ. **222**, 109352 (2022)
106. A. Rodríguez, S. Becerril, S. Ezquerro, L. Mendez-Gimenez, G. Frühbeck, Crosstalk between adipokines and myokines in fat browning. Acta Physiol. **219**(2), 362–381 (2017)
107. S. Qiu, X. Cai, Z. Sun, U. Schumann, M. Zuegel, J.M. Steinacker, Chronic exercise training and circulating irisin in adults: a meta-analysis. Sports Med. **45**, 1577–1588 (2015)
108. Y. Tsuchiya, D. Ando, K. Takamatsu, K. Goto, Resistance exercise induces a greater irisin response than endurance exercise. Metabolism **64**(9), 1042–1050 (2015)
109. M.H. Rokling-Andersen, A.C. Rustan, A.J. Wensaas, O. Kaalhus, H. Wergedahl, T.H. Røst, J. Jensen, B.A. Graff, R. Caesar, C.A. Drevon, Marine n-3 fatty acids promote size reduction of visceral adipose depots, without altering body weight and composition, in male Wistar rats fed a high-fat diet. Br. J. Nutr. **102**(7), 995–1006 (2009)

110. A.M.C. Ooijen, K.R. Westerterp, L. Wouters, P.F. Schoffelen, A.A. Van Steenhoven, W.D.M. van Lichtenbelt, Heat production and body temperature during cooling and rewarming in overweight and lean men. Obesity **14**(11),1914–1920 (2006)
111. M.S. Westerterp-Plantenga, W.D. van Marken Lichtenbelt, H. Strobbe, P. Schrauwen, Energy metabolism in humans at a lowered ambient temperature. Eur. J. Clinical Nutr. **56**(4), 288–296 (2002)
112. X. Zeng, M. Ye, J.M. Resch, M.P. Jedrychowski, B. Hu, B.B. Lowell, D.D. Ginty, B.M. Spiegelman, Innervation of thermogenic adipose tissue via a calsyntenin 3β–S100b axis. Nature **569**(7755), 229–235 (2019)
113. J. Turner, A. Kumar, C. Koch, The effects of indoor and outdoor temperature on metabolic rate and adipose tissue—the Mississippi perspective on the obesity epidemic. Rev. Endocr. Metab. Disord. **17**, 61–71 (2016)
114. F.S. Celi, T.N. Le, B. Ni, Physiology and relevance of human adaptive thermogenesis response. Trends Endocrinol. Metab. **26**(5), 238–247 (2015)
115. M. Luo, X. Zhou, Y. Zhu, J. Sundell, Revisiting an overlooked parameter in thermal comfort studies, the metabolic rate. Energy Build. **118**, 152–159 (2016)
116. F. Johnson, A. Mavrogianni, M. Ucci, A. Vidal-Puig, J. Wardle, Could increased time spent in a thermal comfort zone contribute to population increases in obesity? Obes. Rev. **12**(7), 543–551 (2011)
117. C. Saxton, Effects of severe heat stress on respiration and metabolic rate in resting man. Aviat. Space Environ. Med. **52**(5), 281–286 (1981)
118. M. Luo, Z. Wang, K. Ke, B. Cao, Y. Zhai, X. Zhou, Human metabolic rate and thermal comfort in buildings: the problem and challenge. Build. Environ. **131**, 44–52 (2018)
119. Z. Fang, Z. Lin, C.M. Mak, J. Niu, K.-T. Tse, Investigation into sensitivities of factors in outdoor thermal comfort indices. Build. Environ. **128**, 129–142 (2018)
120. B.E. Ainsworth, W.L. Haskell, S.D. Herrmann, N. Meckes, D.R. Bassett Jr., C. Tudor-Locke, J.L. Greer, J. Vezina, M.C. Whitt-Glover, A.S. Leon, 2011 Compendium of physical activities: a second update of codes and MET values. Med. Sci. Sports Exerc. **43**(8), 1575–1581 (2011)
121. ASHRAE, *ASHRAE Handbook: Fundamentals* (2021)
122. L.W. Ludlow, P.G. Weyand, Energy expenditure during level human walking: seeking a simple and accurate predictive solution. J. Appl. Physiol. **120**(5), 481–494 (2016)
123. P.S. Freedson, E. Melanson, J. Sirard, Calibration of the computer science and applications, inc. accelerometer. Med. Sci. Sports Exerc. **30**(5), 777–781 (1998)
124. P. Freedson, D. Pober, K.F. Janz, Calibration of accelerometer output for children. Med. Sci. Sports Exerc. **37**(11), S523–S530 (2005)
125. D. Hendelman, K. Miller, C. Baggett, E. Debold, P. Freedson, Validity of accelerometry for the assessment of moderate intensity physical activity in the field. Med. Sci. Sports Exerc. **32**(9 Suppl), S442–S449 (2000)
126. S.E. Crouter, E. Kuffel, J.D. Haas, E.A. Frongillo, D.R. Bassett Jr., A refined 2-regression model for the actigraph accelerometer. Med. Sci. Sports Exerc. **42**(5), 1029 (2010)
127. A. Yngve, A. Nilsson, M. Sjostrom, U. Ekelund, Effect of monitor placement and of activity setting on the MTI accelerometer output. Med. Sci. Sports Exerc. **35**(2), 320–326 (2003)
128. J.V.G.A. Durnin, R. Passmore, Energy, work and leisure, in *Energy, Work and Leisure* (1967)
129. N.M. Byrne, A.P. Hills, G.R. Hunter, R.L. Weinsier, Y. Schutz, Metabolic equivalent: one size does not fit all. J. Appl. Phys. (2005)
130. E.E. Broday, A.A. de Paula Xavier, R. de Oliveira, Comparative analysis of methods for determining the metabolic rate in order to provide a balance between man and the environment. Int. J. Ind. Ergon. **44**(4), 570–580 (2014)
131. Y. Zhai, M. Li, S. Gao, L. Yang, H. Zhang, E. Arens, Y. Gao, Indirect calorimetry on the metabolic rate of sitting, standing and walking office activities. Build. Environ. **145**, 77–84 (2018)
132. A. Nomoto, R. Hisayama, S. Yoda, M. Akimoto, M. Ogata, H. Tsutsumi, S.-I. Tanabe, Indirect calorimetry of metabolic rate in college-age Japanese subjects during various office activities. Build. Environ. **199**, 107909 (2021)

133. P.G. Weyand, B.R. Smith, N.S. Schultz, L.W. Ludlow, M.R. Puyau, N.F. Butte, Predicting metabolic rate across walking speed: one fit for all body sizes? J. Appl. Physiol. **115**(9), 1332–1342 (2013)
134. W. Sirichana, B.A. Dolezal, E.V. Neufeld, X. Wang, C.B. Cooper, Wrist-worn triaxial accelerometry predicts the energy expenditure of non-vigorous daily physical activities. J. Sci. Med. Sport **20**(8), 761–765 (2017)
135. B. Jeng, K.L. Cederberg, B. Lai, J.E. Sasaki, M.M. Bamman, R.W. Motl, Accelerometer output and its association with energy expenditure in persons with mild-to-moderate Parkinson's disease. PLoS ONE **15**(11), e0242136 (2020)
136. A.T. Allred, P. Choi, S. Agiovlasitis, Triaxial accelerometer output predicts oxygen uptake in adults with down syndrome. Disabil. Rehabil. **43**(18), 2602–2609 (2021)
137. R. Takigawa, T. Inayama, K. Yamada, Y. Muraoka, K. Ohkawara, Assessment of daily physical activities with sensors attached to the upper limbs in healthy adults using a manual wheelchair. J. Phys. Fitness Sports Med. **7**(3), 193–202 (2018)
138. K. Lyden, S.L. Kozey, J.W. Staudenmeyer, P.S. Freedson, A comprehensive evaluation of commonly used accelerometer energy expenditure and MET prediction equations. Eur. J. Appl. Physiol. **111**, 187–201 (2011)
139. E. Valanou, C. Bamia, A. Trichopoulou, Methodology of physical-activity and energy-expenditure assessment: a review. J. Public Health **14**, 58–65 (2006)
140. J. Lighton, Limitations and requirements for measuring metabolic rates: a mini review. Eur. J. Clin. Nutr. **71**(3), 301–305 (2017)
141. W. Ji, M. Luo, B. Cao, Y. Zhu, Y. Geng, B. Lin, A new method to study human metabolic rate changes and thermal comfort in physical exercise by $CO_2$ measurement in an airtight chamber. Energy and Buildings **177**, 402–412 (2018)
142. D. Teli, M.F. Jentsch, P.A. James, Naturally ventilated classrooms: an assessment of existing comfort models for predicting the thermal sensation and preference of primary school children. Energy Build. **53**, 166–182 (2012)
143. G. Havenith, Metabolic rate and clothing insulation data of children and adolescents during various school activities. Ergonomics **50**(10), 1689–1701 (2007)
144. D.M. Rogers, B.L. Olson, J.H. Wilmore, Scaling for the VO2-to-body size relationship among children and adults. J. Appl. Physiol. **79**(3), 958–967 (1995)
145. R.M. Almeida, N.M. Ramos, V.P. De Freitas, Thermal comfort models and pupils' perception in free-running school buildings of a mild climate country. Energy Build. **111**, 64–75 (2016)
146. M. Ito, H. Nishi, Real-time METs estimation for effective indoor amenity control in HEMS. IEEE Int. Symp. Ind. Electronics **2013**, 1–6 (2013)
147. W. Li, Y. Zhang, L. Zhang, Z. Li, The research of the sport burden grade standard corresponding to the heart rate. Nat. Sci. J. Harbin Normal Univ **24**(5), 99–102 (2008)
148. J. Staudenmayer, D. Pober, S. Crouter, D. Bassett, P. Freedson, An artificial neural network to estimate physical activity energy expenditure and identify physical activity type from an accelerometer. J. Appl. Physiol. (2009)
149. K.N. Wines, A.K. Rzepecki, A.L. Andrews, R.E. Dechert, Validation of the Vmax metabolic cart in a simulated pediatric model. J. Parenter. Enter. Nutr. **39**(3), 353–358 (2015)
150. C. Baldari, M. Meucci, F. Bolletta, M. Gallotta, G. Emerenziani, L. Guidetti, Accuracy and reliability of COSMED K5 portable metabolic device versus simulating system. Sport Sci. Health **11**(1), 58 (2015)
151. R. Duffield, B. Dawson, H. Pinnington, P. Wong, Accuracy and reliability of a Cosmed K4b2 portable gas analysis system. J. Sci. Med. Sport **7**(1), 11–22 (2004)
152. I. Ruiz, M. Sprowls, Y. Deng, D. Kulick, H. Destaillats, E.S. Forzani, Assessing metabolic rate and indoor air quality with passive environmental sensors. J. Breath Res. **12**(3), 036012 (2018)
153. K. Fabbri, Indoor thermal comfort perception, in *A questionnaire approach focusing on children* (Springer, New York City, NY, USA, 2015)
154. W.L. Khong, M. Mariappan, The evolution of heart beat rate measurement techniques from contact based photoplethysmography to non-contact based photoplethysmography imaging, in *2019 IEEE International Circuits and Systems Symposium (ICSyS)* (IEEE, 2019), pp. 1–4

155. S. Madihally, *Principles of Biomedical Engineering*, Artech House (2019)
156. A. Javaloyes, J.M. Sarabia, R.P. Lamberts, M. Moya-Ramon, Training prescription guided by heart-rate variability in cycling. Int. J. Sports Physiol. Perform. **14**(1), 23–32 (2019)
157. P. Jayathissa, M. Quintana, T. Sood, N. Narzarian, C. Miller, *Is Your Clock-Face Cozie, A Smartwatch Methodology for the In-situ Collection of Occupant Comfort Data* (2019)
158. C.-C. Yang, Y.-L. Hsu, A review of accelerometry-based wearable motion detectors for physical activity monitoring. Sensors **10**(8), 7772–7788 (2010)
159. S.C. Mukhopadhyay, Wearable sensors for human activity monitoring: a review. IEEE Sens. J. **15**(3), 1321–1330 (2014)
160. J. Yao, Y. Li, X. Tang, H. Hu, Usability evaluation of finger clip oximeter. Packaging Engineering **41**(10), 95–100 (2020)
161. Z. Fang, T. Tang, Z. Zheng, X. Zhou, W. Liu, Y. Zhang, Thermal responses of workers during summer: an outdoor investigation of construction sites in South China. Sustain. Cities Soc. **66**, 102705 (2021)
162. M.H. Hasan, F. Alsaleem, M. Rafaie, Sensitivity study for the PMV thermal comfort model and the use of wearable devices biometric data for metabolic rate estimation. Build. Environ. **110**, 173–183 (2016)
163. G.M. Revel, M. Arnesano, F. Pietroni, Integration of real-time metabolic rate measurement in a low-cost tool for the thermal comfort monitoring in AAL environments, ambient assisted living: Italian forum. Springer **2015**, 101–110 (2014)
164. T. Chakraborti, K.D. Sharma, A. Chatterjee, A novel local extrema based gravitational search algorithm and its application in face recognition using one training image per class. Eng. Appl. Artif. Intell. **34**, 13–22 (2014)
165. H.F.T. Ahmed, H. Ahmad, C. Aravind, Device free human gesture recognition using Wi-Fi CSI: a survey. Eng. Appl. Artif. Intell. **87**, 103281 (2020)
166. E.J. Choi, Y. Yoo, B.R. Park, Y.J. Choi, J.W. Moon, Development of occupant pose classification model using deep neural network for personalized thermal conditioning. Energies **13**(1), 45 (2019)
167. B.R. Park, E.J. Choi, Y.J. Choi, J.W. Moon, Accuracy analysis of DNN-based pose-categorization model and activity-decision algorithm. Energies **13**(4), 839 (2020)
168. H. Na, J.-H. Choi, H. Kim, T. Kim, Development of a human metabolic rate prediction model based on the use of Kinect-camera generated visual data-driven approaches. Build. Environ. **160**, 106216 (2019)
169. H. Na, H. Choi, T. Kim, Metabolic rate estimation method using image deep learning, in *Building Simulation* (Springer, 2020), pp. 1077–1093
170. P. Lu, J. Li, Acceptable temperature steps for occupants moving between air-conditioned main space and naturally ventilated transitional space of building. Build. Environ. **182**, 107150 (2020)
171. Z. Fang, Z. Zheng, X. Feng, D. Shi, Z. Lin, Y. Gao, Investigation of outdoor thermal comfort prediction models in South China: a case study in Guangzhou. Build. Environ. **188**, 107424 (2021)
172. J. Niu, B. Hong, Y. Geng, J. Mi, J. He, Summertime physiological and thermal responses among activity levels in campus outdoor spaces in a humid subtropical city. Sci. Total Environ. **728**, 138757 (2020)
173. J. Niu, J. Xiong, H. Qin, J. Hu, J. Deng, G. Han, J. Yan, Influence of thermal comfort of green spaces on physical activity: empirical study in an urban park in Chongqing, China. Build. Environ. **219**, 109168 (2022)
174. M. Nikolopoulou, S. Lykoudis, Thermal comfort in outdoor urban spaces: analysis across different European countries. Build. Environ. **41**(11), 1455–1470 (2006)
175. A.H.A. Mahmoud, Analysis of the microclimatic and human comfort conditions in an urban park in hot and arid regions. Build. Environ. **46**(12), 2641–2656 (2011)
176. N.A. Kenny, J.S. Warland, R.D. Brown, T.G. Gillespie, Part A: assessing the performance of the COMFA outdoor thermal comfort model on subjects performing physical activity. Int. J. Biometeorol. **53**, 415–428 (2009)

177. B.E. Ainsworth, W.L. Haskell, M.C. Whitt, M.L. Irwin, A.M. Swartz, S.J. Strath, W.L. O Brien, D.R. Bassett, K.H. Schmitz, P.O. Emplaincourt, Compendium of physical activities: an update of activity codes and MET intensities. Med. Sci. Sports Exerc. **32**(9; SUPP/1), S498–S504 (2000)
178. J.K. Vanos, J.S. Warland, T.J. Gillespie, G.A. Slater, R.D. Brown, N.A. Kenny, Human energy budget modeling in urban parks in Toronto and applications to emergency heat stress preparedness. J. Appl. Meteorol. Climatol. **51**(9), 1639–1653 (2012)
179. L.M. Monteiro, Thermal comfort index for sustainable urban spaces, in *Central Europe Towards Sustainable Building* (2013)
180. M. Fletcher, D. Glew, A. Hardy, C. Gorse, A modified approach to metabolic rate determination for thermal comfort prediction during high metabolic rate activities. Build. Environ. **185**, 107302 (2020)
181. A. Mark, *The Impact of the Individual Layers in Multi-layer Clothing Systems on the Distribution of the Air Gap Thickness and Contact Area* (Albstadt-Sigmaringen University, Albstadt, 2013)
182. K.B. Pandolf, B. Givoni, R.F. Goldman, Predicting energy expenditure with loads while standing or walking very slowly. J. Appl. Physiol. **43**(4), 577–581 (1977)
183. B. Givoni, R.F. Goldman, Predicting metabolic energy cost. J. Appl. Physiol. **30**(3), 429–433 (1971)
184. Y. Epstein, L. Stroschein, K. Pandolf, Predicting metabolic cost of running with and without backpack loads. Eur. J. Appl. Physiol. **56**(5), 495–500 (1987)
185. J.R. Drain, B. Aisbett, M. Lewis, D.C. Billing, The Pandolf equation under-predicts the metabolic rate of contemporary military load carriage. J. Sci. Med. Sport **20**, S104–S108 (2017)
186. C.C. Moore, E.J. Aguiar, S.W. Ducharme, C. Tudor-Locke, Development of a cadence-based metabolic equation for walking. Med. Sci. Sports Exerc. **53**(1), 165 (2021)
187. D. Fiala, K.J. Lomas, M. Stohrer, Computer prediction of human thermoregulatory and temperature responses to a wide range of environmental conditions. Int. J. Biometeorol. **45**(3), 143–159 (2001)
188. J.L. Threlkeld, *Thermal Environmental Engineering* (Prentice Hall, 1970)
189. C. Yang, T. Yin, M. Fu, Study on the allowable fluctuation ranges of human metabolic rate and thermal environment parameters under the condition of thermal comfort. Build. Environ. **103**, 155–164 (2016)
190. H. Zhang, M. Tang, W. Zhao, Investigation of the urban walking space sun-shading, in *Proceedings of the National Conference on Building Environment and Building Energy Conservation* (2007)
191. A. Uğursal, C.H. Culp, The effect of temperature, metabolic rate and dynamic localized airflow on thermal comfort. Appl. Energy **111**, 64–73 (2013)
192. T. Tang, X. Zhou, Y. Zhang, X. Feng, W. Liu, Z. Fang, Z. Zheng, Investigation into the thermal comfort and physiological adaptability of outdoor physical training in college students. Sci. Total Environ. **839**, 155979 (2022)
193. S. Lu, H. Xia, S. Wei, K. Fang, Y. Qi, Analysis of the differences in thermal comfort between locals and tourists and genders in semi-open spaces under natural ventilation on a tropical island. Energy Build. **129**, 264–273 (2016)
194. J. Huang, S. Wu, T. Yuan, D. Shi, Y. Gao, Analysis of equivalent metabolic rate (EMT) used for predicted mean vote: a case study in Chongqing, China. Case Stud. Thermal Eng. **27**, 101230 (2021)
195. Z. Fang, Y. Mao, Y. Zhu, J. Lu, Z. Zheng, X. Chen, Human thermal physiological response of wearing personal protective equipment: an educational building semi-open space experimental investigation. Sci. Total Environ. **877**, 162779 (2023)
196. J.W. Castellani, A.J. Young, Human physiological responses to cold exposure: acute responses and acclimatization to prolonged exposure. Auton. Neurosci. **196**, 63–74 (2016)
197. R. De Dear, G.S. Brager, *Developing an Adaptive Model of Thermal Comfort and Preference* (1998)

198. T. Goto, J. Toftum, R. de Dear, P.O. Fanger, Thermal sensation and thermophysiological responses to metabolic step-changes. Int. J. Biometeorol. **50**, 323–332 (2006)
199. Z. Fang, P. Zhang, P. Wu, L. Li, Z. Zheng, Effects of different metabolic rates on thermal comfort in a warmer thermal environment. J. Build. Eng. **80**, 108030 (2023)
200. H. Wang, S. Hu, Experimental study on thermal sensation of people in moderate activities. Build. Environ. **100**, 127–134 (2016)
201. Y. Zhai, S. Zhao, Y. Gao, W. Song, L. Yang, H. Zhang, E. Arens, Preferred temperatures with and without air movement during moderate exercise. Energy Build. **207**, 109565 (2020)
202. S. Gao, Y. Zhai, L. Yang, H. Zhang, Y. Gao, Preferred temperature with standing and treadmill workstations. Build. Environ. **138**, 63–73 (2018)
203. D. Fiala, G. Havenith, P. Bröde, B. Kampmann, G. Jendritzky, UTCI-Fiala multi-node model of human heat transfer and temperature regulation. Int. J. Biometeorol. **56**(3), 429–441 (2012)
204. A.P. Gagge, A. Fobelets, L. Berglund, A standard predictive index of human response to the thermal environment. ASHRAE Trans **92**(2), 709–731 (1986)
205. M.A. Humphreys, J.F. Nicol, The validity of ISO-PMV for predicting comfort votes in every-day thermal environments. Energy Build. **34**(6), 667–684 (2002)
206. F.R.A. Alfano, B.I. Palella, G. Riccio, The role of measurement accuracy on the thermal environment assessment by means of PMV index. Build. Environ. **46**(7), 1361–1369 (2011)
207. L.M. Chamra, W.G. Steele, K. Huynh, The uncertainty associated with thermal comfort. ASHRAE Trans. **109**, 356 (2003)
208. G. Havenith, I. Holmér, K. Parsons, Personal factors in thermal comfort assessment: clothing properties and metabolic heat production. Energy Build. **34**(6), 581–591 (2002)
209. S. Ter Mors, J.L. Hensen, M.G. Loomans, A.C. Boerstra, Adaptive thermal comfort in primary school classrooms: creating and validating PMV-based comfort charts. Build. Environ. **46**(12), 2454–2461 (2011)
210. N. Hema Latha, *Effects of Different Phases of Menstrual Cycle on Basal Metabolic Rate* (2006)
211. D.S. Molliet, C.E.K. Mady, Exergy analysis of the human body to assess thermal comfort conditions: comparison of the thermal responses of males and females. Case Stud. Thermal Eng. **25**, 100972 (2021)
212. K. Parsons, *Human Thermal Environments: The Effects of Hot, Moderate, and Cold Environments on Human Health, Comfort and Performance* (CRC Press, 2007).
213. W. Ji, Y. Zhu, H. Du, B. Cao, Z. Lian, Y. Geng, S. Liu, J. Xiong, C. Yang, Interpretation of standard effective temperature (SET) and explorations on its modification and development. Build. Environ. **210**, 108714 (2022)
214. J. Lee, Y. Ham, Physiological sensing-driven personal thermal comfort modelling in consideration of human activity variations. Build. Res. Inf. **49**(5), 512–524 (2021)
215. V. Anand, D. Sendhil, E. Rajasekar, Estimating the metabolic rate and associated physiological response for Indian subjects through climate chamber experiments. Build. Environ. **207**, 108466 (2022)

# Impact of Seasonal Variations on Human Outdoor Thermal Comfort

**Abstract** This chapter investigates how seasonal variations shape human thermal comfort in outdoor environments. A field study in Guangzhou, a city with hot summers and mild winters, was conducted to assess the sensitivity of thermal perception to key microclimatic parameters. Using standardized regression coefficients, the influence of air temperature, mean radiant temperature, wind speed, and relative humidity was examined across different seasons, alongside the performance of three thermal indices: PET, UTCI, and SET. Results demonstrate that human thermal sensitivity is season-dependent, with air temperature exerting the strongest effect and relative humidity playing a minimal role. Among the indices, SET provided the most consistent alignment with observed seasonal responses. These findings highlight the importance of considering seasonal dynamics in outdoor thermal comfort research and provide evidence for climate-responsive urban planning and design.

## 1 Introduction

Given the accelerated progression of global warming and urban expansion, heightened attention is increasingly focused on outdoor thermal environments. Contemporary research efforts have expanded significantly toward developing mitigation strategies for improving thermal conditions in open spaces [1–3]. Practical implementation of these approaches demands incorporation of regionally-specific thermal sensitivity metrics, highlighting essential factors that shape human thermal experience outdoors while carefully evaluating economic and functional trade-offs. Human thermal comfort perception is principally controlled by physiological heat equilibrium mechanisms. Established thermal assessment frameworks, including PMV [4], SET* [5, 6], PET [7], and UTCI [8, 9], were formulated based on fundamental heat balance principles. Energy exchange between individuals and their immediate surroundings occurs predominantly through convective and radiative pathways. Thus, the interconnection between human comfort and microclimatic elements may display non-linear patterns [10–12]. Accordingly, outdoor thermal comfort is

 
Z. Fang et al., *Outdoor Thermal Comfort and Heat Stress Mitigation*, Environmental Science and Engineering, https://doi.org/10.1007/978-981-95-8899-2_6

governed by the combined influence of multiple elements, particularly the convective heat transfer described by equations $T_a$ and $V_a$. Bröde et al. [8] have established the correlation between $T_a$ and $V_a$ as substantiation for UTCI's thermal responsiveness. Nonetheless, due to physiological acclimatization and prevailing meteorological conditions, people's susceptibility to different thermal parameters may demonstrate considerable variation across diverse climatic regions and seasonal contexts [11, 13].

Despite considerable research attention on outdoor thermal sensitivity (Table 1), notable inconsistencies persist across study results. Scholarly work has predominantly recognized parameter $T_{mrt}$ as a crucial factor in outdoor thermal comfort; however, Peng et al. [14] noted that $T_{mrt}$ demonstrated only marginal influence on thermal perception. Further contrasting evidence was presented by Chan and Chau [15] and Liu et al. [16], who revealed opposing thermal sensitivity trends under different seasonal conditions within the same geographical context. These patterns indicate that investigations carried out in the same area may generate divergent findings. To illustrate, while Li et al. [17] established parameter $T_{mrt}$ as the primary summertime sensitivity indicator in Hong Kong, Chan and Chau [15] reported parameter $T_a$ as the most influential element.

Apart from qualitative sensitivity studies, quantitative approaches have received considerable focus in multiple research initiatives. A prevalent method for analyzing connections between human thermal perception and particular microclimatic elements employs regression techniques using experimental datasets, while controlling other variables at fixed values or within specified limits [13]. Nonetheless, this methodology shows notable drawbacks; notwithstanding the implementation of comprehensive survey studies, the impact of additional variables in exterior environments cannot be completely removed. A different tactic for evaluating sensitivity focuses on creating models of human thermal comfort. Several investigations have produced such frameworks for various climate zones (Table 2). These analytical tools, however, mainly depend on multiple linear regression approaches, consequently presuming linear correlations between separate variables and thermal comfort. While these models show effectiveness in forecasting thermal comfort in localized areas, they are unable to precisely measure the distinct contribution of every microclimatic element to human thermal perception.

Determining exact quantitative connections between thermal perception and microclimatic factors using established outdoor comfort indices represents a crucial research objective. Nevertheless, investigations have shown that such models generate systematically skewed outcomes when employed across different climate zones, mainly because of variations in human thermal acclimatization. Earlier studies conducted by the research team [18] revealed uncertain interrelationships between SET*, PET, UTCI and the mean thermal sensation vote (MTSV) during Guangzhou's periods of high outdoor heat, particularly when the operative temperature exceeded 34 °C. Thus, existing outdoor thermal comfort frameworks need revisions that incorporate location-specific adaptive traits. Different modeling approaches may also display distinct responsive behaviors. For instance, while PET maintained a

**Table 1** Studies of outdoor thermal sensitivity

| Location | Season/time | The most sensitive parameters (in order of highest to lowest) |
|---|---|---|
| Hongkong, China [17] | Summer, autumn, winter | $T_{mrt}$, sun and wind desirability, $V_a$, |
| Hongkong, China [15] | Summer, winter | Summer: $T_a$;<br>Winter: $T_a$, SR[a] |
| Hongkong, China [26] | Summer, winter | $T_a$, $V_a$, SR |
| Guangzhou, China [25] | Summer | $T_a$, $T_{mrt}$, $V_a$, RH |
| Guangzhou, China [13] | | $T_a$ |
| Nanjing, China [29] | All year | SR, atmospheric pressure, maximum temperature, $V_a$, RH |
| Changsha, China [16] | Spring, summer, autumn, winter | Spring, summer, autumn: $T_a$;<br>Winter: thermal radiation |
| Mianyang, China [30] | Summer, autumn, winter | $T_a$, SR |
| Shanghai, China [6] | Winter | $T_a$, SR |
| Xian, China [31] | Winter | SR, $T_a$, RH, $V_a$ |
| Tian, China [1] | All year | $T_a$ |
| Harbin, China [32] | All year | $T_a$ |
| Wellington, New Zealand [33] | – | Gustiness, $V_a$ |
| Europe [34] | 2001.7–2002.9 | $T_a$, SR |
| Birmingham, UK [35] | 2004.4/2005.2 | $T_a$, $V_a$ |
| Melbourne, Australia [36] | 2014.1, 2015.2, 2015.5 | SVF, time of exposure |
| Isfahan, Iran [37] | 2014.7.12–18, 2014.11.29, 2014.12 | $T_a$ |

[a]SR = Solar Radiation

pronounced linear connection with operative temperature, UTCI showed an exponential relationship with operative temperature [19]. Hence, the sensitivity characteristics of geographically adapted models demand thorough verification.

Accordingly, a field-based study was performed in Guangzhou, China, exemplifying a characteristic hot summer and warm winter climatic area. The methodology integrated simultaneous monitoring of microclimatic parameters with distributed thermal comfort questionnaires. Thermal reactivity in this specific zone was then evaluated by means of standardized regression coefficients. The PET, UTCI, and SET* indices were subsequently adjusted using linear regression methods applied to the gathered research data. The highest-performing model underwent sensitivity assessment, enabling detailed analysis of quantitative relationships between human thermal comfort and key microclimatic drivers.

**Table 2** Regional thermal comfort regression models

| Location (Author) | | Research time | Empirical TSV model | $R^2$ |
|---|---|---|---|---|
| Yokohama, Japan [38] | | 1994–1995 | $TSV = 0.1115T_a + 0.0019SR - 0.3185V_a + 1.2$ (All year) | 0.871 |
| Hong Kong, China [9] | | Summer and Winter 2003 | $TSV = 0.1895T_a + 0.0028SR - 0.7754V_a + 0.1953RH - 2.3$ (Summer) | 0.870 |
| | | | $TSV = 0.1185T_a + 0.0025SR - 0.6019V_a + 0.1155RH - 4.77$ (Summer and winter) | 0.910 |
| Birmingham, UK [36] | | 2004/8–2005/2 | $TSV = 0.0665T_a + 0.0011SR - 0.2256V_a - 0.0079RH + 3.2464$ (All year) | 0.250 |
| Curitiba, Brazil [40] | | 2009/1–2009/8 | $TSV = 0.1293T_a + 0.0007SR - 0.2744V_a - 2.26$ | 0.665 |
| Changsha, China | [40] | 2010/6–2010/8 | $TSV = 0.313T_a + 0.030T_{mrt} - 0.304V_a + 0.026RH - 11.622$(Summer) | 0.552 |
| Singapore, Singapore | | 2010/8–2011/5 | $TSV = 0.398T_a + 0.038T_{mrt} - 0.329V_a + 0.023RH - 14.061$ | 0.801 |
| Wuhan, China [41] | | 20/8/2011–27/11/2011 | $TSV = 0.0643T_a + 0.00076SR - 0.161V_a - 0.00376RH - 1.382$(Summer and autumn) | 0.670 |
| Changsha, China [16] | | 2012/3–2013/2 | $TSV = 0.146T_a - 0.007A_h - 0.308V_a + 0.097T_{ga} - 2.819$(All year) | 0.794 |
| | | | $TSV = 0.172T_a - 0.041A_h - 0.191V_a + 0.042T_{ga} - 2.779$(Spring) | 0.630 |
| | | | $TSV = 0.236T_a + 0.015A_h - 0.325V_a + 0.091T_{ga} - 5.886$(Summer) | 0.777 |
| | | | $TSV = 0.112T_a - 0.011A_h - 0.373V_a + 0.110T_{ga} - 2.297$(Autumn) | 0.608 |
| | | | $TSV = 0.121T_a + 0.102A_h - 0.464V_a + 0.111T_{ga} - 2.882$(Winter) | 0.718 |
| Guangzhou, China [42] | | 2014/8–2014/10 | $TSV = 0.245T_a + 0.059T_{mrt} - 0.457V_a + 0.013RH - 8.527$(Summer) | 0.598 |
| Athens, Greece | [34] | – | $TSV = 0.034T_a + 0.0001SR - 0.086V_a - 0.001RH - 0.412$ | 0.270 |
| Thessaloniki, Greece | | – | $TSV = 0.036T_a + 0.0013SR - 0.038V_a + 0.011RH - 2.197$ | 0.510 |
| Milan, Italy | | – | $TSV = 0.049T_a - 0.0002SR + 0.006V_a + 0.002RH - 0.920$ | 0.440 |
| Fribourg, Switzerland | | – | $TSV = 0.068T_a + 0.0006SR - 0.107V_a - 0.002RH - 0.69$ | 0.680 |
| Kassel, Germany | | – | $TSV = 0.043T_a + 0.0005SR - 0.077V_a + 0.001RH - 0.876$ | 0.480 |

(continued)

**Table 2** (continued)

| Location (Author) | Research time | Empirical TSV model | $R^2$ |
|---|---|---|---|
| Cambridge, UK | – | $TSV = 0.113T_a + 0.0001SR - 0.05V_a - 0.003RH - 1.74$ | 0.570 |
| Sheffield, UK | – | $TSV = 0.07T_a + 0.0012SR - 0.057V_a - 0.003RH - 0.855$ | 0.580 |
| Combined model for Europe | | $TSV = 0.049T_a + 0.001SR - 0.051V_a + 0.014RH - 2.079$ | 0.780 |
| Mendoza, Argentina [44] | Winter 2010, Summer 2011 | $TSV = 0.0621T_a - 0.3257V_a + 0.0079RH - 0.9796$ | 0.720 |
| Rome, Italy [45] | 2014/2–2015/1 | $TSV = 0.058T_a + 0.083T_{mrt} - 0.272V_a + 0.005RH + 0.264I_{cl} - 4.068$ | 0.395 |
| Tehran, Iran [46] | – | $TSV = 0.101T_a + 0.028T_{mrt} - 0.233V_a - 0.002RH - 2.681$ (All year) | 0.848 |

# 2 Methodology

## 2.1 Study Design

This study was executed within a university setting located in Guangzhou, China (112° – 114 °E, 22° – 23° N) (MOHURD 2016). The local climate demonstrates sustained thermal elevation through much of the annual cycle, displaying mean maximum air temperatures exceeding 29 °C between May and October, and achieving peak mean values of 34 °C. Minimum monthly average temperatures remain above 10 °C even throughout the winter season. Resulting from its subtropical maritime monsoon climate regime, the urban area maintains considerable atmospheric humidity, with yearly relative humidity (RH) levels persistently above 65%, and regularly exceeding 90% in summer. Accordingly, the climate pattern presents distinctly hot summers and temperate winters.

Meteorological and questionnaire surveys were implemented from June 2016 to May 2017 encompassing all seasonal variations, with daily operations scheduled from 7:30 to 18:00 whilst omitting periods of intense weather activity such as thunderstorms. To ensure observational variety, measurement sites were determined according to building configuration, terrain coverage, and shade distribution. Four designated monitoring points were established (Fig. 1): Sites A and C were located in building shadows with artificial surfacing; Site B combined shaded conditions with natural vegetation; Site D was situated within an urban canyon morphology.

## 2.2 Measurement Parameters

The collected climatic variables encompassed air temperature ($T_a$, measured with testo175H1), relative humidity (RH, obtained via testo175H1), mean radiant temperature ($T_{mrt}$), and wind speed ($V_a$, recorded using Kanomax Model KA22) (consult Table 3). $T_{mrt}$ was determined according to eq. (1) (ISO 1998):

$$T_{mrt} = \left[ \left(T_g + 273\right)^4 + \frac{1.1 \times 10^8 {V_a}^{0.6}}{\varepsilon D^{0.4}} \left(T_g - T\right)_a \right]^{\frac{1}{4}} - 273 \quad (1)$$

Where $T_g$ refers to globe temperature (measured by JTR04), D indicates the black globe diameter (D = 0.15 m), and ε denotes the emissivity of the black globe (ε = 0.97). Data from the black globe thermometer were compiled as 10 min averages. Marino et al. [20] verified that when variable means were calculated across 10 min spans, results from the globe method corresponded well with those from the more accurate six-directional radiation procedure. Though short-term fluctuations in radiation levels were inevitably reduced, the instrument showed satisfactory tracking performance [21].

**Fig. 1** Survey sites located in Guangzhou, China

**Table 3** Instruments specifications in this investigation

| Measured parameters | Instruments | Measuring range | Accuracy | Sampling interval (min) |
|---|---|---|---|---|
| $T_a$ | testo 175H1 | − 20 to + 55 °C | ± 0.4 °C | 5 |
| RH | | 0 to 100% | ± 2% RH (2–98% RH) at + 25 °C | |
| $T_g$ | JTR04 | − 20 to 125 °C | ± 0.5 °C | 5 |
| $V_a$ | Kanomax KA22 | 0 to 50 m/s | ± 2% full scale | 5 |

Sensors were installed at 1.1 m above ground, representing the approximate center of mass for both sitting and standing occupants [2]. A sun shield was employed during measurements to reduce solar radiation effects on temperature sensors.

$T_{mrt}$ is known to be associated with air temperature, typically elevating with higher $T_a$ values. This interrelationship impedes separate evaluation of $T_a$ or $T_{mrt}$ for examining mean thermal sensation vote (MTSV) fluctuations. Accordingly, operative temperature was utilized in this research. Operative temperature represents the uniform temperature of a hypothetical black enclosure in which heat transfer by radiation and convection equals that in the actual environment. It is calculated using Eq. (2), with coefficient A assigned values of 0.5 ($V_a < 0.2$ m/s), 0.6 ($0.2 \leq V_a \leq 0.6$ m/s), and 0.7 ($V_a > 0.6$ m/s) as specified [22].

$$T_{op} = At_a + (1 - A)T_{mrt} \tag{2}$$

## 2.3 Questionnaire

Questionnaires were distributed during environmental parameter recordings, collecting subjects' demographic data (age, gender, height, weight), clothing thermal resistance, metabolic rates, thermal sensation assessments (TSV), and other relevant metrics (Appendix A). The study included five activity types: sitting (1 Met, 60 W/m$^2$), standing (1.2 Met, 70 W/m$^2$), slow walking (2.0 Met, 115 W/m$^2$), fast walking (2.6 Met, 150 W/m$^2$), and jogging (3.8 Met, 220 W/m$^2$) [2]. The TSV scale spanned nine levels from − 4 (very cold) to 4 (very hot), including intermediate values: − 3 (cold), − 2 (cool), − 1 (slightly cool), 0 (neutral), 1 (slightly warm), 2 (warm), and 3 (hot).

Following Eq. (3) [23], the target sample size was established as 384 surveys, based on Guangzhou's permanent resident population of 18,676,600 recorded at the end of 2020 (GZSB 2020).

$$s = \frac{X^2NP(1 - P)}{d^2(N - 1) + X^2P(1 - P)} \tag{3}$$

Where:

- $s$ required sample size.
- $X^2$ the table value of chi-square for 1 degree of freedom at the desired confidence level;
- $N$ the population size.
- $P$ the population proportion (assumed to be 0.5) because this would provide the maximum sample size.
- $d$ degree of accuracy expressed as a proportion (0.05).

## 2.4 Standardized Regression Coefficient

Thermal sensitivity for four microclimatic parameters—$T_a$, $T_{mrt}$, $V_a$, and RH—was analyzed using the standardized regression coefficient (SRC) method [24] within a multiple linear regression framework. SRC denotes regression coefficients that are standardized to remove scale differences between predictors and the response variable; the absolute value reflects the relative importance of each independent variable in influencing the dependent measure [24]. This coefficient expresses relative weight, which depends on the dispersion of explanatory variables. As such, SRC is particularly suitable for thermal sensitivity studies, since sensitivity itself is closely linked to the variability of ambient conditions. For instance, even when a parameter has a strong theoretical effect on thermal sensation, low actual variation in a given setting may lead to reduced perceived sensitivity, owing to spatial and seasonal differences in thermal adaptation. It must be noted, however, that because SRC relies on linear approximation, it may not fully represent the precise quantitative relationship between thermal sensation and microclimatic parameters.

# 3 Results and Discussion

## *3.1 Data Statistics*

A total of 4675 valid questionnaires were obtained through randomized seasonal distribution: 985 in spring, 1283 in summer, 728 in autumn, and 1679 in winter. Each seasonal sample size exceeded the minimum requirement of 384, confirming statistical reliability (Table 4). The male-female ratio remained approximately 1:1 during all seasons, indicating minimal gender-specific effects. The representativeness of the survey was verified through comparison between measured air temperature ($T_a$) and meteorological data from 2009 to 2018 (Fig. 2), demonstrating consistency with contemporary climate conditions. Seasonal clothing insulation values (Table 4) displayed distinct clustering, with averages of 0.51 clo (spring), 0.42 clo (summer), 0.59 clo (autumn), and 0.99 clo (winter). "Slow walking" emerged as the dominant activity type, accounting for 72% of all documented movements (Fig. 3).

## *3.2 Qualitative Sensitivity*

### 3.2.1 Sensitivity in Guangzhou

To quantify the influence of microclimatic parameters on TSV, standardized regression coefficients (SRC) were computed via multiple linear regression using SPSS software (https://www.ibm.com/products/spss-statistics), with $T_a$, $T_{mrt}$, $V_a$, and RH as independent variables. As defined by Bring [24], the SRC provides a direct interpretation of the relationship: a "+" or "−" sign signifies a positive or negative correlation, while the absolute value indicates the strength of sensitivity. Results in Table 5 show that the SRC varied by both meteorological factor and season. $T_a$ consistently yielded the highest coefficient in each season and annually, confirming its role as the primary driver of thermal sensation. $T_{mrt}$ emerged as the second most sensitive

**Table 4** Detail information of measurement parameters and questionnaire

| Season | Research time | No. of questionnaire | Male to female ratio | Parameter | Minimum | Maximum | Mean value | Standard deviation |
|---|---|---|---|---|---|---|---|---|
| Spring | 2017/4–2017/5 | 985 | 435/550 (1/1.26) | $T_a$ (°C) | 20.6 | 34.0 | 27.7 | 2.70 |
| | | | | $V_a$ (m/s) | 0.0 | 3.1 | 0.4 | 0.50 |
| | | | | $T_{mrt}$ (°C) | 13.4 | 56.9 | 33.6 | 6.10 |
| | | | | RH (%) | 45.5 | 89.0 | 69.3 | 9.35 |
| | | | | $I_{clo}$ (clo) | 0.25 | 0.87 | 0.51 | 0.10 |
| Summer | 2016/6–2016/7 | 1283 | 675/608 (1/0.90) | $T_a$ (°C) | 26.8 | 37.2 | 31.8 | 2.25 |
| | | | | $V_a$ (m/s) | 0.0 | 2.5 | 0.3 | 0.40 |
| | | | | $T_{mrt}$ (°C) | 20.8 | 70.4 | 37.0 | 8.73 |
| | | | | RH (%) | 49.9 | 94.1 | 71.7 | 11.46 |
| | | | | $I_{clo}$ (clo) | 0.21 | 0.78 | 0.42 | 0.11 |
| Autumn | 2016/11 | 728 | 321/407 (1/1.27) | $T_a$ (°C) | 19.1 | 28.6 | 24.3 | 2.14 |
| | | | | $V_a$ (m/s) | 0.0 | 3.4 | 0.7 | 0.75 |
| | | | | $T_{mrt}$ (°C) | 19.0 | 64.0 | 31.7 | 9.30 |
| | | | | RH (%) | 47.4 | 78.4 | 58.6 | 7.51 |
| | | | | $I_{clo}$ (clo) | 0.26 | 1.21 | 0.59 | 0.16 |
| Winter | 2016/12–2017/1 | 1679 | 770/909 (1/1.18) | $T_a$ (°C) | 9.90 | 21.90 | 16.00 | 2.77 |
| | | | | $V_a$ (m/s) | 0.00 | 4.70 | 0.60 | 0.71 |
| | | | | $T_{mrt}$ (°C) | 10.30 | 46.30 | 19.80 | 5.77 |
| | | | | RH (%) | 35.80 | 86.60 | 55.20 | 19.08 |
| | | | | $I_{clo}$ (clo) | 0.33 | 1.75 | 0.99 | 0.24 |
| Total | 2016/6–2017/5 | 4675 | 2201/2472 (1/1.12) | Male height (m) | 1.47 | 1.90 | 1.72 | 0.06 |
| | | | | Male weight (kg) | 40.0 | 96.0 | 61.4 | 7.35 |
| | | | | Male age | 13 | 52 | 21.5 | 2.19 |
| | | | | Female height (m) | 1.20 | 1.83 | 1.60 | 0.05 |
| | | | | Female weight (kg) | 15.0 | 60.0 | 20.8 | 5.45 |
| | | | | Female age | 15 | 60 | 20.8 | 2.23 |

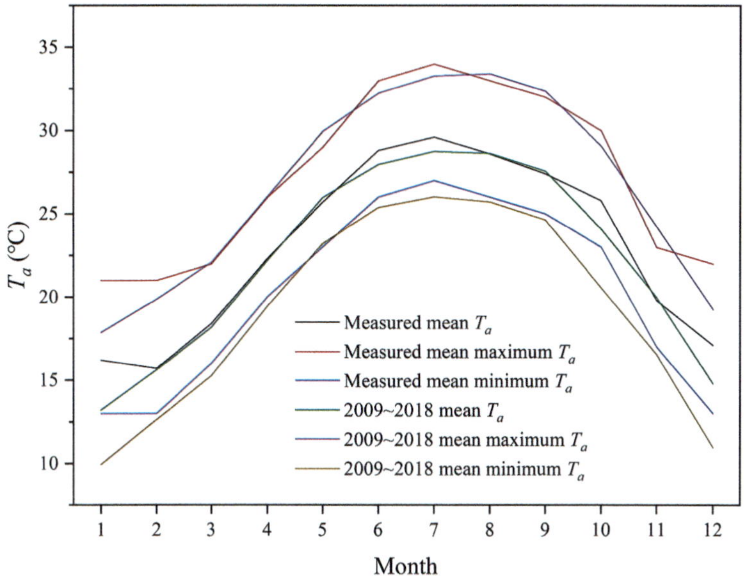


**Fig. 2** Comparison between measured and historical $T_a$

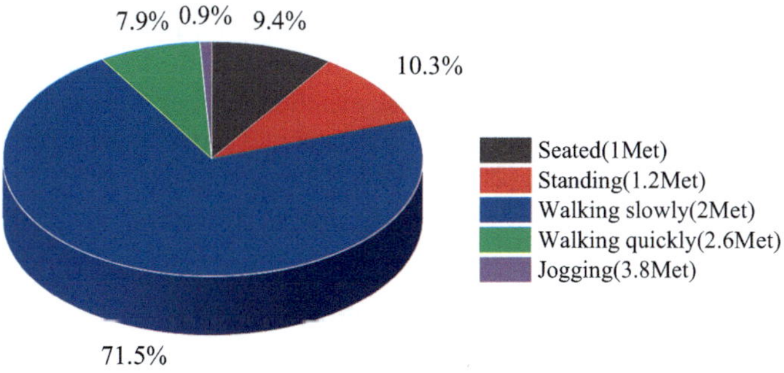


**Fig. 3** Activity pattern distribution

variable in spring and summer, whereas $V_a$ ranked second during autumn, winter, and over the entire year. RH consistently exhibited the lowest sensitivity across all time scales. Overall, in Guangzhou's outdoor environments, $T_a$, $T_{mrt}$, and $V_a$ had a more substantial influence on TSV compared to RH. A positive correlation was observed between TSV and $T_a$, $T_{mrt}$, and RH. Conversely, TSV demonstrated a negative correlation with $V_a$.

**Table 5** Seasonal SRC and weight

| Season | $T_a$ | $V_a$ | $T_{mrt}$ | RH |
|---|---|---|---|---|
| | SRC/weight | | | |
| Spring | 0.406/63.3 | −0.087/13.6 | 0.141/22 | – 0.0063636 |
| Summer | 0.446/29.6 | −0.205/21.4 | 0.215/22.5 | 0.093/9.7 |
| Autumn | 0.406/45.4 | −0.185/20.6 | 0.163/18.2 | 0.141/15.8 |
| Winter | 0.368/81.1 | −0.068/15.1 | 0.014/3.1 | 0.003/0.7 |
| All year | 0.679/77.5 | −0.111/12.7 | 0.07/8 | 0.016/1.8 |

### 3.2.2 Sensitivity in Hot Summer and Warm Winter Zone

A range of thermal sensitivity studies has been performed in multiple urban areas with hot-summer and cold-winter climate patterns (Table 6). Li et al. [25] reported that summer outdoor thermal sensitivity in Guangzhou was influenced by $T_a$, $T_{mrt}$, $V_a$, and RH, which is consistent with the results obtained in this study. Research by Ng et al. [26] established that $T_a$, $V_a$, and solar radiation (SR) served as the most critical factors during summer and winter in Hong Kong. Similarly, Chan and Chau [15] emphasized that Ta represented the most influential thermal parameter in both seasons within Hong Kong. Despite certain differences among these studies, Table 6 illustrates identifiable consistencies. For instance, all cited studies recognized $T_a$ as the primary sensitive variable across summer and autumn, although the second most significant parameter was not the same. This study determined $T_{mrt}$ as the secondary sensitive factor in summer, a finding that matches Li et al. [25] but differs from Ng and Cheng [26] In winter, however, our outcomes are in accordance with those documented by Ng and Cheng [26].

## 3.3 *Modified Outdoor Thermal Comfort Models*

PET, UTCI, and SET* are extensively employed in the evaluation of outdoor thermal comfort, with corresponding thermal categories and sensation ranges detailed in Table 7. Nonetheless, adaptation may be necessary for Guangzhou's specific conditions, considering the thermal adaptation of local inhabitants and the region's characteristic climate. The calculations for PET, UTCI, and SET* were conducted using RayMan Pro (https://www.urbanclimate.net/rayman/). Input variables comprised $T_a$, $V_a$, $T_{mrt}$, RH, metabolic rate, and $I_{clo}$. Linear regression was utilized to modify PET, UTCI, and SET, a methodology corroborated by prior research [27, 28]. A segmentation interval of 1 °C was implemented for each index, with outliers being excluded following box plot assessments (Figs. 4, 5 and 6). Accordingly, any 1 °C segment with fewer than six data points was omitted. The trimmed averages depicted in Figs. 4, 5 and 6 reflect the mean TSV after outlier removal. By correlating actual TSV (ATSV) with corresponding PET, UTCI, or SET values, recalibrated models

**Table 6** Sensitive studies in hot summer and warm winter zone

| Authors | Cities | Spring | | Summer | | Autumn | | Winter | | All year | |
|---|---|---|---|---|---|---|---|---|---|---|---|
| | | 1st | 2nd | 1st | 2nd | 1st | 2nd | 1st | 2nd | 1st | 2nd |
| Ng and Cheng [26], Cheng et al. [9] | Hong Kong | | | $T_a$ | $V_a$ | | | $T_a$ | $V_a$ | | |
| Chan and Chau [15] | Hong Kong | | | $T_a$ | | | | $T_a$ | SR | | |
| Li et al. [25] | Guangzhou | | | $T_a$ | $T_{mrt}$ | | | | | | |
| This study | Guangzhou | $T_a$ | $T_{mrt}$ | $T_a$ | $T_{mrt}$ | $T_a$ | $V_a$ | $T_a$ | $V_a$ | $T_a$ | $V_a$ |
| Maximum frequency parameter | | $T_a$ | $T_{mrt}$ | $T_a$ | $T_{mrt}$ | $T_a$ | $V_a$ | $T_a$ | $V_a$ | $T_a$ | $V_a$ |

were constructed (Tables 8, 9 and 10). The optimized models—MPET, MUTCI, and MSET*—exhibited superior accuracy in forecasting mean thermal sensation vote (MTSV). However, further sensitivity evaluation remains essential to determine the most appropriate model for quantifying microclimatic parameters.

## 3.4 Sensitivity Test

Defining a quantitative correlation between outdoor thermal comfort and thermal sensitivity has historically presented challenges. To accurately assess the relationship between outdoor thermal comfort and any single variable, it was essential to keep other parameters unchanged. Initial investigations employing survey data were carried out by Fang et al. [13]. Yet, these data exhibit constraints when analyzing the isolated relationship of one parameter under fixed remaining conditions. Thus, the current research proposes a framework incorporating adjusted outdoor thermal comfort models after completing sensitivity assessment.

Among microclimatic factors, $T_a$ emerged as the most influential, followed by $T_{mrt}$ and $V_a$. RH showed only marginal effects on human comfort. Confirming parameters with lower sensitivity through survey data continues to present difficulties. For this reason, the analysis of MPET, MUTCI, and MSET* focused exclusively on $T_a$, $T_{mrt}$, and $V_a$.

**Table 7** Original thermal grade of PET, UTCI, and SET*

| Thermal grade | Thermal sensation | Physical stress | PET/°C | UTCI/°C | SET*/°C |
|---|---|---|---|---|---|
| −4 | Very cold | Extreme cold stress | ≤ 4 | <− 40 | |
| −3.5 | | Very cold stress | | − 40 to − 27 | |
| −3 | Cold | Strong cold stress | 4–8 | − 27 to − 13 | 10.0–14.5 |
| −2 | Cool | Moderate cold stress | 8–13 | − 13 to 0 | 14.5–17.5 |
| −1 | Slightly cool | Slight cold stress | 13–18 | 0–9 | 17.5–22.2 |
| 0 | Comfortable | No thermal stress | 18–23 | 9–26 | 22.2–25.6 |
| 1 | Slightly warm | Slight heat stress | 23–29 | / | 25.6–30.0 |
| 2 | Warm | Moderate heat stress | 29–35 | 26–32 | 30.0–34.5 |
| 3 | Hot | Strong heat stress | 35–41 | 32–38 | 34.5–37.5 |
| 3.5 | | Very heat stress | | 38–46 | |
| 4 | Very hot | Extreme heat stress | ≥ 41 | > 46 | > 37.5 |

### 3.4.1 Operating Temperature

The available dataset proved inadequate for robust analysis under fully controlled conditions with all other variables fixed. This investigation therefore implemented a methodology that limited parameter variations to a narrow spectrum. Although changes within this restricted scope continue to influence TSV, calculating the mean thermal sensation vote (MTSV) helped alleviate these effects. In the same manner, $T_{op}$ was examined at 1 °C intervals, employing box plot techniques to eliminate anomalous values. Proper sensitivity consistency demands that both simulated and observed MTSV display analogous patterns as $T_{op}$. fluctuates. As depicted in Fig. 7, the $T_{op}$. sensitivity of MPET, MUTCI, and MSET* was examined utilizing gathered survey responses. Each model exhibited enhanced concordance with actual MTSV (Fig. 7), with MSET* yielding superior results, evidenced by its regression slope most nearly matching that of the empirical data (Table 11).

### 3.4.2 Wind Speed

As indicated in Fig. 8, MUTCI and MPET exhibit similar levels of responsiveness to changes in wind velocity. The UTCI framework proved inadequate for characterizing airflow conditions under 0.3 m/s measured at 1.1 m, corresponding to 0.5 m/s at

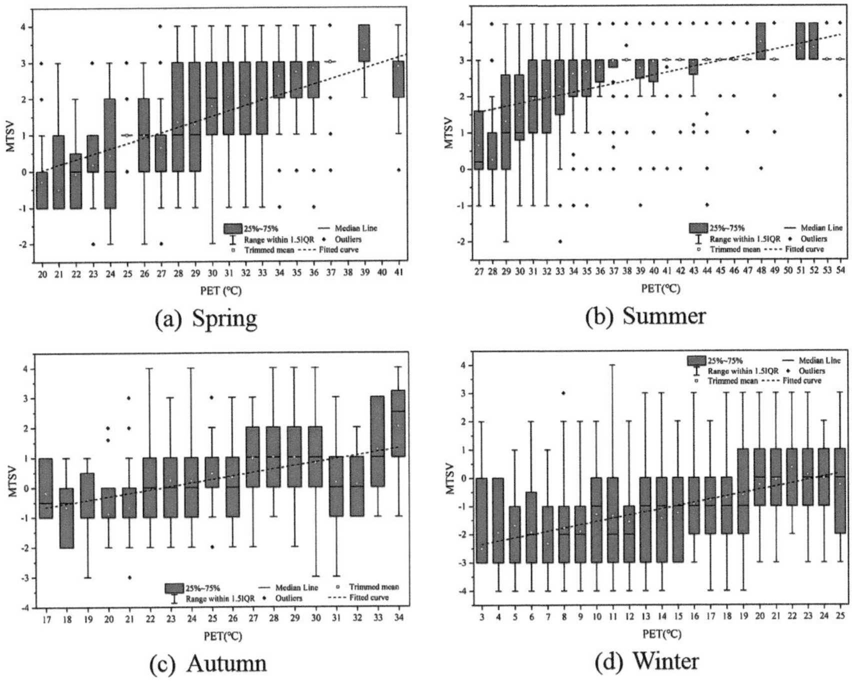


**Fig. 4** Seasonal linear regression between PET and MTSV

10 m height. Consequently, UTCI exhibited constraints when evaluating settings with minimal wind speeds. By contrast, MSET* displayed the highest degree of conformity with empirical measurements (Table 12).

## 3.5 *Quantified Sensitivity*

MSET* showed optimal agreement with measured mean thermal sensation vote (MTSV) under Guangzhou's climatic conditions. As a result, this model was applied to examine the sensitivity of particular parameters while holding other variables constant. Input values consisted of averaged data for the remaining factors, representing typical outdoor environmental characteristics in Guangzhou (Table 4). The study specifically concentrated on "slow walking" as the selected activity, given its status as the most prevalent outdoor movement pattern documented (Fig. 3).

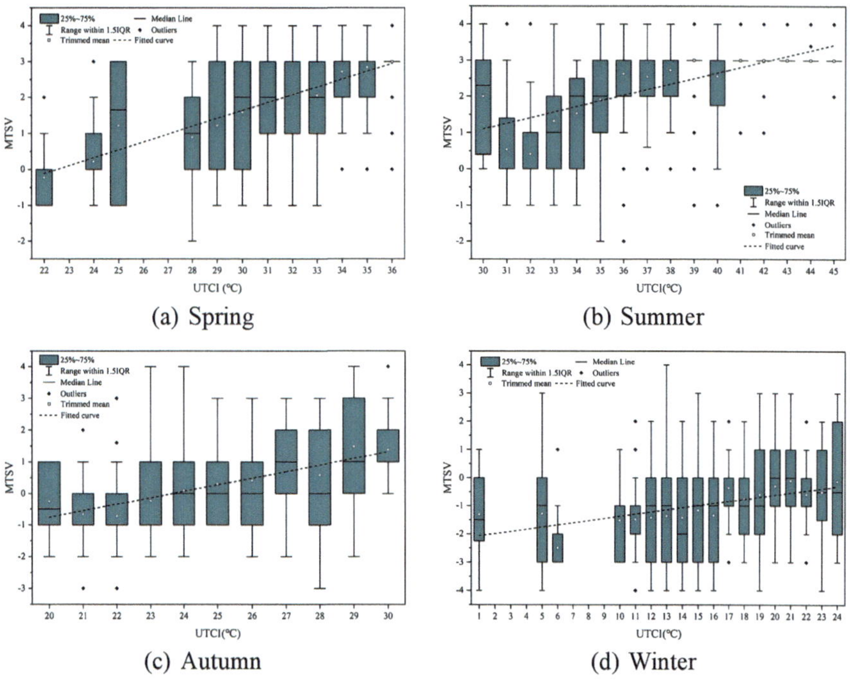


**Fig. 5** Seasonal linear regression between UTCI and MTSV

### 3.5.1 Air Temperature

When evaluating the quantified sensitivity of $T_a$ through mean $T_{mrt}$, it was observed that average $T_{mrt}$ values remained consistently lower than corresponding $T_a$ measurements, with this pattern being most evident during winter months. Although the investigation took place during daytime periods when $T_{mrt}$ characteristically exceeds $T_a$, the study opted to use directly measured $T_{mrt}$ readings rather than averaged values. This analytical strategy mirrors the techniques developed by Bröde et al. [8], Cheng et al. [9]. It is significant that $T_{mrt}$ generally fluctuates in coordination with $T_a$. The computed parameter $T^*_{mrt}$ was derived from Eqs. (4) and (5), in which $\overline{T_{mrt}}$ and $\overline{T_a}$ indicate seasonal mean $T_{mrt}$ and $T_a$ values, respectively.

$$T^*_{mrt} = T_a + \Delta T \tag{4}$$

$$\Delta T = \overline{T_{mrt}} - \overline{T_a} \tag{5}$$

By implementing the MSET* model, alterations in mean thermal sensation vote (MTSV) in response to elevated air temperature ($T_a$) were obtained (Fig. 9). Seasonal $T_a$ values corresponded to measured ranges for each period. Table 13 demonstrates

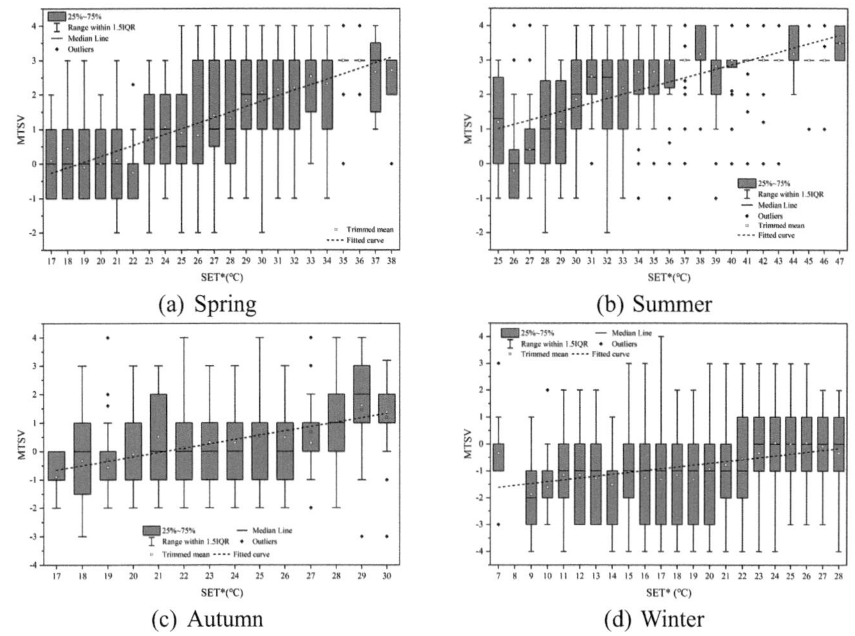


**Fig. 6** Seasonal linear regression between SET* and MTSV

**Table 8** Regression equation of PET

| Season | PET range (°C) | Regression equation | $R^2$ |
|---|---|---|---|
| Spring | 20–41 | $MTSV = 0.187PET - 4.038$ | 0.95 |
| Summer | 27–54 | $MTSV = 0.078PET - 0.537$ | 0.63 |
| Autumn | 17–34 | $MTSV = 0.115PET - 2.619$ | 0.65 |
| Winter | 3–25 | $MTSV = 0.114PET - 2.693$ | 0.90 |

**Table 9** Regression equation of UTCI

| Season | UTCI range (°C) | Regression equation | $R^2$ |
|---|---|---|---|
| Spring | 22–36 | $MTSV = 0.219UTCI - 4.928$ | 0.94 |
| Summer | 30–45 | $MTSV = 0.155UTCI - 3.555$ | 0.69 |
| Autumn | 20–30 | $MTSV = 0.207UTCI - 4.899$ | 0.87 |
| Winter | 1–24 | $MTSV = 0.077UTCI - 2.142$ | 0.62 |

that inverse slope parameters (1/k, where k denotes fitted linear curve slope) for spring and autumn were closely aligned, registering 5.2 and 5.9 respectively. These values signify that a $T_a$ modification of either 5.2 or 5.9 °C during spring or autumn induces a single-grade transition in MTSV (such as moving from MTSV = 2 [warm]

**Table 10** Regression equation of SET*

| Season | SET* range (°C) | Regression equation | $R^2$ |
|---|---|---|---|
| Spring | 17–38 | $MTSV = 0.155SET^* - 2.886$ | 0.79 |
| Summer | 25–47 | $MTSV = 0.189SET^* - 4.174$ | 0.80 |
| Autumn | 17–30 | $MTSV = 0.144SET^* - 3.013$ | 0.80 |
| Winter | 7–28 | $MTSV = 0.072SET^* - 2.309$ | 0.67 |

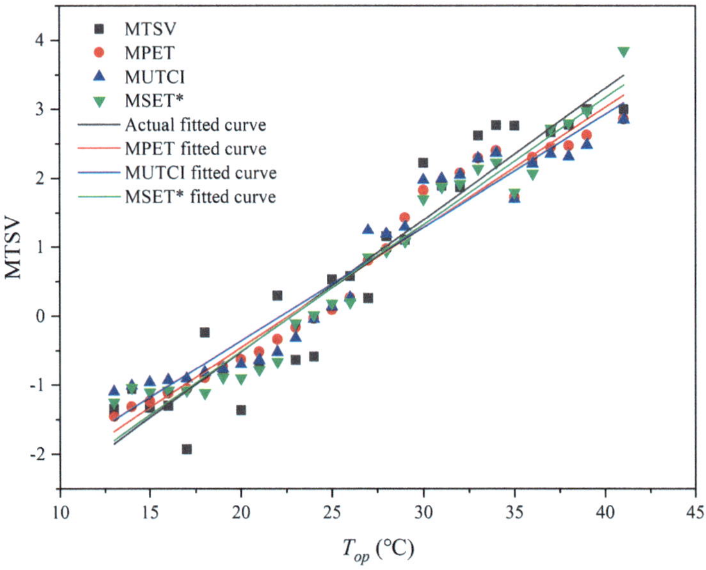


**Fig. 7** Operating temperature sensitivity of MPET, MUTCI, and MSET*. M = 2met, $V_a = 0.3$–0.6 m/s

**Table 11** The MTSV tendency of MPET, MUTCI, and MSET* with $T_{op}$ change

| Model | Slope | $R^2$ |
|---|---|---|
| Actual | 0.191 | 0.91 |
| MPET | 0.174 | 0.96 |
| MUTCI | 0.164 | 0.93 |
| MSET* | 0.184 | 0.96 |

to MTSV = 3 [hot]). Summer conditions revealed heightened MTSV sensitivity to $T_a$, where merely 3.9 °C variation triggered one thermal sensation grade shift. Conversely, winter conditions exhibited a diminished regression slope. This reduction stemmed partially from the weakened $T_{mrt}-T_a$ interdependence during winter compared to other seasons, coupled with increased clothing insulation ($I_{clo}$) that

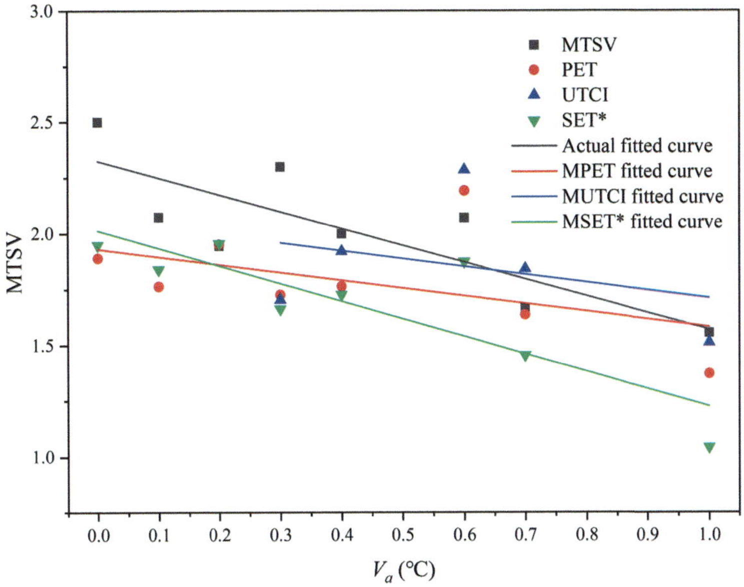


**Fig. 8** Wind speed sensitivity of MPET, MUTCI, and MSET*. M = 2met, $T_{op}$ = 30 °C

**Table 12** The MTSV tendency of MPET, MUTCI, and MSET* with $V_a$ change

| Model | Slope | $R^2$ |
|---|---|---|
| Actual | – 1.219 | 0.81 |
| MPET | – 0.564 | 0.74 |
| MUTCI | – 0.800 | 0.84 |
| MSET* | – 0.610 | 0.83 |

impeded convective thermal exchange between body and environment. The aggregated seasonal 1/k measurement reached 7.6, confirming $T_a$'s role as a determinant microclimatic factor in outdoor thermal comfort assessment.

### 3.5.2 Mean Radiant Temperature

MTSV's response to $T_{mrt}$ exhibited patterns comparable to its relationship with $T_a$, with sensitivity highest in summer, followed sequentially by spring, autumn, and winter (Fig. 10). During summer months, a $T_{mrt}$ fluctuation of 11.9 °C produced a single-grade alteration in MTSV (Table 14). Winter, however, demonstrated diminished responsiveness in MTSV. This phenomenon primarily results from elevated $I_{clo}$ levels, which inhibit heat transfer between the human body and its surroundings.

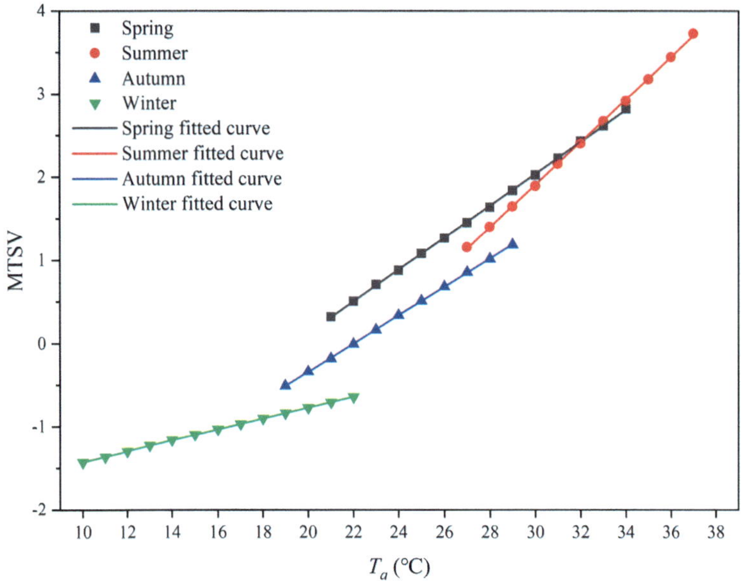


**Fig. 9** MTSV versus $T_a$. RH = *mean* (RH), $V_a = mean\ (V_a)$, $T_{mrt} = T_{mrt*}$, Metabolic rate = 2 Met

**Table 13** The fitting line slope of $T_a$

| Season | Main range (°C) | Mean RH (%) | Mean $V_a$ (m/s) | Mean $T_{mrt}$ (°C) | Metabolic rate (Met) | 1/k | $R^2$ | Mean 1/k |
|---|---|---|---|---|---|---|---|---|
| Spring | 21–34 | 69 | 0.4 | 33.6 | 2 | 5.2 | 1.00 | 7.6 |
| Summer | 27–37 | 72 | 0.3 | 37.0 | 2 | 3.9 | 1.00 | |
| Autumn | 19–29 | 59 | 0.7 | 31.7 | 2 | 5.9 | 1.00 | |
| Winter | 10–22 | 55 | 0.6 | 19.8 | 2 | 15.3 | 1.00 | |

### 3.5.3 Wind Speed

Figure 11 illustrates a non-linear association between MTSV and $V_a$. MTSV altered more rapidly at lower wind speeds than at higher velocities, indicating enhanced sensitivity under low $V_a$ conditions. Most recorded $V_a$ values were concentrated within the $\leq 1$ m/s range. For evaluating sensitivity across this interval, linear regression analyses were performed, with resulting slopes provided in Table 15. The study found uniform $V_a$ sensitivity throughout spring, summer, and autumn, where a 1 m/s $V_a$ modification corresponded to a single-grade MTSV shift. Winter conditions, however, manifested substantially reduced $V_a$ sensitivity relative to other seasons.

The influence of wind speed ($V_a$) on thermal perception demonstrated a strong relationship with air temperature ($T_a$), as $V_a$ modulates convective heat exchange between people and their environment. To explore $V_a$–$T_a$ interactions, seasonal

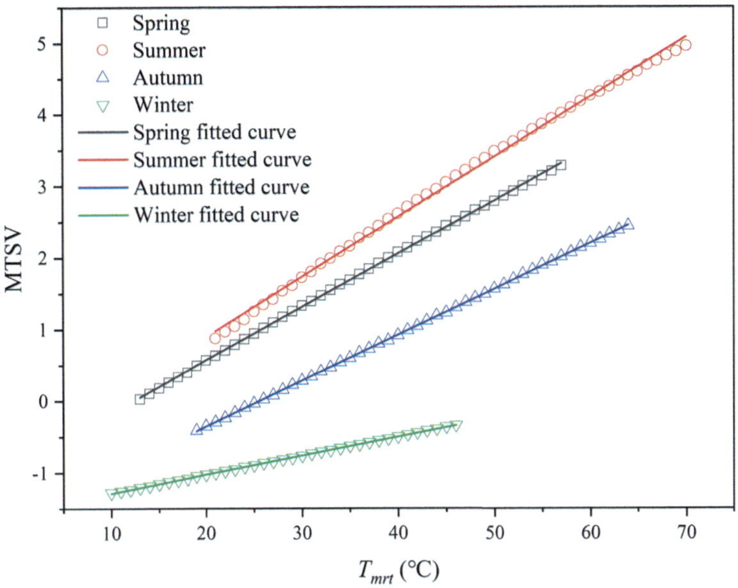


**Fig. 10** MTSV versus $T_{mrt}$. RH = *mean(RH)*, $T_a = mean(T_a)$, $V_a = mean(V_a)$, Metabolic rate = 2 Met

**Table 14** The fitting line slope of $T_{mrt}$

| Season | Main range (°C) | Mean RH (%) | Mean $V_a$ (m/s) | Mean $T_a$ (°C) | Metabolic rate (Met) | 1/k | $R^2$ | Mean 1/k |
|---|---|---|---|---|---|---|---|---|
| Spring | 13–57 | 69 | 0.4 | 27.7 | 2 | 13.5 | 1.00 | 18.8 |
| Summer | 21–70 | 72 | 0.3 | 31.8 | 2 | 11.9 | 1.00 | |
| Autumn | 19–64 | 59 | 0.7 | 24.3 | 2 | 15.6 | 1.00 | |
| Winter | 10–46 | 55 | 0.6 | 16.0 | 2 | 38.0 | 1.00 | |

changes in MTSV relative to $V_a$ at various $T_a$ levels were examined (Fig. 12). Throughout spring, summer, and autumn, decreased Ta resulted in more rapid MTSV variations, particularly in lower $V_a$ ranges. This pattern occurs because reduced $T_a$ enhances the temperature differential between the environment and skin surface, facilitating accelerated heat transfer. In climates characterized by hot summers and mild winters, outdoor temperatures usually remain lower than human body temperature. However, sensitivity variations across different $T_a$ values were less evident in winter. This reduced responsiveness is mainly attributable to clothing layers impeding convective thermal exchange between skin and airflow.

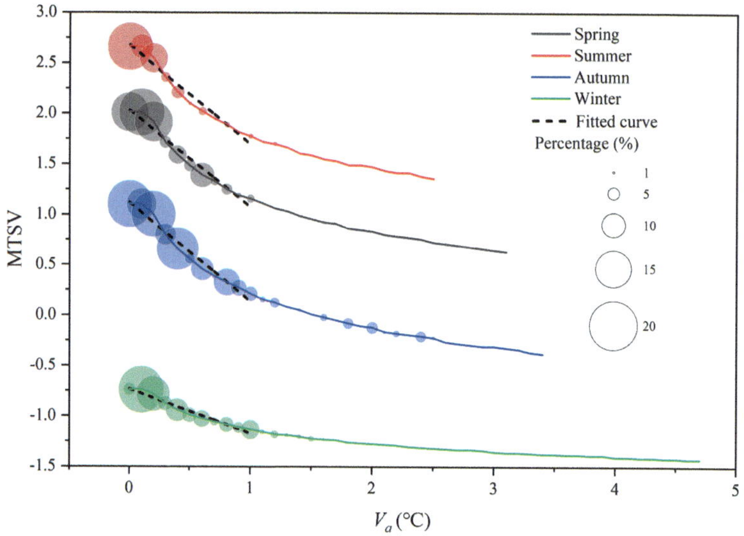


**Fig. 11** MTSV versus $V_a$. RH = *mean*(RH), $T_a = mean(T_a)$, $T_{mrt} = mean(T_{mrt})$, Metabolic rate = 2 Met

**Table 15** The fitting line slope for $V_a$

| Season | Main range (m/s) | Mean RH (%) | Mean $T_{mrt}$(°C) | Mean $T_a$ (°C) | Metabolic rate (Met) | 1/k (0–1 m/s) | $R^2$ | Mean 1/k |
|---|---|---|---|---|---|---|---|---|
| Spring | 0–3.1 | 69 | 33.6 | 27.7 | 2 | − 1.0 | 0.97 | − 1.3 s |
| Summer | 0–2.5 | 72 | 0.3 | 37.0 | 2 | − 1.0 | 0.96 | |
| Autumn | 0–3.4 | 59 | 0.7 | 31.7 | 2 | − 1.0 | 0.97 | |
| Winter | 0–4.7 | 55 | 0.6 | 16.0 | 2 | − 2.3 | 0.96 | |

## 4 Limitations and Future Studies

In the current research, mean radiant temperature ($T_{mrt}$) was evaluated using a standard 150 mm black globe thermometer, a device that exhibits a delay in attaining thermal equilibrium. Although implementing an appropriate sampling interval can help minimize errors associated with this latency, identifying an ideal measurement period remained challenging because of fluctuating microclimatic factors. The duration for the black globe to achieve thermal equilibrium varies significantly across diverse climatic environments. Subsequent investigations should consider utilizing more reliable techniques, such as the six-directional measurement method.

Results pertaining to the second-most sensitive microclimatic parameter displayed considerable variation across different research efforts. Therefore, further verification

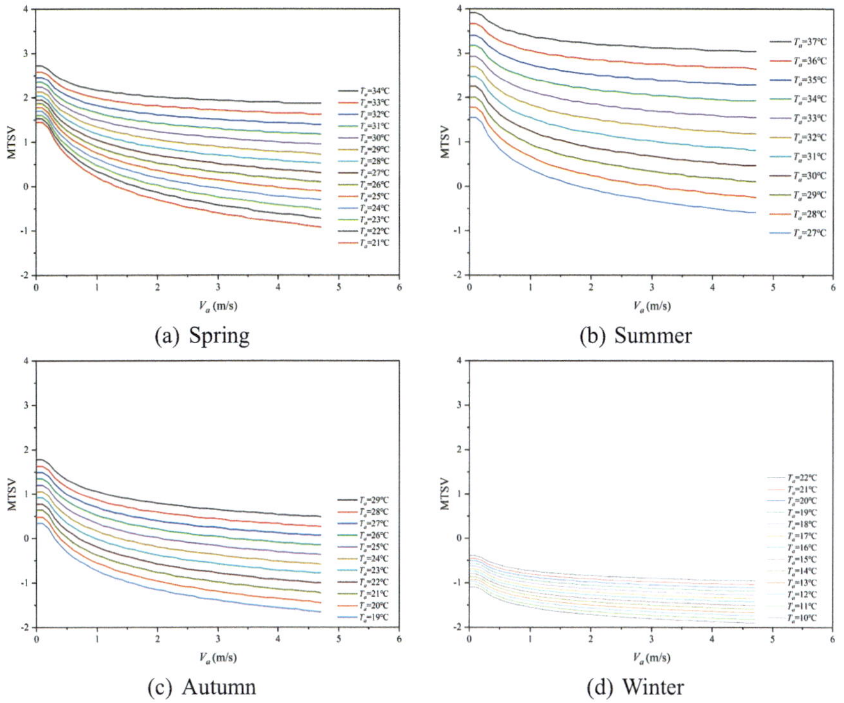


**Fig. 12** MTSV versus $V_a$ at different $T_a$ winter. RH $= mean(\text{RH})$, $T_{mrt} = mean(T_{mrt})$, Metabolic rate = 2 Met

is needed, particularly within the climatic context of hot summer and warm winter zones.

Diverse mitigation strategies may induce distinct changes in microclimatic parameters. The quantification of parameter sensitivity facilitates balanced assessments between the costs and outcomes of mitigation approaches. However, current studies examining the thermal influence of mitigation strategies in regions experiencing hot summers and cold winters are inadequate to support comprehensive effectiveness comparisons. Subsequent research ought to prioritize the empirical quantification of thermal impacts from mitigation interventions, enabling integrated benefit evaluations grounded in quantified sensitivity to support practical urban planning recommendations.

# 5 Conclusion

This chapter analyzed seasonal variations in outdoor thermal comfort in Guangzhou through field measurements and questionnaire surveys. Using standardized regression coefficients, the study identified the relative influence of microclimatic parameters and refined PET, UTCI, and SET* models. The main findings are:

(1) Thermal sensitivity exhibits distinct seasonal patterns in Guangzhou. Ta consistently ranked as the dominant parameter in every season and over the entire year, followed by $T_{mrt}$ and $V_a$. The second most influential parameter varied seasonally: $T_{mrt}$ held this role during spring and summer, while $V_a$ assumed the position in autumn and winter. RH uniformly exhibited minimal sensitivity.

(2) SET* demonstrated better concordance with Guangzhou residents' thermal sensitivity than both PET and UTCI.

(3) Generally, MTSV change one grade with the $\Delta T_a = 5.2$ °C, $\Delta T_{mrt} = 13.5$ °C or $\Delta V_a = 1$ m/s in spring; $\Delta T_a = 3.9$ °C, $\Delta T_{mrt} = 11.9$ °C or $\Delta V_a = 1$ m/s in summer; $\Delta T_a = 5.9$ °C, $\Delta T_{mrt} = 15.6$ °C or $\Delta V_a = 1$ m/s in autumn; $\Delta T_a = 15.3$ °C, $\Delta T_{mrt} = 38$ °C or $\Delta V_a = 2.3$ m/s in winter.These results highlight the necessity of integrating seasonal sensitivity patterns into outdoor comfort assessment and provide a scientific basis for climate-responsive urban design in hot–humid regions.

**Acknowledgements** Funding: This work was supported by the National Natural Science Foundation of China (Project No. 51978180), Research and Development Project of the Ministry of Housing, and Urban-Rural Development of the People's Republic of China (No. 2019-K-003).

# Appendix A

## Questionnaire for outdoor thermal comfort study

- **Location:** ☐ Site A ☐ Site B ☐ Site C ☐ Site D
- **Date:** ____________ **Time:** ____________

### 1. Personal information

**Gender:** ☐ Male ☐ Female **Age:** ________

**Weight:** ________ **Height:** ________

### 2. Dress (please choose the clothes you are currently wearing):

| | |
|---|---|
| Coats and jackets: | □ down jacket coat; □ cotton-padded coat; □ windbreaker; □ sports jacket; □ business suit; |
| Upper clothes: | □ sweater; □ wool vest; □ underwear sweater; □ long-sleeved shirts; □short-sleeved shirts; |
| Lower clothes: | □ business pants; □ woolen trousers; □ sweat pants; □ dress; □cotton-padded trousers; □ underwear trousers; □ shorts; □ skirt; |
| Shoes and socks: | □ socks; □ stockings; □ shoes; □ cotton slippers; □ sandals |
| Others: | □ please specify:________ |

### 3. Your main activity over the past 20 minutes is:

☐ Seated and quiet ☐ Standing and relaxed ☐ Walking at 3.2 km/h ☐ Walking at 4.8 km/h

### 4. Your current thermal sensation is:

☐-4 Very cold ☐-3 Cold ☐-2 Cool ☐-1 Slightly cool ☐0 Neutral ☐1 Slightly warm ☐2 Warm ☐3 Hot ☐4 Very hot

### 5. Your current thermal preference is:

☐−1 Cooler ☐0 No change ☐1 Warmer

### 6. Your current humidity sensation is:

☐-3 Very dry ☐-2 Too dry ☐-1 Slightly dry ☐0 Neutral ☐1 Slightly humid ☐2 Too humid ☐3 Very humid

## References


1. D. Lai, W. Liu, T. Gan, K. Liu, Q. Chen, A review of mitigating strategies to improve the thermal environment and thermal comfort in urban outdoor spaces. Sci. Total Environ. **661**, 337–353 (2019)
2. ASHRAE, *Thermal environmental conditions for human occupancy* (ASHRAE, 2020)
3. J. Li, N. Liu, The perception, optimization strategies and prospects of outdoor thermal comfort in China: a review. Build. Environ. **170**, 106614 (2020)
4. P.O. Fanger, Thermal comfort: analysis and applications in environmental engineering, in *Thermal Comfort*, (1970)
5. A.P. Gagge, A. Fobelets, L. Berglund, A standard predictive Index of human reponse to thermal enviroment. Trans. Am. Soc. Heat. Refrig. Air-Cond. Eng. **92**, 709–731 (1986)
6. L. Chen, Y. Wen, L. Zhang, W.-N. Xiang, Studies of thermal comfort and space use in an urban park square in cool and cold seasons in Shanghai. Build. Environ. **94**, 644–653 (2015)
7. P. Höppe, The physiological equivalent temperature–a universal index for the biometeorological assessment of the thermal environment. Int. J. Biometeorol. **43**, 71–75 (1999)
8. P. Bröde, D. Fiala, K. Błażejczyk, I. Holmér, G. Jendritzky, B. Kampmann, et al., Deriving the operational procedure for the Universal Thermal Climate Index (UTCI). Int. J. Biometeorol. **56**, 481–494 (2012)
9. V. Cheng, E. Ng, C. Chan, B. Givoni, Outdoor thermal comfort study in a sub-tropical climate: a longitudinal study based in Hong Kong. Int. J. Biometeorol. **56**, 43–56 (2012)
10. Y. Zhang, J. Liu, Z. Zheng, Z. Fang, X. Zhang, Y. Gao, et al., Analysis of thermal comfort during movement in a semi-open transition space. Energy Build. **225**, 110312 (2020a)
11. Z. Fang, X. Feng, X. Xu, X. Zhou, Z. Lin, Y. Ji, Investigation into outdoor thermal comfort conditions by different seasonal field surveys in China. Guangzhou. Int. J. Biometeorol. **63**, 1357–1368 (2019)
12. Y. Zhang, X. Zhou, Z. Zheng, M.O. Oladokun, Z. Fang, Experimental investigation into the effects of different metabolic rates of body movement on thermal comfort. Build. Environ. **168**, 106489 (2020b)
13. Z. Fang, Z. Zheng, X. Feng, D. Shi, Z. Lin, Y. Gao, Investigation of outdoor thermal comfort prediction models in South China: a case study in Guangzhou. Build. Environ. **188**, 107424 (2021)
14. Y. Peng, T. Feng, H. Timmermans, A path analysis of outdoor comfort in urban public spaces. Build. Environ. **148**, 459–467 (2019)
15. S.Y. Chan, C.K. Chau, Development of artificial neural network models for predicting thermal comfort evaluation in urban parks in summer and winter. Build. Environ. **164**, 106364 (2019)
16. W. Liu, Y. Zhang, Q. Deng, The effects of urban microclimate on outdoor thermal sensation and neutral temperature in hot-summer and cold-winter climate. Energ. Build. **128**, 190–197 (2016)
17. J. Li, J. Niu, C.M. Mak, T. Huang, Y. Xie, Exploration of applicability of UTCI and thermally comfortable sun and wind conditions outdoors in a subtropical city of Hong Kong. Sustain. Cities Soc. **52**, 101793 (2020)
18. Z. Fang, X. Feng, J. Liu, Z. Lin, C.M. Mak, J. Niu, et al., Investigation into the differences among several outdoor thermal comfort indices against field survey in subtropics. Sustain. Cities Soc. **44**, 676–690 (2019a)
19. Z. Fang, Z. Lin, C.M. Mak, J. Niu, K.-T. Tse, Investigation into sensitivities of factors in outdoor thermal comfort indices. Build. Environ. **128**, 129–142 (2018)
20. C. Marino, A. Nucara, M. Pietrafesa, E. Polimeni, S. Costanzo, Outdoor mean radiant temperature estimation: is the black-globe thermometer method a feasible course of action? in *2018 IEEE International Conference on Environment and Electrical Engineering and 2018 IEEE Industrial and Commercial Power Systems Europe (EEEIC/I&CPS Europe)*, (IEEE, 2018), pp. 1–7
21. S. Thorsson, F. Lindberg, I. Eliasson, B. Holmer, Different methods for estimating the mean radiant temperature in an outdoor urban setting. Int. J. Climatol. **27**, 1983–1993 (2007)

22. ASHRAE, *ASHRAE Handbook: Fundamentals* (ASHRAE, 2017)
23. R.V. Krejcie, D.W. Morgan, Determining sample size for research activities. Educ. Psychol. Meas. **30**, 607–610 (1970)
24. J. Bring, How to standardize regression coefficients. Am. Stat. **48**, 209–213 (1994)
25. L. Li, X. Zhou, L. Yang, The analysis of outdoor thermal comfort in Guangzhou during summer. Procedia Eng. **205**, 1996–2002 (2017)
26. E. Ng, V. Cheng, Urban human thermal comfort in hot and humid Hong Kong. Energ. Buildings **55**, 51–65 (2012)
27. T.-P. Lin, Thermal perception, adaptation and attendance in a public square in hot and humid regions. Build. Environ. **44**, 2017–2026 (2009)
28. O. Potchter, P. Cohen, T.-P. Lin, A. Matzarakis, Outdoor human thermal perception in various climates: a comprehensive review of approaches, methods and quantification. Sci. Total Environ. **631**, 390–406 (2018)
29. J. Yin, Y. Zheng, R. Wu, J. Tan, D. Ye, W. Wang, An analysis of influential factors on outdoor thermal comfort in summer. Int. J. Biometeorol. **56**, 941–948 (2012)
30. B. Cheng, G. Gou, F. Zhang, Q. Feng, Z. Huang, Thermal comfort in urban mountain parks in the hot summer and cold winter climate. Sustain. Cities Soc. **51**, 101756 (2019)
31. M. Xu, B. Hong, J. Mi, S. Yan, Outdoor thermal comfort in an urban park during winter in cold regions of China. Sustain. Cities Soc. **43**, 208–220 (2018)
32. X. Chen, P. Xue, L. Liu, L. Gao, J. Liu, Outdoor thermal comfort and adaptation in severe cold area: a longitudinal survey in Harbin, China. Build. Environ. **143**, 548–560 (2018)
33. D. Walton, V. Dravitzki, M. Donn, The relative influence of wind, sunlight and temperature on user comfort in urban outdoor spaces. Build. Environ. **42**, 3166–3175 (2007)
34. M. Nikolopoulou, S. Lykoudis, Thermal comfort in outdoor urban spaces: analysis across different European countries. Build. Environ. **41**, 1455–1470 (2006)
35. N. Metje, M. Sterling, C. Baker, Pedestrian comfort using clothing values and body temperatures. J. Wind Eng. Ind. Aerod. **96**, 412–435 (2008)
36. S. Shooshtarian, I. Ridley, The effect of individual and social environments on the users thermal perceptions of educational urban precincts. Sustain. Cities Soc. **26**, 119–133 (2016)
37. S. Kariminia, S. Motamedi, S. Shamshirband, D. Petković, C. Roy, R. Hashim, Adaptation of ANFIS model to assess thermal comfort of an urban square in moderate and dry climate. Stoch. Environ. Res. Risk Assess. **30**, 1189–1203 (2016)
38. B. Givoni, M. Noguchi, H. Saaroni, O. Pochter, Y. Yaacov, N. Feller, et al., Outdoor comfort research issues. Energ. Build. **35**, 77–86 (2003)
39. E.L. Krüger, B. Givoni, F.A. Rossi, Outdoor comfort study in Curitiba, Brazil: effects of gender, Body weight and age on the thermal preference. Proceedings of the Conference Adapting to change: new thinking on comfort. Network for Comfort and Energy Use in Buildings, London, (2010)
40. W. Yang, N.H. Wong, G. Zhang, A comparative analysis of human thermal conditions in outdoor urban spaces in the summer season in Singapore and Changsha, China. Int. J. Biometeorol. **57**, 895–907 (2013)
41. D. Lai, D. Guo, Y. Hou, C. Lin, Q. Chen, Studies of outdoor thermal comfort in northern China. Build. Environ. **77**, 110–118 (2014)
42. L. Zhao, X. Zhou, L. Li, S. He, R. Chen, Study on outdoor thermal comfort on a campus in a subtropical urban area in summer. Sustain. Cities Soc. **22**, 164–170 (2016)
43. M.A. Ruiz, E.N. Correa, Adaptive model for outdoor thermal comfort assessment in an Oasis city of arid climate. Build. Environ. **85**, 40–51 (2015)
44. F. Salata, I. Golasi, R. de Lieto Vollaro, A. de Lieto Vollaro, Outdoor thermal comfort in the Mediterranean area. A transversal study in Rome, Italy. Build. Environ. **96**, 46–61 (2016)
45. M. Hadianpour, M. Mahdavinejad, M. Bemanian, F. Nasrollahi, Seasonal differences of subjective thermal sensation and neutral temperature in an outdoor shaded space in Tehran, Iran. Sustain. Cities Soc. **39**, 751–764 (2018)
46. MOHURD, Code for thermal design of civil buildding GB50176-2016, (2016)

47. ISO, Ergonomics of the thermal environment–Instruments for measuring physical quantities. Geneva. Int. Stand. Organ. (1998)
48. GZSB, Population size and distribution in Guangzhou in 2020. GZSB, (2020)

# Exploring Thermal Comfort and Pleasure in Outdoor Thermal Environments

**Abstract** Outdoor recreational spaces play an essential role in supporting both physical activity and psychological well-being, making thermal comfort and thermal pleasure critical aspects of outdoor experience. During hot summer conditions, shaded environments are often preferred because they moderate surface temperatures and provide more favorable microclimates. This study conducted walking experiments in elevated passages and tree-shaded areas to explore how thermal comfort and thermal pleasure vary with microclimatic conditions and activity intensity. Results indicate that tree canopies reduced mean radiant temperature by about 3 °C and wind speed by nearly 1 m/s compared with elevated passages. Across all scenarios, the lowest thermal comfort and pleasure levels were observed at a walking speed of 1.80 m/s. Findings further reveal that thermal pleasure is influenced not only by measurable environmental factors and physiological heat balance but also by psychological expectations, reflecting the interplay of metabolic heat production, external climate, and subjective perception. These outcomes highlight the necessity of integrating both physiological and psychological dimensions into thermal comfort models. By linking objective indicators with subjective thermal pleasure, this work provides insights for refining outdoor comfort evaluation and informs urban planning strategies aimed at designing shaded, enjoyable, and thermally responsive outdoor environments.

## 1 Introduction

Engaging in outdoor recreation delivers important advantages for physical and mental health [1], highlighting the significance of encouraging public use of open-air settings to promote wellness [2]. Nevertheless, climate change has led to increased urban temperatures and amplified heat island impacts, causing more common and extended extreme heat events [3, 4]. The escalating intensity of heatwaves and their negative effects on outdoor thermal environments warrant considerable focus [5–7]. Urban heat islands reduce urban livability by diminishing environmental quality, increasing thermal discomfort, and heightening heat-related health concerns [8]. Intense heat island conditions may even result in higher mortality [8]. Initiatives to mitigate

 
Z. Fang et al., *Outdoor Thermal Comfort and Heat Stress Mitigation*, Environmental Science and Engineering, https://doi.org/10.1007/978-981-95-8899-2_7

urban heat islands and protect people from severe heat form part of comprehensive approaches to combat climate change [9, 10]. Hence, recognizing the importance of outdoor comfort and safety is crucial in urban design projects aimed at improving thermal settings and protecting participants in outdoor pursuits.

During warm conditions, people often prefer shaded routes for transit or leisure [11]. These areas have cooler surfaces than sunlit spaces, substantially enhancing thermal comfort [12]. Consequently, they are more commonly included in urban designs for hot regions [13–15]. Standard examples are elevated passages and tree-covered spaces. Each provides shade from direct sunlight and supports better thermal comfort in summer. Their main difference involves airflow permeability, another key element affecting outdoor comfort [16, 17]. Although wind improvement has limited capacity to reduce solar heat stress in hot seasons [18], research verifies that superior wind conditions boost comfort for pedestrians [19, 20]. To study comfort differences among shade types, this project performed group walking tests in zones between covered and open shaded areas—frequently encountered during daily travel. Varied walking paces simulated different activity levels to assess the joint effect of thermal factors and effort on comfort.

Thermal sensation vote (TSV) is a recognized tool for measuring outdoor thermal perception through verbal and numeric feedback, per established standards [21, 22]. It supplies immediate feedback on personal reactions to outdoor thermal conditions. Based on "Alliesthesia" theory, changing thermal conditions activate the body's temperature regulation systems. If external cues assist in returning to thermal balance, individuals experience positive sensations [23]. Therefore, thermal pleasure ($T_p$) originates from bodily regulation and is viewed as a psychophysiological element in comfort analysis. $T_p$ and TSV capture environmental perception at different time scales, with $T_p$ being a higher-level response [24]. During walking, wind changes caused by movement serve as external thermal cues during physiological stages [25], possibly modifying walkers' moods (producing pleasure or unease). This psychophysiological process additionally influences satisfaction with the thermal setting. Given that outdoor comfort assessment seeks to provide contentment, overlooking emotional factors is insufficient. Currently, limited studies in outdoor thermal comfort have included Alliesthesia, and existing applications predominantly address major, short-term spatial changes [26, 27]. The psychophysiological impact on comfort from self-generated activity over brief durations is still not well grasped.

Considerable study has focused on predicting outdoor thermal comfort, yielding multiple indices that anticipate human physiological responses outdoors. Over recent decades, thermal indices have been created that merge environmental and physiological metrics to determine an equivalent temperature (°C) typifying a reference setting that induces similar bodily responses as real conditions. Popular indices in outdoor thermal science encompass physiological equivalent temperature (PET) [28] and universal thermal climate index (UTCI) [29, 30]. A thermal index model usually links the index to TSV, enabling appraisal of thermal feeling in specific contexts. Models might employ basic linear relations [31], sophisticated polynomial curves [32, 33], exponential functions [34], or other formats, varying with local climate and comfort considerations. Yet, the role of psychological dimensions in the precision of these models is not clearly defined. Hence, persistent advancement and tuning of situation-adapted thermal index models are needed. Aside from PET and UTCI, other models such as standard effective temperature (OUT_SET) [35, 36] and wet bulb globe temperature (WBGT) [37] see some use. However, OUT_SET is rarely applied owing to intricate calculations and data needs, and WBGT—intended for hot workplaces—relies on a basic four-category effort system, rendering it less suitable for changing metabolic states. In contrast, PET and UTCI offer integrated assessment by including environmental and physiological factors, standing as the most extensively used and reliable outdoor thermal indices today [29, 38]. Thus, this research emphasizes these two models.

Consequently, the primary objectives of this work are to assess thermal comfort disparities among walkers at different speeds in shaded outdoor locations under varying thermal situations, and to study emotional changes tied to thermal settings across velocities and durations. It also probes the connection between emotional components and thermal sensation votes. Lastly, the project intends to create customized thermal index models for gauging comfort in shaded environments—encompassing both elevated and natural shaded spaces—by integrating physiological and psychological aspects.

# 2 Methods

## 2.1 Experimental Design

### 2.1.1 Experimental Conditions

Field studies were carried out from August 7 to 19, 2023, at a university in Guangzhou (112°E–114.2°E and 22.3°N–24.1°N). August is one of the hottest months in this area, with average peak temperatures between 31.00 °C and 35.00 °C, and relative humidity commonly ranging from 75.00% to 90.00%. Two sites were designated for testing: case 1 in an elevated shaded structure, and case 2 below a dense tree canopy (Fig. 1). Both settings are notable for forming extensive, enclosed shaded zones that offer protection from solar exposure. The elevated shelter consists of an overhead covering but has open sides that allow substantial natural airflow. In the tree-covered site, overlapping leaves provide shading and facilitate heat exchange between people and surrounding vegetation, although the height and density of branches greatly limit wind movement. Accordingly, thermal environments and comfort expectations differ significantly between these two configurations.

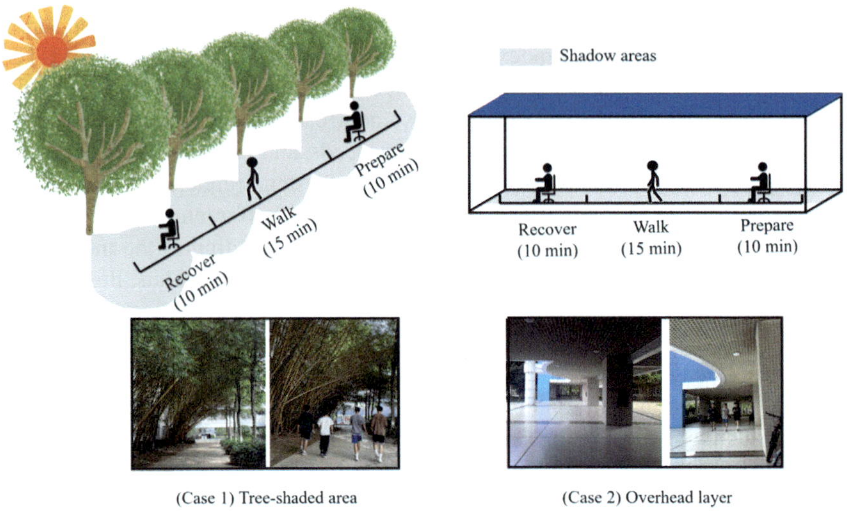


**Fig. 1** Experimental process

### 2.1.2 Experimental Protocol

During the research period, synchronized data acquisition was carried out at both selected locations. Testing was conducted daily from 8:30 to 11:30 in the morning and from 2:00 to 5:00 in the afternoon. To preserve comparability, all simultaneous experiments at different sites used consistent walking speeds. Volunteers joined based on their timetable, with eight subjects taking part in each half-day session. Each half-day cohort was randomly separated into two smaller groups, with every participant completing both the elevated shelter and tree shade components within two half-day intervals. The experimental protocol included three sequential phases. First, a 10 min preparatory stage helped stabilize initial activity rates. Next came a 15 min walking interval. Owing to the restricted pathway length (approximately 50 m), participants performed shuttle walking while research staff moved between locations to monitor and maintain procedural consistency. The subjects walked at five predetermined velocities—1.20, 1.40, 1.60, 1.80, and 2.00 m/s—each matching a particular metabolic demand. Pace regulation was achieved through metronome guidance. Prior research confirms that individuals attain metabolic equilibrium in outdoor transitional areas within 9–11 min during steady-speed walking [32], and within 5–11 min under tree canopies [11]. Thus, a 15 min walking duration was implemented to ensure thorough metabolic stabilization. After completing the walking session, participants underwent a 10 min seated recuperation phase. A schematic representation of the field arrangement and experimental sequence is provided in Fig. 1.

Five separate walking speeds were examined in successive order, with each pace group undergoing a 35 min testing session. Participants received no less than 30 min of rest between consecutive experiments to permit their heart rates to recover to baseline values. To reduce possible effects of daytime environmental changes, the testing sequence for different velocities was strategically alternated daily. As an illustration, on Day 1, speeds increased incrementally from 1.20 to 2.00 m/s, while on Day 2, assessments commenced at 1.60 m/s, moved to 2.00 m/s, and finished with the 1.20 m/s test. This systematic rotation approach verified that all speeds were evaluated during different daytime periods, thereby enhancing the reliability and credibility of the experimental methodology and its conclusions.

### 2.1.3 Testing Instruments

A pair of thermal comfort data acquisition units were implemented during the study to automatically document meteorological measurements, consisting of air temperature ($T_a$), relative humidity (RH), black globe temperature ($T_g$), and air velocity ($V_a$). These measurements are extensively accepted in exterior thermal comfort investigations as decisive elements influencing human thermal perception and constitute primary data for thermal comfort index determinations. The devices were installed 1.1 m above the ground surface, corresponding to the recognized center of gravity elevation for a standing human [39]. All data gathering operations conformed to ASHRAE Standard 55–2020 [40]. Throughout walking experiments, subjects operated a wireless combined anemometer-thermometer unit at umbilical level to record frontal wind velocity. The apparatus was positioned 80–115 cm above ground level, with minor modifications based on subject height, to diminish respiratory influence on air movement measurements. Metabolic rates were determined employing a portable digital pulse oximeter using the cardiac frequency technique, with corresponding mathematical formulations presented in Eqs. (1)–(4).

$$MR = \frac{HR - HR_0}{RM} + M_0 \tag{1}$$

$$RM = \frac{HR_{\max} - HR_0}{MWC - M_0} \tag{2}$$

$$MWV_{male} = (41.7 - 0.22 \times A) \times W^{0.666} \tag{3}$$

$$MWV_{female} = (35.0 - 0.22 \times A) \times W^{0.666} \tag{4}$$

where MR signifies the metabolic rate (W/m$^2$), $M_0$ refers to the metabolic rate at rest (W/m2), HR represents the real-time heart rate, $HR_0$ indicates the resting heart rate under thermally comfortable conditions, and RM expresses the incremental change in heart rate for each unit increase in metabolic rate. HRmax describes the peak heart rate, computed as $HR_{max} = 205 - 0.62 \times A$, and MWC stands for the maximum working capacity. In these formulas, A corresponds to age in years, and W signifies body weight measured in kilograms.

**Table 1** Details of instruments used in this study

| Instrument | Type | Parameter | Measuring range | Accuracy | Sampling rate (s) |
|---|---|---|---|---|---|
| Thermal comfort level recorder | SSDZY-1 | $T_a$ (°C) | − 20.0 °C– + 80.0 °C | ± 0.3 °C | 2 |
| | | *RH* (%) | 0.01–99.9% RH | ± 2.0% RH (10.0–90.0% RH) | 2 |
| | | $T_g$ (°C) | − 20.0–80.0 °C | ± 0.3 °C | 2 |
| Wireless universal wind speed and temperature recorder | WFWZY-1 | $V_a$ (m/s) | 0.05–5.00 m/s | 5% ± 0.05 m/s | 2 |
| Finger clip heart rate meter | YX102 | *HR* (bpm) | 25–250 bpm | ± 5% | 8 |

Table 1 outlines the full technical characteristics of the apparatus used in the experimental studies.

## 2.2 *Subjective Survey and Measurement*

Prior to field data collection, G*Power [41] was used to validate whether the sample size met statistical requirements. Using an effect size of 0.25, significance level of 0.05, and power of 0.95, the two-way ANOVA test required a minimum sample size of 302. Accounting for the five walking speeds and two test environments within the subjective comfort assessment, the study's necessary sample size was defined as no fewer than 31 participants. Exactly 31 individuals were included in the experiment. Table 2 summarizes the participants' background information. All selected subjects were university students who had lived in Guangzhou for at least one year (with enrollment dating to September 2022 or earlier) and were in healthy condition. Information about remuneration and study procedures was openly communicated. Recruitment followed a voluntary basis, and compensation adhered to the university's employment policies.

**Table 2** Information on the subjects

| Gender | Sample size | Age (years) | Height (cm) | Weight (kg) | BMI (kg/m$^2$) | BMR (kcal) |
|---|---|---|---|---|---|---|
| Male | 21.00 | 21.90 ± 2.20 | 172.60 ± 8.50 | 66.30 ± 8.00 | 22.20 ± 2.90 | 1688.20 ± 208.00 |
| Female | 10.00 | 21.10 ± 2.10 | 158.40 ± 8.80 | 50.00 ± 10.10 | 20.00 ± 2.80 | 1321.30 ± 209.50 |
| Average | – | 21.60 | 168.30 | 61.40 | 21.60 | 1577.70 |

**Table 3** Summary of the thermal response vote scale

| TSV | $T_{pref}$ | $T_p$ |
|---|---|---|
| − 3 cold<br>− 2 cool<br>− 1 slightly cool<br>0 neutral<br>+ 1 slightly warm<br>+ 2 warm<br>+ 3 hot | − 1 prefer cooler<br>0 no change<br>+ 1 prefer warmer | − 2 unpleasant<br>− 1 slightly unpleasant<br>0 neutral<br>+ 1 slightly pleasant<br>+ 2 pleasant |

Before testing began formally, subjects were trained in appropriate questionnaire procedures. This introduction covered both the structure of the survey and the use of subjective rating systems to ensure data validity. The digital questionnaire collected personal information, thermal feedback, and physiological data. During preparation and recovery periods, participants answered the questionnaire every two minutes. Throughout the walking phase, surveys were distributed at the 1st, 3rd, 5th, 7th, 9th, 11th, 13th, and 15th min. Consequently, each 35 min test session resulted in 18 completed questionnaires per individual. When accounting for all five walking speeds tested across both locations, 180 questionnaires were generated per participant. With 31 subjects involved, the study gathered a total of 2790 questionnaires. The thermal evaluation component emphasized four aspects: Thermal Sensation Vote (TSV) providing immediate reaction to thermal conditions; Thermal Preference ($T_{pref}$) and Thermal Pleasure ($T_p$) serving as affective measures of thermal adaptation. The corresponding assessment scales are shown in Table 3. The TSV and $T_{pref}$ scales complied with ISO 7730 [22] and ASHRAE Standard 55 [40], while the thermal pleasure scale drew upon research by Cabanac et al. [23].

## 2.3 Data Analysis

Initial questionnaire data and environmental parameter records were compiled using Microsoft Excel. Statistical analysis, incorporating regression techniques and significance testing, was conducted with IBM SPSS Statistics 25. Figures and charts were created employing both Excel and Origin2021 from Origin Lab Corporation. PET was calculated using the RayMan application [42], and UTCI was formulated based on the multi-node human thermal regulation model [29, 43], which can be accessed via the project's official website (www.utci.org).

# 3 Results

## 3.1 *Analysis of Microclimate Exposure During the Field Study*

Figure 2 shows environmental measurements obtained during the field study. The research occurred under typical summer conditions in Guangzhou, with an average air temperature of 31.58 °C and mean relative humidity exceeding 70%. Although data were collected simultaneously in shaded areas without direct sunlight, comparative analysis shows evident differences between tree-shaded locations and overhead layer environments. Areas under tree cover demonstrated significantly lower air temperatures and increased relative humidity compared to the overhead layer. Evaluation of mean radiant temperature ($T_{mrt}$) reveals that tree shading considerably reduces radiation intensity, producing an average $T_{mrt}$ decrease of approximately 3 °C relative to the overhead layer. Additionally, tree canopies buffer wind flow, maintaining stable, low wind speeds in shaded zones (mean $V_a = 0.30$ m/s) with daily standard deviations remaining below 0.50 units. By comparison, the overhead layer showed pronounced wind speed variability, with peaks nearing 4 m/s and daily standard deviations surpassing 0.50 units.

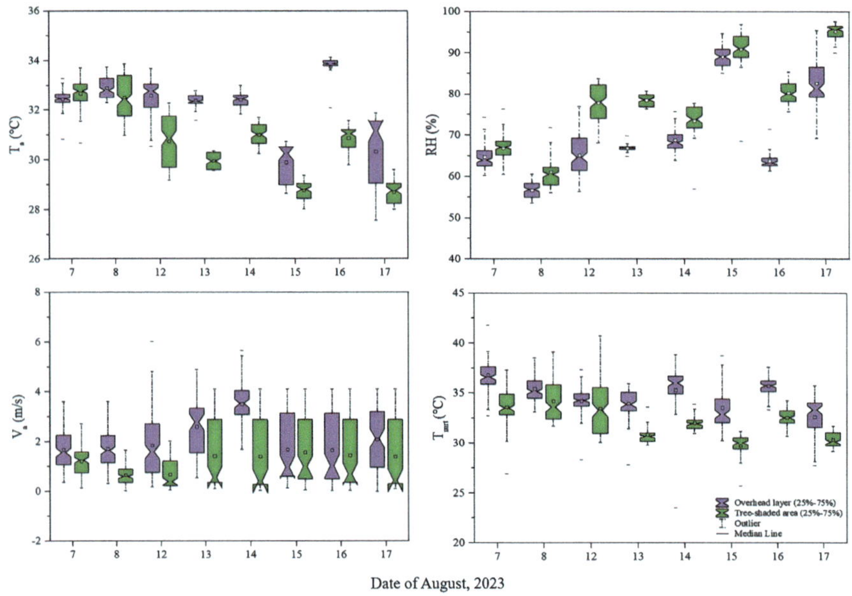


**Fig. 2** Daily environmental parameters in the site studied

## 3.2 Comparison of Mean Thermal Sensation Vote (MTSV) in Two Spaces

Figure 3 depicts the temporal dynamics of the mean thermal sensation vote (MTSV) throughout the study, integrating results from participants categorized by walking speed and environmental conditions. The data demonstrate that MTSV values for stationary outdoor subjects in summer conditions generally range from 0.80 to 1.30. A small decrease in MTSV, limited to 0.20 units, is observed after a ten-minute seated recovery interval. These results confirm that participants had attained a relatively stable psychophysiological state before formal experimentation commenced. Additionally, when moving from walking to recovery, pedestrians' MTSV can recover to near-initial levels within a 10 min period.

Regarding the walking phase, within the initial minute after commencing walking, the mean thermal sensation vote (MTSV) of pedestrians in both environments declined to some degree, with a more pronounced change observed in the tree-covered group compared to the overhead shelter group. The head-on wind speed at the onset of walking influences the average thermal sensation in the shaded group, leading to a stronger shift toward the cooler end of the scale. This is due to the relatively low natural wind speed in tree-shaded areas, where changes occur gradually.

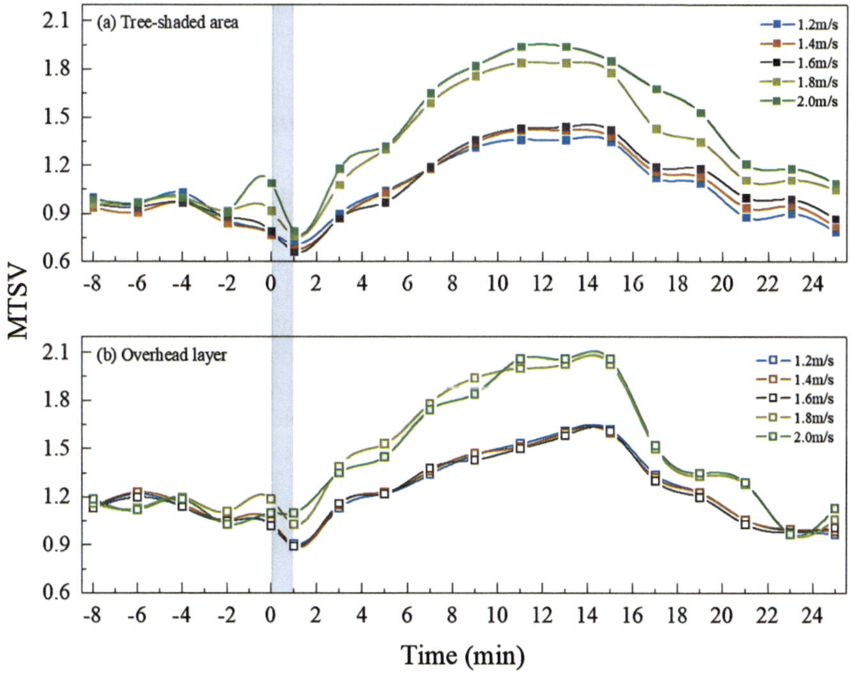


**Fig. 3** Variation of mean thermal sensation vote with times

In contrast, the overhead shelter group did not exhibit a similar reduction in MTSV, as the higher natural wind speed reduces the perceived change in wind sensation at the start of walking compared to that under trees.

From the 3rd minute onward, the MTSV changes in a favorable direction over time and begins to stabilize around the 11th minute. Clearly, at this stage, walking speed shows a stronger correlation with the magnitude of change in pedestrians' MTSV. Whether under overhead cover or tree shade, the MTSV values of pedestrians walking at 1.80 and 2.00 m/s were significantly higher than those walking at the other three speeds. Previous studies indicate that the effect of wind speed disturbance is comparable at steady walking speeds of 1.20, 1.40, and 1.60 m/s, but becomes markedly more pronounced at 2.00 m/s [25]. Furthermore, the $\beta$ value (average negative slope of logarithmic power spectrum curves) of the wind speed spectrum diverges most strongly from that of natural wind at a walking pace of 2.00 m/s, resulting in noticeable deviations from thermal comfort benchmarks [44]. This suggests that the critical range of outdoor walking speeds associated with shifts in thermal comfort lies between 1.60 m/s and 2.00 m/s. To investigate this further, the present study incorporated a walking speed of 1.80 m/s. The resulting data refine the upper threshold for abrupt changes in outdoor walking comfort to 1.80 m/s.

Additionally, under identical walking speed conditions, the thermal sensation vote of pedestrians in overhead shelters is higher than that of those in tree-shaded areas, indicating that walking speed (which correlates with metabolic activity level) is the primary factor influencing MTSV, followed by the location of walking.

## 3.3 *Comparison of Thermal Pleasure ($T_p$) in Two Spaces*

### 3.3.1 Distribution of Thermal Pleasure at Different Walking Speeds

Figure 4 illustrates the thermal pleasure evaluations at various walking speeds observed in two distinct outdoor settings throughout the 15 min experimental walking duration. During low-paced walking (1.2 m/s and 1.4 m/s), subjects in tree-covered zones reported higher thermal pleasure compared to those in elevated shelters, with an average thermal pleasure rating maintained at or above zero. This indicates that under gentle walking conditions in such contexts, people's subjective experience of thermal comfort was somewhat positive and agreeable, even while the mean thermal sensation vote surpassed the neutral thermal threshold. The findings suggest that at reduced metabolic rates, climatic conditions strongly enhance human thermal pleasure, and the effect of lowered radiant and air temperatures is more substantial than that of wind speed. From Fig. 2, it is evident that the average wind speed in the overhead layer is 1.14 m/s greater than in shaded regions, while both air temperature and radiative intensity are diminished under tree cover, with quantified reductions of $\Delta T_{mrt} = 3.03$ °C, $\Delta T_a = 1.70$ °C.

When walking speed reaches or surpasses 1.6 m/s (corresponding to a brisk walking mode), the average thermal pleasure experienced by individuals in elevated

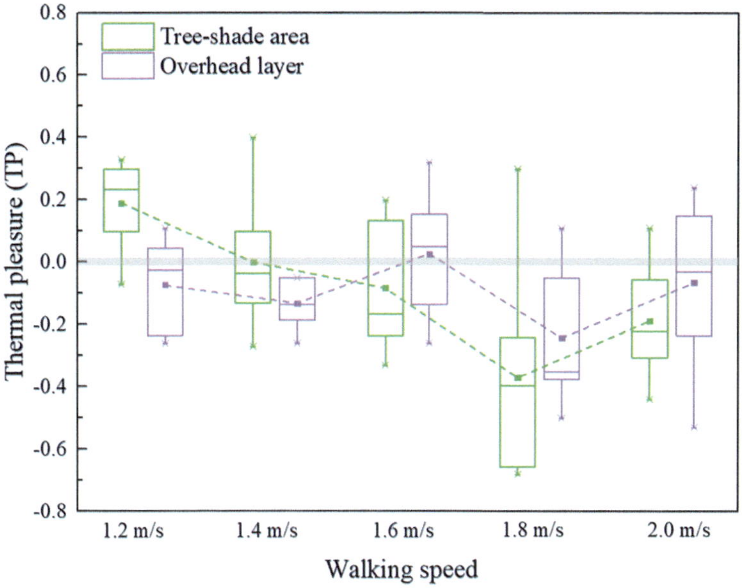


**Fig. 4** Thermal pleasure at different walking speeds

shelters exceeds that reported by those in tree-shaded spaces. In this regime, the heightened frontal disturbance wind speed induced by locomotion reinforces the beneficial role of natural wind on thermal perception, exceeding the effects of other environmental variables. Therefore, under consistent climatic conditions, pedestrians can enhance their thermal pleasure by choosing shaded paths during leisurely walks and elevated structures during faster walking.

As gait velocity increases, thermal pleasure in vegetated areas shows a declining tendency. Conversely, in elevated walkways, it registers a slight initial rise, reaching a minor maximum at 1.60 m/s (above $T_p = 0$), after which it decreases, paralleling the trend seen in shaded settings. Particularly noteworthy is that both environments exhibited the most negative thermal pleasure values at 1.80 m/s, implying that metabolic heat output then overrides other external factors. A shift in the governing factor and subsequent improvement in thermal pleasure manifest at 2.00 m/s, driven by elevated disturbance wind speed. In general, the least favorable thermal pleasure conditions occur at 1.80 m/s, independent of spatial configuration.

### 3.3.2 *Analysis of Influencing Factors of $T_p$*

This study applied Spearman's rank correlation test, a distribution-free statistical technique, to determine the strength and directional association between pairs of variables without presupposing linearity. The test yields a correlation coefficient that captures the monotonic dependence between variables using ranked values

[45]. Significance testing was conducted to verify that observed correlations represented substantive, non-random relationships. In order to explore determinants of thermal pleasure, Spearman correlation analysis was used to investigate associations among $T_p$, diverse environmental factors, walking speeds, and MTSV across various experimental stages in two contrasting environments, as presented in Table 4.

Distinct variations can be observed among the environmental parameters showing significant correlations with thermal pleasure ($T_p$) in the two types of spaces. In the overhead shelter environment, $T_p$ correlated significantly only with wind speed (denoted as surrounding air velocity $V_s$ during walking, and $V_a$ in other stages). A significant relationship was also found between $T_p$ and air temperature specifically during walking. On the other hand, in the shaded area, $T_p$ showed statistically significant associations with all listed environmental factors—relative humidity (RH), mean radiant temperature ($T_{mrt}$), air temperature, ($T_a$), and wind speed—throughout all experimental phases. While the correlation values were not high, these significant

**Table 4** Results of Spearman's rank correlation test

| Site | Categories | | RH | $T_a$ | $V_a/V_s$ | $T_{mrt}$ | $T_p$ | $W_s$ | MTSV |
|---|---|---|---|---|---|---|---|---|---|
| Overhead layer | Preparing period | $T_p$ | 0.06 | – 0.10 | 0.33** | – 0.04 | 1.00 | – | – 0.64** |
| | | MTSV | – 0.06 | 0.14 | – 0.11 | – 0.05 | – 0.64** | – | 1.00 |
| | Walking period | $T_p$ | 0.11 | – 0.14* | – 0.22** | – 0.10 | 1.00 | – 0.19** | – 0.67** |
| | | MTSV | – 0.10 | 0.15* | 0.35** | 0.04 | – 0.67** | 0.37** | 1.00 |
| | Recovery period | $T_p$ | – 0.01 | – 0.08 | 0.21** | – 0.01 | 1.00 | – 0.03 | – 0.64** |
| | | MTSV | – 0.14 | 0.22** | – 0.16 | 0.02 | – 0.64** | 0.09 | 1.00 |
| Tree-shaded area | Preparing period | $T_p$ | – 0.22** | 0.20* | 0.21* | 0.24** | 1.00 | – | – 0.37** |
| | | MTSV | – 0.16* | 0.34** | – 0.20* | 0.30** | – 0.37** | – | 1.00 |
| | Walking period | $T_p$ | – 0.20** | 0.19** | – 0.31** | 0.17** | 1.00 | – 0.38** | – 0.33** |
| | | MTSV | 0.11 | 0.15* | 0.39** | 0.22** | – 0.38** | 0.42** | 1.00 |
| | Recovery period | $T_p$ | – 0.19* | 0.33** | 0.21** | 0.26** | 1.00 | – 0.43** | – 0.18* |
| | | MTSV | – 0.11 | 0.18* | – 0.14 | 0.23** | – 0.43** | 0.24** | 1.00 |

** The correlation is statistically significant at the 0.01 level
*. The correlation is statistically significant at the 0.05 level
The absence of "*" signifies a non-significant correlation

relationships indicate connections that might be accounted for by intricate patterns, stage transitions, or alternative nonlinear models.

Furthermore, during the walking phase, $T_p$ displayed statistically significant correlations with walking speed ($W_s$) and mean thermal sensation vote (MTSV) in both spaces, even though the correlation coefficients were relatively low. Walking speed influences not only the oncoming air movement around the pedestrians but also alters their internal heat generation, both of which considerably impact thermal perception and contribute psychological signals affecting thermal anticipation during motion, which in turn influences thermal pleasure responses. In a dynamic context, $T_p$—which integrates emotional aspects and mirrors psychological condition—becomes more prominent compared to MTSV, which only reflects the strength of thermal stimulation [46]. Accordingly, $T_p$ may be viewed as a more comprehensive expression of thermal perception compared to MTSV [24], as it is shaped not just by external thermal conditions but also by internal emotional mechanisms. These two facets represent separate dimensions with substantial differences in interpretive value, which partly clarifies why, despite a significant linear relationship between $T_p$ and MTSV, the correlation strength is limited.

The findings also affirm that, alongside objective environmental variables, walking speed and MTSV are the two most strongly influential parameters affecting $T_p$.

### 3.3.3 Analysis of Change in Thermal Pleasure with Thermal Sensation

Marked differences in subjective ratings emerged between the walking and recovery phases across the three experimental stages. In comparison, the preparatory stage showed more stable environmental and internal thermal conditions, leading to only minimal variations in MTSV (Fig. 3). Although a significant correlation was identified between $T_p$ and MTSV throughout the entire experiment (Table 4), the analysis specifically targeted data from the 1st to 30th minutes to study temporal patterns in thermal pleasure and its relationship with thermal sensation dynamics.

Changes in thermal pleasure (dP) and thermal sensation (dT) were defined as differences between two consecutive questionnaire responses, computed using Eqs. (5) and (6).

$$dP = P_t - P_{t-1} \tag{5}$$

$$dT = T_t - T_{t-1} \tag{6}$$

Thermal pleasure ($P_t$) and thermal sensation ($T_t$) values were gathered from the nth questionnaire administration ($n = 2, 3, 4 \ldots 23$). The temporal change in thermal sensation (dT) was assessed according to whether subjects reported feeling warmer ($dT > 0$) or cooler ($dT < 0$) between successive surveys conducted at 2–5 min intervals. Similarly, thermal affective changes (dP) were categorized as positive alliesthesia when subjects experienced increased pleasure ($dP > 0$), or negative alliesthesia for

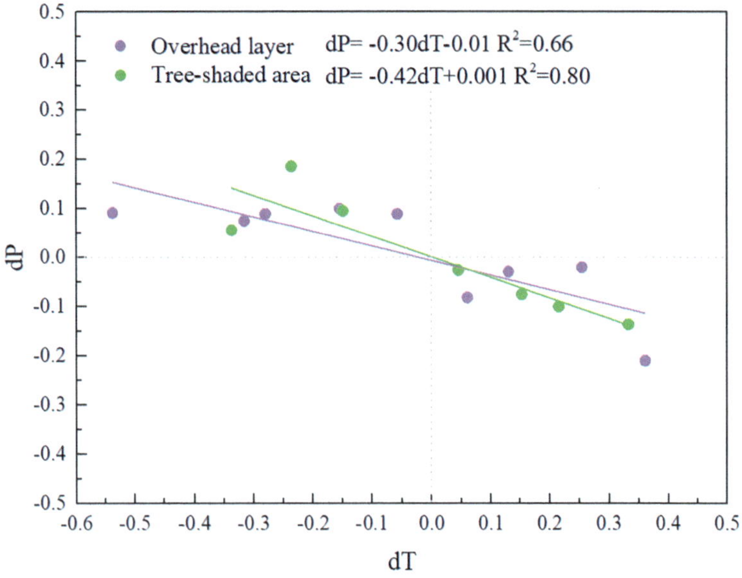


**Fig. 5** Linear regression analysis between dP and dT

diminished pleasure (dP < 0) across consecutive measurements. To evaluate the dP-dT relationship, linear regression analysis was conducted using averaged dP values against binned dT values (bin size = 0.1 dT), as shown in Fig. 5.

Under summer conditions (representative of South China's typical weather), all thermal sensation ratings were situated toward the warmer segment of the thermally neutral range. A clear inverse correlation was found between thermal pleasure—considered as a psychological metric—and momentary fluctuations in thermal sensation: increases in thermal sensation are associated with decreases in thermal pleasure. When thermal sensation exhibits a positive change (dT > 0), thermal pleasure is reduced, as signaled by a negative dP value. On the other hand, unchanged thermal sensation across two intervals leads to largely consistent thermal pleasure, a finding corroborated by the convergence of the two regression curves near the (0, 0) point in the graph. This model serves as a predictive tool for estimating moment-to-moment variations in pedestrians' thermal pleasure.

## *3.4 Analysis of Metabolic Changes*

As shown in Fig. 6, shifts in metabolic rate are predominantly tied to physical exertion and reveal only slight variations across both environments. Metabolic activity exhibits a short-term surge at the onset of walking, continues a steady climb through the exercise period (1–15 min), and eventually nears a dynamically balanced state.

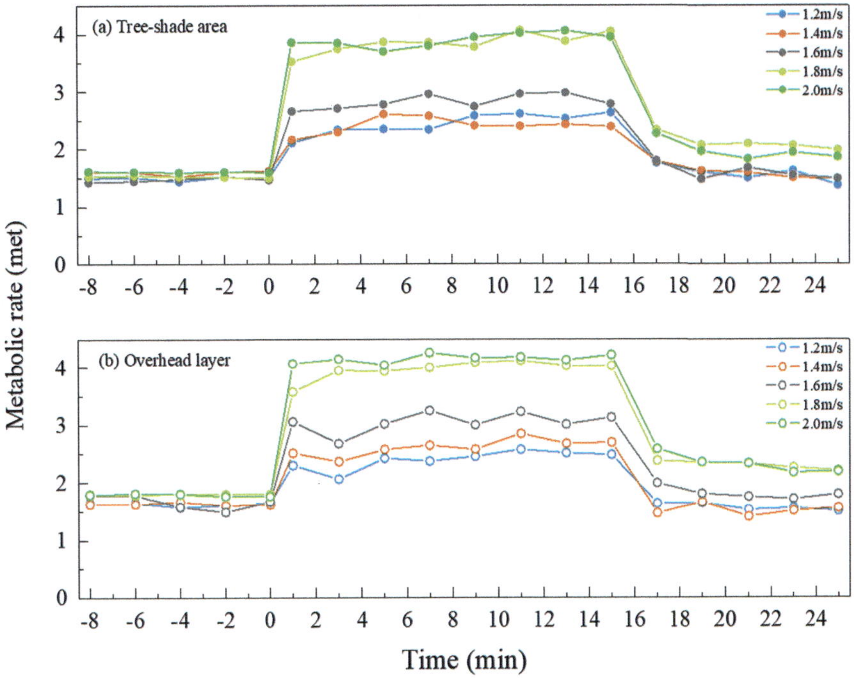


**Fig. 6** Variation of metabolic rate in different walking speed groups with time

After walking ended, participants moving at speeds of 1.20 m/s, 1.40 m/s, and 1.60 m/s restored metabolic readings close to their original values within a brief span (about 2–4 min). However, although those walking at 1.80 m/s and 2.00 m/s also reached a stable level following a 10 min rest, their finalized metabolic rates stayed elevated compared to slower paces. This finding suggests that faster-walking individuals need longer recuperation to regain a baseline metabolic condition. One prominent conclusion from this figure is that metabolic rate varies continually during walking, and both the range of variation and the final equilibrium level are influenced by walking speed.

## 4 Establishment and Improvement of Thermal Index Models

Prior studies have established the effectiveness of outdoor thermal index models for people performing relatively sedentary activities like sitting or standing [34, 47]. This research focuses on examining how well these thermal indices perform during walking. Thus, all data presented and examined in this chapter—unless specifically noted—were gathered throughout walking sessions (1–15 min).

PET and UTCI are broadly accepted as standard thermal indicators suitable for assessing outdoor thermal comfort [48]. They represent the equivalent temperature (°C) in a controlled reference setting that elicits the same physiological response as the real outdoor environment in a typical person, serving as a measure of thermal stress. In practical thermal evaluations, these indices are often correlated with subjective thermal sensation votes to develop predictive thermal models. Using relatively simple calculations, such models translate intangible thermal stress into tangible and adjustable thermal perceptions, supplying actionable information for urban planners to adopt targeted measures in climate-aware urban design.

However, it has been observed that single-predictor regression models linking MTSV with thermal indices possess certain drawbacks in real-world thermal environment evaluation [25, 32]. As a result, the goal of this study is to improve the adaptability of thermal index models to specific local conditions and activity types. The work concentrates on optimizing the model's applicability rather than altering the fundamental thermal index formulas. The structure for constructing and refining these thermal index models is presented in Fig. 7.

## *4.1 Linear Regression*

The established methodology for thermal index modeling centers on constructing a direct linear relationship between the thermal index and mean thermal sensation vote. Research shows that thermal index models created through this procedure may achieve better adaptability in particular outdoor activity contexts [49–51]. Accordingly, following classification by exercise phase and spatial condition, MTSV was determined with the thermal index (PET and UTCI) as the predictor variable, using a bin width of 0.50 °C. A scatter diagram was produced to represent the dataset, and linear regression was utilized to evaluate the association. The derived outcomes are provided in Fig. 8.

As shown in Fig. 8, UTCI and PET both demonstrated a clear linear relationship with MTSV in the preparation phase, where the $R^2$ values were above 0.6. However, during walking—whether in shaded or overhead shelter environments—thermal indices relying on direct linear correlation showed poor applicability, indicated by low $R^2$ measures. The weak performance of the linear model may be partly due to its failure to adequately represent high-temperature outdoor scenarios, as supported by earlier studies [33, 34]. Additional constraints are inherent to the thermal indices. For instance, UTCI uses a constant metabolic rate of 2.30 Met [38], which does not account for metabolic fluctuations during activity or their influence on thermal sensation, and focuses mainly on physiological reactions to environmental factors like $T_a$, RH, V, and $T_{mrt}$. Thus, using UTCI alone to link thermal sensation responses and evaluate thermal comfort during exercise is insufficient. Likewise, while PET integrates metabolic rates adjusted for walking speed [52], Fig. 6 indicates that metabolic rate varies dynamically during walking and is not static even

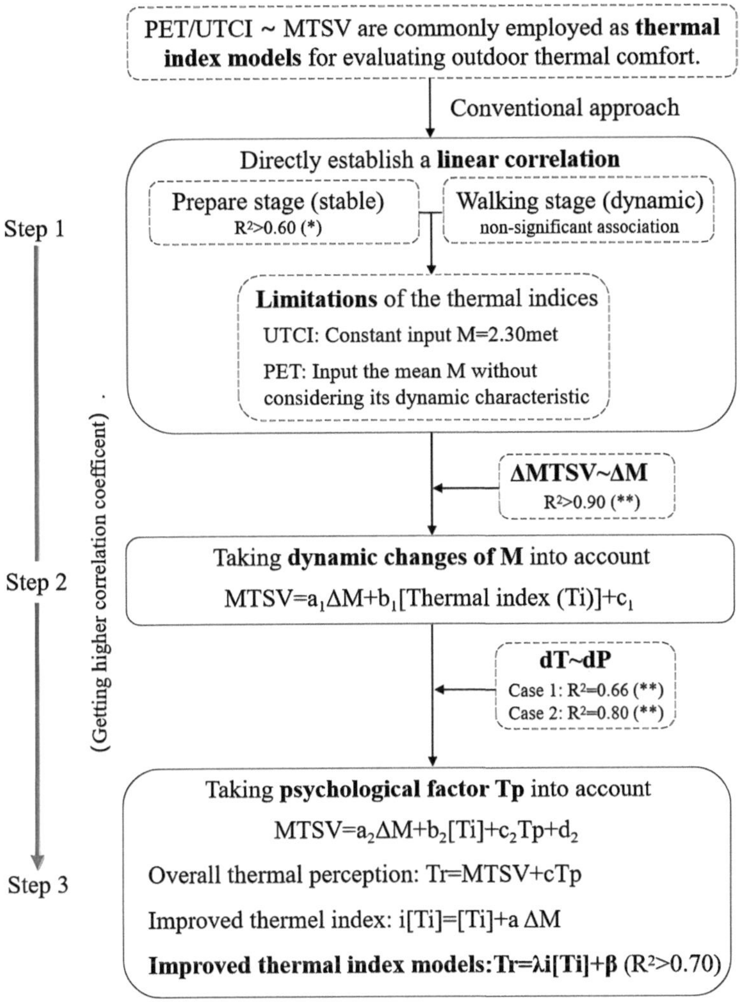


**Fig. 7** Research framework of Sect. 4

under consistent intensity. Therefore, PET-based linear regression models are also unsuitable for accurately assessing pedestrian thermal comfort in real-world settings.

## 4.2 *Taking Dynamic Changes of M into Account*

Existing research confirms that omitting metabolic rate variability during walking results in predictive inaccuracies within linear thermal index models [25, 32]. Accordingly, this investigation examined the association between ΔMTSV and ΔM—where

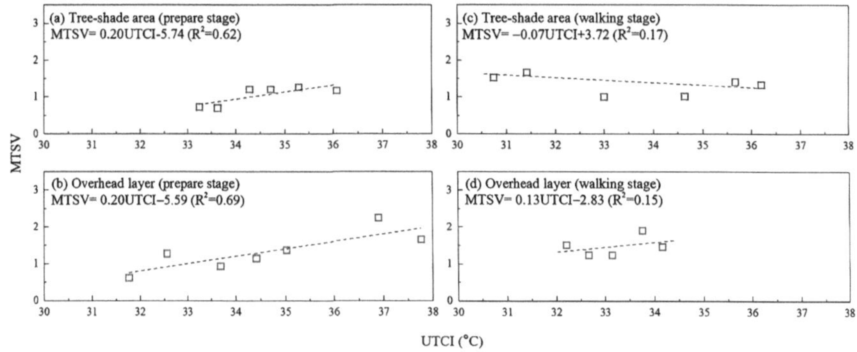


(i) UTCI

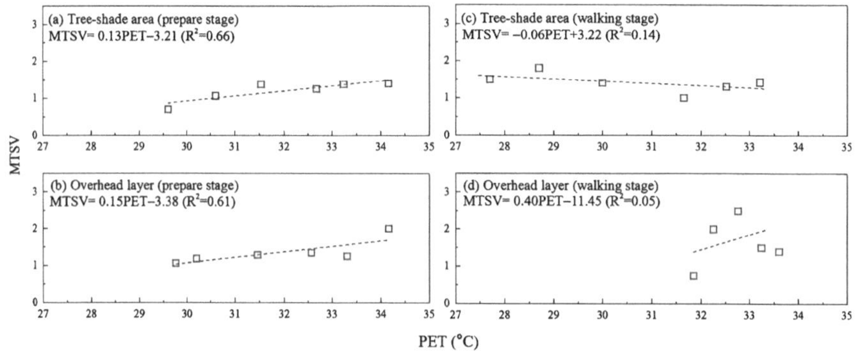


(ii) PET

**Fig. 8** Variation of MTSV with thermal indices, **i** UTCI and **ii** PET: (a) (b) prepare stage, (c) (d) walking stage

these variables reflect differential values between actual walking and the preparatory phase. As depicted in Fig. 9, increases in metabolic activity coincided with proportional alterations in thermal sensation. A distinct linear correlation was identified, thereby corroborating the proposed hypothesis. These outcomes demonstrate that fluctuating metabolic rates during walking substantially contribute to the limited agreement between thermal indices and MTSV.

Real-time metabolic rate was incorporated into the thermal index models using multivariate linear regression, leading to observable performance gains as formulated in Eqs. 7–10 (Table 5). With this initial modification, correlation coefficients for both the PET and UTCI models demonstrated substantial increases. These findings indicate enhanced MTSV prediction accuracy in the revised models, most notably in overhead shelter conditions—a result consistent with existing literature [11, 32]. Despite these improvements, the general robustness of the thermal index models remains unsatisfactory, as three models maintain $R^2$ values below 0.6.

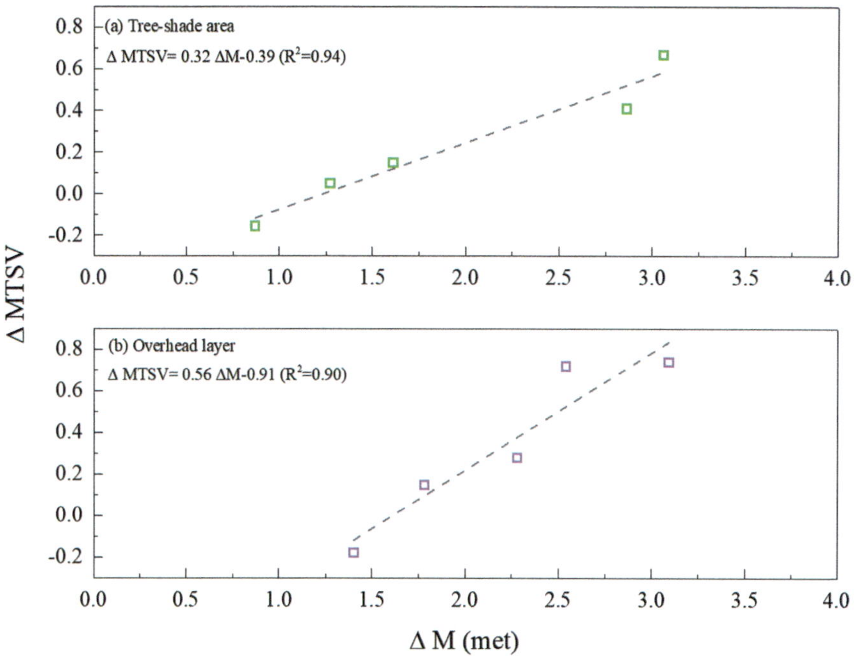


**Fig. 9** The relationship between ΔMTSV and ΔM

**Table 5** Summary table of fitting results (1)

| Tree-shaded area | | Overhead layer | |
|---|---|---|---|
| MTSV = 0.68 ΔM + 0.16UTCI–5.44 ($R^2 = 0.53$, sig. < 0.05) | (7) | MTSV = 0.52 ΔM + 0.21UTCI–6.59 ($R^2 = 0.73$, sig. < 0.05) | (9) |
| MTSV = 0.67 ΔM + 0.15PET–4.65 ($R^2 = 0.58$, sig. < 0.05) | (8) | MTSV = 0.73 ΔM + 0.17PET–5.44 ($R^2 = 0.59$, sig. < 0.10) | (10) |

## 4.3 *Taking Psychological Factor ($T_p$) into Account*

In dynamic thermal environments, psychological factors such as thermal preference and pleasure carry greater weight than thermal sensation votes, which principally reflect the strength of thermal stimuli [46]. This investigation shows that wind speed variations caused by different walking rates elicit differentiated affective states along temporal and spatial dimensions, assessed through graded thermal pleasure scales. Outcomes detailed in Sect. 3.3.2 reaffirm a consistent connection between dT and dP across successive time points. As displayed in Fig. 4, $T_p$ exhibits marked fluctuations under diverse metabolic intensities and related wind disturbances, confirming that the emotional component $T_p$ plays a definable role within the thermal sensing structure.

Therefore, $T_p$ has been included as an auxiliary variable in the continued development of thermal index models. Multivariate fitting outputs are shown in Eqs. (11)–(12) and (15)–(16) (Table 6). The refined models display elevated R2 values without loss of statistical significance. In line with earlier works [11, 25, 32], composite expressions merging thermal indices and metabolic rate are classified as improved UTCI (iUTCI) and improved PET (iPET). The labels "t" and "o" indicate tree-shaded and overhead shelter settings, respectively, following Eqs. (13)–(14) and (17)–(18).

The overall thermal perception (Tr) is defined by combining MTSV and $T_p$, as represented in Eq. (19) of Table 7, where *a* is a constant. This representation leads to upgraded thermal models based on the UTCI and PET frameworks. In these models, $Tr_{UTCI\text{-}t}$, $Tr_{PET\text{-}t}$, $Tr_{UTCI\text{-}o}$ and $Tr_{PET\text{-}o}$ denote thermal perceptions in tree-shaded and overhead shelter environments, respectively, according to Eqs. (20)–(23).

As presented in Fig. 10, the iUTCI and iPET models offer consistent prediction accuracy for holistic thermal perception. Correlation analysis further substantiates that the modified thermal index models under overhead shelter conditions demonstrate higher dependability, with correlation coefficients above 0.84. These developed models also achieve acceptable predictive precision, as correlation measures exceed 0.74. Considering these results, reassessing the performance of both original and refined thermal index models across diverse outdoor scenarios should be a key objective in future research.

The revised thermal indices and their associated heat stress levels are shown in Table 8. Enhancing the quantitative estimation of thermal comfort across diverse

**Table 6** Summary table of fitting results (2)

| Tree-shaded area | | Overhead layer | |
|---|---|---|---|
| $MTSV + 0.71Tp = 0.10(UTCI + 3.80\Delta M)–2.84$ ($R^2 = 0.68$, sig. $< 0.05$) | (11) | $MTSV + 0.57Tp = 0.24(UTCI + 1.71\Delta M)–7.28$ ($R^2 = 0.78$, sig. $< 0.05$) | (15) |
| $MTSV + 0.70Tp = 0.11(PET + 4.10\Delta M)–2.75$ ($R^2 = 0.71$, sig. $< 0.05$) | (12) | $MTSV + 0.32Tp = 0.14(PET + 4.93\Delta M)–4.62$ ($R^2 = 0.66$, sig. $< 0.1$) | (16) |
| $iUTCI_t = UTCI + 3.80\ \Delta M$ | (13) | $iUTCI_o = UTCI + 1.71\ \Delta M$ | (17) |
| $iPET_t = PET + 4.10\ \Delta M$ | (14) | $iPET_o = PET + 4.39\ \Delta M$ | (18) |

**Table 7** Summary table of fitting results (3)

| Tree-shaded area | | Overhead layer | |
|---|---|---|---|
| $Tr = MTSV + aT_p$ (19) | | | |
| $Tr_{UTCI\text{-}t} = 0.12\ iUTCI_o–3.63$ ($R^2 = 0.74$, sig. $< 0.05$) | (20) | $Tr_{UTCI\text{-}o} = 0.24\ iUTCI_o–7.53$ ($R^2 = 0.88$, sig. $< 0.05$) | (22) |
| $Tr_{PET\text{-}t} = 0.10\ iPET–2.42$ ($R^2 = 0.78$, sig. $< 0.05$) | (21) | $Tr_{PET\text{-}o} = 0.15\ iPET–4.91$ ($R^2 = 0.84$, sig. $< 0.1$) | (23) |

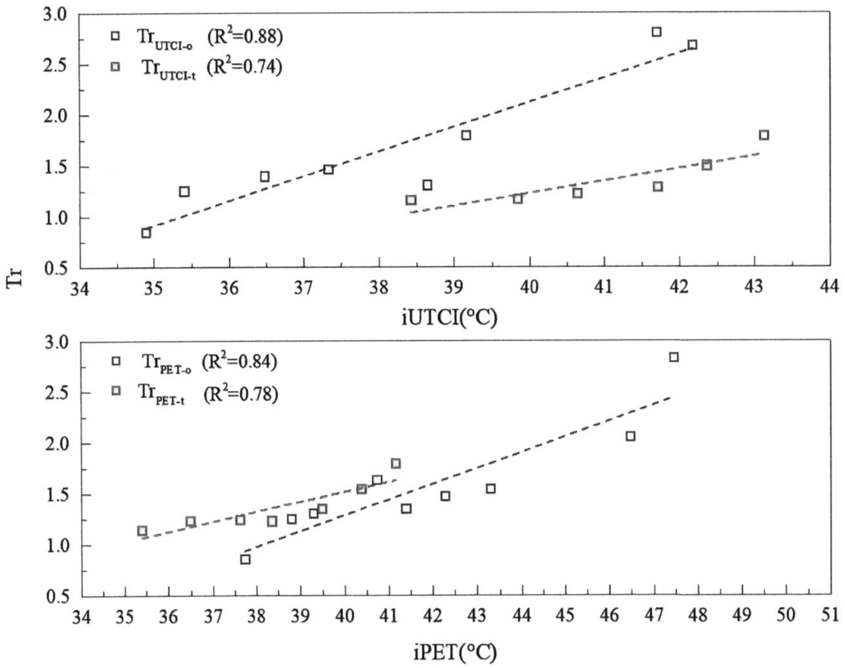


**Fig. 10** The improved thermal index models

activity intensities and increasing the predictive reliability of thermal index frameworks in multifaceted urban outdoor contexts can offer evidence-based insights for urban planning professionals. The fundamental objective of current thermal index model refinement efforts continues to be the improvement of thermal environments in urban exterior areas from the standpoint of user experience.

**Table 8** The heat stress categories of the improved thermal indices

| Stress category | | No thermal stress | Moderate heat stress | Strong heat stress | Very strong heat strong |
|---|---|---|---|---|---|
| Overall thermal perception | | − 0.5–0.5 | 0.5–1.5 | 1.5–2.5 | 2.5–3.0 |
| iUTCI (°C) | Tree-shaded area | 26.08–34.42 | 34.42–42.75 | 42.75–51.08 | 51.08–55.25 |
| | Overhead layer | 29.29–33.46 | 33.46–37.63 | 37.63–41.79 | 41.79–43.86 |
| iPET (°C) | Tree-shaded area | 19.20–29.20 | 29.20–39.20 | 39.20–49.20 | 49.20–54.20 |
| | Overhead layer | 29.40–36.07 | 36.07–42.73 | 42.73–49.40 | 49.40–52.73 |

# 5 Discussion

## *5.1 The Necessity of Investigating the Impact of Psychological Factors on Outdoor Dynamic Thermal Comfort*

The perception of a stimulus as pleasant or unpleasant depends not on environmental changes per se, but on the internal state of the individual. Cabanac coined the term "Alliesthesia" to describe this psychophysiological mechanism [23]. In thermal contexts, the affective quality of transient skin-detected thermal signals—pleasant or unpleasant—is determined by whether they drive the body toward or away from thermal neutrality. Movement toward neutrality elicits thermal pleasure, while deviation from it causes displeasure.

Thermally neutral environments represent the ideal psychophysiological state for individuals under steady thermal conditions. The unchanging nature of a neutral indoor temperature offers little sensory gratification, owing to the absence of dynamic thermal cues. In comparison, outdoor settings exhibit continual variations in temperature, humidity, wind, solar radiation, activity levels, and environmental interruptions. These ongoing changes in space and time function as external activators. Per alliesthesia theory, whether such stimuli aid in reinstating the body's internal stability governs their ability to induce pleasure. Within variable outdoor surroundings, personal emotional conditions—spanning from pleasant to unpleasant—frequently oscillate, producing a gap between neutral and desirable thermal states, a conclusion supported by earlier research [26, 27, 53]. Throughout this research, the dominant external influence during walking is the shifting wind velocity around pedestrians across different moments. Alliesthesia suggests that wind speeds associated with distinct walking rhythms elicit diverse affective reactions through time and space, evaluated using thermal pleasure ($T_p$) metrics. Findings reveal that $T_p$ fluctuates substantially under assorted disturbances and temporal changes, verifying its meaningful contribution to the comprehensive thermal perceptual structure.

Research on thermal environment evaluation regarding scale hypothesis [54] suggests that the spacing between verbal anchors on subjective thermal sensation scales (for instance, the ASHRAE 7-point scale) and affective scales (implemented as thermal comfort scales in that work) is inconsistent. Although thermal sensation vote (TSV) indicates thermal stimulus intensity, it fails to express an individual's comfort condition. Sensation refers to registering external stimuli, whereas perception involves interpreting such signals—a higher-level cognitive function [24]. Despite identifying a notable correlation between $T_p$ and MTSV in this research (Sect. 3.3), in non-steady settings, emotional aspects like thermal preference and pleasure assume greater importance compared to TSV, which only measures stimulus strength [46]. Hence, neglecting psychological (i.e., emotional) components and using thermal sensation votes alone for comfort evaluation may lead to inaccuracies. Unlike conventional steady-state models, obtaining both sensory and emotional thermal assessments is vital for a complete understanding of outdoor thermal comfort, particularly among mobile individuals. Alliesthesia provides a useful theoretical basis for examining thermal perception in open and semi-open spaces. Dzyuban et al. also reported a weak association between TSV and thermal pleasure in their exploration of pedestrian alliesthesia, pointing to the complexity of psychological factors and the value of targeted investigation under fluctuating thermal conditions [55]. Although a large portion of outdoor thermal comfort studies has focused on reducing TSV/MTSV scores to enhance comfort—using them as the central criterion for thermal index model validation—emotional influences were recognized in earlier works [56, 57] and have regained traction in recent publications [58]. These studies assert that excluding higher-order psychological factors results in an incomplete representation of thermal comfort.

Guided by the previous analysis, Sect. 4 of this work introduced psychological evaluation using the Thermal Pleasure Scale. This portion concentrates on appraising holistic thermal comfort experienced by people outdoors under specific physical states and environmental layouts in non-steady scenarios. A structured evaluation system for outdoor thermal indices was further established and upgraded, yielding initial support for its applicability. Within urban external design, walking behavior and related psychological and physiological changes should be incorporated, since they tangibly affect human comfort in thermal surroundings. However, tactically deployed shading interventions to regulate microclimate factors persist as the most effective strategy for raising thermal comfort in unenclosed spaces.

## 5.2 *Implications for Adding Psychological Adaptability to Improvements of Thermal Index Models*

Individuals experience their environment differently, and their responses to physical stimuli depend not just on strength but also on situational "cues". Hence, psychological factors play a major role in shaping the perception of thermal conditions and ensuing reactions. The concept of "naturalness", introduced by Griffiths et al. [59], describes environments without artificial modifications. In such natural settings, people exhibit higher tolerance to environmental fluctuations. Yet, individuals frequently hold preset beliefs about climatic attributes—for example, thinking, "It is summer in Guangzhou, so it must be very hot outside". This brings in key components influencing psychological thermal adjustment: "anticipations" and "previous exposure".Nikolopoulou and Steemers developed a framework incorporating environmental psychology to explain human engagement with outdoor spaces [60], underscoring how past experiences directly form expectations and modify thermal perception. The subjects in this study were college students with at least one year of residency in Guangzhou, providing them with particular exposure to the city's summer thermal climate. The test situation mirrored a typical student transit context. Accordingly, their prolonged thermal adaptation likely shaped their anticipations for current conditions, which then influenced subjective measures including MTSV, $T_{pref}$, and $T_p$. On a representative summer test day, more than 10% of subjects favored no modification to the ambient setting and expressed positive thermal pleasure, although their thermal sensation votes lay beyond the comfort interval ($\pm$ 0.5). This points to a steady perception and psychological anticipation of the prevailing thermal environment. At this point, subjective thermal sensation votes result from both sensory and psychological adaptive processes. In summer, this typically dampens the positive growth tendency in MTSV levels. Lam et al. corroborated this outcome in their study examining the influence of short-term physiological and psychological adaptation on thermal comfort among local and non-local students during summer exercise [61].

Psychological factors substantially influence how individuals perceive thermal comfort within a specific environment. Current thermal comfort models mainly emphasize thermophysiological mechanisms involving body heat management and energy exchange (PET derives from the Munich Energy Balance Model for Individuals, and UTCI applies the Fiala multi-node system). Yet these approaches often disregard psychological components of thermal comfort. As a result, conventional thermal indices may offer reduced accuracy in predicting thermal conditions. Introducing psychological elements into modernized thermophysiological frameworks may improve the forecasting ability of thermal models [11, 32, 33, 62]. Incorporating measures such as thermal pleasure offers innovative avenues for creating interventions that positively influence warmth perception [63]. However, integrating psychological adaptation into urban planning presents substantial practical difficulties. It introduces the strategic question of whether urban policies should support enduring psychological adaptation [64] or transient physiological adaptation [65] across varied populations. Identifying a suitable thermal index model to direct urban development is therefore essential. This is particularly important when designating functional districts—for instance, emphasizing commercial zones with high numbers of international tourists or residential areas largely populated by long-term inhabitants. Enhancing urban planning protocols for these specific contexts may serve as a goal for subsequent research. Figure 11 suggests a conceptual research structure for advancing thermal index models through psychophysiological approaches, outlining a trajectory for further inquiry.

## 6 Conclusion

This chapter explored how outdoor thermal comfort and pleasure are shaped by both environmental conditions and psychological responses during walking in shaded spaces. Field trials in tree canopies and elevated shelters demonstrated that, although both provide protection from direct solar radiation, they produce distinct thermal experiences: vegetation more effectively lowers radiant temperature, while elevated covers better preserve airflow.

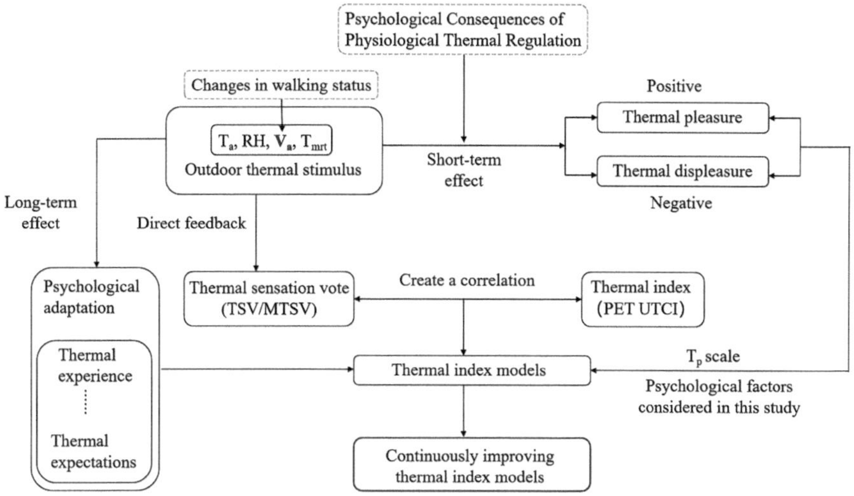


**Fig. 11** A future research framework for improving thermal index models

The findings highlight that thermal pleasure is not determined solely by measurable thermal parameters but emerges from the interaction of external stimuli, bodily heat regulation, and psychological anticipation. Walking intensity further modifies this process, with slower movement under trees and faster movement under overhead covers yielding higher pleasure levels. Conversely, the lowest comfort consistently occurred at moderate walking speed.

Importantly, by integrating thermal pleasure into assessment frameworks, traditional indices can be enhanced to more accurately capture the lived experience of pedestrians. This psychophysiological perspective moves outdoor thermal comfort research beyond physical parameters alone, underscoring the need to incorporate both physiological and emotional dimensions. Such an approach provides a richer foundation for urban planning, enabling the design of outdoor environments that are not only thermally comfortable but also pleasurable and inviting.

**Acknowledgements** The work described in this paper is supported by the National Natural Science Foundation of China (Project No. 52278097) and a Theme-based Research Scheme Grant from the Research Grants Council of the Hong Kong Special Administrative Region, China (Project No. T22-504/21-R).

## References

1. P.A. Coventry, J.E. Brown, J. Pervin, S. Brabyn, R. Pateman, J. Breedvelt, S. Gilbody, R. Stancliffe, R. McEachan, P.L. White, Nature-based outdoor activities for mental and physical health: systematic review and meta-analysis. SSM-Popul. Health **16**, 100934 (2021)

2. M.P. White, I. Alcock, J. Grellier, B.W. Wheeler, T. Hartig, S.L. Warber, A. Bone, M.H. Depledge, L.E. Fleming, Spending at least 120 minutes a week in nature is associated with good health and wellbeing. Sci. Rep. **9**(1), 1–11 (2019)
3. W. Fei, S. Zhao, Urban land expansion in China's six megacities from 1978 to 2015. Sci. Total Environ. **664**, 60–71 (2019)
4. M. Taleghani, Outdoor thermal comfort by different heat mitigation strategies-a review. Renew. Sustain. Energy Rev. **81**, 2011–2018 (2018)
5. Y. Kim, H. Kim, S.-D. Shin, Y.-C. Hong, Different influence of outdoor temperature on traumatic and nontraumatic injuries. J. Trauma Acute Care Surg. **73**(4), 944–949 (2012)
6. F. Brocherie, O. Girard, G.P. Millet, Emerging environmental and weather challenges in outdoor sports. Climate **3**(3), 492–521 (2015)
7. I.-M. Lee, E.J. Shiroma, F. Lobelo, P. Puska, S.N. Blair, P.T. Katzmarzyk, Effect of physical inactivity on major non-communicable diseases worldwide: an analysis of burden of disease and life expectancy. The Lancet **380**(9838), 219–229 (2012)
8. J. Fahed, E. Kinab, S. Ginestet, L. Adolphe, Impact of urban heat island mitigation measures on microclimate and pedestrian comfort in a dense urban district of Lebanon. Sustain. Cities Soc. **61**, 102375 (2020)
9. V.Q. Doan, H. Kusaka, T.M. Nguyen, Roles of past, present, and future land use and anthropogenic heat release changes on urban heat island effects in Hanoi, Vietnam: numerical experiments with a regional climate model. Sustain. Cities Soc. **47**, 101479 (2019)
10. H. Akbari, D. Kolokotsa, Three decades of urban heat islands and mitigation technologies research. Energy Build. **133**, 834–842 (2016)
11. H.H. Yuchun Zhang, K. Dai, Zhang Lin, Z. Fang, Z. Zheng, Thermal responses of face-masked pedestrians during summer: an outdoor investigation under tree-shaded areas. Build. Environ. **233** 110058 (2023)
12. H. Zhang, M. Tang, W. Zhao, Investigation of the urban walking space sun-shading, in *Proceedings of the National Conference on Building Environment and Building Energy Conservation* (2007)
13. M. Yıldırım, Shading in the outdoor environments of climate-friendly hot and dry historical streets: the passageways of Sanliurfa, Turkey. Environ. Impact Assess. Rev. **80**, 106318 (2020)
14. S. Mazouz, M. Zerouala, Shading as a modulator for the design of urban layouts based on vernacular experiences. Energy Build. **29**(1), 11–15 (1998)
15. D. Shi, J. Song, J. Huang, C. Zhuang, R. Guo, Y. Gao, Synergistic cooling effects (SCEs) of urban green-blue spaces on local thermal environment: a case study in Chongqing, China. Sustain. Cities Soc. **55**, 102065 (2020)
16. Y. Zhang, D. Shi, R. Guo, C. Zhuang, Y. Gao, K. Zhao, Single image modeling (SIM) for predicting the temperature and airflows of outdoor air zones in regional planning. Sustain. Cities Soc. **53**, 101934 (2020)
17. Y. Du, C.M. Mak, Y. Li, A multi-stage optimization of pedestrian level wind environment and thermal comfort with lift-up design in ideal urban canyons. Sustain. Cities Soc. **46**, 101424 (2019)
18. Y. Xie, J. Liu, T. Huang, J. Li, J. Niu, C.M. Mak, T.-C. Lee, Outdoor thermal sensation and logistic regression analysis of comfort range of meteorological parameters in Hong Kong. Build. Environ. **155**, 175–186 (2019)
19. X. Zhang, K.-T. Tse, A.U. Weerasuriya, S. Li, K.C. Kwok, C.M. Mak, J. Niu, Z. Lin, Evaluation of pedestrian wind comfort near 'lift-up' buildings with different aspect ratios and central core modifications. Build. Environ. **124**, 245–257 (2017)
20. X. Zhang, A. Weerasuriya, B. Lu, K.T. Tse, C.H. Liu, Y. Tamura, *Pedestrian-Level Wind Environment Near a Super-Tall Building with Unconventional Configurations in a Regular Urban Area* (Springer, Building Simulation, 2020), pp.439–456
21. ASHRAE, *ASHRAE Handbook: Fundamentals* (2021)
22. ISO, ISO 7730, *Ergonomics of the Thermal Environment—Analytical Determination and Interpretation of Thermal Comfort Using Calculation of the PMV and PPD Indices and Local Thermal Comfort Criteria* (2005)

23. M. Cabanac, Physiological role of pleasure: a stimulus can feel pleasant or unpleasant depending upon its usefulness as determined by internal signals. Science **173**(4002), 1103–1107 (1971)
24. R. De Dear, Revisiting an old hypothesis of human thermal perception: alliesthesia. Build. Res. Inf. **39**(2), 108–117 (2011)
25. Y. Zhang, X. Zhou, Z. Zheng, M.O. Oladokun, Z. Fang, Experimental investigation into the effects of different metabolic rates of body movement on thermal comfort. Build. Environ. **168**, 106489 (2020)
26. S. Liu, N. Nazarian, J. Niu, M.A. Hart, R. de Dear, From thermal sensation to thermal affect: a multi-dimensional semantic space to assess outdoor thermal comfort. Build. Environ. **182**, 107112 (2020)
27. T. Huang, J. Li, Y. Xie, J. Niu, C. Mak, Outdoor thermal comfort study in the underneath-elevated-building (UEB) area: on-site measurements and surveys in Hong Kong. Build. Environ. **125**, 502–514 (2017)
28. P. Höppe, The physiological equivalent temperature–a universal index for the biometeorological assessment of the thermal environment. Int. J. Biometeorol. **43**, 71–75 (1999)
29. D. Fiala, G. Havenith, P. Bröde, B. Kampmann, G. Jendritzky, UTCI-Fiala multi-node model of human heat transfer and temperature regulation. Int. J. Biometeorol. **56**(3), 429–441 (2012)
30. G. Jendritzky, R. de Dear, G. Havenith, UTCI—why another thermal index? Int. J. Biometeorol. **56**, 421–428 (2012)
31. S. Manavvi, E. Rajasekar, Semantics of outdoor thermal comfort in religious squares of composite climate: New Delhi, India. Int. J. Biometeorol. **64**, 253–264 (2020)
32. Y. Zhang, J. Liu, Z. Zheng, Z. Fang, X. Zhang, Y. Gao, Y. Xie, Analysis of thermal comfort during movement in a semi-open transition space. Energy Build. **225**, 110312 (2020)
33. T. Tang, Y. Zhang, Z. Zheng, X. Zhou, Z. Fang, W. Liu, Detailed thermal indicators analysis based on outdoor thermal comfort indices in construction sites in South China. Build. Environ. **205**, 108191 (2021)
34. Z. Fang, X. Feng, J. Liu, Z. Lin, C.M. Mak, J. Niu, K.-T. Tse, X. Xu, Investigation into the differences among several outdoor thermal comfort indices against field survey in subtropics. Sustain. Cities Soc. **44**, 676–690 (2019)
35. J. Pickup, R. de Dear, An outdoor thermal comfort index (OUT_SET*)-part I-the model and its assumptions, biometeorology and urban climatology at the turn of the millenium, in *Selected papers from the Conference ICB-ICUC* (2000), pp. 279–283
36. R. De Dear, *An Outdoor Thermal Comfort Index (OUT_SET*)-Part II-Applications, Biometeorology and Urban Climatology at the Turn of the Millennium, Selected Papers from the ICB-ICUC'99 Conference* (World Meteorological Organization, Sydney, 2000), pp.284–290
37. G.M. Budd, Wet-bulb globe temperature (WBGT)—its history and its limitations. J. Sci. Med. Sport **11**(1), 20–32 (2008)
38. K. Blazejczyk, Y. Epstein, G. Jendritzky, H. Staiger, B. Tinz, Comparison of UTCI to selected thermal indices. Int. J. Biometeorol. **56**(3), 515–535 (2012)
39. ISO, ISO 7726, *Ergonomics of the Thermal Environment-Instruments for Measuring Physical Quantities* (ISO, 1998)
40. ASHRAE, *ASHRAE Standard 55, Thermal Environmental Conditions for Human Occupancy* (ANSI/ASHRAE Standard 55–2021 Atlanta, ASHRAE, 2021)
41. F. Faul, E. Erdfelder, A. Buchner, A.-G. Lang, Statistical power analyses using G* Power 3.1: tests for correlation and regression analyses. Behav. Res. Methods **41**(4), 1149–1160 (2009)
42. A. Matzarakis, F. Rutz, H. Mayer, Modelling radiation fluxes in simple and complex environments—application of the RayMan model. Int. J. Biometeorol. **51**(4), 323–334 (2007)
43. P. Weihs, H. Staiger, B. Tinz, E. Batchvarova, H. Rieder, L. Vuilleumier, M. Maturilli, G. Jendritzky, The uncertainty of UTCI due to uncertainties in the determination of radiation fluxes derived from measured and observed meteorological data. Int. J. Biometeorol. **56**(3), 537–555 (2012)
44. X. Zhou, Q. Ouyang, G. Lin, Y. Zhu, Impact of dynamic airflow on human thermal response. Indoor Air **16**(5) (2006)

45. C. Spearman, *The Proof and Measurement of Association Between Two Things* (1961)
46. Y. Zhang, R. Zhao, Overall thermal sensation, acceptability and comfort. Build. Environ. **43**(1), 44–50 (2008)
47. Y. Wang, R. de Groot, F. Bakker, H. Wörtche, R. Leemans, Thermal comfort in urban green spaces: a survey on a Dutch university campus. Int. J. Biometeorol. **61**(1), 87–101 (2017)
48. O. Potchter, P. Cohen, T.-P. Lin, A. Matzarakis, Outdoor human thermal perception in various climates: a comprehensive review of approaches, methods and quantification. Sci. Total Environ. **631**, 390–406 (2018)
49. J. Huang, C. Zhou, Y. Zhuo, L. Xu, Y. Jiang, Outdoor thermal environments and activities in open space: an experiment study in humid subtropical climates. Build. Environ. **103**, 238–249 (2016)
50. M.H. Elnabawi, N. Hamza, S. Dudek, Thermal perception of outdoor urban spaces in the hot arid region of Cairo, Egypt. Sustain. Cities Soc. **22**, 136–145 (2016)
51. F. Salata, I. Golasi, R. de Lieto Vollaro, A. de Lieto Vollaro, Outdoor thermal comfort in the Mediterranean area. A transversal study in Rome, Italy. Build. Environ. **96**, 46–61 (2016)
52. A. Matzarakis, F. Rutz, H. Mayer, Modelling radiation fluxes in simple and complex environments: basics of the RayMan model. Int. J. Biometeorol. **54**(2), 131–139 (2010)
53. A. Middel, N. Selover, B. Hagen, N. Chhetri, Impact of shade on outdoor thermal comfort—a seasonal field study in Tempe, Arizona. Int. J. Biometeorol. **60**, 1849–1861 (2016)
54. M. Schweiker, M. André, F. Al-Atrash, H. Al-Khatri, R.R. Alprianti, H. Alsaad, R. Amin, E. Ampatzi, A.Y. Arsano, E. Azar, Evaluating assumptions of scales for subjective assessment of thermal environments–do laypersons perceive them the way, we researchers believe? Energy Build. **211**, 109761 (2020)
55. Y. Dzyuban, D.M. Hondula, J.K. Vanos, A. Middel, P.J. Coseo, E.R. Kuras, C.L. Redman, Evidence of alliesthesia during a neighborhood thermal walk in a hot and dry city. Sci. Total Environ. **834**, 155294 (2022)
56. S. Thorsson, T. Honjo, F. Lindberg, I. Eliasson, E.-M. Lim, Thermal comfort and outdoor activity in Japanese urban public places. Environ. Behav. **39**(5), 660–684 (2007)
57. I. Knez, S. Thorsson, Thermal, emotional and perceptual evaluations of a park: cross-cultural and environmental attitude comparisons. Build. Environ. **43**(9), 1483–1490 (2008)
58. T. Zhang, M. Su, B. Hong, C. Wang, K. Li, Interaction of emotional regulation and outdoor thermal perception: a pilot study in a cold region of China. Build. Environ. **198**, 107870 (2021)
59. I. Griffiths, J. Huber, A. Baillie, Integrating the environment, in: *Proc. Eur. Conf. in Arch* (1987)
60. M. Nikolopoulou, K. Steemers, Thermal comfort and psychological adaptation as a guide for designing urban spaces. Energy Build. **35**(1), 95–101 (2003)
61. C.K.C. Lam, J. Hang, D. Zhang, Q. Wang, M. Ren, C. Huang, Effects of short-term physiological and psychological adaptation on summer thermal comfort of outdoor exercising people in China. Build. Environ. **198**, 107877 (2021)
62. Z. Fang, Y. Zhang, Y. Yang, X. Zhou, Z. Zheng, Experimental investigation of standard effective temperature (SET*) adapted for human walking in an indoor and transitional thermal environment. Sci. Total Environ. **793**, 148421 (2021)
63. T. Parkinson, R. De Dear, Thermal pleasure in built environments: physiology of alliesthesia. Build. Res. Inf. **43**(3), 288–301 (2015)
64. C.K.C. Lam, K.K.-L. Lau, Effect of long-term acclimatization on summer thermal comfort in outdoor spaces: a comparative study between Melbourne and Hong Kong. Int. J. Biometeorol. **62**, 1311–1324 (2018)
65. C.K.C. Lam, Y. Gao, H. Yang, T. Chen, Y. Zhang, C. Ou, J. Hang, Interactive effect between long-term and short-term thermal history on outdoor thermal comfort: Comparison between Guangzhou, Zhuhai and Melbourne. Sci. Total Environ. **760**, 144141 (2021)

# Modification Calibration Methods of Outdoor Thermal Comfort Model

# Exploring the Applicability of the Classic PMV Model in Outdoor Thermal Environments and Proposing Correction Suggestions

**Abstract** Outdoor thermal environments strongly influence human energy balance and subjective comfort, requiring robust indices for reliable evaluation. The Predicted Mean Vote (PMV) model, though widely used in outdoor studies, was originally derived from controlled indoor experiments, raising concerns about its applicability in exterior contexts. This chapter examines the performance of PMV through field investigations conducted in Guangzhou during the summer. Results demonstrate a clear linear relationship between PMV and operative temperature, while the corresponding link between Thermal Sensation Votes (TSV) and operative temperature is less consistent. TSV increases steadily within the range of 28–36 °C, but tends to plateau at higher operative temperatures, remaining within 2.5–4.0 on the thermal sensation scale. In moderate conditions (28–34 °C), PMV aligns closely with TSV, whereas at elevated temperatures, PMV consistently overestimates thermal sensation. Based on statistical analysis, an adaptive correction model was developed, reducing discrepancies and offering improved reliability for outdoor thermal comfort evaluation. These findings provide both empirical validation and methodological refinement, contributing to the advancement of outdoor comfort assessment tools.

## 1 Introduction

Within industrialized regions, an average individual passes over 90% of their day inside built structures [1, 2], contributing to a considerable growth in energy use related to indoor environmental regulation. In recent years, increased attention has been directed toward encouraging the public to engage more frequently in outdoor activities. However, in these open settings, individuals come into direct contact with natural surroundings wherein the interplay of atmospheric heat, air movement, humidity, and solar radiation influences the human thermoregulatory system, characterized through equivalent temperature, which profoundly shapes thermal comfort. To cultivate favorable and health-supportive microclimatic conditions, urban planning—as a pivotal task—should embrace multiple considerations covering tangible, ecological, financial, and communal facets [3, 4]. Strategic planning of exterior

 
Z. Fang et al., *Outdoor Thermal Comfort and Heat Stress Mitigation*, Environmental Science and Engineering, https://doi.org/10.1007/978-981-95-8899-2_8

zones can not only improve metropolitan habitability but also diminish the energy required for interior heating and air conditioning by fostering increased outdoor occupancy. Consequently, it is imperative to determine fitting outdoor thermal comfort frameworks for appraising open environmental areas.

Over recent years, multiple thermoregulatory models have been proposed to enhance predictive accuracy regarding human thermal responses [5]. One of the most widely utilized thermal metrics is the Predicted Mean Vote (PMV), which is based on the human body's heat balance mechanism. This model is designed to predict general thermal perception and discomfort under moderate climatic conditions, though its application is advised only for PMV values within the range of $-2$ to $+2$ and when critical variables remain within specified bounds—such as ambient air temperature falling between 10 and 30 °C [6]. Employing the 7-point ASHRAE thermal sensation scale, PMV is widely endorsed and applied in studies of indoor environments [7]. It has also been regularly adopted in analyses of outdoor thermal comfort [8–10]. However, given that the PMV model was originally derived from indoor chamber studies, a comprehensive comparative evaluation is required to assess its validity for evaluating outdoor thermal conditions in southern China.

This study aimed to elucidate the divergences between PMV and MTSV. A field survey was first conducted in Guangzhou during summer months. Several key outdoor thermal parameters were recorded concurrently with a thermal comfort questionnaire. Subsequently, PMV values were calculated using the gathered field measurements. Finally, discrepancies between PMV and MTSV were analyzed, and revisions to the PMV model were proposed to improve its applicability to open outdoor settings.

## 2 Methods

### *2.1 Location of Study*

This investigation took place at Guangzhou University, situated in the southern part of China within the longitudinal range of 112.8–114.2 °E and latitudinal bounds of 22.3–24.1 °N. Meteorological records spanning 2010–2015 [11] indicate that the mean radiant temperature ($T_{mrt}$) attained 28.4 °C during July. Relative humidity (RH) levels averaged around 83% in summer months and approximately 70% throughout winter [12]. From a climatic perspective, Guangzhou exemplifies a typical subtropical urban environment, marked by sustained high thermal conditions, elevated moisture content in the atmosphere, and considerable rainfall in summer.

To assess human subjective perceptions and analyze the impacts of microclimatic parameters, multiple sites representing varied environmental settings—incorporating regional climate patterns, terrain characteristics, and built form—were selected. Within these areas, locations featuring different microclimate conditions including sheltered, exposed, landscaped, and hard-surfaced spaces were identified and

**Fig. 1** Sites of survey

employed. The investigated environments encompassed public squares, institutional building groups, and an expansive open plaza, as depicted in Fig. 1. The methodology incorporated two essential elements: systematic measurement of exterior thermal parameters and implementation of thermal perception questionnaires.

## 2.2 *Protocol of Field Data*

A set of on-site surveys and instrumental measurements were performed from July through September. Each day's investigation and monitoring covered a 9-h period starting at 8:30 and concluding at 18:30. Three thermal parameters—air temperature ($T_a$), relative humidity (RH), and globe temperature ($T_g$)—were continually measured with automatic recording at intervals of one minute. Wind speed ($V_a$) was regularly captured and documented every five minutes. While microclimate data were being gathered, participants in proximity to the measurement locations were randomly invited to respond to the questionnaire. In total, 323 validated questionnaires were collected (Table 1).

The operative temperature ($T_{op}$), which takes the impact of air temperature, mean radiation temperature and air velocity on thermal comfort into consideration, is calculated by the following equation [6, 13]:

**Table 1** Instruments for micrometeorological measurements

| Name | Parameters | Measurement range | Accuracy |
|---|---|---|---|
| ZDR-20 | $T_a$ | – 40–100 °C | ± 0.5 °C |
| | RH | 0–100% | ± 3% |
| JTR10 | $T_g$ | 5–120 °C | ± 0.2 °C |
| Kanomax model KA22 | $V_a$ | 0–4.99 m/s | ± 2% |

$$T_{op} = AT_a + (1 - A)\,T_{mrt} \tag{1}$$

In Eq. (1), the coefficient (A) only depends on $V_a$, $T_{mrt}$, which considers both short-wave and long-wave radiation and represents the weighted average temperature of an imaginary enclosure that gives the same radiation as the complex urban environment, has a strong influence on human thermal comfort [14]. $T_{mrt}$ was calculated according to ISO 7726 [15] for forced convection from the measured $T_g$, $V_a$, $T_a$ and globe emissivity ($\varepsilon_g$, assumed to be 0.95) and diameter (D, approximately 110 mm):

$$T_{mrt} = \left\{ \left(T_g + 273\right)^4 + \left[ \frac{(1.1 \times 10^8 \times V_a^{0.6}}{\varepsilon_g \times D^{0.4}} \right] \times \left(T_g - T_a\right) \right\}^{\frac{1}{4}} - 273 \tag{2}$$

## 2.3 *Survey Questionnaire*

During this investigation, respondents were invited to complete survey forms concurrently with physical data acquisition. All subjects participated on a voluntary basis. The questionnaire was structured into three segments. The opening segment gathered basic demographic data such as age, gender, stature, and mass. The subsequent segment captured information on thermal adaptive measures, covering historical thermal experiences, forms of exertion, and attire selection. The final segment examined subjects' instantaneous thermal comfort perceptions. The questionnaire framework adhered to thermal environment specifications outlined in ASHRAE Standard 55 [13] and ISO 7730 [6]. Multiple earlier studies have utilized a 9-point thermal perception scale for interior environment evaluation (– 4: very cold, – 3: cold, – 2: cool, – 1: slightly cool, 0: neutral, 1: slightly warm, 2: warm, 3: hot, 4: very hot) [16]. Given the anticipated prevalence of hot and humid conditions during data collection, the current investigation integrated this 9-point scaling system into its survey instrument.

# 3 Results and Discussion

## *3.1 PMV and Thermal Sensation*

The Predicted Mean Vote (PMV), acknowledged as a leading thermal comfort index, was first developed to estimate indoor thermal sensation. It has sometimes been extended to open-air thermal assessments. As displayed in Fig. 1, the relationships between PMV, mean thermal sensation vote (MTSV), and operative temperature ($T_{op}$) are presented. A strong linear correlation between PMV and $T_{op}$ was identified, aligning with previous research findings [13]. When PMV is used to assess indoor thermal environments, the following constraints must be observed: PMV must range from − 2 to + 2, and air temperature ($T_a$) should not exceed 30 °C. Figure 2 indicates that PMV predictions remained reasonably accurate at $T_{op}$ values below 34 °C. However, the divergence between PMV and MTSV intensified with rising $T_{op}$, especially under high $T_a$ scenarios outdoors. Furthermore, the association of MTSV with $T_{op}$ was ambiguous. Although MTSV increased together with $T_{op}$ below 34 °C, beyond this threshold PMV consistently overpredicted thermal sensation.

Under high $T_{op}$ conditions, MTSV converged around 3.0. Most subjects indicated hot or very hot sensations once $T_{op}$ surpassed 34 °C. Thus, if PMV is applied to assess outdoor thermal conditions, adaptation of the model appears essential.

Figure 3 illustrates the regression relationship between PMV and MTSV. When MTSV values were below 2.5, they exhibited an increasing trend as PMV rose. However, once PMV exceeded 2.5, MTSV became concentrated within the range

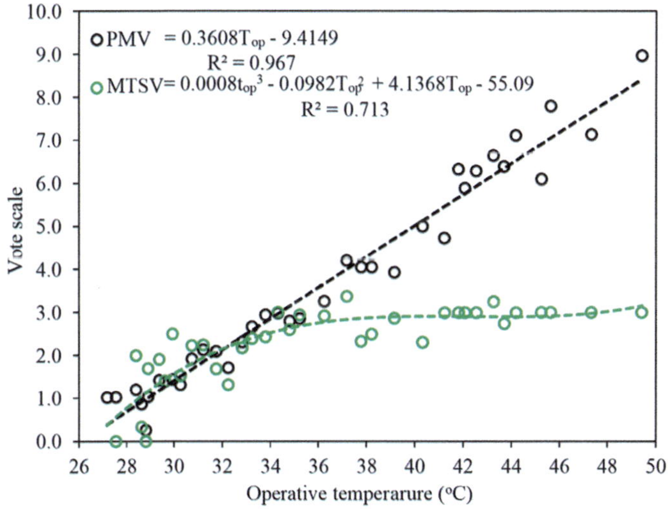


**Fig. 2** Difference between the PMV and MTSV against operative temperature

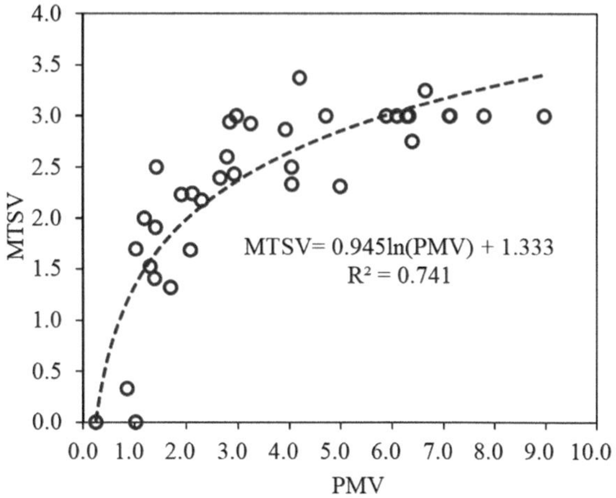


**Fig. 3** Regression relationship between the MTSV and PMV

of 2.5–3.5. MTSV showed predictability through PMV, with an $R^2$ value of 0.741 indicating a statistically significant correlation.

# 4 Conclusion

This chapter explored the applicability of the classic PMV model for outdoor thermal comfort evaluation. Field data collected in Guangzhou demonstrated that PMV maintains a strong correlation with operative temperature and aligns reasonably well with subjective thermal sensation within moderate temperature ranges. However, under high-temperature conditions, PMV consistently overestimates thermal sensation, highlighting its limited validity in outdoor environments.

To overcome these shortcomings, an adaptive correction model was developed and validated with survey data. The refined model improves predictive accuracy, offering a more reliable framework for outdoor comfort assessments. These results emphasize the importance of adjusting classical indoor-based indices for exterior contexts, and provide a basis for future applications in urban climate research and sustainable outdoor design.

**Acknowledgements** The work described in this paper was fully supported by a grant from the Research Grants Council of the Hong Kong Special Administrative Region, China (Project No. C5002-14G).

# References


1. P. Hoppe, Different aspects of assessing indoor and outdoor thermal comfort. Energy Build. **34**, 661–665 (2002)
2. J.A. Leech, R. Burnett, W. Nelson, S.D. Aaron, M. Raizenne, Outdoor air pollution epidemiologic studies. Am. J. Respir. Crit. Care Med. **161**(3), A308 (2003)
3. C. Hass-Klau, Impact of pedestrianisation and traffic calming on retailing: a review of the evidence from Germany. Transp. Policy **1**, 21–31 (1993)
4. A.A. Hakim, H. Petrovitch, C.M. Burchfiel, G.W. Ross, B.L. Rodriguez, L.R. White et al., Effects of walking on mortality among nonsmoking retiredmen. N. Engl. J. Med. **338**, 94–99 (1998)
5. K. Kati, R. Li, W. Zeiler, Thermophysiological models and their applications: a review. Build. Environ. **106**, 286–300 (2016)
6. ISO, Ergonomics of the thermal environment—analytical determination and interpretation of thermal comfort using calculation of the PMV and PPD indices and local thermal comfort criteria, ISO 7730, (2005)
7. L. Schellen, M.G.L.C. Loomans, B.R.M. Kingma, M.H. de Wit, A.J.H. Frijns, W.D. Van Marken Lichtenbelt, The use of a thermophysiological model in the built environment to predict thermal sensation. Build. Environ. **59**, 10–22 (2013)
8. S. Berkovic, A. Yezioro, A. Bitan, Study of thermal comfort in courtyardsin a hot arid climate. Sol. Energy **86**, 1173–1186 (2012)
9. V. Cheng, E. Ng, C. Chan, B. Givoni, Outdoor thermal comfort study insub-tropical climate: a longitudinal study based in Hong Kong. Int. J. Biometeorol. **56**, 43–56 (2012)
10. E. Johansson, S. Thorsson, R. Emmanuel, E. Krüger, Instruments andmethods in outdoor thermal comfort studies-the need for standardization. Urban Clim. **10**, 346–366 (2014)
11. http://tianqi.2345.com/wea_history/59287.htm
12. Y. Zhang, H. Chen, Q. Meng, Thermal comfort in buildings with split air-conditioners in hot-humid area of China. Build. Environ. **64**, 2 (2013)
13. ANSI/ASHRAE. *Standard 55–2013: Thermal Environmental Conditions for Human Occupancy*. Atlanta, Georgia: American Society of Heating Refrigerating and Air-Conditioning Engineers (ASHRAE) (2013)
14. M.W. Yahia, E. Jojansson, Evaluation the behavior of different thermal indices by investigating various outdoor urban environments in the hot dry city of damascus, Syria. Int. J. Biometeorol. **57**, 615–630 (2013)
15. ISO 7726 Ergonomics of the thermal environment—Instruments for measuring physical quantities (1998)
16. H. Zhang, *Human Thermal Sensation and Comfort in Transient and Non-uniform Thermal Environments*. Ph. D. Thesis, University of California Berkeley (2003)

# Comparative Analysis of Outdoor Thermal Indices and Their Applicability in Evaluating Outdoor Thermal Sensations

**Abstract** Comfortable and healthy outdoor microclimates are essential for advancing sustainable urban development. This chapter presents a comparative evaluation of widely applied outdoor thermal comfort indices, including Predicted Mean Vote (PMV), Wet-Bulb Globe Temperature (WBGT), Physiological Equivalent Temperature (PET), Standard Effective Temperature (SET*), and the Universal Thermal Climate Index (UTCI). Each index differs in its energy balance assumptions, mathematical structure, and applicable boundary conditions, which affects their performance in real-world environments. Field surveys and on-site meteorological measurements conducted at a university campus in Guangzhou, China, were used to assess the indices under hot summer conditions. Results indicate strong correlations between operative temperature and thermal indices such as WBGT, PET, SET*, UTCI, $T_{mrt}$, and PMV. However, when operative temperatures exceeded 34 °C, their predictive relationships with mean thermal sensation vote (MTSV) became less consistent. These findings highlight both the value and the limitations of current indices, and underscore the need for refined classification schemes to better address outdoor heat stress in warm and humid climates.

## 1 Introduction

Outdoor environments play a vital role in enhancing the sustainability and habitability of cities. In developed nations, individuals are increasingly choosing to spend time outside engaging in recreational activities, taking advantage of sunlight and natural ventilation. As a key element influencing the usability and attractiveness of open areas, outdoor thermal comfort is receiving growing interest within urban planning and design [1]. Improved microclimatic conditions offer cities multifaceted advantages—encompassing physical, environmental, economic, and social dimensions—by encouraging greater outdoor activity [2, 3]. This can also contribute to energy conservation by reducing dependence on air conditioning systems [4]. As a result of the pronounced Urban Heat Island (UHI) effect, the energy required for cooling in cities has risen substantially. According to data, the global energy penalty

 
Z. Fang et al., *Outdoor Thermal Comfort and Heat Stress Mitigation*, Environmental Science and Engineering, https://doi.org/10.1007/978-981-95-8899-2_9

per unit surface area per degree of UHI intensity was estimated at 0.74 kWh/($m^2$/K), and the average total energy demand for heating and cooling in typical buildings grew by 11% from 1970 to 2010 [5]. Furthermore, the impact of thermal conditions on public health requires careful consideration. Research such as Salata et al. [6] indicated a correlation between elevated air temperatures and higher mortality rates. Hence, it is essential for designers and urban planners to account for microclimatic variability when creating comfortable outdoor built environments.

In recent decades, numerous studies have been conducted to evaluate outdoor thermal conditions. Multiple thermal comfort indices, founded on human energy balance principles and heat exchange mechanisms, have been established to quantify thermal perception. These indices differ significantly—ranging from steady-state to dynamic modeling approaches, and from single-node, two-node, to multi-node and multi-segment systems [7]. The single-node model, for instance, relies solely on the heat balance equation, accounting for internal metabolic production and losses through skin and respiration, with heat transferred via clothing to the surroundings. It incorporates six fundamental parameters: air temperature ($t_a$), mean radiant temperature ($T_{mrt}$), relative humidity (RH), air velocity ($v_a$), clothing insulation ($I_{clo}$), and metabolic rate (Met). In contrast, the two-node model accounts for thermal resistance from clothing and differentiates between the core and skin layers, incorporating skin and core temperatures as central variables. More complex multi-node models—used in dynamic simulations—segment the body into multiple compartments and nodes, integrating skin temperature, core temperature, and the rate of change in skin temperature to predict comfort responses.

A frequently used index in outdoor settings, often applied without extensive calibration, is the Predicted Mean Vote (PMV) [8], originally devised for indoor environments [9]. PMV employs a seven-point scale to gauge the average thermal sensation of people and is included in the ISO standard 7730 [10]. It has been utilized in several earlier outdoor thermal comfort studies [11, 12].

The Physiological Equivalent Temperature (PET) is derived from the Munich Energy-balance Model for Individuals (MEMI) [13]. PET is employed to represent the air temperature under standard indoor settings at which an individual's thermal sensation matches that experienced in a more complex outdoor environment [14]. This index incorporates independent physiological variables—such as height, age, and activity level—into the calculation of metabolic and sweat rates. Matzarakis et al. [15] established PET thresholds for varying levels of thermal perception across Europe, while Lin and Matzarakis [16] later identified distinct PET neutral temperatures and perceptual ranges in subtropical climates, underscoring the influence of climatic adaptation on human thermal response. The index has been widely applied in assessing outdoor thermal conditions across diverse climates [17–21].

The Wet Bulb Globe Temperature (WBGT) is among the most widely used thermal indices, introduced over five decades ago [22]. It was initially adopted by the U.S. Army and Marine Corps for evaluating training conditions and has since been utilized to assess outdoor thermal environments [23].

The Standard Effective Temperature (SET*) simplifies complex radiant conditions into a standardized setting based on the Pierce two-node model [24]. This metric represents the human thermoregulatory system as two isothermal compartments—skin and core—enabling the calculation of core, skin, and mean body temperatures based on deviations from set points. SET* has been employed in numerous empirical studies on outdoor thermal comfort [25–27].

A more recent development in outdoor thermal comfort modeling is the Universal Thermal Climate Index (UTCI), which employs a multi-node thermodynamic approach and is increasingly adopted in biometeorological research [28]. Building upon Fiala's multi-node physiological model [29, 30] and integrating the clothing model by Havenith et al. [31], UTCI adopts an equivalent temperature concept—similar to SET*—to characterize outdoor thermal conditions [28]. The index accounts for the collective effects of air temperature, wind, solar radiation (direct, diffuse, and reflected), long-wave radiation, and humidity, and computes the corresponding temperature in a reference environment that would elicit the same physiological strain as the actual outdoor setting.

In recent years, a team of scientists at UC-Berkeley developed a computational model of the human thermoregulatory system to quantify comfort conditions, drawing on a range of empirical measurements. This model was introduced to evaluate complex thermal settings while accounting for personal physiological variations. It also elaborated on the connections between local thermal sensation and local comfort of specific body segments, as well as overall sensation and comfort under uniform, non-uniform, and transient conditions [32]. Nonetheless, its application has not extended to evaluating and forecasting outdoor thermal comfort.

Much of the early research on human thermal comfort was carried out in indoor settings [33], despite notable disparities between indoor and outdoor thermal environments. Outdoor conditions typically involve fluctuations in solar radiation and shade, along with variations in air temperature, wind speed, and direction. Moreover, metabolic rates and clothing insulation differ significantly from those indoors. In addition, studies on thermal comfort have shown that an approach based solely on physiology is insufficient for describing outdoor comfort conditions. Thermal adaptation—encompassing behavioral adaptations (personal, environmental, technological, or cultural), physiological elements (genetic or acclimatization factors), and psychological aspects (such as habituation or expectation) [33]—should also be taken into account [34]. At the same time, relatively limited research has been done on outdoor thermal comfort in subtropical China, and comparative analyses of adaptations among thermal comfort indices remain scarce.

Based on the above, most thermal comfort indices were established in temperate regions using experimental data gathered under specific climatic conditions. Consequently, discrepancies may arise when comparing outcomes from thermal comfort studies in southern China with the predictions of these indices. In this research, several thermal comfort indicators—including PMV, SET*, WBGT, UTCI, and PET—are compared with data obtained from field surveys. The aim of this work is to examine the adaptability of various thermal comfort indices for assessing outdoor thermal conditions in built environments within China's subtropical region.

# 2 Methodology

## *2.1 Field Measurement and Survey Sites*

This investigation was conducted on the campus of Guangzhou University, situated in southern China within the geographical coordinates of 112.8–114.2 °E longitude and 22.3–24.1 °N latitude. According to meteorological records from 2010 to 2015 [35], the average air temperature ($T_a$) in July reached 28.4 °C. Relative humidity (RH) averaged approximately 83% during summer and 70% in winter [36] (Fig. 1). Climatically, Guangzhou is characterized as a typical subtropical urban area, featuring consistently high temperatures, elevated humidity levels, and substantial summer precipitation. Guangzhou University is one of ten institutions forming the Guangzhou University City cluster, located on Xiaoguwei Island in Panyu District, in the south-central part of Guangzhou.

To capture human subjective responses for assessing microclimatic influences, several sites representing varied microenvironments—such as shaded, unshaded, grassy, and concrete-paved surfaces—were selected. These areas encompassed public squares, academic building complexes, and an extensive open ground floor, as illustrated in Fig. 2. The research comprised two main components: monitoring parameters of the outdoor thermal environment and administering thermal comfort questionnaires.

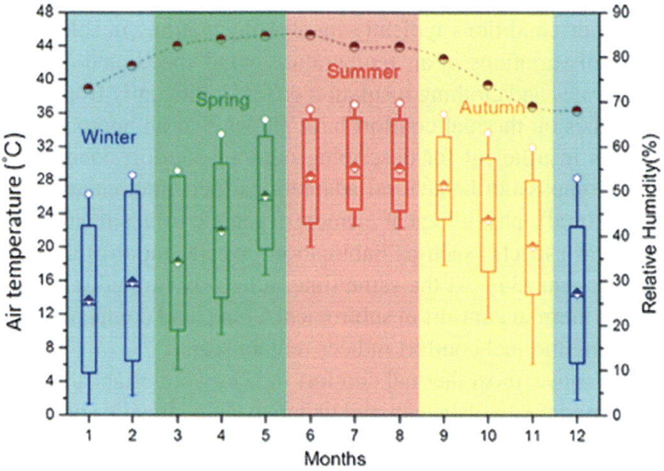


**Fig. 1** Monthly variation of air temperature and relative humidity

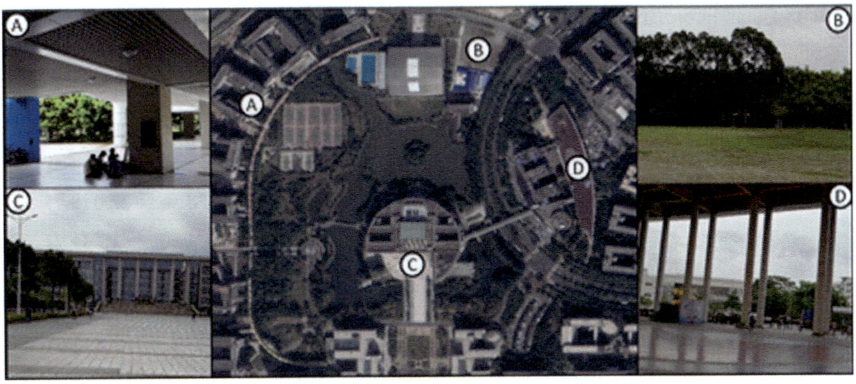


**Fig. 2** Sites of survey, **a** open area underneath the elevated building, **b** open area (lawn), **c** square, **d** open area (concrete-paved)

## 2.2 *Environmental Measurement and Survey Method*

### 2.2.1 Overall Introduction

A series of field surveys and measurements were conducted from June to July in 2016. The investigations spanned 9 h daily, from 8:30 to 18:30. Three thermal parameters—air temperature ($T_a$), relative humidity (RH), and globe temperature ($T_g$)—were continuously monitored and logged automatically at one-minute intervals. Air velocity ($V_a$) was also continuously measured, with recordings taken every five minutes. During the collection of microclimate data, individuals near the measurement sites were randomly invited to participate in the survey. Each subject first received an explanation of the questionnaire items. After fully understanding all questions, the subject completed the form. A total of 674 questionnaires were gathered in this field campaign. Four distinct locations (A–D), indicated in Fig. 2, were selected for the survey. The numbers of days over which surveys were conducted at these sites were 6, 3, 4, and 6, respectively. Detailed information regarding the survey days is presented in Table 1. The corresponding numbers of participants were 58, 85, 82, and 97. During screening, 30 questionnaires were excluded due to incomplete responses, resulting in 644 valid submissions. Further details are provided in Table 1.

### 2.2.2 Measurement Parameters and Instruments

The primary microclimatic parameters gathered in accordance with ASHRAE Standard 55-2017 comprised air temperature ($T_a$), relative humidity (RH), globe temperature ($T_g$), and air velocity ($V_a$) [37]. Measurements were performed at a height of 0.6 m for seated subjects and at 1.1 m for those standing [38], in close proximity to the participants. $T_a$ and RH were measured using a temperature and humidity

**Table 1** Detail information of field data

| Location | Date | Number of subjects | | Age |
|---|---|---|---|---|
| A | 13, 15, 20, 22, 23, 24 July, 2016 | Male | 54 | 21.8 (1.1) |
| | | Female | 62 | 20.7 (1.6) |
| | | Male + Female | 116 | 21.2 (1.4) |
| B | 15, 20, 22, 29 July, 2016 | Male | 132 | 22.0 (3.1) |
| | | Female | 45 | 21.2 (1.7) |
| | | Male + Female | 177 | 21.9 (2.9) |
| C | 21, 22, 28 July, 2016 | Male | 103 | 21.8 (1.5) |
| | | Female | 52 | 21.5 (2.3) |
| | | Male + Female | 155 | 21.7 (1.7) |
| D | 12, 13, 28, 29, 30 July; 1 June, 2016 | Male | 108 | 21.3 (1.1) |
| | | Female | 91 | 21.0 (1.0) |
| | | Male + Female | 199 | 21.2 (1.0) |
| Total number | | | 644 | |

sensor, model ZDR-20. $T_g$ was obtained with a globe thermometer, model JTR10. The JTR10 device is a standard black globe thermometer with a diameter of 150 mm and an emissivity ($\varepsilon_g$) of 0.95. Its response time is approximately 15 min to reach equilibrium, which satisfies the requirements of this study. In contrast, smaller globe thermometers with a diameter of 38 mm are influenced more by convective heat transfer than radiation, leading to reduced accuracy in mean radiant temperature ($T_{mrt}$) estimation [19]. The performance of the JTR10 complies with ISO 7726 [38]. Wind speed was measured using a Kanomax Model KA22 hot wire anemometer. This instrument offers two measurement ranges: a low range from 0 to 4.99 m/s and a high range from 5 to 50 m/s, with an accuracy of ± 2%. Since wind speeds at the study sites consistently remained below 5 m/s, the low range (0–4.99 m/s) was employed for all recordings. All sensors were calibrated prior to use. A summary of the micrometeorological measurement details, including ranges and accuracies, is provided in Table 2.

The operative temperature ($T_{op}$), with consideration of the effects of air temperature, mean radiation temperature, and air velocity, is calculated by the following

**Table 2** Instruments used in micrometeorological measurements

| Sensor | Meteorological parameters | Measuring range | Accuracy |
|---|---|---|---|
| ZDR-20 | $T_a$ | – 40 to 100 °C | ± 0.5 °C |
| | RH | 0–100% | ± 3% |
| JTR10 | $T_g$ | 5–120 °C | ± 0.2 °C |
| Kanomax model | $V_a$ | 0–4.99 m/s,5 | ± 2% |
| KA22 | | 50 m/s | |

equation [10, 37]:

$$T_{op} = AT_a + (1 - A)T_{mrt} \tag{1}$$

In Eq. (1), A is the mean weight coefficient of the air temperature and mean radiant temperature, which depends only on $V_a$, shown in Table 3.

$T_{mrt}$, which considers both short-wave and long-wave radiation and represents the weighted average temperature of an imaginary enclosure that gives the same radiation as the complex urban environment, has a strong influence on human thermal comfort [39]. According to ISO 7726 [38], $T_{mrt}$ was calculated by Eq. (2). In Eq. (2), $\varepsilon_g$ was assumed to be 0.95, and black globe thermometer diameter (D) was 150 mm.

$$T_{mrt} = \left\{ \left(T_g + 273\right)^4 + \left[ \frac{\left(1.1 \times 10^8 \times V_a^{0.6}\right)}{\left(\varepsilon_g \times D^{0.4}\right)} \right] \times \left(T_g - T_a\right) \right\}^{1/4} - 273 \tag{2}$$

In this investigation, the clothing insulation value referred to the total insulation. According to the ASHRAE Handbook [40], the most precise methods for determining clothing insulation involve measurements using heated mannequins [39, 41] and tests with active human subjects [42]. However, in most engineering applications, directly measuring clothing insulation is often not practical. Instead, insulation values for individual garments commonly worn can be obtained from reference tables in the ASHRAE Handbook. The total insulation of a clothing ensemble can then be estimated from these individual values by applying a summation formula as:

$$I_{cl} = 0.835 \sum_i I_{clu,i} + 0.161 \tag{3}$$

where $I_{clu,i}$ was the effective insulation of garment i, and $I_{cl}$, as before, was the insulation for the entire ensemble. A simpler and nearly accurate summation formula [43] was:

$$I_{cl} = \sum_i I_{clu,i} \tag{4}$$

In this survey, the total clothing insulation was calculated by Eq. (4). The detail of the information of the clothing insulation was collected in the field survey.

**Table 3** Coefficient a indifferent air velocity ranges

| $V_a$ (m/s) | < 0.2 | 0.2–0.6 | > 0.6 |
|---|---|---|---|
| A | 0.5 | 0.6 | 0.7 |

### 2.2.3 Questionnaire

During the on-site environmental data collection, each participant was asked to complete a questionnaire. All respondents were volunteers. The survey was structured into three parts. The first section gathered basic personal details such as age, gender, height, and weight. The second part documented adaptive behaviors related to thermal conditions, including recent thermal history, types of activities performed, and clothing worn. The third segment focused on subjective thermal assessment. All items were reviewed for relevance, and the questionnaire received approval from the University Research Office. The design of the survey adhered to the thermal environment guidelines outlined in ASHRAE Standard 55 [40] and ISO 7730 [10]. The conventional ASHRAE 7-point thermal sensation scale was defined as: − 3: cold, − 2: cool, − 1: slightly cool, 0: neutral, 1: slightly warm, 2: warm, and 3: hot. Only a limited number of studies have employed a 9-point thermal sensation scale to assess indoor environments, which includes: − 4: very cold, − 3: cold, − 2: cool, − 1: slightly cool, 0: neutral, 1: slightly warm, 2: warm, 3: hot, and 4: very hot [33]. Zhang et al. also applied the 9-point scale to investigate thermal comfort in buildings equipped with split air-conditioning systems in hot-humid climates [36]. Considering the potential for extreme heat and humidity during the survey period, the 9-point scale was chosen for use in this study.

Throughout the field survey, microclimate parameters were recorded simultaneously as individuals near the measurement locations were randomly selected to participate. Each respondent received a clear explanation of the questions. Participants were instructed to "indicate your current thermal perception by marking the appropriate box on the scale." Questionnaires were filled out based on their own interpretation of each item.

## 2.3 *Thermal Comfort Assessing Indices*

This study evaluated multiple thermal indices, namely WBGT, PMV, SET*, PET, and UTCI, as summarized in Table 4. These indices exhibit considerable differences in their underlying principles and applications. The PMV index, characterized as a "static" or "constancy" model, was introduced by Fanger in 1970. It is derived from the human body's heat balance and incorporates four environmental parameters—air temperature ($T_a$), air velocity ($V_a$), mean radiant temperature ($T_{mrt}$), and relative humidity (RH)—along with two individual factors: clothing insulation and metabolic rate. PMV was formulated based on thermal sensation feedback from over 1300 participants [8]. In contrast, WBGT, proposed by Yaglou and Minard [22], was initially applied in military contexts by the U.S. Army for evaluating training safety and was later endorsed by the World Health Organization [44]. It is computed using three climatic variables: $T_a$, black globe temperature ($T_g$), and wet bulb temperature ($T_w$), as indicated in Table 4.

**Table 4** Detail information of different thermal comfort indices

| Year | Index | Author(s) | Model (model type) | Description | Addition |
|---|---|---|---|---|---|
| 1957 | WBGT | Yaglou and Minard [22] | Statistics | $WBGT = 0.7T_w + 0.1T_a + 0.2T_g$ (outdoorair)<br>$WBGT = 0.7T_w + 0.3T_g$ (indoorair) | |
| 1970 | PMV | Fanger [8] | Steady-state energy balance (one node model) | $H\text{-}ED\text{-}E_{sw}\text{-}E_{Re}\text{-}L = R + C$ $PMV = (0.028 + 0.3033e^{-0.036\,M}).\ \{(M\text{-}W)\text{-}3.05$ $[5.733\text{–}0.000699(M\text{-}W)\text{-}P_a]\text{-}0.42[(M\text{-}W)\text{-}58.15]\text{-}0.0173\ M(5.867\text{-}P_a)\text{-}0{,}0014\ M(34\text{-}T_a)\text{-}3.96.10\text{-}8f_{cl}[(T_{cl} + 273)4\text{-}(T_{mrt} + 273)4]\text{-}f_{cl}\cdot h_c(T_{cl}-T_a)\}$<br>$T_{cl} = 35.7\text{–}0.28(M\text{-}W)\text{-}0.155_{Icl}[3.9610\text{-}3f_{cl}[(T_{cl} + 273)4\text{-}(T_{mrt} + 273)4]\text{-}\ f_{cl}\cdot h_c(T_{cl}-T_a)]$<br>$h_c = 2.38\ (T_{cl}-T_a)^{0.25}$ for $2.38\ (T_{cl}-T_a)\ 0.25 \geq 12.1\ (v_a)^{0.5}$ or<br>$h_c = 12.1v_{\varepsilon}{}^{0.5}$ for $2.38\ (T_{cl}-T_a)\ 0.25 \leq 12.1\ (v_a)^{0.5}$<br>$f_{cl} = 1.0 + 0.2\ I_{cl}$ for $I_{cl}<0.5$ clo or $f_{cl} = 1.05 + 0.1I_{cl}$ for $I_{cl} > 0.5$ clo | PMV: − 2 to + 2; Metabolic rate: 46–232W/ $m^2$ (0.8–4 met); Clothing thermal resistance: 0 0.310 $m^2$ °C/ W(0–2clo); Ambient air temperature: 10–30 °C; Mean radiant temperature: 10–40 °C; Air velocity: 0–1 m/s |

(continued)

**Table 4** (continued)

| Year | Index | Author(s) | Model (model type) | Description | Addition |
|---|---|---|---|---|---|
| 1973 | SET* | Gagge et al. [24] | Two node model | $S = M\text{-}E\text{-}R\text{-}C\text{-}W$<br>$S_{cr} = M\text{-}W\text{-}(C_{res} + E_{res})\text{-}(t_{cr} - t_{sk}) \times (5.28 + 1.163 \times skbf)$<br>$S_{sk} = (t_{cr} - t_{sk}) \times (5.28 + 1.163 \times skbf)\text{-}(C + R + E_{sk})$<br>$S_{cr} = (1\text{-}\alpha)\ mc_{p,b}(dt_{cr}/d\theta)/AD$<br>$S_{sk} = \alpha mc_{p,b}(dt_{sk}/d\theta)/AD$<br>$H_{sk} = h_s(t_{sk}\text{-}SET^*) + \omega h_s,\ e(p_{s,sk} - 0.5p_{SET^*})$ | Ambient air temperature equals mean radiation temperature; relative humidity: 50%; Air velocity:0.15 m/s; Clothing thermal resistance: 0.6clo; Metabolic rate: 1.0met; The same mean skin temperature and shin wittedness as the person in the actual complex environment |

(continued)

**Table 4** (continued)

| Year | Index | Author(s) | Model (model type) | Description | Addition |
|---|---|---|---|---|---|
| 1999 | PET | Höppe [13, 14] | Munich energy balance model of individuals (Two–node model) | $H\text{-}C\text{-}R_N\text{-}E_D\text{-}E_{SW}\text{-}E_{Re} = S = C_k \cdot GE.(dT_M/dt)$ $F_{cs} = v_b \cdot \rho_b \cdot c_b \cdot A_{Du} \cdot (T_c - T_{sk})$ $F_{sc} = (A_{Be}/I_{cl})(T_{sk} - T_{cl})$<br>$SW = 8.47 \times 10^{-5}.((0.1T_{sk} + 0.9T_c) - 36.6)\ A_{Du}$<br>$E_{SW} = SW \times r$<br>$M_{shiv} = 19.4 \times (34.0 – T_{sk}) \times (37.0 – T_c)\ A_{Du}$<br>$E_{SW} = h_e\ (VP_a\text{-}SVPT_{sk}).A_{Du}$ | Ambient air temperature equals mean radiation temperature; Air temperature: 20 °C<br>Air velocity: 0.1 m/s;<br>The vapor pressure of the ambient air: 12 hPa<br>Relative humidity: 50%;<br>Metabolism rate: 80 W;<br>Clothing thermal resistance: 0.9clo; |

(continued)

**Table 4** (continued)

| Year | Index | Author(s) | Model (model type) | Description | Addition |
|---|---|---|---|---|---|
| 2001 | UTCI | Jendritzky et al. [28, 45] | UTCI-Fiala model (multi-node) | $UTCI = f(T_a; T_{mrt}; v_a; p_a) = T_a + Offset(T_a; T_{mrt}; v_a; p_a)$ | Ambient air temperature equals mean radiation temperature; Metabolic rate: 135 $W/m^2$; Walking speed of1.1 m/s Air velocity: 0.5 m/s (10/m above ground). Relative humidity: 50% |

A key distinction between WBGT and other indices lies in its exclusion of human energy balance, relying solely on ambient conditions. Both SET* and PET are grounded in human energy equilibrium, employing respectively the two-node model [24] and the Munich Energy Balance Model for Individuals [13]. These indices diverge primarily in their computation of physiological sweat rate and surface heat transfer (Table 4). SET* assumes a reference person with specific mean skin temperature and skin wettedness under actual conditions, adopts a default RH of 50%, and uses a vapor pressure fixed at 1.2 kPa in PET. The reference air velocity in SET* is set to 0.15 m/s, slightly above PET's 0.1 m/s. Further variations exist in default garment insulation and metabolic rate: SET* uses 0.6 clo and 1.0 met, whereas PET applies 0.9 clo and 80 W, respectively. Comprehensive details are provided in Table 4.

UTCI, representing an equivalent temperature, is founded on a multi-node thermoregulatory model [28]. In contrast to two-node models, multi-node frameworks simulate the human body with greater anatomical precision, predicting both systemic and localized physiological reactions. The original 19-compartment, 342-node structure was streamlined into a symmetric 12-compartment, 187-node system by merging bilateral and sectional body segments into consolidated units [44]. Key parameters encompass mean skin temperature ($T_{sk}$), core temperature ($T_c$), and various modes of heat exchange. UTCI's reference conditions differ from those of SET* and PET, incorporating a 10-m height wind speed of 0.5 m/s, equivalent mean radiant and air temperatures, and 50% humidity. Moreover, it employs a metabolic rate of 2.3 met (135 W), exceeding values used in other indices [45]. A full summary of these thermal comfort metrics is presented in Table 4, while the influences of diverse parameters across indices are compared in Table 5.

In this study, the WBGT was determined using the formula provided in Table 4. To account for adjustments in PMV and SET*, these indices were computed with the Center for the Built Environment (CBE) Thermal Comfort Tool for ASHRAE-55 (http://comfort.cbe.berkeley.edu/). The RayMan software [46], a widely-used application, was employed to estimate PET. UTCI equivalent temperature values were

**Table 5** The parameters' effect on different thermal comfort indices (1999) [8, 13, 22, 24, 28, 45, 84]

| Thermal comfort indices | $T_a$ | RH/$T_w$ | $V_a$ | $T_{mrt}$/$T_g$ | I | Met | Skin wetting | $T_{sk}$ | $T_c$ |
|---|---|---|---|---|---|---|---|---|---|
| WBGT | √ | √ | | √ | | | | | |
| PMV (one node model) | √ | √ | √ | √ | √ | √ | | | |
| SET* (two node model) | √ | √ | √ | √ | √ | | √ | √ | √ |
| PET (two node model) | √ | √ | √ | √ | √ | √ | | √ | √ |
| UTCI (multi-node model) | √ | √ | √ | √ | √ | √ | | √ | √ |

derived through an operational procedure accessible both as source code and an executable program on the official website (www.utci.org).

To define thermal sensation ranges for various comfort indices—including WBGT, SET*, PET, $T_{mrt}$, and UTCI—within China's humid subtropical region, this research adopted a methodology consistent with earlier studies [4, 6, 16, 47]. The mean thermal sensation vote (MTSV) for respondents within each 0.5 °C interval of the aforementioned indices was computed using Eq. (5).

$$MTSV = \frac{\sum_{n}^{j} TSV}{n} \tag{5}$$

In Eq. (5), n was the number of TSV in each 0.5 °C thermal indices. Thus, considering the differences among these different thermal comfort indices, the WBGT, PMV, SET*, PET, and UTCI were compared and analyzed in this investigation.

# 3 Results

## *3.1 Meteorological Parameters of Outdoor Thermal Environment*

Air velocity ($V_a$) has been established as one of the key parameters of the thermal environment affecting human thermal comfort. In this research, $V_a$ was monitored, and its distribution is presented in Fig. 3. The majority of measurements were within the range of 0–1 m/s, with a maximum value of approximately 2.1 m/s. This suggests that the studied sites predominantly experienced low-wind conditions. A contributing factor may be that several sites are enclosed by buildings, where sheltering effects can reduce local wind speed.

In regions with high temperature and humidity, the influence of relative humidity (RH) on thermal comfort is particularly notable [48]. As shown in Fig. 4, RH levels were consistently high, with most readings above 60%, indicating its important role in thermal perception. Air temperature, also illustrated in Fig. 4, varied between 27 °C and 39 °C, peaking at 39 °C, which underscores that the outdoor thermal conditions could at times be extreme. For comfort and health considerations, it is advisable to avoid intense physical activities—such as outdoor exercise or manual labor—under such conditions [40]. Additionally, RH tended to decrease as air temperature rose, highlighting the necessity to account for the interactive effects of air temperature and humidity in outdoor thermal comfort assessment. Figure 5 displays instantaneous records of air temperature and RH at Location D on July 30, 2016. When air temperature peaked, RH reached its lowest level. Specifically, between 10:30 and 12:00, Ta exceeded 36 °C, peaking at 37.8 °C, while RH remained above 60%. This pattern further confirms the typical subtropical summer climate of Guangzhou.

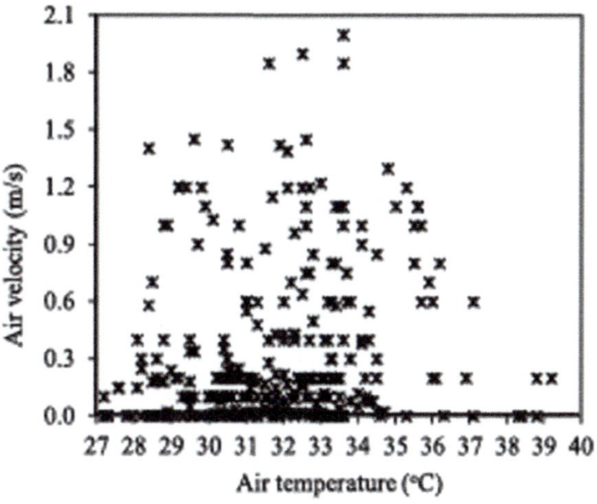


**Fig. 3** Air temperature against air velocity

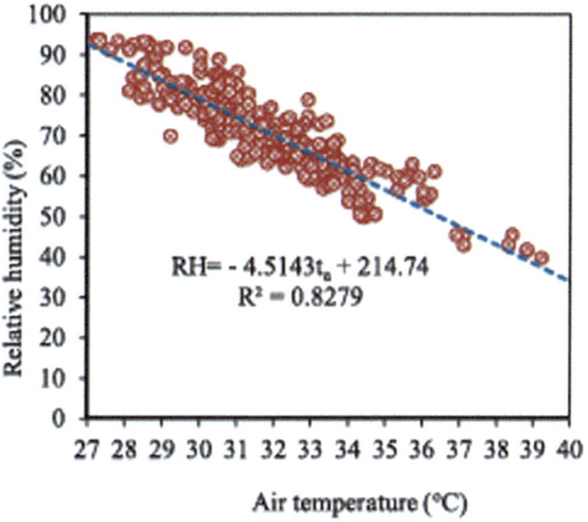


**Fig. 4** Air temperature against relative humidity

## 3.2 Differences in Terms of Clothing Insulation

Clothing insulation significantly influences thermal sensation. As an effective means of regulating thermal comfort, clothing insulation typically varies with air temperature ($T_a$) [8]. In this study, the clothing worn by each participant was documented through questionnaires. Total clothing insulation was subsequently calculated using Eq. (4). This method has been applied in previous research to estimate clothing insulation [4, 5, 49]. As shown in Fig. 6, clothing insulation values ranged from 0.3 to 0.6 clo, with a maximum near 0.8 clo and a minimum around 0.25 clo. The average insulation was 0.425 clo, which is lower than the standard value of 0.57 clo recommended

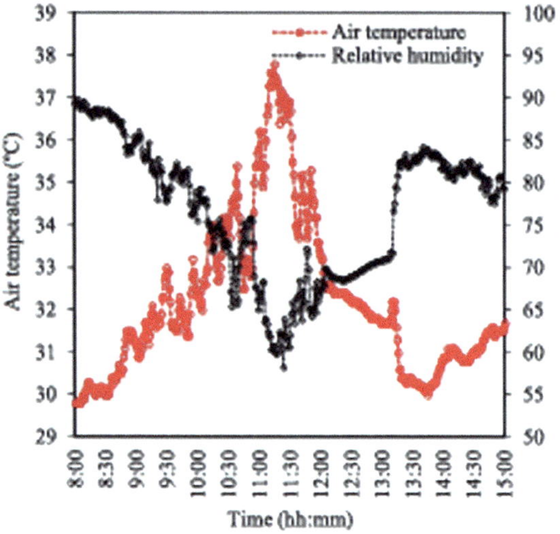


**Fig. 5** Monitored values of air temperature and RH at Location C on 30th July 2016

for summer in ASHRAE 55 [37]. This difference is mainly attributable to the fact that all participants were students at Guangzhou University, who generally dressed more casually compared to office workers. Similarly, Zhang et al. [36] reported that during hot summer conditions, subjects in offices wore lightweight clothing with an average insulation of 0.43 clo.

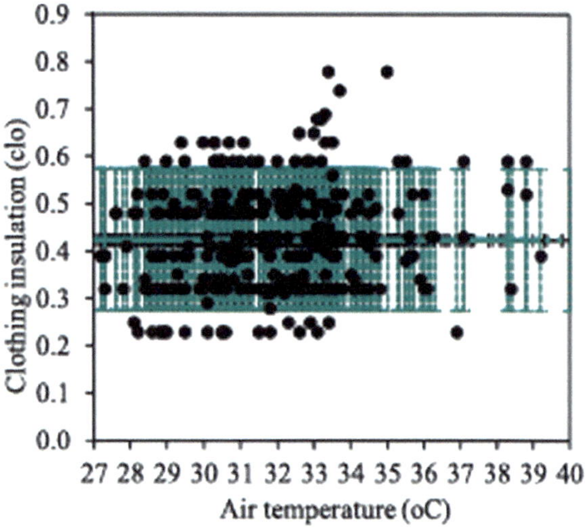


**Fig. 6** Clothing insulation against air temperature

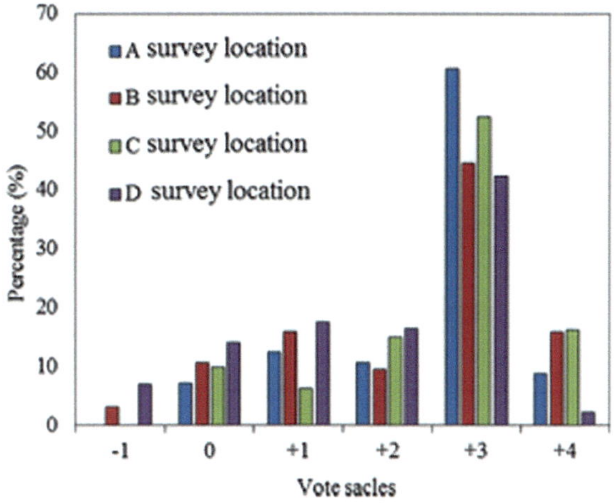


**Fig. 7** Percentage of scale voted at different locations

## *3.3 Thermal Sensation Vote Distribution*

Figure 7 presents the distribution of thermal sensation votes. The warm sensation (+ 3) was the most frequently selected option, exceeding 40% across all locations. Approximately 10% of participants reported feeling hot (+ 4). A contributing factor may be that the immediate air temperature in the outdoor environment occasionally surpassed 36 °C during the survey period, as illustrated in Fig. 4. The combined percentages for votes + 2, + 3, and + 4 approached 70%. Only a small proportion of respondents indicated a neutral sensation. Therefore, within subtropical climatic conditions, enhancing the outdoor thermal environment is essential to better support human outdoor activities.

## *3.4 The Variations of Thermal Indices*

Based on the collected survey data, a range of thermal comfort indices—including WBGT, SET*, PET, $T_{mrt}$, $T_{op}$, and UTCI—were computed. Figure 8 illustrates the variations among these thermal comfort indices in relation to $T_{op}$. All indices exhibited an upward trend as $T_{op}$ increased, though their rates of change differed considerably. At a given $T_{op}$ value, the various indices yielded divergent predictions of thermal comfort levels. As shown in Fig. 8, $T_{op}$ values covered a broad spectrum from 27 °C to 50 °C, with most data points clustering between 27 and 38 °C. Analysis of the measurement data revealed a high average summer outdoor temperature of 31.8 °C. To examine the correlation between operative temperature and the different thermal comfort indices, a simple regression approach was employed.

Figure 9 presents the fitted linear relationships between each thermal comfort index and $T_{op}$. Consistent with earlier analysis, linear regression was used to facilitate

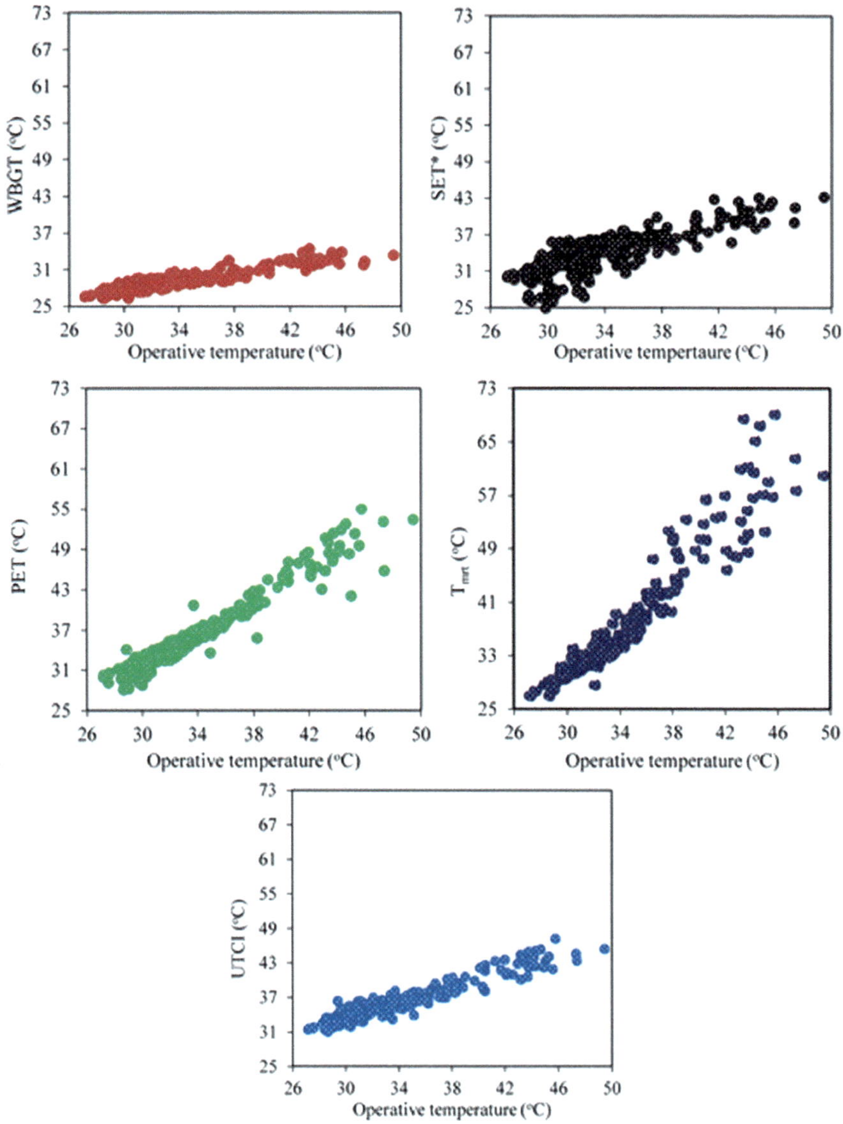


**Fig. 8** Different thermal comfort indices versus operative temperature: (1) WBGT, (2) SET*, (3) PET, (4) $T_{mrt}$ and (5) UTCI

comparison among the indices. As depicted in Fig. 9, all thermal comfort metrics demonstrated strong linear associations with $T_{op}$, with all $R^2$ values exceeding 0.8 except for that of SET*. Furthermore, the slopes of the regression lines varied, as also observed in Fig. 8. The slope for $T_{mrt}$ was the steepest at 1.847, markedly higher than those of the other indices. In outdoor settings with intense solar radiation, $T_{mrt}$ can

**Fig. 9** Relationships between different thermal indices and operative tem perature

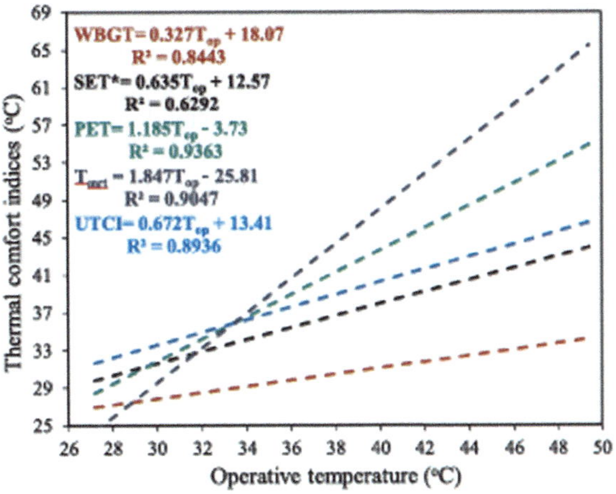


remain elevated even when air temperature is relatively low, particularly in unshaded areas. In contrast, WBGT displayed the gentlest slope of only 0.327. This can be attributed to its formulation, in which wet-bulb temperature ($T_w$) carries a weight of 0.7, while air temperature ($T_a$) and globe temperature ($T_g$) are assigned weights of 0.2 and 0.1, respectively. Since $T_w$ is generally lower than $T_a$ and $T_g$, WBGT values tend to be lower compared to other indices. The slope of $T_{mrt}$, which reached 1.847, is influenced by five variables: $T_g$, $T_a$, $V_a$, $\varepsilon_g$, and D. Under strong solar radiation, $T_{mrt}$ increases significantly. The slopes for SET* and UTCI were nearly identical, resulting in parallel regression lines. The slope of PET fell between those of $T_{mrt}$ and UTCI. When $T_{op}$ was below 34 °C, the differences among SET*, PET, and UTCI were minimal; however, these differences became more pronounced at higher $T_{op}$ values. Thus, the discrepancies among thermal comfort indices are influenced by the level of $T_{op}$.

## 3.5 *Outdoor Thermal Comfort Indices Versus Thermal Sensation*

Figure 10 illustrates the relationships between the mean thermal sensation vote (MTSV) and various thermal comfort indices. When indices such as SET*, PET, $T_{mrt}$, and UTCI were below 38 °C, MTSV increased at a rate consistent with these indices. However, once the indices exceeded 38 °C, MTSV values concentrated between 3 and 3.5. This phenomenon may be attributed to the upper limit of the thermal sensation scale being "very hot" (+ 4), with most participants voting either "very hot" (+ 4) or "hot" (+ 3). Consequently, the relationship between MTSV and thermal comfort indices became less distinct under high outdoor temperatures. Many previous studies [6, 50–57] employed the ASHRAE 55-2017 7-point scale to evaluate outdoor thermal environments. Strong linear correlations have been reported between MTSV and PET [4, 6, 39, 50–53, 58, 59], SET* [54–56], and UTCI [4, 57, 60, 61]. However, these linear models are inadequate for capturing responses at high outdoor temperatures, suggesting a need to extend the thermal sensation scale, particularly at the high-temperature end. Furthermore, the slope of MTSV relative to WBGT was the steepest among all indices. MTSV exceeded 3.0 when WBGT reached approximately 32 °C, and the linear segment of the MTSV-WBGT relationship was shorter compared to other indices. The regression models for each thermal comfort index are summarized in Table 6. Most $R^2$ values exceeded 0.7, with a maximum of 0.745. All *p*-values were below 0.01, indicating that the regression models are statistically significant. PMV and Thermal Sensation.

As one of the most widely used thermal comfort indices, the Predicted Mean Vote (PMV) was initially developed for indoor thermal environments and has occasionally been applied outdoors. Figure 11 displays the relationships among PMV, MTSV, and $T_{op}$. A strong linear correlation was observed between PMV and $T_{op}$, consistent with prior studies [37]. As indicated in Table 4, the application of PMV indoors requires specific conditions: the PMV should range between − 2 and + 2,

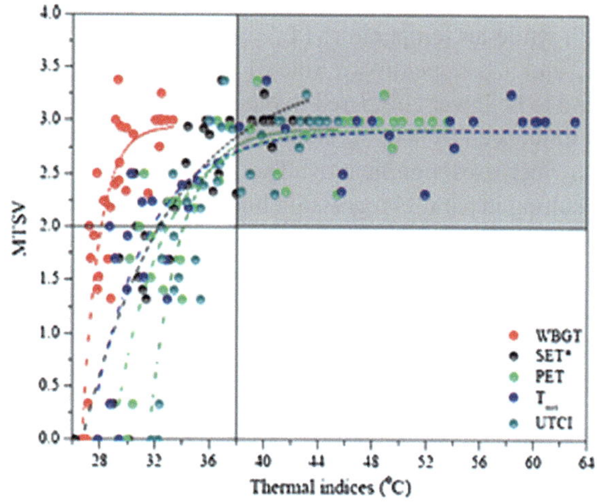


**Fig. 10** MTSV versus different thermal comfort indices

**Table 6** Detail information of the regression models

| Equation: | $MTSV = A_1 \cdot Exp(-x/\beta) + y0$ | | | | |
|---|---|---|---|---|---|
| x | $A_1$ | β | $y_0$ | $R^2$ | *p*-value |
| WBGT | 2.94 | − 8.02·Exp(9) | 1.23 | 0.737 | < 0.01 |
| SET* | 3.46 | − 1.4·Exp(5) | 6.53 | 0.745 | < 0.01 |
| PET | 2.93 | − 1.41·Exp(9) | 3.50 | 0.726 | < 0.01 |
| $T_{mrt}$ | 2.9 | − 2.31·Exp(7) | 4.01 | 0.684 | < 0.01 |
| UTCI | 2.93 | − 4.73·Exp(6) | 2.21 | 0.721 | < 0.01 |

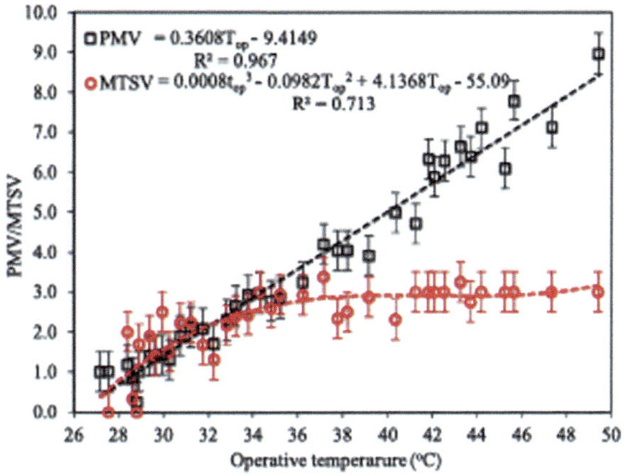


**Fig. 11** Difference between PMV and MTSV respectively versus operative temperature

and air temperature should not exceed 30 °C. Figure 11 shows that PMV predictions were accurate when $T_{op}$ was below 34 °C, but beyond this point, the discrepancy between PMV and MTSV grew significantly, particularly under high outdoor air temperatures. Additionally, the relationship between MTSV and $T_{op}$ was not well-defined. While MTSV increased with $T_{op}$ below 34 °C, PMV overestimated thermal sensation at higher $T_{op}$ values. A key reason is that PMV increases uniformly with $T_{op}$ without accounting for human thermal adaptation in hot environments. In reality, as Humphreys (1994) noted, people employ various adaptation strategies—physiological, psychological, and behavioral—to adjust to thermal conditions. Individuals are not passive recipients of environmental conditions but actively interact with their surroundings to improve comfort. From Fig. 11, it is evident that MTSV plateaued around 3.0 under high $T_{op}$ conditions, with most respondents feeling "hot" or "very hot" once operative temperature surpassed 34 °C. Therefore, if PMV is used to assess outdoor environments where $T_{op}$ exceeds 34 °C, modifications to the model are likely necessary.

# 4 Discussion

## *4.1 Thermal Sensation Vote Scale*

In the context of thermal comfort studies, subjective thermal sensation votes hold considerable importance, and the selection of an appropriate sensation scale is essential. Previous studies, as summarized in Table 7, most frequently employed the ASHRAE 7-point thermal sensation scale, defined as: − 3 (cold), − 2 (cool), − 1 (slightly cool), 0 (neutral), 1 (slightly warm), 2 (warm), and 3 (hot) [37]. Both ASHRAE 55 [37] and ISO 7730 [10] recommend this scale alongside the PMV index for assessing indoor thermal conditions, under constrained ranges of air temperature.

Specifically, ISO 7730 [10] stipulates that indoor air temperature ($T_a$) should lie between 10 and 30 °C, and mean radiant temperature ($T_{mrt}$) between 10 and 40 °C. The comfort criteria in ASHRAE standards trace back to climate chamber studies by Nevins et al. [62] and Robles and Nevins [63], which identified a comfortable air temperature range of 16.7–36.6 °C across all subjects [64]. Hence, the 7-point scale has been applied even in environments where $T_a$ approaches or exceeds typical indoor limits. Under such warm conditions, sweating often occurs across most of the body surface, accompanied by hot sensations. Experimental work by Song et al. [65], conducted at $T_a = 34.0$ °C, RH = 65%, and $V_a = 0.15$ m/s, reported maximum overall thermal sensation votes reaching 3.1 and wetness sensation up to 2.4. While personal cooling systems can help maintain comfort in warm environments without HVAC (e.g., $T_a < 32.0$ °C) [66], they become less effective in hotter settings where $T_a$ surpasses 34.0 °C [67]. Thus, the 7-point scale remains adequate for most indoor assessments and has also been used outdoors [57].

As indicated in Table 7, all referenced surveys were conducted during hot summers. In subtropical regions, $T_a$ frequently exceeded 35 °C [1, 39, 52, 58, 60, 61, 68–70], while $T_{mrt}$ often rose above 40 °C due to intense solar radiation [1, 27, 39, 52, 53, 57, 58, 60, 61, 68–72]. Under these conditions, participants often reported feeling extremely hot, suggesting that the 7-point scale may be insufficient to capture the full range of thermal sensation. Consequently, several studies have adopted an extended 9-point scale for outdoor thermal evaluation across diverse climates, including cool desert, subtropical, Mediterranean, and transitional climates. Although a 5-point scale is still occasionally used [11, 73], results from Fig. 10 show that average sensation votes exceeded 3.0 once operative temperature rose above 34 °C, supporting the use of the more sensitive 9-point scale for evaluating outdoor thermal conditions in this investigation.

## 4.2 Differences in Current Thermal Indices

As defined in ASHRAE 55 [37], $T_{op}$ is calculated as the average of air temperature ($T_a$) and mean radiant temperature ($T_{mrt}$), weighted by the convective and linearized radiant heat transfer coefficients, respectively, as indicated in Eq. (1). It is widely adopted for assessing indoor thermal environments within ASHRAE Standard, ISO-7730 [10], and GB/T50785 [38]. He et al. [74] identified a strong linear correlation between $T_{op}$ and thermal sensation within the range of 22–30 °C. Similarly, Wang et al. [53], in a study of thermal comfort in urban green spaces on a Dutch university campus, demonstrated that mean thermal sensation vote (MTSV) could be effectively predicted using a linear model based on $T_{op}$. As illustrated in Fig. 11, various thermal comfort indices exhibit strong linear relationships with $T_{op}$, with an average $R^2$ value close to 0.85, supporting the applicability of $T_{op}$ in evaluating outdoor thermal environments. To facilitate comparison across heat stress categories of different thermal indices, the corresponding $T_{op}$ values were derived using the linear regression equations presented in Fig. 9.Based on extensive survey

**Table 7** Thermal sensation scale used in surveys

| Place | Climate characters | Thermal sensation scale | Range of $T_a$ (°C) | Rang or $T_{mrt}/T_g$ (°C) | Authors and year |
|---|---|---|---|---|---|
| Sydney | Humid subtropical,mild and cool in winter to warm and hot in the summer | ASHRAE7-point | 20.4–43.3 | 20.9–67.9 | Spagnolo and de Dear [61] |
| Taichung | Warm humid subtropical climate | ASHRAE7-point | 14–39 | / | Lin et al. [16] |
| Guangzhou | Typical subtropical climate, hot summer and warm winter | ASHRAE7-point | 32–39 | 32–51 | Xi et al. [56] |
| Szeged | Transitional climate between oceanic(Marine West Coast Climate/Oceanic climate)and continental climate | ASHRAE9-point | 12.5–28.5 | 10–60 | Kántor et al. [71] |
| Damascus | Cool desert climate | ASHRAE9-point | 17.3–39.6 | 27.4–71.4 | Yahia and Joiansson [39] |
| Tel Aviv | Mediterranean climate,hot,humid yet rainless summers | ASHRAE9-point | 21.4–33.7 | / | Cohena et al. [59] |
| Nagoya | Humid subtropical climate, hot summer and cold winter | ASHRAE7-point | 31.9–32.9 | 31.9–48.4 | Watanabe et al. [57] |
| Tianjin | Temperate, continental-type monsoon climate, cool zone | ASHRAE7-point | 25.6–30.7 | / | Lai et al. [4] |
| Rio de Janeiro | Tropical savanna climate | ASHRAE7-point | 24.8–37.0 | / | Krüger et al. [79] |
| Wuhan | Humid subtropical climate, hot summer and cold winter | ASHRAE7-point | 31–38 | / | Huang et al. [68] |

(continued)

**Table 7** (continued)

| Place | Climate characters | Thermal sensation scale | Range of $T_a$ (°C) | Rang or $T_{mrt}/T_g$ (°C) | Authors and year |
|---|---|---|---|---|---|
| Rome | Mediterranean climate, with mild winters and hot, dry summers | ASHRAE7-point | 3.2–35.9 | / | Salata et al. [6] |
| Changsha | Humid subtropical climate, hot summer and cold winter | ASHRAE9-point | 18.7–39.8 | 18.8–57.8 | Liu et al. [52] |
| Guangzhou | Subtropical climate zone | ASHRAE7-point | 24.7–34.6 | 25.6–65.93 | Zhao et al. [27] |
| Brazil | Tropical climate zone | ASHRAE7-point | / | / | Hirashima et al. [80] |
| Campo Grande | Highland tropical climate | ASHRAE7-point | 15.3–32.3 | 17.5–33.6 | Lucchese et al. [81] |
| Italy | Typical Mediterranean climate | ASHRAE7-point | 3.2–35.9 | 7.6–48.1 | Salata et al. [6] |
| HongKong | Subtropical climate zone | ASHRAE7-point | / | / | Huang et al. [68] |
| Dar es Salaam | Hot-humid climate | ASHRAE7-point | 20.0–33.2 | 27.0–52.4 | Ndetto and Matzarakis [72] |
| Rio de Janeiro | Tropical savanna climate | ASHRAE7-point | 24.9–36.8 | 13.1–76.4 | Kruger et al. [69] |
| Brazil | Highland tropical climate | ASHRAE7-point | 15.3–34.4 | / | Lucchese and Andreasi [82] |
| Groningen | Cool temperate climate, summers are somewhat warm and humid | ASHRAE7-point | 21.3–33.5 | 22.3–40.4 | Wang et al. [53] |
| Mediterranean | Climate of mediterranean | ASHRAE7-point | 10.47–35.86 | 9.98–48.07 | Golasi et al. [1] |
| Cuenca | Typical continental mediterranean | ASHRAE7-point | / | / | Galindoa and Hermida (2018) |
| Tehran | Cold semi-arid climate | ASHRAE7-point | 2.8–37.5 | 0.1–49.1 | Hadianpour et al. [70] |

data, Table 8 summarizes the $T_{op}$ intervals corresponding to different heat stress categories for each thermal index. It is evident that the same stress category corresponds to varying operative temperatures across indices, highlighting considerable discrepancies. These differences suggest that existing thermal indices may require adjustment when applied to outdoor thermal conditions in Guangzhou. Notably, the $T_{op}$ ranges reported by Lin et al. [6] align closely with those in the current study. For instance, in Taiwan, the "hot" category corresponds to a $T_{op}$ range of 35.2–38.6 °C, while "warm" and "slightly warm" correspond to 31.84–35.2 °C and 28.5–31.8 °C, respectively. As shown in Fig. 10 of this study, a $T_{op}$ above 34 °C corresponds to a "hot" sensation, 30–34.0 °C to "warm", and 28–30 °C to "slightly warm"—showing good agreement with the Taiwan data. This consistency may be attributed to similar climatic conditions, as well as comparable psychological, social, economic, and cultural factors between Taichung and Guangzhou. Nevertheless, significant deviations are observed when comparing these ranges with originally proposed heat stress categories. Similar inconsistencies have been noted in previous studies [4, 12, 53]. Other thermal comfort indices, including UTCI, SET*, and WBGT, were also applied to outdoor thermal evaluation, and their corresponding $T_{op}$ ranges are included in Table 8. Inter-study comparisons reveal notable variations in these ranges [75]. For example, based on field data from Tianjin [4], UTCI stress categories deviated significantly from those in the UTCI-Fiala multi-node model [76]. Pantavou et al. [5] also reported differences in UTCI stress thresholds compared to the original scale—for instance, the "no thermal stress" range was 17.4–24.5 °C, narrower than the original 9–26 °C [77]. These findings indicate that index modifications may be necessary in future applications. It should be noted that the current regression models are derived solely from summer warm-thermal environment data and do not account for seasonal variations throughout the year. Therefore, future predictive models linking MTSV to thermal comfort indices should incorporate data across different seasons. Subsequent research will include investigations during spring, autumn, and winter to improve model comprehensiveness and accuracy.

## 4.3 Different Performances for Thermal Comfort Indices

The use of reference conditions was considered reasonable, since thermal comfort indices under actual conditions—defined by air temperature ($T_a$), mean radiant temperature ($T_{mrt}$), vapor pressure ($P_a$), and air velocity ($V_a$)—were expressed as equivalent temperatures representing the reference environment that would elicit the same physiological response. Thus, to facilitate practical assessment of thermal environments, it was necessary to define categorization scales for thermal stress. Previous research often employed the Bin method, which aggregates mean thermal sensation votes (MTSV) per 1 K interval [6, 45]. This approach involves: (a) grouping observations into half-degree Kelvin increments and using the mean vote of each bin (MTSV) rather than individual responses, and (b) fitting a linear regression model between these binned sensation votes and a thermal index (e.g., PET, SET*, or

**Table 8** Heat stress categories

| UTCI [75] | | | WBGT [74] | | | PET [78, 15] | | | | | SET* [83] | | | Present study |
|---|---|---|---|---|---|---|---|---|---|---|---|---|---|---|
| Thermal sensation | EU | $T_{op}$ | Thermal sensation | USA | $T_{op}$ | Thermal sensation | EU | Taiwan | $T_{op}$ (EU) | $T_{op}$ (Taiwan) | Thermal sensation | USA | $T_{op}$ | $T_{op}$ |
| Extreme heat stress | > 46 | > 48.5 | Black (all training should be stopped) | > 32.2 | > 43.21 | Extreme heat stress | > 41 | > 42 | > 37.8 | > 38.6 | Very hot | > 37.5 | > 39.26 | |
| Very strong heat stress | 38–46 | 36.6–48.5 | Red (active exercises for all but the well acclimated should be curtailed) | 31.1–32.2 | 39.9–43.21 | | | | | | | | | |
| Strong heat stress | 32–38 | 27.7–36.6 | Yellow (active exercise for un-acclimated persons should be curtailed) | 29.4–31.0 | 34.7–39.5 | Strong heat stress | 35–41 | 38–42 | 32.7–37.8 | 35.2–38.6 | Hot | 34.5–37.5 | 39.3–34.5 | > 34.0 |
| Moderate heat stress | 26–32 | 18.7–27.7 | Green (keep alert for possible increases in the index and for symptoms of heat stress) | 27.8–29.3 | 29.8–34.3 | Moderate heat stress | 29–35 | 34–38 | 27.6–32.7 | 31.8–35.2 | Warm | 30.0–34.5 | 27.5–34.5 | 30.0–34.0 |
| | | | | | | Slight heat stress | 23–29 | 30–34 | 22.6–27.6 | 28.5–31.8 | Slightly warm | 25.6–30.0 | 20.5–27.4 | 28–30 |

(continued)

**Table 8** (continued)

| UTCI [75] | | | WBGT [74] | | | PET [78, 15] | | | | | SET* [83] | | | Present study |
|---|---|---|---|---|---|---|---|---|---|---|---|---|---|---|
| Thermal sensation | EU | $T_{op}$ | Thermal sensation | USA | $T_{op}$ | Thermal sensation | EU | Taiwan | $T_{op}$ (EU) | $T_{op}$ (Taiwan) | Thermal sensation | USA | $T_{op}$ | $T_{op}$ |
| No thermal stress | 9–26 | – 6.6–18.7 | No flag (unlimited) | 25.6–27.7 | 23.0–29.5 | No thermal stress | 18–23 | 26–30 | 18.4–22.6 | 25.1–28.5 | Neutral | 22.2–25.6 | 15.2–20.5 | / |
| Slight cold stress | 0–9 | – 20.0–(– 6.6) | / | / | / | Slight cold stress | 13–18 | 22–26 | 14.1–18.3 | 21.7–25.1 | Slight cool | 17.5–22.2 | 7.8–15.2 | / |
| Moderate cold stress | – 13–0 | – 39.1–(– 20.0) | / | / | / | Moderate cold stress | 8–13 | 18–22 | 9.9–14.1 | 18.3–21.7 | Cool | 14.5–17.5 | 3.0–7.8 | / |
| Strong cold stress | – 27–(– 13) | – 60.1–(– 39.3) | / | / | / | Strong cold stress | 4–8 | 14–18 | 6.5–9.9 | 15.0–18.3 | Cold | 10.0–14.5 | – 4.1–3.0 | / |
| Very strong cold stress | – 27–(– 40) | – 79.5–(– 60.1) | / | / | / | Extreme cold stress | <4 | <14 | <6.5 | <15.0 | Very hot | <10.0 | <– 4.05 | / |
| Extreme cold stress | <– 40 | <– 79.5 | / | / | / | | | | | | | | | |

UTCI), with each point weighted by the number of observations within the bin. In the current study, observations were first grouped into 1 K intervals, and the mean thermal sensation per bin was computed. Regression models were then developed between MTSV and various thermal indices. Most previous surveys (Table 7) utilized the ASHRAE 7-point scale for collecting thermal sensation votes. Applying the Bin method to such data, linear regression models linking MTSV with thermal indices were derived (Table 9). The coefficients of determination ($R^2$) varied considerably, with some near unity and others below 0.5. The lowest $R^2$, reported by Lin et al. [55], was only 0.305, indicating a weak correlation between subjective thermal sensation and comfort indices. Elnabawi et al. [52], in a study of thermal comfort in an urban park in Cairo, Egypt, observed that thermal responses clustered around "hot" (+ 3) on the sensation scale when PET exceeded 40 °C, including near seating areas and fountains—a pattern consistent with the results shown in Fig. 10 of this study. This suggests that at extreme temperatures, subjective thermal responses exceed conventional heat stress categories, leaving participants little option but to select the highest available rating. Brode et al. [75] also demonstrated that the relationship between dynamic thermal sensation and UTCI was not strongly linear across a wide range (−50 to 50 °C UTCI), a finding applicable to other thermal indices. Thus, when outdoor temperatures surpass 40 °C, correlations between sensation votes and thermal indices become less distinct. In this investigation, a 9-point scale—extending the ASHRAE 7-point system—was used to evaluate outdoor thermal conditions. Under extreme heat, MTSV values were concentrated between + 3 and + 4, implying that the expanded scale helped subjects express perceptions near the upper limit of their experience. Consequently, the design of thermal sensation scales should reflect regional climatic characteristics. In tropical regions, where unbearably hot conditions may occur, scale modification is advised. Since magnitude estimation has no upper bound, regression models also indicate that heat stress categories for different thermal indices may require adjustment according to climate zone.

## 5 Conclusions

This chapter presented a comparative analysis of several widely used outdoor thermal indices, including PMV, WBGT, PET, SET*, and UTCI, using field surveys and on-site measurements in Guangzhou. The results confirmed that these indices exhibit strong correlations with operative temperature, but their predictive accuracy for thermal sensation weakens under high-temperature conditions, particularly when operative temperatures exceed 34 °C. While PMV and related models remain valuable references, their performance outdoors diverges from indoor-based assumptions, highlighting the need for adaptation. Non-linear regression approaches provided better fits in extreme heat, and discrepancies between current heat stress categories and local thermal responses further underscore the necessity for refined classification frameworks in warm and humid climates. As outdoor comfort is shaped by more complex and dynamic factors than indoor conditions, future research should

**Table 9** Linear models in some previous investigations

| Reference | Linear models | $R^2$ |
|---|---|---|
| Lin et al. [54] | MTSV = 0.1302SET*– 3.8142 (Hotseason) | 0.919 |
| | MTSV = 0.0739SET*– 2.0657 (Coolseason) | 0.945 |
| Hwang and Lin [55] | MTSV = 0.087SET*– 2.248(Semi-outdoor environments) | 0.970 |
| | MTSV = 0.116SET*– 3.156(Foroutdoor environments) | 0.980 |
| Lin et al. [58] | MTSV = 0.199PET–4.722(Cool season) | 0.890 |
| | MTSV = 0.118PET–3.025(Hot season) | 0.960 |
| Mahmoud [50] | MTSV = 0.206PET–6.680(Peak hot season) | 0.953 |
| | MTSV = 0.099PET–3.009(Peak cool season) | 0.348 |
| | MTSV = 0.145PET–3.625(Enrance hot season) | 0.944 |
| | MTSV = 0.071PET–2.479(Enrance cool season) | 0.768 |
| | MTSV = 0.52PET–4.222(Spine hot season) | 0.719 |
| | MTSV = 0.139PET–3.009(Spine cool season) | 0.768 |
| | MTSV = 0.106PET–2.117(Fountain hot season) | 0.606 |
| | MTSV = 0.074PET–2.253(Fountain cool season) | 0.768 |
| | MTSV = 0.12PET–2.156(Lake hot season) | 0.651 |
| | MTSV = 0.087PET–2.698(Lake cool season) | 0.654 |
| | MTSV = 0.129PET–2.845(Pavement hot season) | 0.863 |
| | MTSV = 0.087PET–2.363(Pavement cool season) | 0.447 |
| | MTSV = 0.118PET–2.399(Canopy hot season) | 0.784 |
| | MTSV = 0.127PET–3.236(Canopy cool winter) | 0.727 |
| | MTSV = 0.146PET–3.869(Seating hot season) | 0.876 |
| | MTSV = 0.093PET–2.603(Seating cool season) | 0.703 |
| | MTSV = 0.211PET–6.346(Cascade hot season) | 0.985 |
| | MTSV = 0.073PET–1.889(Cascade cool winter) | 0.663 |
| Xi et al. [56] | TSV = 0.1382SET*–3.3469 0.305 | 0.305 |
| Cohena et al. [59] | MTSV = 0.3292PET–5.9692(UPK hot season) | 0.885 |
| | MTSV = 0.0.2146PET–3.5737(UPK cool season) | 0.963 |
| | MTSV = 0.2078PET–3.6741(UST hot season) | 0.965 |
| | MTSV = 0.2363PET–3.9149(UST cool season) | 0.965 |
| | MTSV = 0.2198PET–3.9077(USQ hot season) | 0.966 |
| | MTSV = 0.2111PET–3.6226(USQ cool season) | 0.966 |

(continued)

**Table 9** (continued)

| Reference | Linear models | $R^2$ |
|---|---|---|
| Yahia and Johansson [39] | MTSV = 0.060PET–0.941(Hot season) | 0.420 |
| | MTSV = 0.114PET–2.755(Cool season) | 0.600 |
| Lai et al. [4] | MTSV = 0.101PET–1.571(Hot season) | 0.893 |
| | MTSV = 0.188PET–1.73(Cool season) | 0.752 |
| | MTSV = 0.13UTCI–2.273(Hot season) | 0.876 |
| | MTSV = 0.183UTCI–0.392(Cool season) | 0.946 |
| Watanabe et al. [57] | MTSV = 0.271UTCI–9.237 | 0.665 |
| Krüger et al. [79] | MTSV = 0.1404UTCI–3.2933 | 0.940 |
| Elnabawi et al. [51] | MTSV = 0.0998(PET)–2.947(Hot season) | 0.830 |
| | MTSV = 0.0881(PET)–2.1411(Cool season) | 0.811 |
| Huang et al. [68] | MTSV = 0.123UTCI–2.362 | 0.401 |
| Liu et al. [52] | MTSV = 0.131PET–2.296(Spring) | 0.585 |
| | MTSV = 0.188PET–4.386(Summer) | 0.778 |
| | MTSV = 0.112PET–2.232(Autumn) | 0.521 |
| | MTSV = 0.163PET–2.431(Winter) | 0.663 |
| Salata et al. [6] | MTSV = 0.17PET–4.575 Hot season) | 0.847 |
| | MTSV = 0.118PET–4.575(Cool season) | 0.949 |
| Wang et al. [53] | MTSV = 0.058PET–0.696 | 0.680 |
| Hadianpour et al. [70] | TSV = 0.11PET–2.42(Wholeyear) | 0.846 |
| | TSV = 0.11PET–2.36(Spring) | 0.814 |
| | TSV = 0.17PET–4.26(Summer) | 0.648 |
| | TSV = 0.11PET–2.41(Autumn) | 0.658 |
| | TSV = 0.15PET–2.58(Winter) | 0.562 |
| | TSV = 0.64PMV–0.16(Wholeyear) | 0.841 |
| | TSV = 0.61PMV–0.14(Spring) | 0.809 |
| | TSV = 0.63PMV–0.15(Summer) | 0.624 |
| | TSV = 0.55PMV–0.04(Autumn) | 0.736 |
| | TSV = 0.62PMV–0.34(Winter) | 0.495 |
| | TSV = 0.13UTCI–2.70(Wholeyear) | 0.845 |
| | TSV = 0.12UTCI–2.50(Spring) | 0.804 |
| | TSV = 0.22UTCI–5.68(Summer) | 0.653 |
| | TSV = 0.13UTCI–2.80(Autumn) | 0.657 |
| | TSV = 0.17UTCI–2.92(Winter) | 0.588 |

incorporate variations in clothing insulation, metabolic rate, and seasonal adaptation, together with physiological monitoring of subjects, to enhance the robustness of thermal indices. By refining and validating these models in diverse climates, more accurate and context-sensitive tools can be developed to support sustainable urban design and improve outdoor livability.

**Acknowledgements** The work described in this paper was fully supported by a grant from the Research Grants Council of the Hong Kong Special Administrative Region, China (Project No. C5002-14G). The authors would like to express our gratitude to Dr. Tsz-cheung Lee of Hong Kong Observatory for his advice and all the students who participated in the surveys. Special thanks are also given to the anonymous reviewers for their valuable comments on the manuscript.

## References

1. I. Golasi, F. Salata, V.E. de Lieto, M. Coppi, Complying with the demand of standardization in outdoor thermal comfort: a first approach to the Global Outdoor Comfort Index (GOCI). Build. Environ. **130**, 104–119 (2018)
2. A.A. Hakim, H. Petrovitch, C.M. Burchfiel, G.W. Ross, B.L. Rodriguez, L.R. White, Effects of walking on mortality among nonsmoking retired men. New Engl. J. Med. **338**, 94–99 (1998)
3. E. García-López, C. Heard, A study of the social acceptability of a proposal to improve the thermal comfort of a traditional dwelling. Appl. Therm. Eng. **75**(22), 1287–1295 (2015)
4. D. Lai, D. Guo, Y. Hou, C. Lin, Q. Chen, Studies of outdoor thermal comfort in northern China. Build. Environ. **77**, 110–118 (2014)
5. K. Pantavou, G. Theoharatos, M. Santamouris, D. Asimakopoulos, Outdoor thermal sensation of pedestrians in a Mediterranean climate and a comparison with UTCI. Build. Environ. **66**, 82–95 (2013)
6. F. Salata, I. Golasi, R.L. Vollaro, A.L. Vollaro, Outdoor thermal comfort in the Mediterranean area. A transversal study in Rome Italy. Build. Environ. **96**, 46–61 (2016)
7. K. George, H.A. Flocas, M. Andreas, Dynamic modeling of human thermal comfort after the transition from an indoor to an outdoor hot environment. Int. J. Biometeorol. **59**, 205–216 (2015)
8. P.O. Fanger, *Thermal comfort* (Analysis and application in environment engineering. Danish Technical Press, Copenhagen, 1970)
9. S. Gao, Y. Li, Y. Wang, X. Meng, Y. Zhang, C. Yang, L. Jin, A human thermal balance based evaluation of thermal comfort subject to radiant cooling system and sedentary status. Appl. Therm. Eng. **122**(25), 461–472 (2017)
10. ISO 7730, Ergonomics of the Thermal Environment—Analytical Determination and Interpretation of Thermal Comfort using Calculation of the PMV and PPD Indices and Local Thermal Comfort Criteria. International Standard Organization: Geneva (2005)
11. M. Nikolopoulou, N. Baker, K. Steemers, Thermal comfort in outdoor urban spaces: understanding the human parameter. Sol. Energy **70**, 227–235 (2001)
12. V. Cheng, E. Ng, C. Chan, B. Givoni, Outdoor thermal comfort study in sub-tropical climate: a longitudinal study based in Hong Kong. Int. J. Biometeorol. **56**(1), 43–56 (2012)
13. P. Höppe, *Die Energiebilanz des Menschen* (München Universitäts Schriften, Fachbereich Physic, Wissenschaftliche Mitteilungen, 1984), p.49
14. P. Höppe, The physiological equivalent temperature- a universal index for the biometeorological assessment of the thermal environment. Int. J. Biometeorol. **43**, 71–75 (1999)
15. A. Matzarakis, H. Mayer, M.G. Iziomon, Applications of a universal thermal index: physiological equivalent temperature. Int. J. Biometeorol. **43**, 76–84 (1999)

16. T.P. Lin, A. Matzarakis, Tourism climate and thermal comfort in Sun Moon Lake Taiwan. Int. J. Biometeorol. **52**, 281–290 (2008)
17. H. Mayer, J. Holst, P. Dostal, F. Imbery, D. Schindler, Human thermal comfort in summer within an urban street canyon in Central Europe. Meteorol. Zeitschrif. **17**, 241–250 (2008)
18. J. Niu, J. Liu, T. Lee, Z. Lin, C.M. Mak, K.T. Tse, B.S. Tang, K.C.S. Kwok, A new method to assess spatial variations of outdoor thermal comfort: onsite monitoring results and implications for precinct planning. Build. Environ. **91**, 263–270 (2015)
19. B. Yang, T. Olofsson, G. Nair, A. Kabansi, Outdoor thermal comfort under subarctic climate of north Sweden—A pilot study in Umea. Sustain. Cities Soc. **28**, 387–397 (2017)
20. S. Amindeldar, S. Heidari, M. Khalili, The effect of personal and microclimatic variables on outdoor thermal comfort: a field study in Tehran in cold season. Sustain. Cities Soc. **32**, 153–159 (2017)
21. K. Pantavou, S. Lykoudis, M. Nikolopoulou, L.X. Tsiros, Thermal sensation and climate: a comparison of UTCI and PET thresholds in different climates. Int. J. Biometeorol. **7**, 1–14 (2018)
22. C.P. Yaglou, D. Minard, Control of heat casualties at military training centers. Arch. Ind. Health **16**, 302–316 (1957)
23. M.B. Grahame, Wet-bulb globe temperature (WBGT)——its history and its limitations. J. Sci. Med. Sport **11**, 20–32 (2008)
24. A.P. Gagge, A.P. Fobelets, P.E. Berglund, A standard predictive index of human response to the thermal environment. ASHRAE Trans. **92**, 709–731 (1986)
25. Y. Zhu, J. Liu, A. Hagishima, J. Tanimoto, Y. Yao, Z. Ma, Evaluation of coupled outdoor and indoor thermal comfort environment and anthropogenic heat. Build. Environ. **2**(42), 1018–1025 (2007)
26. T. Tian, Y. Xi, Q. Li, A. Mochida, Q. Meng, Study on the outdoor thermal environment and thermal comfort around campus clusters in subtropical urban areas. Build. Environ. **52**, 162–170 (2010)
27. L. Zhao, X. Zhou, L. Li, S. He, R. Chen, Study on outdoor thermal comfort on a campus in a subtropical urban area in summer. Sustain. Cities Soc. **22**, 164–170 (2016)
28. J.J. Gerd, R. de Dear, G. Havenith, UTCI-Why another thermal index? Int. J. Biometeorol. **56**, 421–428 (2012)
29. D. Fiala, K.J. Lomas, M. Stohrer, A computer model of human thermoregulation for a wide range of environmental conditions: the passive system. J. Appl. Physiol. **87**, 1957–1972 (1999)
30. D. Fiala, K.J. Lomas, M. Stohrer, Computer prediction of human thermoregulatory and temperature responses to a wide range of environmental conditions. Int. J. Biometeorol. **45**, 143–159 (2001)
31. G. Havenith, D. Fiala, K. Błazejczyk et al., The UTCI-clothing model. Int. J. Biometeorol. **56**, 461–470 (2012)
32. H. Zhang, Human thermal sensation and comfort in transient and non-uniform thermal environments. Ph. D. Thesis, University of California Berkeley (2003)
33. Z. Fang, Z. Lin, C.M. Mak, J. Niu, K.T. Tse, Investigation into sensitivities of factors in outdoor thermal comfort indices. Build. Environ. **128**, 129–142 (2018)
34. N. Marialena, S. Koen, Thermal comfort and psychological adaptation as a guide for designing urban spaces. Energy Build. **35**, 95–101 (2003)
35. 2345 weather report, http://tianqi.2345.com/wea_history/59287.htm
36. Y. Zhang, H. Chen, Q. Meng, Thermal comfort in buildings with split air-conditioners in hot-humid area of China. Build. Environ. **64**, 213–224 (2013)
37. ANSI/ASHRAE. Standard 5. Thermal environmental conditions for human occupancy. Atlanta, Georgia: American Society of Heating Refrigerating and Air-Conditioning Engineers (ASHRAE) (2017)
38. ISO 7726, Ergonomics of the thermal environment—Instruments for measuring physical quantities (1998)
39. M.W. Yahia, E. Johansson, Evaluation the behavior of different thermal indices by investigating various outdoor urban environments in the hot dry city of Damascus, Syria. Int. J. Biometeorol. **57**, 615–630 (2013)

40. ASHRAE Handbook-Fundamentals, Atlanta, Georgia: American Society of Heating Refrigerating and Air-Conditioning Engineers (ASHRAE) (2017)
41. E.A. McCullough, B.W. Jones, A comprehensive data base for estimating clothing insulation. IER Technical Report 84–01, Institute for Environmental Research, Kansas State University, Manhattan. ASHRAE Research Project RP-411, Final Report (1984)
42. Y. Nishi, R.R. Gonzalez, A.P. Gagge, Direct measurement of clothing heat transfer properties during sensible and insensible heat exchange with thermal environment. ASHRAE Trans. **81**(2), 183 (1975)
43. B.W. Olesen, A new and simpler method for estimating the thermal insulation of a clothing ensemble. ASHRAE Trans. **91**(2), 478–492 (1985)
44. D.S. Moran, K.B. Pandolf, Y. Shapiro, Y. Heled, Y. Shani, W.T. Mathew, R.R. Gonzale, An environmental stress index (ESI) as a substitute for the wet bulb globe temperature (WBGT). J. Therm. Biol. **26**, 427–431 (2001)
45. D. Fiala, G. Havenith, B. Peter, K. Bernhard, J. Gerd, UTCI-Fiala multi-node model of human heat transfer and temperature regulation. Int. J. Biometeorol. Spec. Issue **56**(3), 429–441 (2011)
46. A. Matzarakis, F. Rutz, H. Mayer, Modelling radiation fluxes in simple and complex environemnts—application of the RayMan model. Int. J. Biometeorol. **51**, 323–334 (2007)
47. Y. Wang, Z. Ni, Y. Peng, B. Xia, Local variation of outdoor thermal comfort in different urban green spaces in Guangzhou, a subtropical city in south China. Urban For. Urban Greening **32**, 99–112 (2018)
48. A. Alahmer, M.A. Omar, A. Mayyas, D. Shan, Effect of relative humidity and temperature control on in-cabin thermal comfort state: thermodynamic and psychometric analyses. Appl. Therm. Eng. **31**(14–15), 2636–2644 (2011)
49. K. Li, Z. Zhang, L. Zhao, Outdoor thermal comfort and activities in the urban residential community in a humid subtropical area of China. Energy Build. **133**, 498–511 (2016)
50. A.H.A. Mahmoud, Analysis of the microclimatic and human comfort conditions in an urban park in hot and arid regions. Build. Environ. **46**, 2641–2656 (2011)
51. M.H. Elnabawi, N. Hamza, S. Dudek, Thermal perception of outdoor urban spaces in the hot arid region of Cairo Egypt. Sustain. Cities Soc. **22**, 136–145 (2016)
52. W. Liu, Y. Zhang, Q. Deng, The effects of urban microclimate on outdoor thermal sensation and neutral temperature in hot-summer and cold-winter climate. Energy Build. **128**, 190–197 (2016)
53. Y. Wang, R. Groot, F. Bakker, H. Wörtche, R. Leemans, Thermal comfort in urban green spaces: a survey on a Dutch university campus. Int. J. Biometeorol. **61**(1), 87–101 (2017)
54. T.P. Lin, R. de Dear, R.L. Hwang, Effect of thermal adaptation on seasonal outdoor thermal comfort. Int. J. Climatol. **31**, 302–312 (2011)
55. R.L. Hwang, T.P. Lin, Thermal comfort requirements for occupants of semi-outdoor and outdoor environments in hot-humid regions. Archit. Sci. Rev. **50**(4), 357–364 (2007)
56. T. Xi, Q. Li, A. Mochida, Q. Meng, Study on the outdoor thermal environment and thermal comfort around campus clusters in subtropical urban areas. Build. Environ. **52**, 162–170 (2012)
57. S. Watanabe, K. Nagano, J. Ishii, T. Horikoshi, Evaluation of outdoor thermal comfort in sunlight, building shade, and pergola shade during summer in a humid subtropical region. Build. Environ. **82**, 556–565 (2014)
58. T.P. Lin, Thermal perception, adaptation and attendance in a public square in hot and humid regions. Build. Environ. **44**, 2017–2026 (2009)
59. P. Cohena, O. Potchter, A. Matzarakis, Human thermal perception of Coastal Mediterranean outdoor urban environments. Appl. Geogr. **37**, 1–10 (2013)
60. E. Kriuger, P. Drach, P. Brode, Implications of air-conditioning use on thermal perception in open spaces: a field study in downtown Rio de Janeiro. Build. Environ. **94**, 417–425 (2015)
61. J. Spagnolo, R. de Dear, A field study of thermal comfort in outdoor and semi-outdoor environments in subtropical Sydney Australia. Build. Environ. **38**, 721–738 (2003)
62. R.G. Nevins, F.H. Rohles, W. Springer, A.M. Feyerherm, A temperature humidity chart for thermal comfort of seated persons. ASHRAE Trans. **72**(1), 283–291 (1966)

63. F.H. Rohles, R.G. Nevins, The natural of thermal comfort for sedentary man. ASHRAE Trans. **77**(1), 239–246 (1971)
64. K. Parsons, *Human Thermal Environments* (CRC Press Taylor & Francis Group, Boca Raton London New York, 2002)
65. W. Song, F. Wang, F. Wei, Hybrid cooling clothing to improve thermal comfort of office workers in a hot indoor environment. Build. Environ. **100**, 92–101 (2016)
66. D. Scheatzle, H. Wu, J. Yellott, Extending the summer comfort envelope with ceiling fans in hot, acrid climates. ASHRAE Trans. **95**, 269–280 (1989)
67. S. Atthajariyakul, C. Lertsatittanakorn, Small fan assisted air conditioner for thermal comfort and energy saving in Thailand. Energy Convers. Manage. **49**, 2499–2504 (2008)
68. J. Huang, C. Zhou, Y. Zhuo, L. Xu, Y. Jiang, Outdoor thermal environments and activities in open space: an experiment study in humid subtropical climates. Build. Environ. **103**, 238–249 (2016)
69. E. Kruger, P. Drach, P. Broede, Outdoor comfort study in Rio de Janeiro: site-related context effects on reported thermal sensation. Int. J. Biometeorol. **61**, 463–475 (2017)
70. M. Hadianpour, M. Mahdavinejad, M. Bemanian, F. Nasrollahi, Seasonal differences of subjective thermal sensation and neutral temperature in an outdoor shaded space in Tehran Iran. Sustain. Cities Soc. **39**, 751–764 (2018)
71. N. Kántor, L. Égerházi, J. Unger, Subjective estimation of thermal environment in recreational urban spaces—Part 1: investigations in Szeged Hungary. Int. J. Biometeorol. **56**(6), 1075–1088 (2012)
72. E.L. Ndetto, A. Matzarakis, Assessment of human thermal perception in the hot-humid climate of Dar es Salaam, Tanzania. Int. J. Biometeorol. **6**, 69–85 (2017)
73. M. Nikolopoulou, S. Lykoudis, Thermal comfort in outdoor urban spaces: Analysis across different European countries. Build. Environ. **41**, 1455–1470 (2006)
74. ISO 7243, International Organization for Standardization, Hot Environments—Estimation of the Heat Stress on Working Man, Based on the WBGT-Index (Wet Bulb Globe Temperature) (1989)
75. P. Brode, D. Fiala, K. Blazejczyk, I. Holmer, G. Jendritzky, B. Kampmann, B. Tinz, G. Havenith, Deriving the operational procedure for the Universal Thermal Climate Index (UTCI). Int. J. Biometeorol. **56**, 481–494 (2012)
76. G. Jendritzky, R. de Dear, G. Havenith, UTCI—Why another thermal index? Int. J. Biometeorol. **56**, 421–428 (2012)
77. R. de Dear, M.E. Fountain, Field experiments on occupant comfort and office thermal environments in a hot-humid climate. ASHRAE Trans. **100**(2), 457–474 (1994)
78. T. Marianna, T. Theocharis, D. Tryfon, Evaluation of comfort conditions in urban open spaces. Application in the island of Crete. Energy Convers. Manage. **86**, 250–258 (2014)
79. E. Krüger, P. Drach, P. Bröde, Implications of air-conditioning use on thermal perception in open spaces: A field study in downtown Rio de Janeiro. Build. Environ. 94417–425 (2015). https://doi.org/10.1016/j.buildenv.2015.07.024/
80. S.Q. da Silveira Hirashima, E.S. de Assis, M. Nikolopoulou, Daytime thermal comfort in urban spaces: A field study in Brazil. Build. Environ. 107245–253 (2016). https://doi.org/10.1016/j.buildenv.2016.08.006
81. J.R. Lucchese, L.P. Mikuri, N.V.S. Freitas, W.A. Andreasi, Application of selected indices on outdoor thermal comfort assessment in Midwest Brazil. Int. J. Energ. Environ. **7**(4), 291–302 (2016)
82. J.R. Lucchese, W.A. Andreasi, Designing Thermally Pleasant Open Areas: The Influence of Microclimatic Conditions on Comfort and Adaptation in Midwest Brazil. J. Sust. Devel. **10**(4) (2017). 11-10.5539/jsd.v10n4p11
83. D.A. McIntyre, Indoor climate. London: Applied Science Publishers Ltd. (1980)
84. A. Psikuta, D. Fiala, G. Laschewski, G. Jendritzky, M. Richards, K. Błażejczyk, I. Mekjavič, H. Rintamäki, R. de Dear, G. Havenith, Validation of the Fiala multi-node thermos-physiological model for UTCI application. Int. J. Int. J. Biometeorol. **56**, 443–460 (2012)

# Developing an Adaptive-Rational Outdoor Thermal Comfort Model for Feedback on Outdoor Thermal Adaptations

**Abstract** Models for outdoor thermal comfort are essential for guiding urban design and promoting the effective use of open spaces. Since thermal equilibrium alone cannot fully explain human comfort outdoors, it is necessary to incorporate both rational heat balance models and adaptive mechanisms. This chapter develops adaptive-rational models that integrate two established indices—Physiological Equivalent Temperature (PET) and Universal Thermal Climate Index (UTCI)—with adaptive and extension strategies. Field validation demonstrates that the adaptive-rational approach substantially improves predictive accuracy and robustness compared to conventional rational models. Specifically, PET-based adaptive-rational models achieve accuracy gains of up to 83% and robustness improvements of 86%, while UTCI-based adaptive-rational models reach improvements of 87%. These results highlight the effectiveness of combining adaptive principles with rational formulations, providing a more reliable framework for evaluating outdoor thermal comfort and supporting the design of thermally livable urban environments.

## 1 Introduction

Sustainable development in built environments, which includes key elements such as thermal comfort [1], health [2], energy efficiency [3], and lower carbon emissions, presents both significant challenges and urgent demands [4]. Achieving outdoor thermal comfort is a vital aspect of advancing sustainable urban development [5]. Comfortable outdoor spaces encourage individuals to engage in outdoor activities, which not only promote healthier lifestyles but also help to reduce energy consumption within buildings [6]. However, the impacts of global climate change and the urban heat island effect create substantial obstacles to maintaining thermally comfortable outdoor environments [7]. To address these issues, researchers are increasingly focused on studying and modeling outdoor thermal comfort. For example, Potchter et al. [8] analyzed 254 studies examining outdoor thermal conditions across varying climatic regions. Outdoor thermal comfort models are essential for designing urban spaces that are both climate-responsive and livable [9]. Lai et al. [10] highlighted

 
Z. Fang et al., *Outdoor Thermal Comfort and Heat Stress Mitigation*, Environmental Science and Engineering, https://doi.org/10.1007/978-981-95-8899-2_10

the importance of these models in guiding urban planning decisions, such as the integration of vegetation, reflective surfaces, water features, and the layout of public spaces.

The task of modeling outdoor thermal comfort is highly complex [11]. A major factor influencing thermal comfort outdoors is the heat balance of the human body [12]. Environmental variables, including air temperature, mean radiant temperature, relative humidity, and wind speed, significantly affect this balance and, in turn, determine comfort levels [13]. Changes in mean radiant temperature due to solar radiation and variations in wind speed introduce additional challenges to the modeling process [6, 13]. Furthermore, personal attributes, such as clothing insulation and metabolic rates, also play a crucial role in thermal comfort outcomes [14]. Differences in metabolic rates caused by varying outdoor activities add another layer of complexity to outdoor comfort modeling [15]. To address these challenges, rational thermal indices have been developed. These indices are based on heat balance Eqs derived from human thermoregulation models and are used to predict outdoor thermal comfort [6]. Two well-established indices in this field are the Physiological Equivalent Temperature (PET), which is calculated using the Munich Energy-Balance Model for Individuals [16], and the Universal Thermal Climate Index (UTCI), which is developed from the multi-node dynamic thermo-physiological UTCI-Fiala model [17].

In addition to the heat balance of the human body, thermal adaptations play a vital role in accurately modeling outdoor thermal comfort [18]. These adaptations are generally divided into three categories: physiological adaptation (the body's responses to alleviate heat strain), psychological adaptation (changes in perception influenced by prior thermal experiences and expectations), and behavioral adaptation (modifications in behavior to reduce thermal discomfort). These adaptations lead to significant differences in outdoor thermal comfort across various climate zones [5]. For example, in various parts of China, thermal neutrality has been documented to range from 11.0 to 32.4 °C [19]. A common approach for calibrating rational thermal indices associated with these adaptations involves establishing a linear relationship between the indices and subjective assessments of thermal sensation [20, 21]. Nevertheless, there is an immediate necessity to enhance the precision of this conventional calibration technique [22]. Therefore, new methodologies are essential for developing outdoor thermal comfort models that simultaneously account for both heat equilibrium and thermal adjustments [23].

Adaptive-rational thermal comfort modeling holds promise for integrating the human body's heat balance with thermal adaptations [9]. This modeling approach has been applied in indoor thermal environments [24, 25]. Fanger and Toftum [26] introduced a fixed extension factor to enable rational indoor thermal comfort models to incorporate thermal adaptations, while Yao et al. [27] proposed a fixed adaptive coefficient for similar purposes. Recognizing the variability of thermal adaptations, Zhang and Lin developed adaptive-rational indoor thermal comfort models utilizing extension methods with a variable extension factor [28] and adaptive methods with a variable adaptive factor [29]. These adaptive-rational models have demonstrated effective performance in air-conditioned, naturally ventilated, and mixed-mode buildings across various climate zones [29], significantly enhancing the accuracy of indoor thermal comfort predictions by as much as 90% [30].

However, current adaptive-rational indoor thermal comfort models are predominantly tailored for indoor environments [24], with a lack of specifically designed adaptive-rational models for outdoor settings. The intricacies of outdoor thermal comfort modeling surpass those of indoor environments [8], as evidenced by the limited accuracy of applying indoor comfort models outdoors [31]. The challenges associated with outdoor thermal comfort modeling can be attributed in part to the fluctuating solar radiation [13], wind speed [32], and metabolic rates [15]. This emphasizes the importance of dedicated rational thermal indices like PET and UTCI, specifically formulated for evaluating outdoor thermal comfort [33]. Therefore, the development of adaptive-rational outdoor thermal comfort models that consider both heat equilibrium and thermal adjustments is critical for precise outdoor thermal comfort prediction.

This research introduces adaptive-rational models for outdoor thermal comfort that integrate two primary outdoor thermal indices: PET and UTCI. These indices are among the most prominent out of a pool of over 160 available outdoor thermal indices [5, 34]. Four adaptive-rational models for outdoor thermal comfort have been established: two based on PET—one incorporating an adaptive technique and the other an extension technique—and two based on UTCI, each utilizing adaptive and extension methods (Sect. 2.1 and 2.2). The models have undergone validation using real-world data, with a detailed analysis of their applicability and efficacy provided in Sect. 4. Further validation outcomes can be found in the Appendix.

## 2 Methodology

### *2.1 Adaptive-Rational Outdoor Thermal Comfort Models with Adaptive Method*

The adaptive method formulates the arPTS (adaptive arPTS) by integrating an adaptive coefficient into the rational model. The rational model, which represents the predicted thermal sensation (PTS) in Eq. (1), considers the heat equilibrium within the human body. The adaptive coefficient ($\lambda$ in Eq. 1) reflects thermal adjustments. This method draws inspiration from the negative feedback theory in cybernetics (refer to Fig. 1) [27]. The body's responses (G in Figs. 1) to thermal stimuli from the outdoor environment ($\delta$ in Fig. 1) generate the initial thermal sensation (PTS). Subsequently, thermal adaptations (K in Fig. 1) act as negative feedback to reduce thermal discomfort. The final thermal sensation, termed the adaptive arPTS, is derived from the initial sensation and the negative feedback (Eq. 3). The adaptive coefficient is defined as the ratio of thermal adaptations to thermal signals (Eq. 4). By integrating Eqs. (2)–(4), the complete formulation of the adaptive-rational outdoor thermal comfort model using the adaptive approach is established in Eq. (1).

$$arPTS_a = \frac{PTS}{1 + \lambda \times PTS} \tag{1}$$

$$PTS = G \times \delta \tag{2}$$

$$arPTS_a = G \times \delta - arPTS_a \times K \times G \tag{3}$$

$$\lambda = \frac{K}{\delta} \tag{4}$$

where $arPTS_a$ is the adaptive-rational outdoor thermal comfort model with the adaptive method; G is the human body responses; K is the thermal adaptations; PTS is the predicted thermal sensation; δ is the thermal stimuli; λ is the adaptive coefficient.

The PTS is computed based on outdoor thermal indices. This research emphasizes the two most utilized outdoor thermal indices, PET and UTCI. The PET denotes the air temperature in a standard environment where the core and skin temperatures match those of the actual environment [6]. Core and skin temperatures are computed using the Munich Energy-Balance Model for Individuals, which is founded on the heat equilibrium of the human body [35]. The PET considers multiple factors, including air temperature, mean radiant temperature, wind speed, relative humidity, metabolic rate, and clothing insulation [36]. Detailed theoretical insights on PET can be found in Höppe [16], and PET computations can be easily executed with Software RayMan [5]. The PTS in Eq. (1) can be inferred from PET by establishing a relationship between PET and thermal sensation. This correlation varies by climate zone and is well-documented [36, 37]. In this research, the relationship between PET and thermal sensation is established based on field data from Hirashima et al. [38], yielding a coefficient of determination of 0.94 (Eq. 5 and Fig. 2).

$$PTS_{PET} = 0.17PET - 4.16 \tag{5}$$

where $PTS_{PET}$ is the thermal sensation predicted from the Physiological Equivalent Temperature (PET).

Bröde et al. [39] introduced the Universal Thermal Climate Index (UTCI), a one-dimensional metric based on the concept of equivalent temperature, designed to evaluate physiological responses to complex outdoor thermal conditions. This index assesses physiological factors, including core body temperature, sweat rate, skin

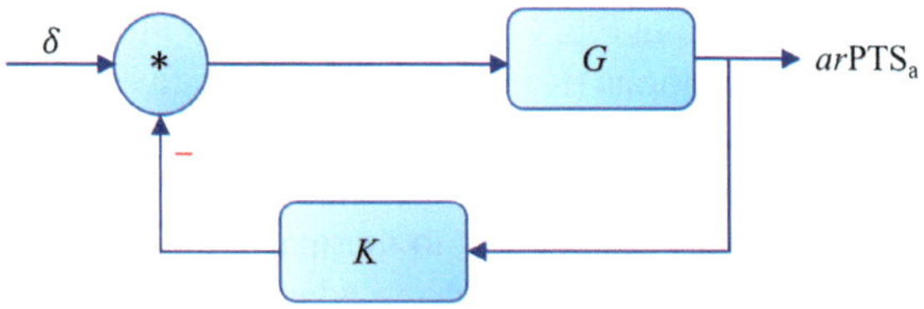


**Fig. 1** Negative feedback theory of cybernetics for adaptive method [27]

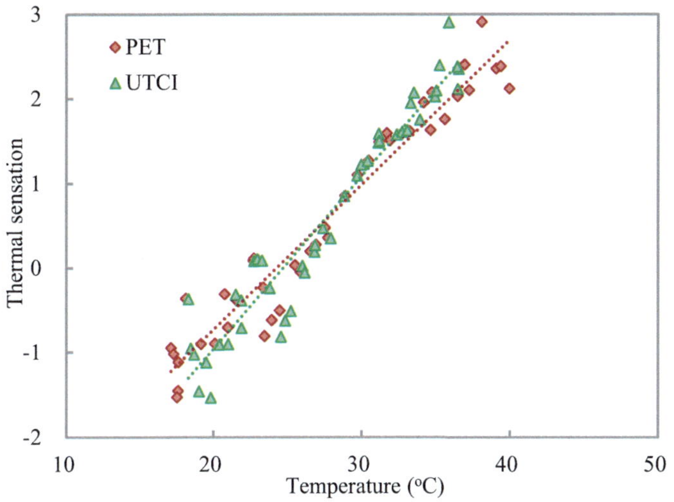


**Fig. 2** Variations of predicted thermal sensation (PTS) with PET and UTCI

wittedness, and skin temperature, using the UTCI-Fiala multi-node model, which relies on the heat balance of the human body [40]. UTCI computes equivalent temperature by considering aspects such as air temperature, mean radiant temperature, wind speed, and relative humidity [41]. Additionally, it incorporates adaptive clothing insulation [42]. Detailed theoretical insights into UTCI are available in Bröde [39], and calculations can be performed using RayMan Software [43]. The Physiological Thermal Sensation (PTS) in Eq. (1) can be derived from UTCI by establishing a relationship between UTCI and thermal sensation. This correlation varies across climate zones and has been well-documented [44]. In this research, the relationship between UTCI and thermal sensation is based on field data from Hirashima et al. [38] (Eq. 6) and Fig. 2), yielding a high coefficient of determination of 0.93.

$$PTS_{\mathrm{UTCI}} = 0.20UTCI - 5.05 \tag{6}$$

where $PTS_{\mathrm{UTCI}}$ is the predicted thermal sensation (PTS) from the Universal Thermal Climate Index (UTCI).

Apart from PTS, an essential element in calculating the adaptive arPTS is the adaptive coefficient ($\lambda$ in Eq. 1). Initially considered fixed [27, 45], field data indicate that thermal adaptations can vary [18]. To address these variations, a linear relationship is proposed between thermal adaptations and outdoor thermal stimuli (Eq. 7), consistent with the linear adaptive model commonly cited in thermal comfort standards [24]. By integrating Eqs. (1), (4), and (7), the adaptive arPTS is reformulated in Eq. (8). The constants p and q in Eq. (8) are determined to minimize the disparity between thermal sensation ratings from field surveys and those predicted by the adaptive arPTS (Eq. 9). This can be rearranged into Eq. (10). Using Eq. (8),

Eq. (10) can be expressed as Eq. (11). To reduce the difference between observed thermal sensation ratings and those predicted by the adaptive arPTS, the derivatives of this difference with respect to Constants p and q must equal zero Eqs. (12) and (13). Solving these Equations provides values for Constants q and p, as shown in Eqs. (14) and (15).

$$K = p\delta + q \tag{7}$$

$$arPTS_{\mathrm{a}} = \frac{PTS}{1 + \left(\frac{q}{\delta} + p\right)PTS} \tag{8}$$

$$\varepsilon_{\mathrm{a}} = \sum_{i=1}^{n} \frac{TSV_{\mathrm{i}}^{2}}{arPTS_{\mathrm{a,i}}^{2}} \left(arPTS_{\mathrm{a,i}} - TSV_{\mathrm{i}}\right)^{2} \tag{9}$$

$$\varepsilon_{\mathrm{a}} = \sum_{i=1}^{n} TSV_{\mathrm{i}}^{4} \left(\frac{1}{arPTS_{\mathrm{a,i}}} - \frac{1}{TSV_{\mathrm{i}}}\right)^{2} \tag{10}$$

$$\varepsilon_{\mathrm{a}} = \sum_{i=1}^{n} TSV_{\mathrm{i}}^{4} \left(\frac{1}{PTS_{\mathrm{i}}} + \frac{q}{\delta_{\mathrm{i}}} + p - \frac{1}{TSV_{\mathrm{i}}}\right)^{2} \tag{11}$$

$$\frac{\partial \varepsilon_{\mathrm{a}}}{\partial p} = 2\sum_{i=1}^{n} TSV_{\mathrm{i}}^{4} \left(\frac{1}{PTS_{\mathrm{i}}} + \frac{q}{\delta_{\mathrm{i}}} + p - \frac{1}{TSV_{\mathrm{i}}}\right) = 0 \tag{12}$$

$$\frac{\partial \varepsilon_{\mathrm{a}}}{\partial q} = 2\sum_{i=1}^{n} \frac{TSV_{\mathrm{i}}^{4}}{\delta_{\mathrm{i}}} \left(\frac{1}{PTS_{\mathrm{i}}} - \frac{1}{TSV_{\mathrm{i}}} + \frac{q}{\delta_{\mathrm{i}}} + p\right) = 0 \tag{13}$$

$$q = \frac{\sum_{i=1}^{n} \frac{TSV_{\mathrm{i}}^{4}}{\delta_{\mathrm{i}} PTS_{\mathrm{i}}} \sum_{i=1}^{n} TSV_{\mathrm{i}}^{4} + \sum_{i=1}^{n} \frac{TSV_{\mathrm{i}}^{4}}{\delta_{\mathrm{i}}} \sum_{i=1}^{n} TSV_{\mathrm{i}}^{3} - \sum_{i=1}^{n} \frac{TSV_{\mathrm{i}}^{4}}{PTS_{\mathrm{i}}} \sum_{i=1}^{n} \frac{TSV_{\mathrm{i}}^{4}}{\delta_{\mathrm{i}}} - \sum_{i=1}^{n} TSV_{\mathrm{i}}^{4} \sum_{i=1}^{n} \frac{TSV_{\mathrm{i}}^{3}}{\delta_{\mathrm{i}}}}{\sum_{i=1}^{n} \frac{TSV_{\mathrm{i}}^{4}}{\delta_{\mathrm{i}}} \sum_{i=1}^{n} \frac{TSV_{\mathrm{i}}^{4}}{\delta_{\mathrm{i}}} - \sum_{i=1}^{n} \frac{TSV_{\mathrm{i}}^{4}}{\delta_{\mathrm{i}}^{2}} \sum_{i=1}^{n} TSV_{\mathrm{i}}^{4}} \tag{14}$$

$$p = \frac{\sum_{i=1}^{n} TSV_{\mathrm{i}}^{3} - \sum_{i=1}^{n} \frac{TSV_{\mathrm{i}}^{4}}{PTS_{\mathrm{i}}} - q\sum_{i=1}^{n} \frac{TSV_{\mathrm{i}}^{4}}{\delta_{\mathrm{i}}}}{\sum_{i=1}^{n} TSV_{\mathrm{i}}^{4}} \tag{15}$$

where p and q are two constants; i is the ith data set from field studies and there are n data sets in total; $\varepsilon_{\mathrm{a}}$ is the difference between the thermal sensation votes from field studies and those predicted by the adaptive arPTS.

As Eqs. (14) and (15) suggest that a predicted thermal sensation of zero would render Constants p and q undefined, the thermal sensation is adjusted to be greater than zero by introducing Constant $\gamma$ Eqs. (16) and (17). The general thermal sensation scale has a minimum value above $-5$; hence, a Constant $\gamma$ of 5 is utilized in this research. Outdoor thermal stimuli can be described using parameters such as outdoor air temperature, mean radiant temperature, wind speed, relative humidity, and other factors [36, 41]. This research utilizes both PET and UTCI to represent the overall outdoor thermal stimuli for the PET-based adaptive arPTS and the UTCI-based adaptive arPTS, respectively. Consequently, the formulations for the PET-based adaptive arPTS and the UTCI-based adaptive arPTS are presented in Eqs. (18) and (19) (Table 1). Constants q and p for the PET-based adaptive arPTS are calculated using Eqs. (20) and (21), while those for the UTCI-based adaptive arPTS are derived from Eqs. (22) and (23).

$$\overline{PTS} = PTS + \gamma \tag{16}$$

$$\overline{TSV} = TSV + \gamma \tag{17}$$

$$arPTS_{\mathrm{a,PET}} = \frac{\overline{PTS_{\mathrm{PET}}}}{1 + \left(\frac{q_{\mathrm{PET}}}{PET} + p_{\mathrm{PET}}\right)\overline{PTS_{\mathrm{PET}}}} - \gamma \tag{18}$$

$$arPTS_{\mathrm{a,UTCI}} = \frac{\overline{PTS_{\mathrm{UTCI}}}}{1 + \left(\frac{q_{\mathrm{UTCI}}}{UTCI} + p_{\mathrm{UTCI}}\right)\overline{PTS_{\mathrm{UTCI}}}} - \gamma \tag{19}$$

$$q_{\mathrm{PET}} = \frac{\sum_{i=1}^{n} \frac{TSV_{\mathrm{i}}^4}{PET_{\mathrm{i}} PTS_{\mathrm{i}}} \sum_{i=1}^{n} TSV_{\mathrm{i}}^4 + \sum_{i=1}^{n} \frac{TSV_{\mathrm{i}}^4}{PET_{\mathrm{i}}} \sum_{i=1}^{n} TSV_{\mathrm{i}}^3 - \sum_{i=1}^{n} \frac{TSV_{\mathrm{i}}^4}{PTS_{\mathrm{i}}} \sum_{i=1}^{n} \frac{TSV_{\mathrm{i}}^4}{PET_{\mathrm{i}}} - \sum_{i=1}^{n} TSV_{\mathrm{i}}^4 \sum_{i=1}^{n} \frac{TSV_{\mathrm{i}}^3}{PET_{\mathrm{i}}}}{\sum_{i=1}^{n} \frac{TSV_{\mathrm{i}}^4}{PET_{\mathrm{i}}} \sum_{i=1}^{n} \frac{TSV_{\mathrm{i}}^4}{PET_{\mathrm{i}}} - \sum_{i=1}^{n} \frac{TSV_{\mathrm{i}}^4}{PET_{\mathrm{i}}^2} \sum_{i=1}^{n} TSV_{\mathrm{i}}^4} \tag{20}$$

$$p_{\mathrm{PET}} = \frac{\sum_{i=1}^{n} TSV_{\mathrm{i}}^3 - \sum_{i=1}^{n} \frac{TSV_{\mathrm{i}}^4}{PTS_{\mathrm{i}}} - q_{\mathrm{PET}} \sum_{i=1}^{n} \frac{TSV_{\mathrm{i}}^4}{PET_{\mathrm{i}}}}{\sum_{i=1}^{n} TSV_{\mathrm{i}}^4} \tag{21}$$

$$q_{\mathrm{UTCI}} = \frac{\sum_{i=1}^{n} \frac{TSV_{\mathrm{i}}^4}{UTCI_{\mathrm{i}} PTS_{\mathrm{i}}} \sum_{i=1}^{n} TSV_{\mathrm{i}}^4 + \sum_{i=1}^{n} \frac{TSV_{\mathrm{i}}^4}{UTCI_{\mathrm{i}}} \sum_{i=1}^{n} TSV_{\mathrm{i}}^3 - \sum_{i=1}^{n} \frac{TSV_{\mathrm{i}}^4}{PTS_{\mathrm{i}}} \sum_{i=1}^{n} \frac{TSV_{\mathrm{i}}^4}{UTCI_{\mathrm{i}}} - \sum_{i=1}^{n} TSV_{\mathrm{i}}^4 \sum_{i=1}^{n} \frac{TSV_{\mathrm{i}}^3}{UTCI_{\mathrm{i}}}}{\sum_{i=1}^{n} \frac{TSV_{\mathrm{i}}^4}{UTCI_{\mathrm{i}}} \sum_{i=1}^{n} \frac{TSV_{\mathrm{i}}^4}{UTCI_{\mathrm{i}}} - \sum_{i=1}^{n} \frac{TSV_{\mathrm{i}}^4}{UTCI_{\mathrm{i}}^2} \sum_{i=1}^{n} TSV_{\mathrm{i}}^4} \tag{22}$$

$$p_{\mathrm{UTCI}} = \frac{\sum_{i=1}^{n} TSV_{\mathrm{i}}^3 - \sum_{i=1}^{n} \frac{TSV_{\mathrm{i}}^4}{PTS_{\mathrm{i}}} - q_{\mathrm{UTCI}} \sum_{i=1}^{n} \frac{TSV_{\mathrm{i}}^4}{UTCI_{\mathrm{i}}}}{\sum_{i=1}^{n} TSV_{\mathrm{i}}^4} \tag{23}$$

## 2.2 *Adaptive-Rational Outdoor Thermal Comfort Models with Extension Method*

The extension method enhances the arPTS, referred to as the extended arPTS, by integrating an expansion factor and a thermal neutrality factor into the rational model. The expansion factor (designated as e in Eq. 24) along with the thermal neutrality factor (c in Eq. 24) allows the rational model (PTS in Eq. 24) to consider thermal adjustments.

**Table 1** Summary of proposed PET-based and UTICI-based adaptive-rational outdoor thermal comfort models

| | | |
|---|---|---|
| Adaptive method | PET-based | $arPTS_{\mathrm{a,PET}} = \frac{PTS_{\mathrm{PET}}}{1+(\frac{q_{\mathrm{PET}}}{PET}+p_{\mathrm{PET}})PTS_{\mathrm{PET}}} - \gamma$ <br> $q_{\mathrm{PET}}$: Eq. (20) <br> $p_{\mathrm{PET}}$: Eq. (21) |
| | UTCI-based | $arPTS_{\mathrm{a,UTCI}} = \frac{PTS_{\mathrm{UTCI}}}{1+(\frac{q_{\mathrm{UTCI}}}{UTCI}+p_{\mathrm{UTCI}})PTS_{\mathrm{UTCI}}} - \gamma$ <br> $q_{\mathrm{UTCI}}$: Eq. (22) <br> $p_{\mathrm{UTCI}}$: Eq. (23) |
| Extension method | PET-based | $arPTS_{\mathrm{e,PET}} = (a_{\mathrm{PET}}PET + b_{\mathrm{PET}})PTS_{\mathrm{PET}} + c_{\mathrm{PET}}$ <br> $a_{\mathrm{PET}}$: Eq. (34) <br> $b_{\mathrm{PET}}$: Eq. (35) <br> $c_{\mathrm{PET}}$: Eq. (36) |
| | UTCI-based | $arPTS_{\mathrm{e,UTCI}} = (a_{\mathrm{UTCI}}UTCI + b_{\mathrm{UTCI}})PTS_{\mathrm{UTCI}} + c_{\mathrm{UTCI}}$ <br> $a_{\mathrm{UTCI}}$: Eq. (38) <br> $b_{\mathrm{UTCI}}$: Eq. (39) <br> $c_{\mathrm{UTCI}}$: Eq. (40) |

Fanger and Toftum [26] initially introduced the expansion method with a fixed expansion factor. This research proposes that the extension factor should be variable to reflect the differing thermal adaptations (Eq. 25). This factor is directly related to outdoor thermal stimuli, consistent with the linear thermal comfort model for adaptations widely acknowledged in thermal comfort norms [30, 46]. Furthermore, various field studies have reported thermal adaptations around thermal neutrality, leading to different thermally neutral outdoor scenarios [19]. Nonetheless, the extension factor does not account for these adaptations surrounding thermal neutrality [29]. Consequently, the thermal neutrality factor is included to address this limitation.

$$arPTS_{\mathrm{e}} = e \times PTS + c \tag{24}$$

$$e = a \times \delta + b \tag{25}$$

where $\mathrm{arPTS_e}$ is the adaptive-rational outdoor thermal comfort model with the extension method; a and b are constants; c is the thermal neutrality factor; e is the extension factor.

The three constants in the extended arPTS are determined to minimize the discrepancy between thermal sensation ratings gathered from field surveys and those estimated by the extended arPTS Eq. (26). This necessitates that the derivatives of the variance between the field thermal sensation ratings and those predicted by the extended arPTS concerning Constants a, b, and c must equal zero Eqs. (27)–(29). By combining Eqs. (27)–(29), the values for Constants a, b, and c are calculated as shown in Eqs. (30)–(32). It is important to highlight that x, y, and z in Eqs. (30)–(32) signify the anticipated (PTS), thermal sensation vote (TSV), and outdoor thermal stimuli (δ), respectively.

$$\varepsilon_{\mathrm{e}} = \sum_{i=1}^{n} ((a\delta_{\mathrm{i}} + b)PTS_{\mathrm{i}} + c - TSV_{\mathrm{i}})^2 \tag{26}$$

$$\frac{\partial \varepsilon_{\mathrm{e}}}{\partial a} = 2\sum_{i=1}^{n} ((a\delta_{\mathrm{i}} + b)PTS_{\mathrm{i}} + c - TSV_{\mathrm{i}})\delta_{\mathrm{i}}PTS_{\mathrm{i}} = 0 \tag{27}$$

$$\frac{\partial \varepsilon_{\mathrm{e}}}{\partial b} = 2\sum_{i=1}^{n} ((a\delta_{\mathrm{i}} + b)PTS_{\mathrm{i}} + c - TSV_{\mathrm{i}})PTS_{\mathrm{i}} = 0 \tag{28}$$

$$\frac{\partial \varepsilon_{\mathrm{e}}}{\partial c} = 2\sum_{i=1}^{n} ((a\delta_{\mathrm{i}} + b)PTS_{\mathrm{i}} + c - TSV_{\mathrm{i}}) = 0 \tag{29}$$

$$a = \frac{\left(\frac{\sum x_{\mathrm{i}}z_{\mathrm{i}}\sum y_{\mathrm{i}}}{n} - \sum x_{\mathrm{i}}y_{\mathrm{i}}z_{\mathrm{i}}\right)\left(\sum x_{\mathrm{i}}^2 - \frac{(\sum x_{\mathrm{i}})^2}{n}\right) - \left(\frac{\sum x_{\mathrm{i}}\sum y_{\mathrm{i}}}{n} - \sum x_{\mathrm{i}}y_{\mathrm{i}}\right)\left(\sum x_{\mathrm{i}}^2 z_{\mathrm{i}} - \frac{\sum x_{\mathrm{i}}z_{\mathrm{i}}\sum x_{\mathrm{i}}}{n}\right)}{\left(\sum x_{\mathrm{i}}^2 z_{\mathrm{i}} - \frac{\sum x_{\mathrm{i}}z_{\mathrm{i}}\sum x_{\mathrm{i}}}{n}\right)^2 - \left(\sum x_{\mathrm{i}}^2 z_{\mathrm{i}}^2 - \frac{\sum x_{\mathrm{i}}z_{\mathrm{i}}\sum x_{\mathrm{i}}z_{\mathrm{i}}}{n}\right)\left(\sum x_{\mathrm{i}}^2 - \frac{(\sum x_{\mathrm{i}})^2}{n}\right)} \tag{30}$$

$$b = \frac{\sum x_{\mathrm{i}}y_{\mathrm{i}} - \frac{\sum x_{\mathrm{i}}\sum y_{\mathrm{i}}}{n} - a\left(\sum x_{\mathrm{i}}^2 z_{\mathrm{i}} - \frac{\sum x_{\mathrm{i}}z_{\mathrm{i}}\sum x_{\mathrm{i}}}{n}\right)}{\sum x_{\mathrm{i}}^2 - \frac{(\sum x_{\mathrm{i}})^2}{n}} \tag{31}$$

$$c = \frac{\sum y_{\mathrm{i}} - a\sum x_{\mathrm{i}}z_{\mathrm{i}} - b\sum x_{\mathrm{i}}}{n} \tag{32}$$

where $\varepsilon_e$ is the difference between the thermal sensation votes from field studies and those predicted by the extended arPTS; x, y, and z are the predicted thermal sensation, thermal sensation vote, and outdoor thermal stimuli respectively.

For the PET-based extended arPTS, the PTS is derived from the PET Eq. (5), as specified in Eq. (33). In this scenario, outdoor thermal stimuli in the PET-based extended arPTS are indicated by the PET. Hence, Constants a, b, and c for the PET-based extended arPTS are computed using Eqs. (34)–(36). Similarly, for the UTCI-based extended arPTS, the PTS is calculated from the UTCI Eq. (6), as described in Eq. (37). In this case, outdoor thermal stimuli are characterized by the UTCI, and Constants a, b, and c for the UTCI-based extended arPTS are determined through Eqs. (38)–(40), respectively (Table 1).

$$arPTS_{\mathrm{e,PET}} = (a_{\mathrm{PET}}PET + b_{\mathrm{PET}})PTS_{\mathrm{PET}} + c_{\mathrm{PET}} \tag{33}$$

$$a_{\mathrm{PET}} = \frac{\left(\frac{\sum x_{\mathrm{i}}PET_{\mathrm{i}}\sum y_{\mathrm{i}}}{n} - \sum x_{\mathrm{i}}y_{\mathrm{i}}PET_{\mathrm{i}}\right)\left(\sum x_{\mathrm{i}}^2 - \frac{(\sum x_{\mathrm{i}})^2}{n}\right) - \left(\frac{\sum x_{\mathrm{i}}\sum y_{\mathrm{i}}}{n} - \sum x_{\mathrm{i}}y_{\mathrm{i}}\right)\left(\sum x_{\mathrm{i}}^2 PET_{\mathrm{i}} - \frac{\sum x_{\mathrm{i}}PET_{\mathrm{i}}\sum x_{\mathrm{i}}}{n}\right)}{\left(\sum x_{\mathrm{i}}^2 PET_{\mathrm{i}} - \frac{\sum x_{\mathrm{i}}PET_{\mathrm{i}}\sum x_{\mathrm{i}}}{n}\right)^2 - \left(\sum x_{\mathrm{i}}^2 PET_{\mathrm{i}}^2 - \frac{\sum x_{\mathrm{i}}PET_{\mathrm{i}}\sum x_{\mathrm{i}}PET_{\mathrm{i}}}{n}\right)\left(\sum x_{\mathrm{i}}^2 - \frac{(\sum x_{\mathrm{i}})^2}{n}\right)} \tag{34}$$

$$b_{\mathrm{PET}} = \frac{\sum x_{\mathrm{i}}y_{\mathrm{i}} - \frac{\sum x_{\mathrm{i}}\sum y_{\mathrm{i}}}{n} - a_{\mathrm{PET}}\left(\sum x_{\mathrm{i}}^2 PET_{\mathrm{i}} - \frac{\sum x_{\mathrm{i}}PET_{\mathrm{i}}\sum x_{\mathrm{i}}}{n}\right)}{\sum x_{\mathrm{i}}^2 - \frac{(\sum x_{\mathrm{i}})^2}{n}} \tag{35}$$

$$c_{\mathrm{PET}} = \frac{\sum y_{\mathrm{i}} - a_{\mathrm{PET}} \sum x_{\mathrm{i}} PET_{\mathrm{i}} - b_{\mathrm{PET}} \sum x_{\mathrm{i}}}{n} \tag{36}$$

$$arPTS_{\mathrm{e,UTCI}} = (a_{\mathrm{UTCI}} UTCI + b_{\mathrm{UTCI}}) PTS_{\mathrm{UTCI}} + c_{\mathrm{UTCI}} \tag{37}$$

$$a_{\mathrm{UTCI}} = \frac{\left(\frac{\sum x_{\mathrm{i}} UTCI_{\mathrm{i}} \sum y_{\mathrm{i}}}{n} - \sum x_{\mathrm{i}} y_{\mathrm{i}} UTCI_{\mathrm{i}}\right)\left(\sum x_{\mathrm{i}}^2 - \frac{(\sum x_{\mathrm{i}})^2}{n}\right) - \left(\frac{\sum x_{\mathrm{i}} \sum y_{\mathrm{i}}}{n} - \sum x_{\mathrm{i}} y_{\mathrm{i}}\right)\left(\sum x_{\mathrm{i}}^2 UTCI_{\mathrm{i}} - \frac{\sum x_{\mathrm{i}} UTCI_{\mathrm{i}} \sum x_{\mathrm{i}}}{n}\right)}{\left(\sum x_{\mathrm{i}}^2 UTCI_{\mathrm{i}} - \frac{\sum x_{\mathrm{i}} UTCI_{\mathrm{i}} \sum x_{\mathrm{i}}}{n}\right)^2 - \left(\sum x_{\mathrm{i}}^2 UTCI_{\mathrm{i}}^2 - \frac{\sum x_{\mathrm{i}} UTCI_{\mathrm{i}} \sum x_{\mathrm{i}} UTCI_{\mathrm{i}}}{n}\right)\left(\sum x_{\mathrm{i}}^2 - \frac{(\sum x_{\mathrm{i}})^2}{n}\right)} \tag{38}$$

$$b_{\mathrm{UTCI}} = \frac{\sum x_{\mathrm{i}} y_{\mathrm{i}} - \frac{\sum x_{\mathrm{i}} \sum y_{\mathrm{i}}}{n} - a_{\mathrm{UTCI}}\left(\sum x_{\mathrm{i}}^2 UTCI_{\mathrm{i}} - \frac{\sum x_{\mathrm{i}} UTCI_{\mathrm{i}} \sum x_{\mathrm{i}}}{n}\right)}{\sum x_{\mathrm{i}}^2 - \frac{(\sum x_{\mathrm{i}})^2}{n}} \tag{39}$$

$$c_{\mathrm{UTCI}} = \frac{\sum y_{\mathrm{i}} - a_{\mathrm{UTCI}} \sum x_{\mathrm{i}} UTCI_{\mathrm{i}} - b_{\mathrm{UTCI}} \sum x_{\mathrm{i}}}{n} \tag{40}$$

## 2.3 *Field Data for Model Validation*

Field studies designed to evaluate outdoor thermal comfort serve to validate the proposed adaptive-rational outdoor thermal comfort models. These investigations were conducted across four construction sites in Guangzhou (see Fig. 3) from July 18 to August 9, 2019. The subtropical climate of Guangzhou is marked by high temperatures and humidity levels throughout the summer. A total of 1063 workers participated in surveys about their thermal sensation. Participants used a nine-point scale to rate their comfort, where 4 indicates very hot, 3 is hot, 2 means warm, 1 is slightly warm, 0 is neutral, − 1 denotes slightly cool, − 2 indicates cool, − 3 represents cold, and − 4 means very cold. The environmental parameters measured are summarized in Table 2. Heart rates were also monitored to determine metabolic rates [47]. Clothing insulation values were calculated to range from 0.43 to 0.58 clo, following guidelines from the ASHRAE Handbook [48]. These data were used to calculate the PET [6] (from 23.6 to 58.5 °C) and UTCI [39] (from 27.8 °C). Objective measurements were taken at one-minute intervals during the workday. During work hours (8:00–11:30 a.m. and 2:30–6:00 p.m.), thermal sensation ratings were collected every 30 min. Additional details on the field studies can be found in Tang et al. [49].

**Fig. 3** Subjective surveys at construction sites

**Table 2** Measured parameters

| Parameter | Equipment | Measuring range | Measuring accuracy |
|---|---|---|---|
| Air temperature | Thermal comfort instrument SSDZY-1 | – 20–80 °C | ± 0.3 °C |
| Globe temperature | Thermal comfort instrument SSDZY-1 | – 20–80 °C | ± 0.3 °C |
| Relative humidity | Thermal comfort instrument SSDZY-1 | 0.01–99.9% | ± 2% |
| Wind speed | Universal wind speed recorder WFWZY-1 | 0.05–5 m/s | 5% ± 0.05 m/s |
| Heart rate | ADZ AV5 | 40–195 beats/min | ± 5% |

# 3 Results

## *3.1 PET-Based Adaptive-Rational Thermal Comfort Models*

As shown in Fig. 4, thermal sensation ratings from the field studies varied from 1.8 (warm) to 4.0 (very hot). The elevated ratings are largely due to the extreme outdoor conditions (with PET values reaching 58.5 °C) and the high activity levels of the construction workers. The PTS derived from the PET Eq. (5) does not precisely forecast these thermal sensation ratings (Fig. 4). When the PET falls below 45 °C, the PTS underestimates the thermal sensation ratings; conversely, it overestimates them when the PET exceeds this threshold. The dataset from the field studies is split into two parts: one segment is used to determine Constants p and q for the PET-based adaptive arPTS and Constants a, b, and c for the PET-based extended arPTS (training data), while the other segment is used for validation of the developed models (validation data) (Fig. 4). The model's performance is evaluated using the root mean square error (RMSE) Eq. (41) and the standard deviation of absolute errors (SDAE)

Eq. (42). A lower RMSE indicates improved accuracy [50], while a smaller SDAE reflects better robustness [51].

$$RMSE = \sqrt{\frac{\sum (TSV_i - prediction_i)^2}{n}} \tag{41}$$

$$SDAE = \sqrt{\frac{\sum \left( |TSV_i - prediction_i| - \frac{1}{n} \sum |TSV_i - prediction_i| \right)^2}{n - 1}} \tag{42}$$

where RMSE is the root mean square error; SDAE is the standard deviation of absolute errors.

Using the training data, Constants p and q for the PET-based adaptive arPTS were determined as 0.0769 and − 3.4325 based on Eqs. (20) and (21), respectively. Figure 4

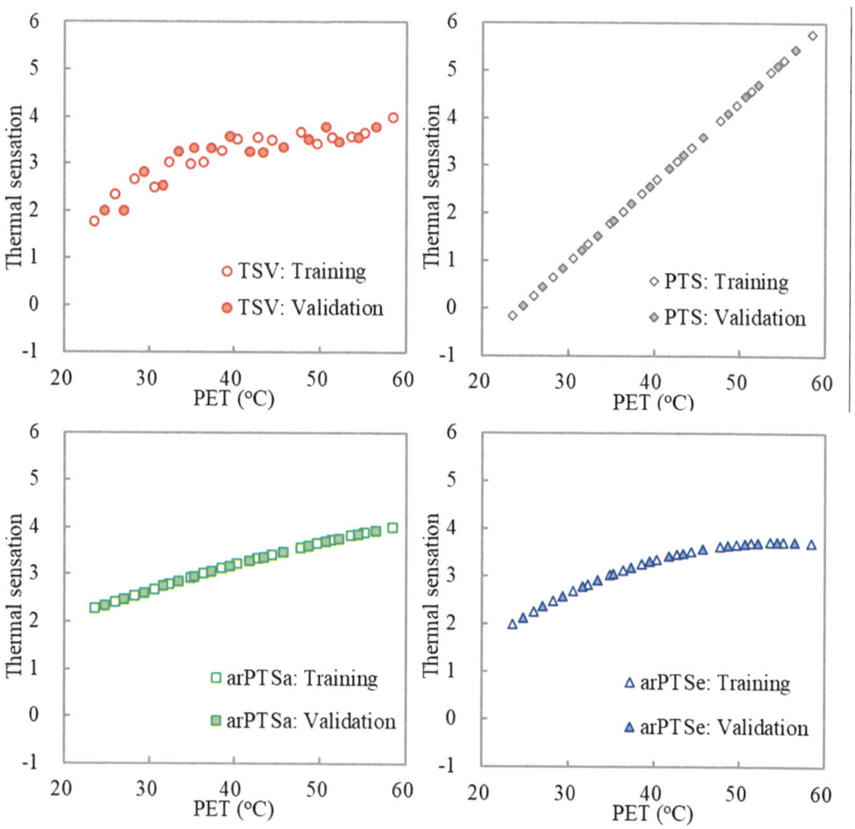


**Fig. 4** Variations of TSV, PTS, adaptive arPMV (arPMVa), and extended arPMV (arPMVe) with PET

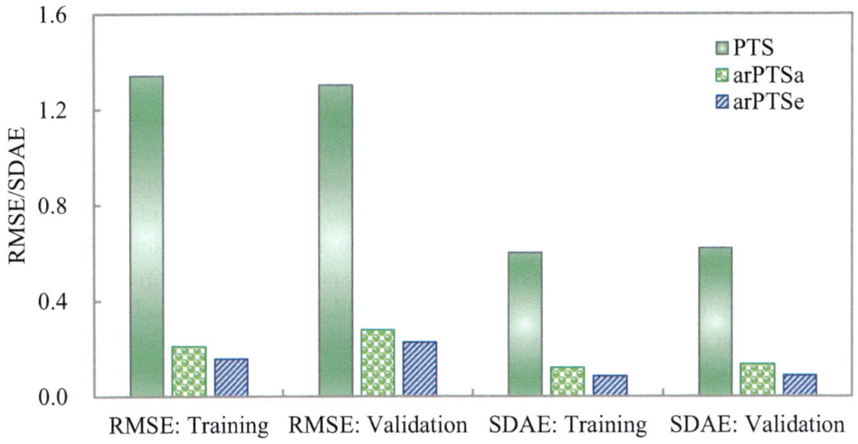


**Fig. 5** Root mean square error (RMSE) and standard deviation of absolute errors (SDAE) of PTS, adaptive *ar*PMV ($ar$PMV$_a$), and extended *ar*PMV ($ar$PMV$_e$): PET-based models

illustrates a strong correlation between the predicted thermal sensations by the PET-based adaptive arPTS and the actual thermal sensation votes in both the training and validation datasets. For the training dataset, the RMSE and SDAE values for the PET-based adaptive arPTS are 0.21 and 0.12, respectively (Fig. 5). In the validation dataset, these values are 0.28 and 0.13, respectively. The low RMSE and SDAE values for both datasets indicate the reliability of the PET-based adaptive arPTS. Compared to the PET-based PTS for the validation data, the PET-based adaptive arPTS enhances accuracy and robustness by 79% (reducing RMSE from 1.30 to 0.28) and 78% (reducing SDAE from 0.62 to 0.13), respectively. Thus, the PET-based adaptive-rational outdoor thermal comfort model employing the adaptive method aligns well with the field study results.

From the training data, Constants a, b, and c for the PET-based extended arPTS were calculated as − 0.0103, 0.8801, and 2.0970 using Eqs. (34)–(36). Figure 4 indicates that thermal sensations predicted by the PET-based extended arPTS closely match the actual thermal sensation votes for both training and validation datasets. For the training data, the RMSE and SDAE for the PET-based extended arPTS are 0.16 and 0.08, respectively (Fig. 5). For the validation data, these values are 0.23 and 0.09. The small RMSE and SDAE values across both datasets indicate the reliability of the PET-based extended arPTS. Compared to the PET-based PTS for the validation data, the PET-based extended arPTS enhances accuracy and robustness by 83% (reducing RMSE from 1.30 to 0.23) and 86% (reducing SDAE from 0.62 to 0.09), respectively. Therefore, the PET-based adaptive-rational outdoor thermal comfort model utilizing the extended method is effectively applicable to the field studies.

### *3.2 UTCI-Based Adaptive-Rational Thermal Comfort Models*

Figure 6 shows that the UTCI-based PTS Eq. (6) exhibits notable discrepancies from the thermal sensation ratings. Specifically, this model tends to underestimate thermal sensation when UTCI falls below 40 °C and overestimate when it exceeds this threshold. To create and validate the UTCI-based adaptive-rational outdoor thermal comfort models, the field studies are segmented into training and validation datasets (Fig. 6). The UTCI-based adaptive arPTS are calculated from the training data, with Constants p and q set at 0.0874 and − 3.5786, correspondingly Eqs. (22) and (23). The thermal sensations predicted by the UTCI-based adaptive arPTS closely correspond to the thermal sensation votes in both datasets (Fig. 6). The RMSE values for the UTCI-based adaptive arPTS are 0.27 for the training set and 0.21 for the validation set, while the SDAE values are 0.21 for training and 0.14 for validation (Fig. 7). In the validation dataset, the UTCI-based adaptive arPTS enhances accuracy and robustness by 87% and 85%, respectively, compared to the UTCI-based PTS, with RMSE and SDAE improving from 1.60 to 0.21 and from 0.96 to 0.14, respectively. Hence, the UTCI-based adaptive-rational outdoor thermal comfort model using the adaptive approach is well-suited for the field studies.

Using the training data from the field studies, the extension method calculates the UTCI-based extended arPTS with Constants a, b, and c set at − 0.0058, 0.6453, and 1.8364, sequentially Eqs. (38)–(40). The thermal sensation votes anticipated by the UTCI-based extended arPTS closely match the actual thermal sensation votes (Fig. 3). The RMSE and SDAE for the UTCI-based extended arPTS are 0.26 and 0.20 for the training data, and 0.20 and 0.13 for the validation data, respectively. When contrasted with the UTCI-based PTS for the validation data, the UTCI-based extended arPTS boosts accuracy by 87%, reducing RMSE from 1.60 to 0.20, and enhances robustness by 87%, lowering SDAE from 0.96 to 0.13. Therefore, the UTCI-based adaptive-rational outdoor thermal comfort model employing the extended method is effectively applicable to the field studies.

## 4 Discussion

Given thermal adaptations, the original PTS developed for one context may not be suitable for others when variables such as climate type and cultural influences vary. This study introduces adaptive-rational outdoor thermal comfort modeling methods that can modify the original PTS for application in various contexts. The original PET-based and UTCI-based PTSs Eqs. (5) and (6) were specifically designed for the Brazilian context [38] and do not apply to the Chinese context. In contrast, the proposed modeling methods effectively adapt the original PTSs to be relevant to the Chinese scenario.

Studies indicate that different thermal indices significantly impact thermal comfort modeling [5, 34]. Among these indices, PET and UTCI are the most frequently

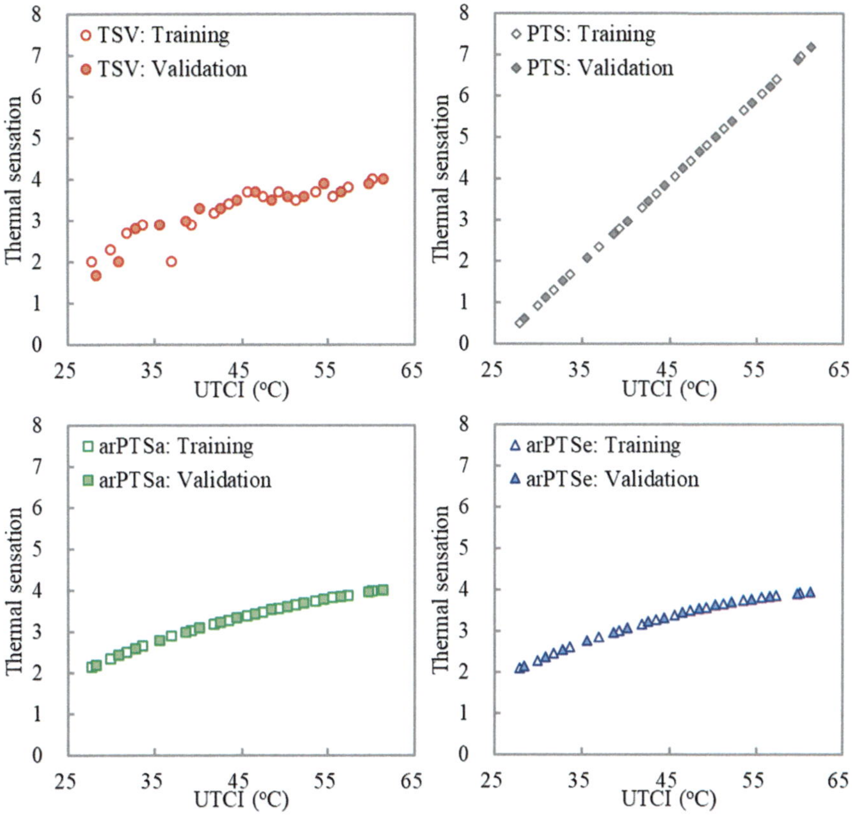


**Fig. 6** Variations of TSV, PTS, adaptive arPMV (arPMVa), and extended arPMV (arPMVe) with UTCI

employed for evaluating outdoor thermal comfort [5, 34]. Since these indices can yield varying results [5, 8], validating modeling methods based on one thermal index does not ensure their effectiveness for another. Therefore, the modeling techniques have been assessed for both PET and UTCI. The findings suggest that the proposed adaptive-rational outdoor thermal comfort models can accurately predict thermal comfort, regardless of the thermal index utilized (Sect. 3). Future research will focus on further developing these models using additional thermal indices and validating them across a range of climate types, thermal sensation scales, and metabolic rates.

Furthermore, the variable thermal adaptations addressed in this study show a linear correlation with both PET and UTCI for their respective adaptive-rational outdoor thermal comfort models (Sects. 2.1 and 2.2). However, other factors, such as outdoor air temperature [24] and relative humidity [52], might provide a more precise depiction of variable thermal adaptations. For instance, standards like GB50785 [53], ASHRAE 55 [54], and EN15251 [55] use different running mean outdoor air temperatures to reflect thermal adaptations. Future investigations will seek to identify

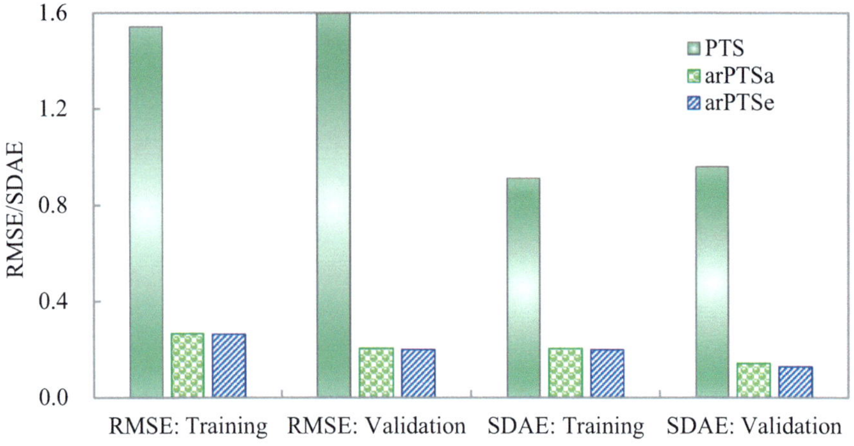


**Fig. 7** Root mean square error (RMSE) and standard deviation of absolute errors (SDAE) of PTS, adaptive arPMV (arPMVa), and extended arPMV (arPMVe): UTCI-based models

the most effective parameters for characterizing these variable adaptations to enhance the adaptive-rational outdoor thermal comfort models.

The effectiveness and dependability of the proposed adaptive-rational models for outdoor thermal comfort are further assessed by contrasting them with traditional methods, including linear regression, commonly used in outdoor thermal comfort modeling. This traditional technique establishes a linear correlation between the thermal index and thermal sensation ratings, a widely acknowledged practice. To demonstrate this, the PET-based traditional model is formulated in Eq. (43), while the UTCI-based traditional model is expressed in Eq. (44). As depicted in Fig. 8, the PET-based adaptive arPTS enhances accuracy and robustness by 6% and 17%, respectively, in comparison to the traditional models for the validation dataset. Similarly, the UTCI-based adaptive arPTS improves accuracy and robustness by 23% and 5%. In contrast, the PET-based extended arPTS enhances accuracy and robustness by 23% and 47%, respectively, compared to the traditional models for the validation dataset. The UTCI-based extended arPTS demonstrates enhancements of 25% in accuracy and 16% in robustness. The superior performance of the extended arPTS is attributed to its capability to consider thermal adaptations around thermal neutrality (Sect. 2.2), a crucial aspect that the adaptive arPTS fails to address (Sect. 2.1). Thermal adaptations close to thermal neutrality play a vital role in outdoor thermal comfort. Future endeavors will focus on refining the adaptive arPTS to integrate these adaptations.

$$PTS_{\mathrm{L,PET}} = 0.0494PET + 1.1704 \quad (43)$$

$$PTS_{\mathrm{L,UTCI}} = 0.0572UTCI + 0.6713 \quad (44)$$

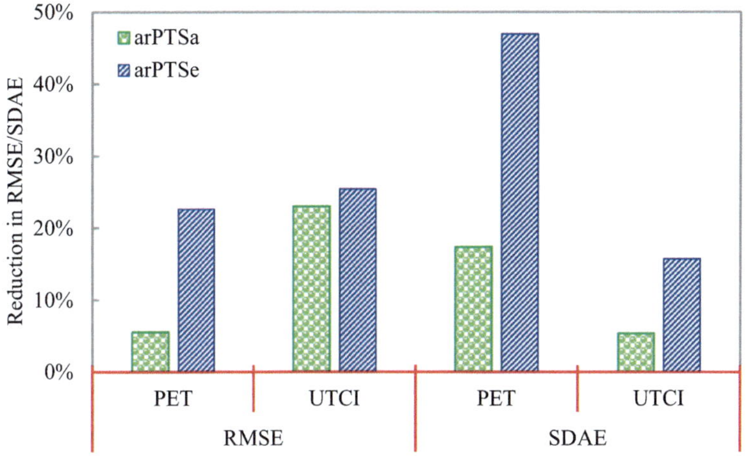


**Fig. 8** Reductions in root mean square error (RMSE) and standard deviation of absolute errors (SDAE) by adaptive *ar*PMV (*ar*$PMV_a$) and extended *ar*PMV (*ar*$PMV_e$) compared with conventional models

where PTSL, PET and PTSL, UTCI are the PET-based and UTCI-based thermal sensation predictions from the conventional linear regressions respectively.

## 5 Conclusions

This chapter introduced adaptive-rational models for outdoor thermal comfort that merge rational heat balance formulations with adaptive and extension strategies, using PET and UTCI as representative indices. Validation through field studies demonstrated that these models significantly enhance predictive accuracy and robustness compared to conventional rational approaches. The PET-based adaptive-rational models improved performance by more than 80%, while the UTCI-based models achieved gains of up to 87%, underscoring the value of explicitly incorporating thermal adaptation processes into comfort prediction. Among the proposed methods, the extension approach provided the greatest improvements, particularly near thermal neutrality, where adaptive behavior exerts a stronger influence on perception. Overall, the adaptive-rational modeling framework offers a more reliable and flexible basis for assessing outdoor comfort across diverse climates and activity contexts. By strengthening the connection between physiological heat balance and adaptive responses, this approach provides valuable feedback for urban planning and design aimed at fostering thermally comfortable and sustainable outdoor environments.

## References


1. M. Mohammadi, P.W. Tien, J.K. Calautit, Numerical evaluation of the use of vegetation as a shelterbelt for enhancing the wind and thermal comfort in peripheral and lateral-type skygardens in highrise buildings. Build. Simul. **16**, 243–261 (2022). https://doi.org/10.1007/s12273-022-0943-7
2. S. Zhang, D. Niu, Z. Lin, Occupancy-aided ventilation for airborne infection risk control: continuously or intermittently reduced occupancies? Build. Simul. **16**, 733–747 (2022). https://doi.org/10.1007/s12273-022-0951-7
3. S. Zhang, Y. Lu, D. Niu, Z. Lin, Energy performance index of air distribution: thermal utilization effectiveness. Appl. Energy **307**, 118122 (2022)
4. Y. Zhang, N.C.F. Tse, H. Ren, Y. Sun, A novel coordinated control for NZEB clusters to minimize their connected grid overvoltage risks. Build. Simul. **15**, 1831–1848 (2022)
5. P. Kumar, A. Sharma, Study on importance, procedure, and scope of outdoor thermal comfort-A review. Sustain. Cities Soc. **61**, 102297 (2020)
6. S. Coccolo, J. Kämpf, J.L. Scartezzini, D. Pearlmutter, Outdoor human comfort and thermal stress: a comprehensive review on models and standards. Urban Climate **18**, 33–57 (2016)
7. C.S. Fong, N. Aghamohammadi, L. Ramakreshnan, N.M. Sulaiman, P. Mohammadi, Holistic recommendations for future outdoor thermal comfort assessment in tropical Southeast Asia: a critical appraisal. Sustain. Cities Soc. **46**, 101428 (2019)
8. O. Potchter, P. Cohen, T.P. Lin, A. Matzarakis, A systematic review advocating a framework and benchmarks for assessing outdoor human thermal perception. Sci. Total Environ. 155128 (2022)
9. S. Khalili, R. Fayaz, S.A. Zolfaghari, Analyzing outdoor thermal comfort conditions in a university campus in hot-arid climate: a case study in Birjand. Iran. Urban Clim. **43**, 101128 (2022)
10. D. Lai, W. Liu, T. Gan, K. Liu, Q. Chen, A review of mitigating strategies to improve the thermal environment and thermal comfort in urban outdoor spaces. Sci. Total. Environ. **661**, 337–353 (2019)
11. J. Zhang, W. Guo, B. Cheng, L. Jiang, S. Xu, A review of the impacts of climate factors on humans' outdoor thermal perceptions. J. Therm. Biol. 103272 (2022)
12. Y. Zhang, X. Zhou, Z. Zheng, M.O. Oladokun, Z. Fang, Experimental investigation into the effects of different metabolic rates of body movement on thermal comfort. Build. Environ. **168**, 106489 (2020)
13. Y. Ji, J. Song, P. Shen, A review of studies and modelling of solar radiation on human thermal comfort in outdoor environment. Build. Environ. 108891 (2022)
14. Y. Shimazaki, A. Yoshida, R. Suzuki, T. Kawabata, D. Imai, S. Kinoshita, Application of human thermal load into unsteady condition for improvement of outdoor thermal comfort. Build. Environ. **46**(8), 1716–1724 (2011)
15. Z. Fang, Z. Lin, C.M. Mak, J. Niu, K.T. Tse, Investigation into sensitivities of factors in outdoor thermal comfort indices. Build. Environ. **128**, 129–142 (2018)
16. P. Höppe, The physiological equivalent temperature–a universal index for the biometeorological assessment of the thermal environment. Int. J. Biometeorol. **43**(2), 71–75 (1999)
17. D. Fiala, G. Havenith, P. Bröde, B. Kampmann, G. Jendritzky, UTCI-Fiala multi-node model of human heat transfer and temperature regulation. Int. J. Biometeorol. **56**, 429–441 (2012)
18. T.P. Lin, R. de Dear, R.L. Hwang, Effect of thermal adaptation on seasonal outdoor thermal comfort. Int. J. Climatol. **31**(2), 302–312 (2011)
19. J. Li, N. Liu, The perception, optimization strategies and prospects of outdoor thermal comfort in China: a review. Build. Environ. **170**, 106614 (2020)
20. P.K. Cheung, C.Y. Jim, Determination and application of outdoor thermal benchmarks. Build. Environ. **123**, 333–350 (2017)
21. M. Haghshenas, M. Hadianpour, A. Matzarakis, M. Mahdavinejad, M. Ansari, Improving the suitability of selected thermal indices for predicting outdoor thermal sensation in Tehran. Sustain. Cities Soc. **74**, 103205 (2021)

22. X. Chen, L. Gao, P. Xue, J. Du, J. Liu, Investigation of outdoor thermal sensation and comfort evaluation methods in severe cold area. Sci. Total. Environ. **749**, 141520 (2020)
23. K. Pantavou, M. Santamouris, D. Asimakopoulos, G. Theoharatos, Empirical calibration of thermal indices in an urban outdoor Mediterranean environment. Build. Environ. **80**, 283–292 (2014)
24. R. Yao, S. Zhang, C. Du, M. Schweiker, S. Hodder, B.W. Olesen, J. Toftum, F. d'Ambrosio, H. Gebhardt, S. Zhou, F. Yuan, B. Li, Evolution and performance analysis of adaptive thermal comfort models–a comprehensive literature review. Build. Environ. 109020 (2022)
25. S. Zhang, Z. Lin, Standard effective temperature based adaptive-rational thermal comfort model. Appl. Energy **264**, 114723 (2020)
26. P.O. Fanger, J. Toftum, Extension of the PMV model to non-air-conditioned buildings in warm climates. Energy Build. **34**(6), 533–536 (2002)
27. R. Yao, B. Li, J. Liu, A theoretical adaptive model of thermal comfort–adaptive predicted mean vote (aPMV). Build. Environ. **44**(10), 2089–2096 (2009)
28. S. Zhang, Z. Lin, Adaptive-rational thermal comfort model: adaptive predicted mean vote with variable adaptive coefficient. Indoor Air **30**(5), 1052–1062 (2020)
29. S. Zhang, Z. Lin, Extended predicted mean vote of thermal adaptations reinforced around thermal neutrality. Indoor Air **31**(4), 1227–1227 (2021)
30. S. Zhang, Z. Lin, Extending predicted mean vote using adaptive approach. Build. Environ. **171**, 106665 (2020)
31. X. Zhou, D. Lai, Q. Chen, Evaluation of thermal sensation models for predicting thermal comfort in dynamic outdoor and indoor environments. Energy Build. **238**, 110847 (2021)
32. J. Li, J. Niu, C.M. Mak, T. Huang, Y. Xie, Exploration of applicability of UTCI and thermally comfortable sun and wind conditions outdoors in a subtropical city of Hong Kong. Sustain. Cities Soc. **52**, 101793 (2020)
33. S. Ghani, A.O. Mahgoub, F. Bakochristou, E.A. ElBialy, Assessment of thermal comfort indices in an open air-conditioned stadium in hot and arid environment. J. Build. Eng. **40**, 102378 (2021)
34. O. Potchter, P. Cohen, T.P. Lin, A. Matzarakis, Outdoor human thermal perception in various climates: a comprehensive review of approaches, methods and quantification. Sci. Total. Environ. **631**, 390–406 (2018)
35. H. Mayer, P. Höppe, Thermal comfort of man in different urban environments. Theoret. Appl. Climatol. **38**(1), 43–49 (1987)
36. E. Krüger, F. Rossi, P. Drach, Calibration of the physiological equivalent temperature index for three different climatic regions. Int. J. Biometeorol. **61**(7), 1323–1336 (2017)
37. K.F.A. Ribeiro, A.C.A. Justi, J.W.Z. Novais, F.M. de Moura Santos, M.C.D.J.A. Nogueira, S.A. Miranda, J.B. Marques, Calibration of the physiological equivalent temperature (PET) index range for outside spaces in a tropical climate city. Urban Clim. **44**, 101196 (2022)
38. S.Q. Hirashima, E.S.D. Assis, M. Nikolopoulou, Dataset on daytime outdoor thermal comfort for Belo Horizonte, Brazil. Data Brief **9**, 530–535 (2016)
39. P. Bröde, D. Fiala, K. Błażejczyk, I. Holmér, G. Jendritzky, B. Kampmann, B. Tinz, G. Havenith, Deriving the operational procedure for the universal thermal climate index (UTCI). Int. J. Biometeorol. **56**(3), 481–494 (2012)
40. A. Psikuta, D. Fiala, G. Laschewski, G. Jendritzky, M. Richards, K. Błażejczyk, I. Mekjavič, H. Rintamäki, R. de Dear, G. Havenith, Validation of the Fiala multi-node thermophysiological model for UTCI application. Int. J. Biometeorol. **56**(3), 443–460 (2012)
41. G. Jendritzky, R. de Dear, G. Havenith, UTCI—why another thermal index? Int. J. Biometeorol. **56**(3), 421–428 (2012)
42. G. Havenith, D. Fiala, K. Błazejczyk, M. Richards, P. Bröde, I. Holmér, H. Rintamaki, Y. Benshabat, G. Jendritzky, The UTCI-clothing model. Int. J. Biometeorol. **56**(3), 461–470 (2012)
43. D. Fröhlich, M. Gangwisch, A. Matzarakis, Effect of radiation and wind on thermal comfort in urban environments-Application of the RayMan and SkyHelios model. Urban Clim. **27**, 1–7 (2019)
44. K. Błażejczyk, UTCI—10 years of applications. Int. J. Biometeorol. **65**(9), 1461–1462 (2021)

45. S. Zhang, Y. Cheng, Z. Fang, Z. Lin, Improved algorithm for adaptive coefficient of adaptive predicted mean vote (aPMV). Build. Environ. **163**, 106318 (2019)
46. R. de Dear, J. Xiong, J. Kim, B. Cao, A review of adaptive thermal comfort research since 1998. Energy Build. **214**, 109893 (2020)
47. P.O. Fanger, Thermal comfort, in *Analysis and Applications in Environmental Engineering* (1970)
48. R.A. Engineers, *ASHRAE Handbook: Fundamentals* (2017)
49. T. Tang, Y. Zhang, Z. Zheng, X. Zhou, Z. Fang, W. Liu, Detailed thermal indicators analysis based on outdoor thermal comfort indices in construction sites in South China. Build. Environ. **205**, 108191 (2021)
50. T.O. Hodson, Root-mean-square error (RMSE) or mean absolute error (MAE): when to use them or not. Geosci. Model Dev. **15**(14), 5481–5487 (2022)
51. S. Guo, X. Bai, Y. Chen, S. Zhang, H. Hou, Q. Zhu, P. Du, An improved approach for soil moisture estimation in gully fields of the Loess Plateau using Sentinel-1A radar images. Remote Sens. **11**(3), 349 (2019)
52. M. Vellei, M. Herrera, D. Fosas, S. Natarajan, The influence of relative humidity on adaptive thermal comfort. Build. Environ. **124**, 171–185 (2017)
53. B. Li, R. Yao, Q. Wang, Y. Pan, An introduction to the Chinese evaluation standard for the indoor thermal environment. Energy Build. **82**, 27–36 (2014)
54. ASHRAE 55, Thermal environmental conditions for human occupancy, ASHRAE Standard 55, in *American Society of Heating, Refrigerating and Air-Conditioning Engineers* (Atlanta, Georgia, 2013)
55. EN 15251, *Indoor Environmental Input Parameters for Design and Assessment of Energy Performance of Buildings: Addressing Indoor Air Quality, Thermal Environment, Lighting and Acoustics* (European Committee for Standardization, Brussels, 2007)

# Improved Thermal Index Models for Outdoor Thermal Comfort Evaluation—Calibration of Simulation Results Using Experimental Data

**Abstract** ENVI-met is a widely applied computational platform for simulating urban microclimates, with its BIO-met module enabling the calculation of thermal comfort indices such as PMV, PET, UTCI, and SET*. These indices provide valuable guidance for assessing outdoor thermal conditions, yet their accuracy depends on calibration against real-world data. This chapter evaluates the precision of ENVI-met simulations by combining field experiments with model-based outputs in a transitional space on a university campus in Guangzhou. Walking trials under controlled metabolic conditions were paired with simultaneous environmental monitoring and three-dimensional ENVI-met modeling of the test site. Results revealed systematic deviations, with simulated air temperatures exceeding measured values by up to 2 °C and wind speed variations being insufficiently captured. To reconcile these differences, experimental data were integrated with BIO-met predictions to generate refined β-model indices, which were then compared with previously developed α-models. The overlapping trends between α- and β-models, together with the formulation of an enhanced calibration algorithm, demonstrate the potential to improve BIO-met's predictive reliability. These advancements support more accurate evaluation of outdoor thermal comfort and provide a practical basis for integrating simulation tools with field data in sustainable urban design.

## 1 Introduction

The pronounced urban heat island effect is increasingly severe in major metropolitan areas [1, 2], and the altered climatic conditions in cities are adversely impacting residents [3, 4]. As outdoor temperatures continue to rise, thermal safety and its implications for public health have garnered significant attention [5–7]. Beyond the high energy demands of indoor climate control [8], prolonged climatic exposures—such as sick building syndrome—may also pose health risks [9, 10]. Consequently, rational urban design is critical to encourage outdoor physical activity.

 
Z. Fang et al., *Outdoor Thermal Comfort and Heat Stress Mitigation*, Environmental Science and Engineering, https://doi.org/10.1007/978-981-95-8899-2_11

Appropriate urban planning and spatial configuration can effectively lower ambient temperatures and enhance thermal comfort. Empirical measurements indicate that expanding vegetated areas can reduce peak air temperatures by as much as 2 °C [11–14]. In recent years, the integration of semi-open transitional spaces has gained traction in urban design [15–17]. These spaces feature distinct architectural forms, often including overhead protections such as elevated floors or eaves [18], which block substantial solar radiation during hot summers and provide comfortable shaded zones. Research demonstrates that such shading can reduce surface temperatures by up to 5.2 °C [19]. Thanks to their semi-open and permeable design, covered areas facilitate shaded and well-ventilated microenvironments that improve human thermal comfort [20, 21]. Thus, the role of transitional spaces deserves greater emphasis in urban planning.

Advances in urban microclimate simulation tools offer considerable benefits for urban design. Numerical modeling is increasingly favored for its efficiency and predictive capabilities. Among small-scale simulation tools are ENVI-met [22], CTTC [23], MISKAM [24], and RayMan [24]. The CTTC model [25] and its subsequent refinements [23] focus primarily on air temperature assessment. The RayMan model [24] emphasizes radiation flux distribution, while MISKAM [26] is used mainly for wind field and pollutant dispersion analysis. Since its introduction in 1998, ENVI-met has been extensively applied in urban microclimate studies worldwide [22]. Rooted in fluid dynamics and thermodynamics, ENVI-met simulates interactions among terrain, vegetation, structures, and the atmosphere at small-to-medium scales. Compared to conventional computational fluid dynamics (CFD) tools, it offers more robust simulations of outdoor thermal conditions and incorporates a detailed vegetation model [27]. With its capacity to simulate diverse microclimates and support high-quality visualization, ENVI-met has become the leading software in this domain. It is widely used by urban designers and landscape architects to enhance air quality [28] and mitigate heat island effects [29–31].

In recent years, urban development has been progressively improved to offer residents more comfortable outdoor living environments [32–34]. In urban planning, the value of computing thermal comfort indices lies in their holistic nature, integrating both outdoor climatic parameters and human physiological factors, thereby enabling more human-centric predictions of environmental comfort. Commonly adopted thermal indicators include predicted mean vote (PMV), standard effective temperature (SET), physiological equivalent temperature (PET), and universal thermal climate index (UTCI). Extensive research has established that PMV is appropriate primarily for indoor thermal assessments [35–38], whereas SET [39, 40], PET [32, 41–44], and UTCI [45, 46] are widely employed in evaluating outdoor environments.

BIO-met serves as a post-processing tool that computes human thermal comfort indices using simulation data [47]. Since version 1.5, it supports the calculation of PMV/PPD (predicted mean vote/predicted percentage of dissatisfied), PET, UTCI, and SET*. However, limited research has been conducted to examine this functionality. Many studies directly extract thermal indices via BIO-met for site evaluation [48–50], with PET being the most frequently selected index as of 2021, accounting for approximately 45.6% of cases [51]. Other studies have utilized different software to derive thermal comfort measures based on ENVI-met meteorological outputs [52–54]. Whether through direct extraction or auxiliary computation, most studies treat these results as final evaluation metrics without further validation.

Some research has compared BIO-met's thermal index outputs with actual thermal sensation votes (TSV) to assess software accuracy. For instance, Salata et al. found generally reliable agreement between simulated PMV and on-site TSV in outdoor microclimates, supporting the software's use in comfort evaluation [50]. Similarly, Elgheznawy et al. reported a strong correlation ($R^2 > 0.8$) between computed PMV and BIO-met simulated values in a study on shaded school courtyards [55].

Nevertheless, few studies have critically examined the feasibility of using ENVI-met meteorological outputs for thermal index computation or the accuracy of BIO-met extractions, resulting in a notable gap in validation research for this software function. A primary reason is that existing studies often focus on specific environmental settings—such as residential areas [52], urban districts [51], green spaces [49], or schoolyards [55]—rather than on people. Consequently, these works often overlook factors influencing dynamic human thermal comfort, including the effect of human presence on the surrounding microenvironment and variations in metabolic rate during physical activity. Ignoring these crucial aspects when evaluating thermal comfort in dynamic settings can lead to deviations between simulated results and actual human experience, limiting the practical relevance of such assessments.

Given that the applicability of the BIO-met software for assessing dynamic human thermal comfort has not been thoroughly examined, this study examines the precision of the thermal indices generated by the program using experimental data from dynamic walking tests performed in a transition space. A preliminary algorithm is proposed to enhance the software's output, aiming to improve the suitability of the derived thermal indices for evaluating thermal comfort under dynamic human conditions.

## 2 Methods

In this study, an approach integrating field measurements and computational simulations was employed. Environmental and physiological parameters of the participants were collected during field tests, and thermal comfort indices were computed using corresponding tools. SET* was computed utilizing the CBE Thermal Comfort Tool, available at: http://comfort.cbe.berkeley.edu/. PET was derived through the RayMan application [24]. UTCI, founded on a multi-node model of human thermal regulation [45, 56], was acquired from the official platform: www.utci.org. By configuring boundary conditions and inputting initial environmental parameters—along with data related to vegetation, ground surfaces, and pollution sources from the surrounding area—the spatial environmental dataset for the test area was generated via ENVI-met software. The BIO-met module was subsequently used to estimate thermal indices at each measurement interval. Finally, the experimentally obtained data were compared against the software-generated results to assess computational accuracy. A graphical abstract of the procedure is provided in Fig. 1.

## 2.1 Experimental Conditions

The field measurement took place on June 23, 2019, within the elevated open area of a university campus in Guangzhou. This structure is representative of a semi-open design, featuring only an overhead roof without peripheral enclosures, as depicted in Fig. 2. Throughout the experiment, two thermal comfort monitoring devices were positioned at a height of 1.1 m [57] to capture continuous climatic measurements, encompassing air temperature ($T_a$), relative humidity (RH), globe temperature ($T_g$), and air velocity ($V_a$). A portable weather station was employed to log wind direction ($W_d$). The two comfort monitors were installed near the front and back edges of the testing zone, within shaded areas situated about 1 m from the walking path. The mobile meteorological unit was positioned adjacent to one of these recorders. All procedures adhered rigorously to the specifications outlined in ASHRAE Standard 55-2020 [58]. Furthermore, wearable smart devices were utilized to monitor the heart rates (HRs) of the participants. A summary of the equipment deployed in this study is presented in Table 1.

Thirty healthy college students (18 males and 12 females) were recruited for this study. The average age was 22.3 years. Each participant was asked to wear clothing with a total thermal resistance of 0.57 during the test. Detailed information about the participants is presented in Table 2.

The TSV for each participant was collected via a questionnaire survey. The scale of the subjective vote, which strictly followed the relevant provisions of ASHRAE standards 55 [58] and ISO7730 [57], is listed in Table 3 for reference.

## 2.2 Experimental Protocol

The experiment employed four fixed walking speeds (1.2, 1.4, 1.6, and 2.0 m/s) to simulate varying metabolic rates. Participants were instructed to provide oral thermal sensation feedback at each interval and to monitor their heart rates (HRs) using smart bracelets. Additionally, during the activity, subjects carried a wireless anemometer (positioned at navel height, between 80 and 110 cm) to measure incoming airflow speeds caused by movement. A schematic of the testing procedure is provided in Fig. 3 [59].

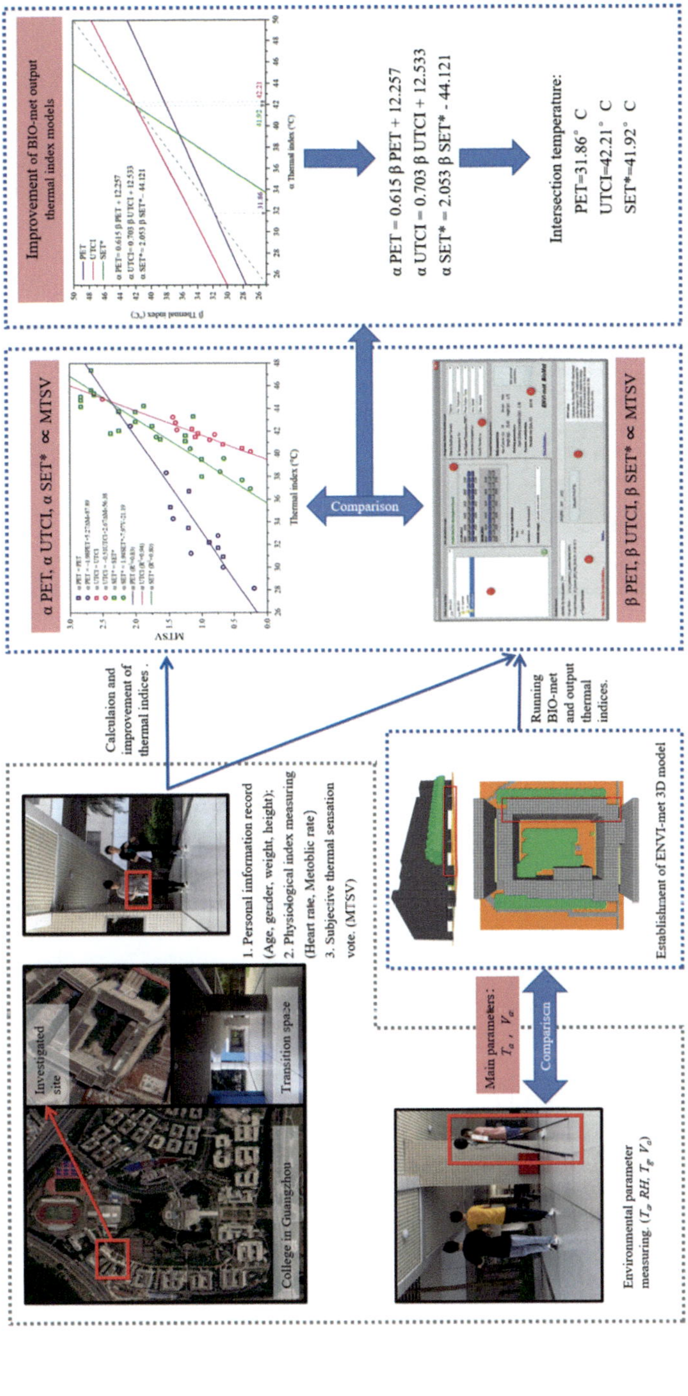


Fig. 1 Graphical abstract

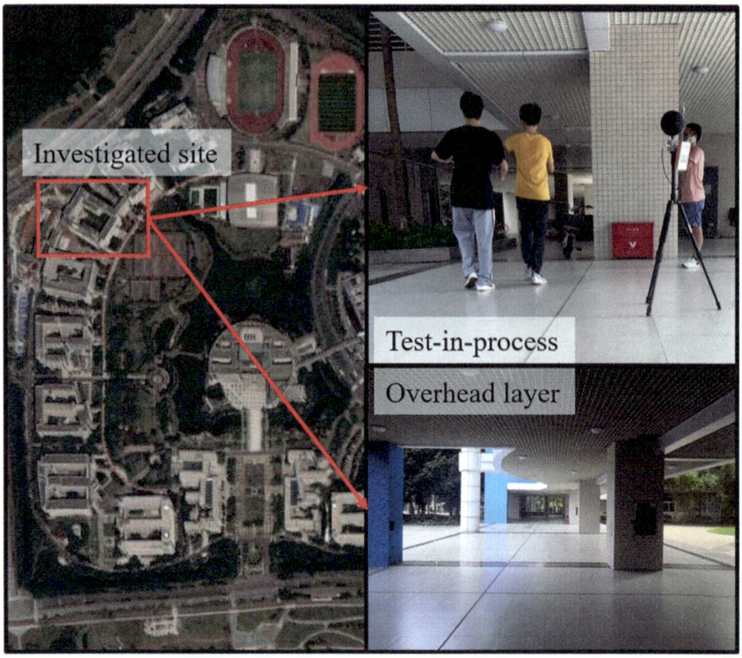


**Fig. 2** Investigation site

**Table 1** Details of instruments used in the study

| Instrument | Type | Parameter | Measuring range | Accuracy | Sampling rate (s) |
|---|---|---|---|---|---|
| Thermal comfort level recorder | SSDZY-1 | $T_a$ (°C) | – 20.0–80.0 °C | ± 0.3 °C | 2 |
| | | RH (%) | 0.01–99.9% RH | ± 2.0% RH (10.0–90.0% RH) | 2 |
| | | $T_g$ (°C) | – 20.0–80.0 °C | ± 0.3 °C | 2 |
| Mobile meteorological station | Vantage pro2 | $W_d$ | 0–360° | ± 7° | 60 |
| Wireless universal wind speed and temperature recorder | WFWZY-1 | $V_a$ (m/s) | 0.05–5.00 m/s | 5% ± 0.05 m/s | 2 |
| Smart bracelet | 3XMSH05HM | HR (bpm) | 30–250 bpm | ± 3 bpm | 2 |

**Table 2** Detailed participant information

| Gender | Sample size | Age (years) | Height (cm) | Weight (kg) | BMI (kg/m$^2$) | BMR (kcal) |
|---|---|---|---|---|---|---|
| Male | 18.0 | 22.5 ± 1.5 | 175.3 ± 5 | 66.0 ± 8.0 | 19.3 ± 1.5 | 1480.2 ± 145.1 |
| Female | 12.0 | 21.5 ± 2.5 | 161.0 ± 3 | 50.2 ± 1.5 | 19.0 ± 1.3 | 1197.0 ± 20.0 |
| Average | – | 22.3 | 171.6 | 57.8 | 19.2 | 1338.4 |

**Table 3** Subjective vote scale

| Thermal sensation | Thermal preference | Humidity preference | Acceptability |
|---|---|---|---|
| − 3 cold<br>− 2 cool<br>− 1 slightly cool<br>0 neutral<br>+ 1 slightly warm<br>+ 2 warm<br>+ 3 hot | − 1 cooler<br>0 no change<br>+ 1 warmer | − 1 lower<br>0 no change<br>+ 1 higher | − 1 unacceptable<br>1 acceptable |

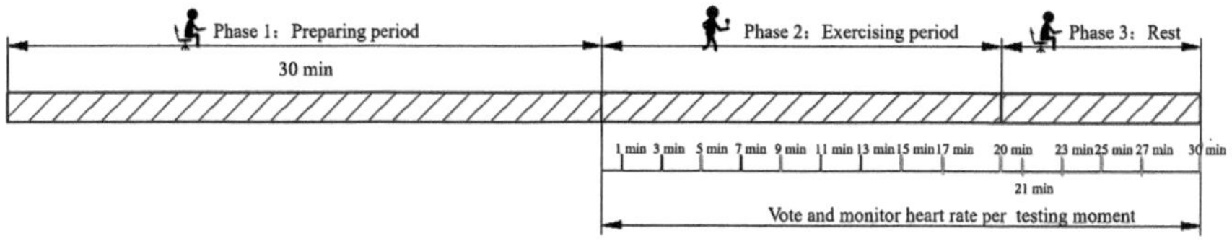


**Fig. 3** Testing process

## 2.3 *ENVI-Met Model*

### 2.3.1 Model Construction

ENVI-met is a simulation tool for urban microclimates initially developed in 1998 by Michael Bruse and Heribert Fleer in Germany [22]. The software employs computational fluid dynamics (CFD) as its foundation, integrating models of airflow, temperature, relative humidity, atmospheric turbulence, and surrounding vegetation to simulate urban microclimate conditions.

Since its introduction, numerous studies have utilized ENVI-met to validate results from empirical measurements [60, 61]. Multiple validation experiments have been conducted at research sites within the Guangzhou region [62–64], confirming the software's reliability for predicting outdoor microenvironments. By configuring plant height and adjusting parameters of surface materials, ENVI-met serves as an effective tool for forecasting urban microclimatic conditions.

In this research, version V4.4.5 of ENVI-met was used to construct a three-dimensional model with a simulation domain of 60 m × 60 m × 90 m and a grid resolution of 2.14 m × 2.14 m × 2.14 m, as illustrated in Fig. 4. The geographic coordinates were set to 113.23° E and 23.12° N, and the time zone was configured as UTC/GMT+08:00. The model accurately represented the building where testing occurred along with adjacent vegetation. The target structure stands 32 m tall, comprising six above-ground floors.

Wind speed and direction at 10 m height, along with hourly temperature and humidity data, were obtained from meteorological station records. Solar radiation was computed using the geographical location data integrated into the ENVI-met model, with a solar correction ratio input set to 1.0. Field surveys identified the surrounding vegetation primarily as *Cinnamomum burmannii*, with trees exhibiting a conical form and an average height of approximately 15 m, as depicted in Fig. 5. Within the ENVI-met model, the plant type was chosen to match field observations,

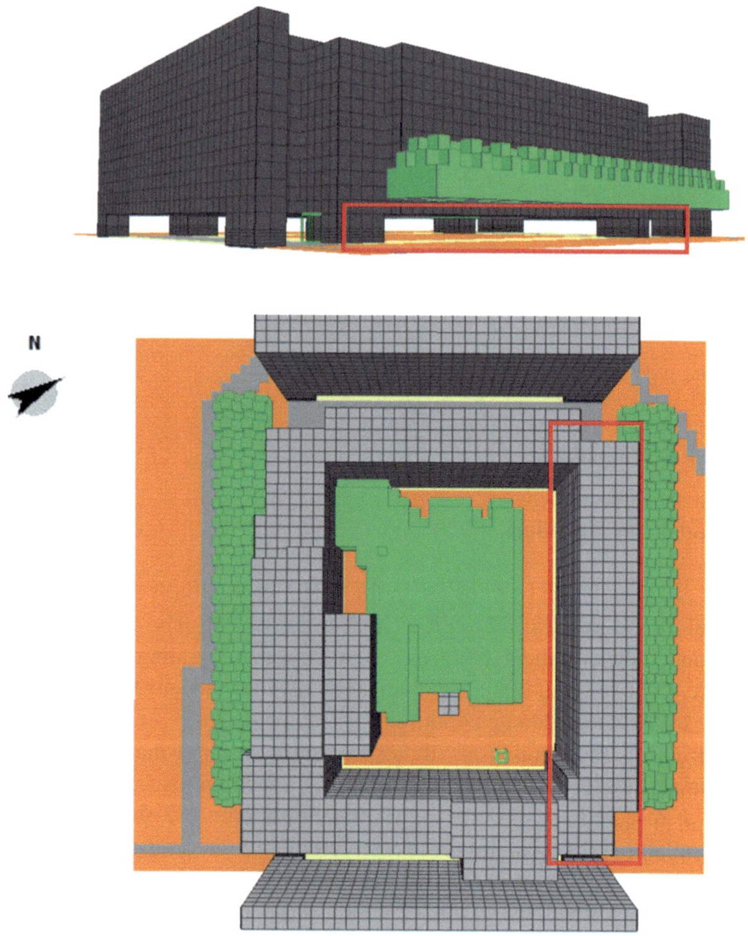


**Fig. 4** Three-dimensional view of simulated model. Area outlined in red represents the study site

**Fig. 5** The rounding plant type and the corresponding 3 D model in ENVI-met

**Table 4** ENVI-met simulation parameter settings

| | |
|---|---|
| Environmental parameters | Hourly $T_a$, hourly RH, and $T_{mrt}$, were taken from data recorded by thermal comfort recorder |
| | $V_a$: 1.9 m/s; $W_d$: 271.71, recorded by mobile meteorological station |
| | Absolute air content value at an altitude of 2500 m above ground: 10.76 g/kg |
| Initial soil state | 0–0.2 m: 305 K/30%; 0.2–0.5 m: 307 K/40%; below 0.5 m: 306 K/50% |
| Ground form | Concrete pavement: 4 cm concrete, 2 cm sand, and soil |
| | Red brick pavement: 3 cm red brick, 1 cm sand, and soil |
| | Ceramic tile: 2 cm ceramic tiles, 1 cm mortar, 3 cm concrete, 2 cm sand and soil |
| Time period | June 23, 2019, from 9:00 to 16:00 |

selecting a medium-sized conical tree from the default vegetation library. The selected tree model was configured with a height of 15 m and a width of 7 m. Leaf area density (LAD) was set to 2.3 $m^2\ m^{-3}$, and the short-wave albedo of the foliage was defined as 0.12 clo. Initial soil temperature and moisture conditions were assigned according to climatic parameters specific to Guangzhou [65]. Key input parameters used in the ENVI-met simulations are summarized in Table 4.

### 2.3.2 BIO-Met Calculation

When utilizing the BIO-met tool for thermal index computation, only individual parameters—such as age, gender, weight, height, static clothing insulation, and metabolic rate—need to be entered, under the condition that environmental simulation outputs from ENVI-met are already available. Subsequently, hourly contour maps for the chosen thermal index can be readily generated [47].

Within current thermal comfort literature, this function has been employed to derive thermal index values [66]. Nonetheless, such simulations typically contrast software-generated results with static environmental measurements. Consequently, these outcomes are applicable only for predicting the average thermal perception of individuals maintaining a fixed metabolic rate over a specified duration in a thermal setting. To date, no studies have addressed thermal indices under dynamically varying conditions.

# 3 Results

## 3.1 Model Validation

In numerous previous studies, the accuracy of outputs from the ENVI-met model is typically validated through comparison with empirically recorded air temperature and relative humidity data [63, 67, 68]. This study employed two widely recognized evaluation metrics: the root-mean-square error (RMSE) and the mean absolute percentage error (MAPE). RMSE is a standard measure for assessing simulation precision. MAPE, in contrast, expresses model deviation in percentage terms, which remains independent of the original data's numerical range and is thus suitable for cross-dataset comparisons. Lower values for both indicators correspond to better model performance (Table 5).

$$\mathrm{RMSE} = \sqrt{\frac{\sum_{\mathrm{i=i}}^{\mathrm{n}} \left(\mathrm{y}_i{}' - \mathrm{y}_i\right)^2}{\mathrm{n}}}, \tag{1}$$

$$\mathrm{MAPE} = \frac{1}{\mathrm{n}} \sum_{\mathrm{i=i}}^{\mathrm{n}} \frac{\left|\mathrm{y_i}' - \mathrm{y_i}\right|}{\mathrm{y}'} \times 100\%, \tag{2}$$

where $\mathrm{y_i}'$ is the simulated value, $\mathrm{y_i}$ is the measured value, and n is the total amount of data (Fig. 6).

Validation outcomes indicated that the RMSE values for both air temperature ($T_a$) and relative humidity (RH) were minimal, while the MAPE remained below 5%, suggesting that discrepancies between simulated and empirically gathered data fell within an acceptable margin. The coefficient of determination $R^2$ between modeled and observed values exceeded 0.90, further reflecting a high level of agreement.

**Table 5** Verification results of ENVI-met environment parameter output

| Parameters | RMSE | MAPE | $R^2$ |
|---|---|---|---|
| $T_a$ | 1.02 °C | 2.50% | 0.98 |
| RH | 3.81% | 3.67% | 0.90 |

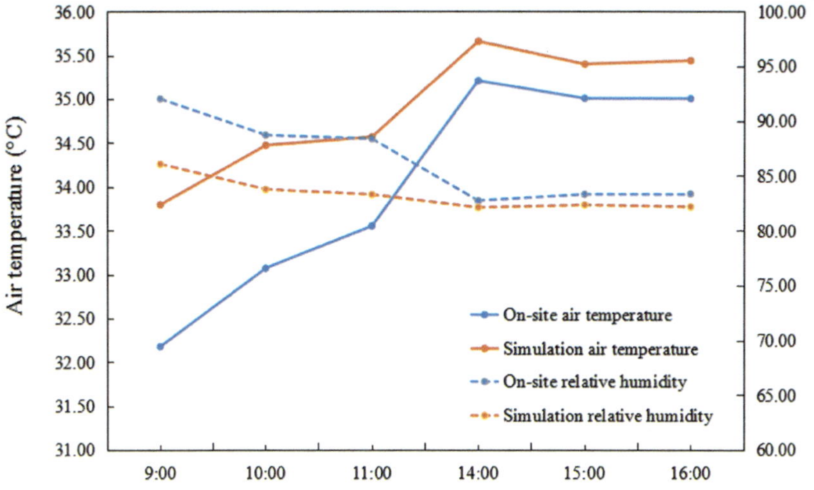


**Fig. 6** Comparison of on-site data and simulation data

Consequently, the ENVI-met model was deemed to possess adequate accuracy for predicting thermal conditions at the test location.

## *3.2 Comparison of ENVI-Met Output with the Experimental Data*

On the test days, the test periods were 9:00–11:00 and 14:00–16:00. According to the field test data, the air temperature range of the investigation site was 34 ± 2.5 °C, and the average air velocity was 1.9 m/s. Detailed information on the environmental parameters is presented in Table 6.

Within the ENVI-met software, contour maps of the most variable environmental parameters—air temperature ($T_a$) and air velocity ($V_a$)—were exported for each test hour (the smallest output unit of the software), as illustrated in Figs. 7 and 8. The cross-sectional height of these contours was set at 1.5 m (z = 1.5 m), aligning with the measurement height used during field tests. The maximum temperature value from

**Table 6** Details of test conditions

| Classification | $T_a$ (°C) | RH (%) | $T_g$ (°C) | $V_a$ (m/s) |
|---|---|---|---|---|
| Max. value | 35.7 | 95.0 | 36.1 | 7.7 |
| Minimum value | 31.6 | 80.2 | 31.8 | 0.0 |
| Average value | 33.9 | 85.9 | 34.1 | 1.9 |
| Standard deviation | 1.1 | 3.8 | 1.2 | 1.2 |

the ENVI-met output reached 36.73 °C. Compared with the field data in Table 4, this represents an overestimation of about 1 °C relative to the measured maximum, and a deviation of nearly 2 °C from the minimum recorded value. These discrepancies indicate that the software's numerical simulation of the temperature field deviates to some extent from actual measurements, consistently predicting higher air temperatures than those observed.

In terms of wind speed, the software-generated values generally underestimated the actual air velocity within the overhead layer. As the transitional space was open on all sides, the real-world wind field experienced substantial real-time fluctuations in both speed and direction. Monitoring wind speed at one-hour intervals inevitably introduced deviations. Furthermore, in dynamic conditions, the wind speed perceived by pedestrians includes not only natural airflow but also disturbance wind generated by walking motion. Consequently, a considerable difference exists between the physically monitored air velocity and the ambient wind speed output from the software, indicating that the model generally underestimates the actual air movement around the human body.

## *3.3 Analysis of BIO-Met Output Thermal Index Model*

Thermal comfort indices—namely PET, UTCI, and SET*—for each test hour within the transitional space were computed using the BIO-met software, with PMV values also generated. Distribution contour plots of each index are provided in Figs. 9, 10, 11, and 12, where the study area is demarcated by a black rectangle at a height of z = 1.5 m. The LEONARDO visualization tool, used to produce these contour maps, allows the grid system defined in the ENVI-met model to be displayed. By positioning the cursor over any location within the map, the corresponding numerical value can be viewed, as illustrated in Fig. 13. Hourly averaged values for each thermal index were derived by accounting for the spatial proportion of colored grid cells and incorporating specific measured data. These results are summarized in Table 6.

The color patterns in the contour maps offered a comprehensive overview of the thermal index values across the target area for each test interval, while the transitions between color zones aided in assessing the uniformity of the thermal comfort distribution (Table 7). By comparing the averaged output thermal index with established thermal comfort thresholds [69–71], it becomes possible to evaluate the thermal conditions of the environment and propose tailored strategies to mitigate localized discomfort. Validating the precision and dependability of the software's computational outcomes will offer valuable insights for urban planning practices.

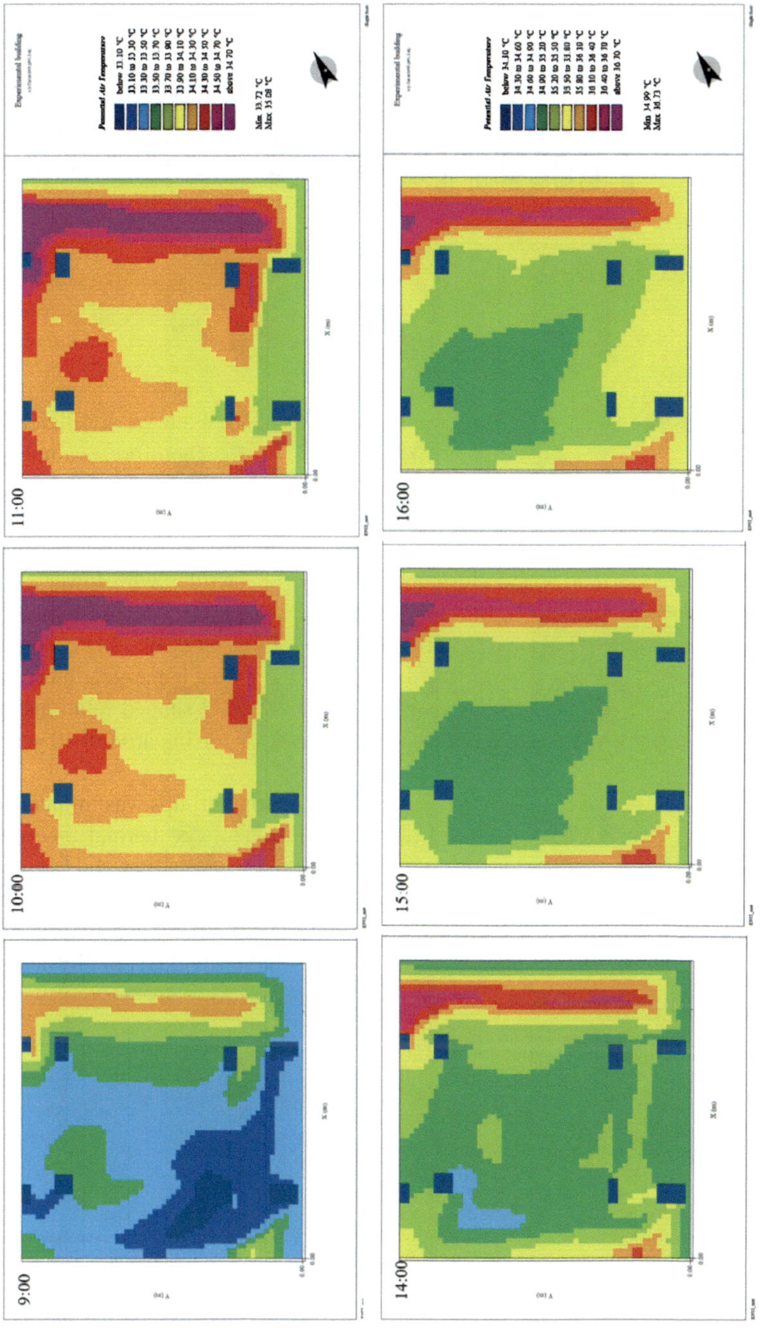


**Fig. 7** Distribution of $T_a$ around buildings during test period

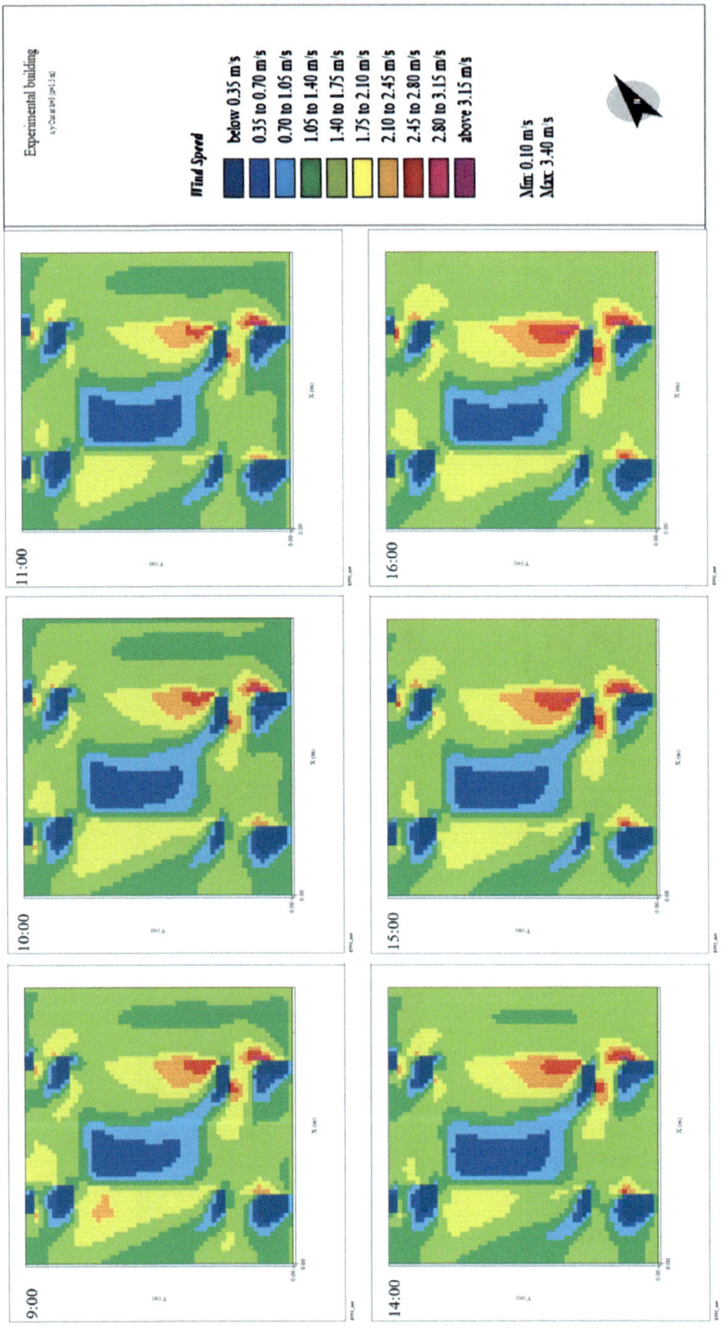


**Fig. 8** Distribution of $V_a$ around buildings during test period

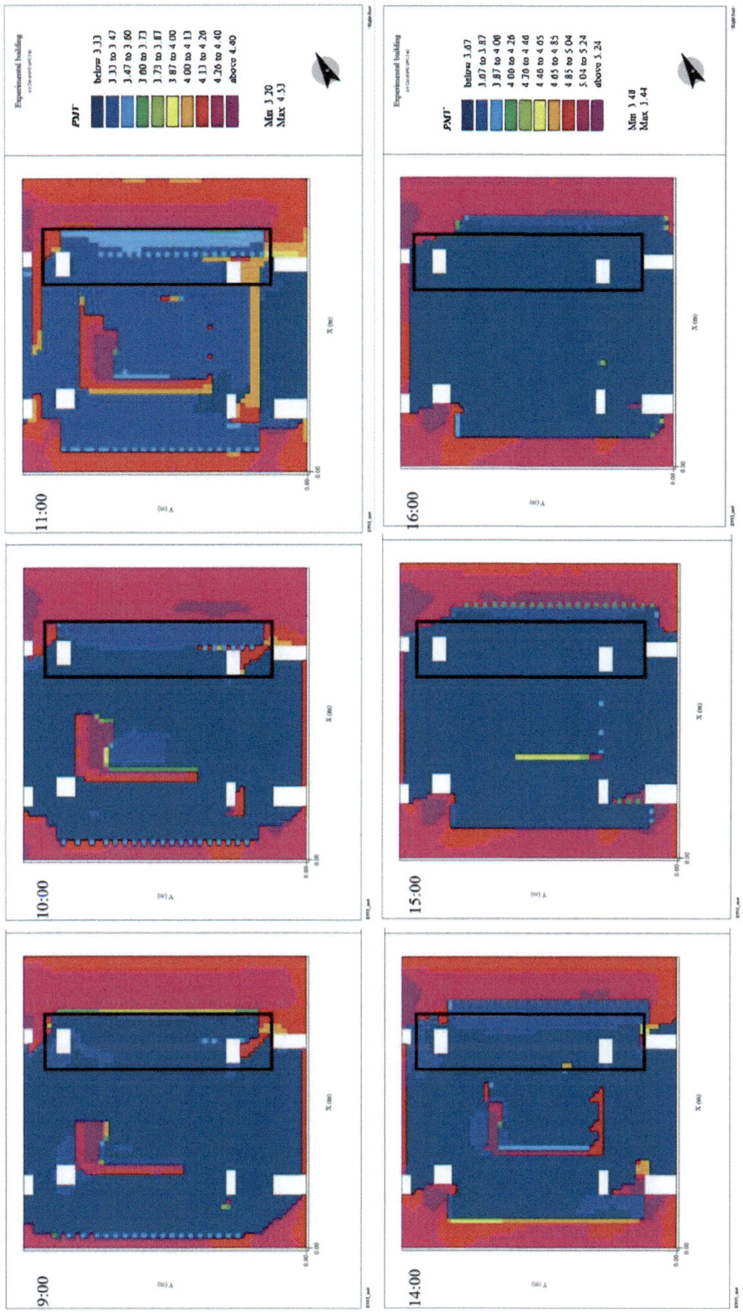


**Fig. 9** Distribution contour map of PMV

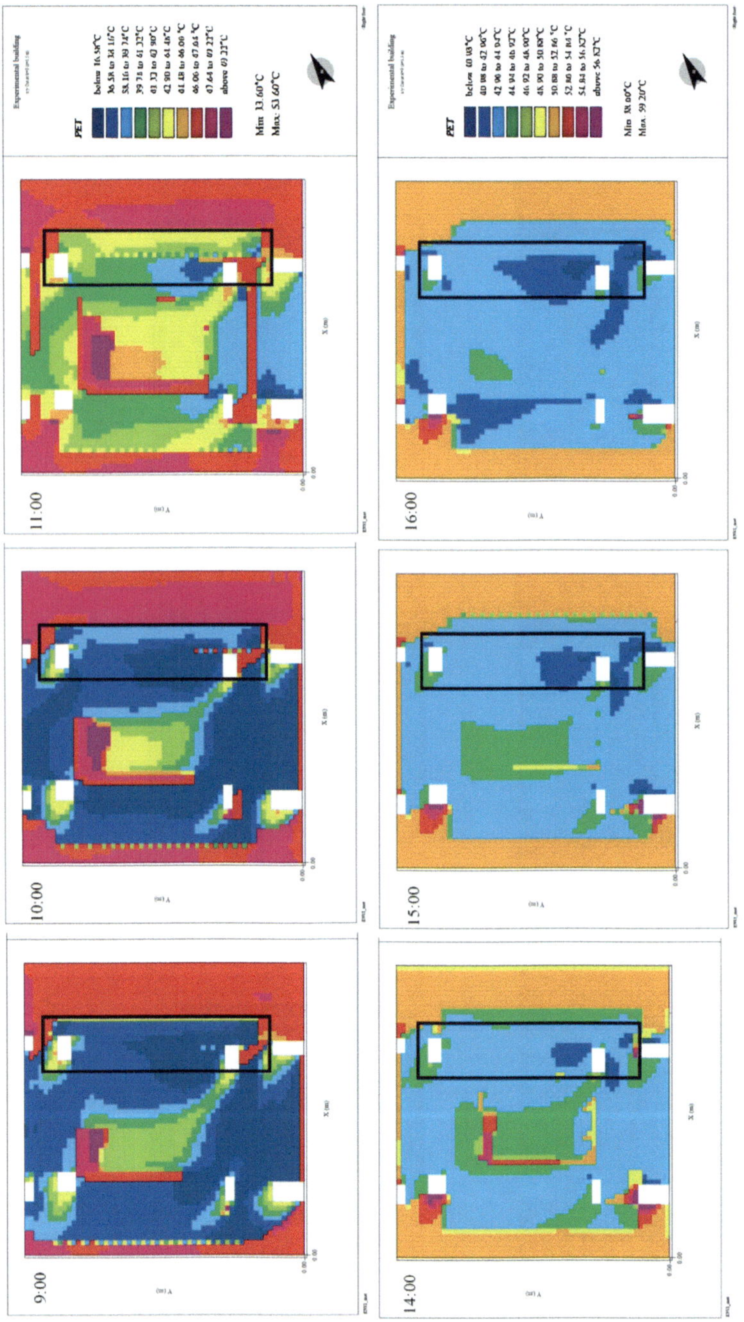


**Fig. 10** Distribution contour map of PET

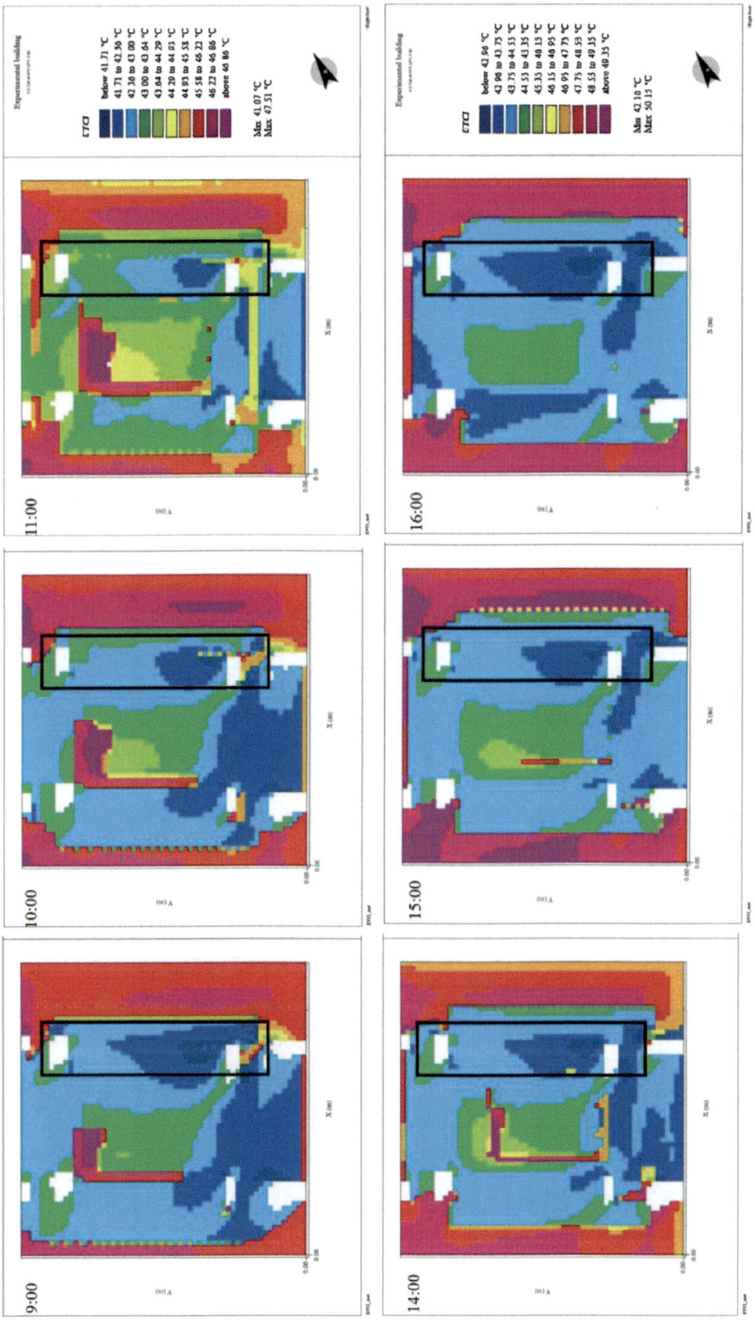


**Fig. 11** Distribution contour map of UTCI

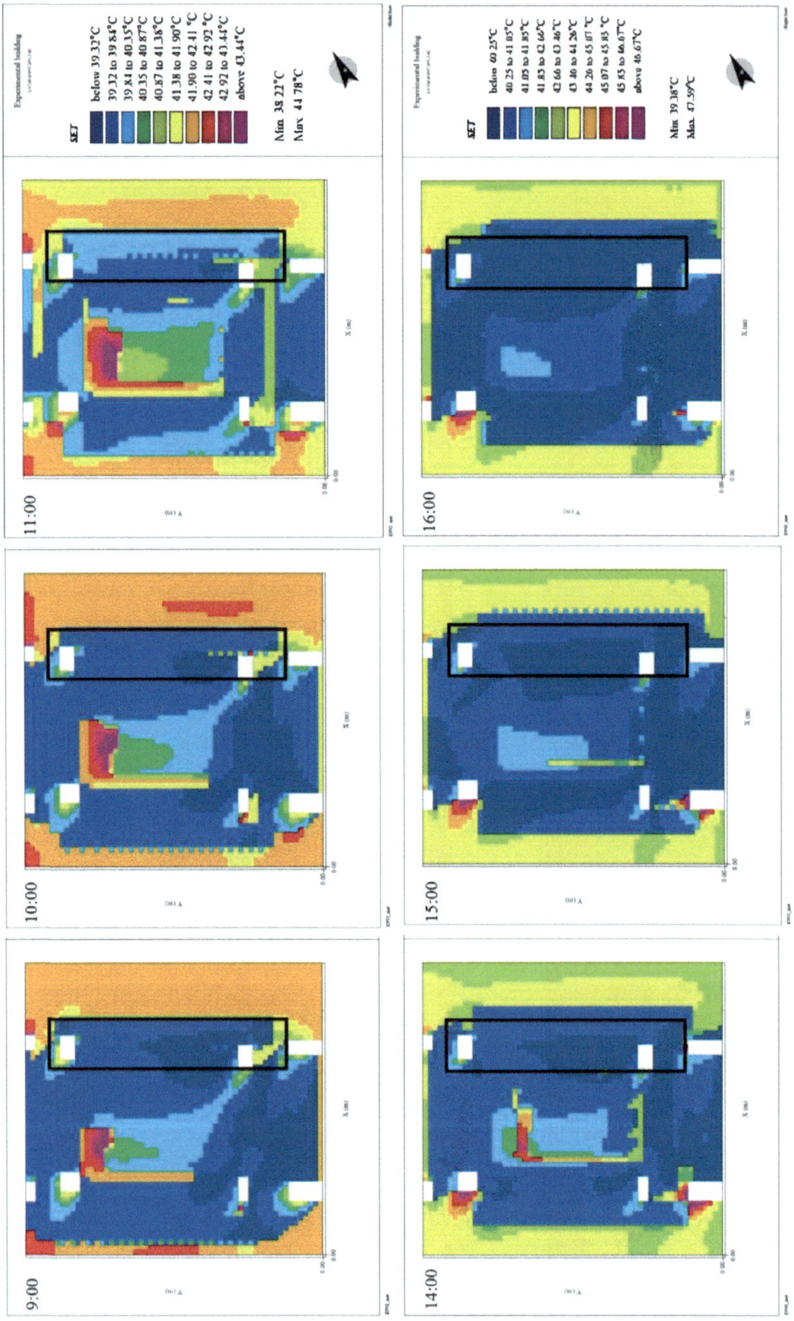


**Fig. 12** Distribution contour map of SET*

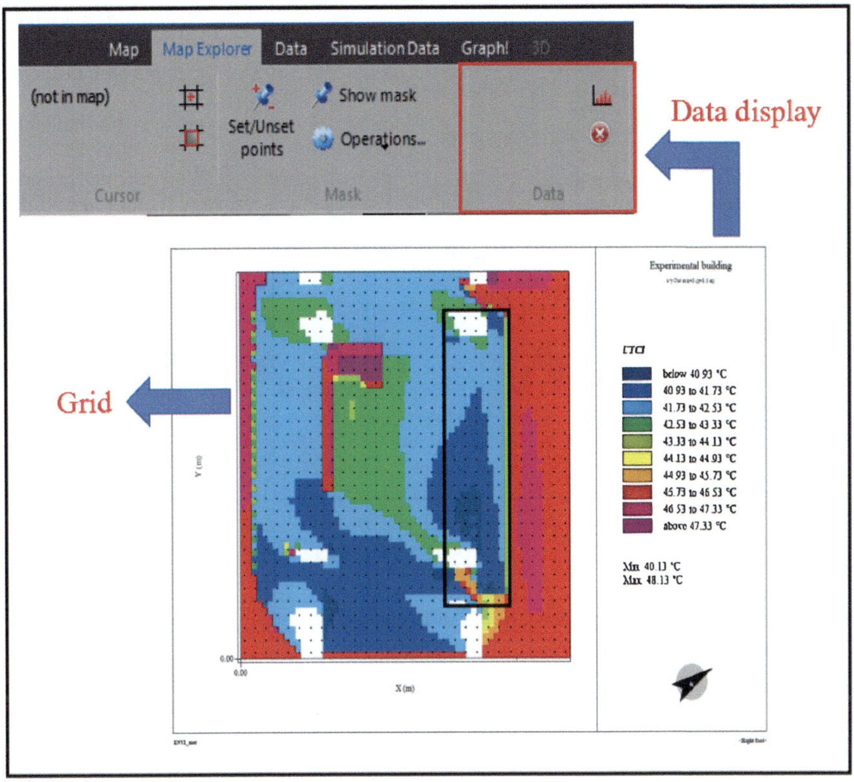


**Fig. 13** Illustration of the calculation method of the average thermal index

**Table 7** Values of thermal indices at each test hour

| Time | 9:00 | 10:00 | 11:00 | 14:00 | 15:00 | 16:00 | Average | Thermal neutral interval |
|---|---|---|---|---|---|---|---|---|
| β PMV | 3.100 | 3.250 | 3.50 | 3.810 | 3.700 | 3.610 | 3.495 | - 0.5–0.5 |
| β PET (°C) | 35.960 | 36.360 | 39.050 | 43.850 | 43.340 | 43.160 | 40.287 | 18–23 |

## *3.4 Initial Improvement of the BIO-Met Calculation Program in Transitional Space*

The correlation equations between the PMV and the other thermal indices output by BIO-met (defined as β PET, β UTCI, and β SET*) were established, as shown in Eqs. (3)–(5).

$$\beta\ \text{PMV} = 0.072\ \beta\ \text{PET} + 0.597\ \left(R^2 = 0.92\right), \tag{3}$$

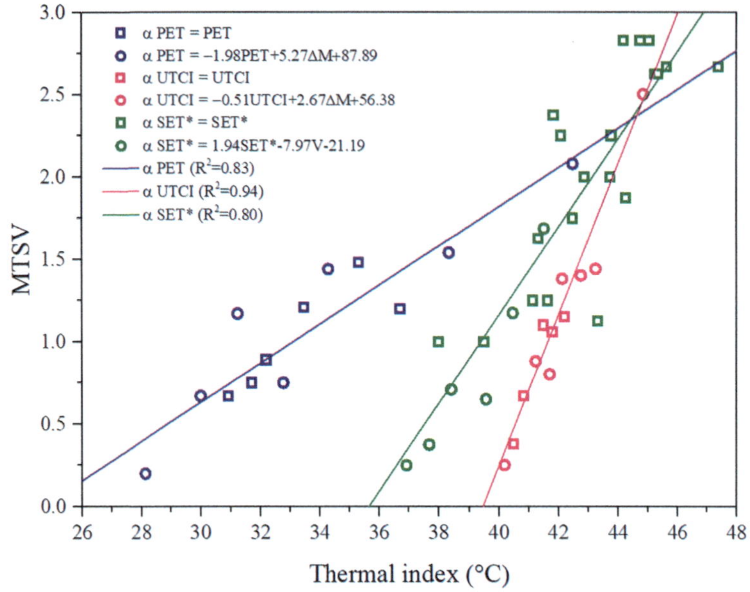


**Fig. 14** Improved thermal index models of transitional space

$$\beta\ \text{PMV} = 0.315\ \beta\ \text{UTCI} - 9.991\ \left(R^2 = 0.91\right), \tag{4}$$

$$\beta\ \text{PMV} = 0.548\ \beta\ \text{SET}^* - 18.256\ \left(R^2 = 0.88\right). \tag{5}$$

In our previous studies, improved thermal index models for transitional spaces were obtained (defined as α PET, α UTCI, and α SET*) [59, 72], as shown in Fig. 14, which are summarized in Eqs. (6)–(8):

$$\text{MTSV} = 0.117\ \alpha\ \text{PET} - 2.872, \tag{6}$$

$$\text{MTSV} = 0.448\ \alpha\ \text{UTCI} - 17.640, \tag{7}$$

$$\text{MTSV} = 0.267\ \alpha\ \text{SET}^* - 8.516. \tag{8}$$

A comparison between the PMV and MTSV is shown in Fig. 15 based on the experimental data statistics.

The average PMV value was 4.765 (defined as α PMV), which was 1.27 grades higher than the average value obtained by ENVI-met (β PMV). The average gap between α-PMV and MTSV was 3.305. Thus, the relationship between MTSV and βPMV can be derived using Eq. (9).

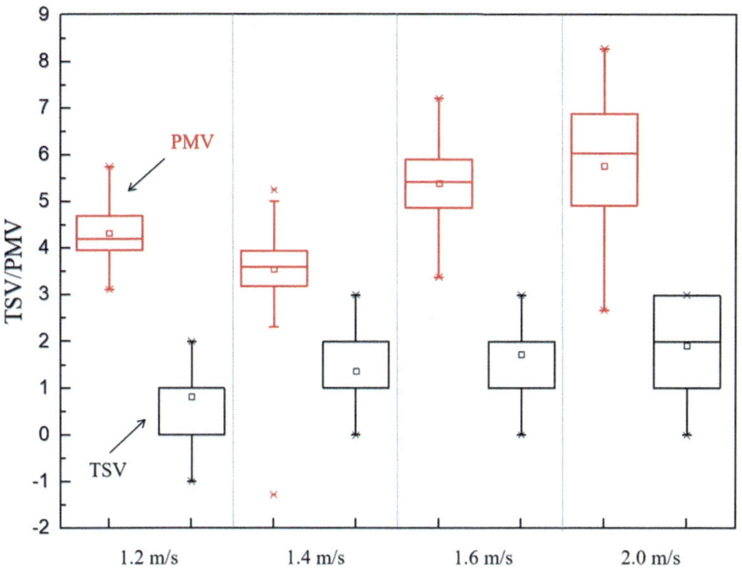


**Fig. 15** Comparison between MTSV and PMV

$$\beta \text{ PMV} = \text{MTSV} + 2.035. \tag{9}$$

Then, Eq. (10) can be converted into Eq. (11):

$$\begin{cases} \beta\, PMV = 0.072\, \beta\, PET + 0.597 = MTSV + 2.035 \\ \beta\, PMV = 0.315\, \beta\, UTCI\ 9.991 = MTSV + 2.035 \\ \beta\, PMV = 0.548\, \beta\ SET^*\ 18.256 = MTSV + 2.035 \end{cases} \tag{10}$$

$$\begin{cases} \text{MTSV} = 0.072\, \beta\, PET\ \ 1.438 \\ \text{MTSV} = 0.315\, \beta\, UTCI\ \ 12.026 \\ \text{MTSV} = 0.267\, \beta\ SET^*\ \ 20.291 \end{cases} \tag{11}$$

A comparison was made between the thermal index models generated by the BIO-met program and those derived from experimental data under the same coordinate framework, as illustrated in Figs. 16, 17 and 18. Discrepancies are evident between the indices produced by the two approaches. As seen in Figs. 16 and 17, prior to the intersection temperature, the PET and UTCI values from BIO-met overestimated the actual thermal sensation votes, while the mean thermal sensation vote (MTSV) was underestimated beyond that point. Agreement between model evaluations and actual conditions occurred solely at the intersection, with coordinates (31.86 °C, 0.86) for PET and (42.21 °C, 1.27) for UTCI. In contrast, the SET* model followed an inverse trend: the β SET* values underestimated subject MTSV until 41.92 °C, where α SET* matched β SET*.

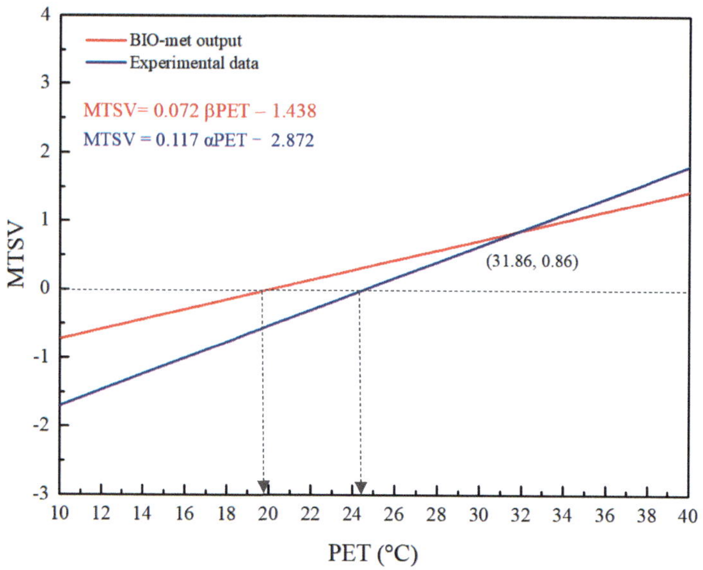


**Fig. 16** Comparison of PET values between BIO-met output and experimental data

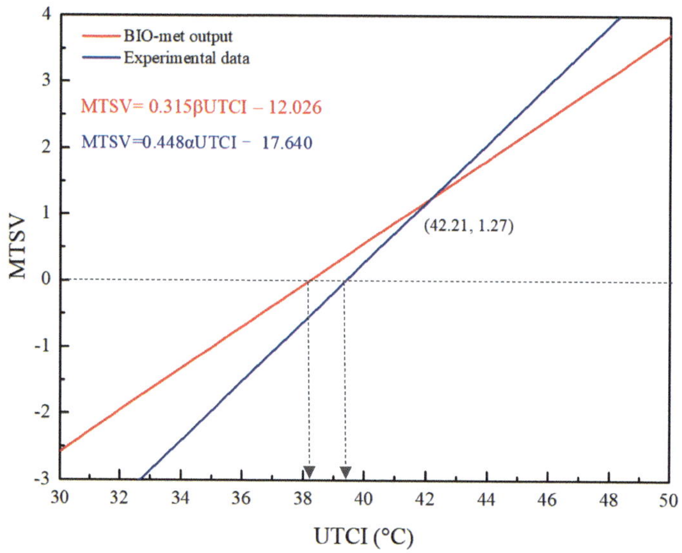


**Fig. 17** Comparison of UTCI values between BIO-met output and experimental data

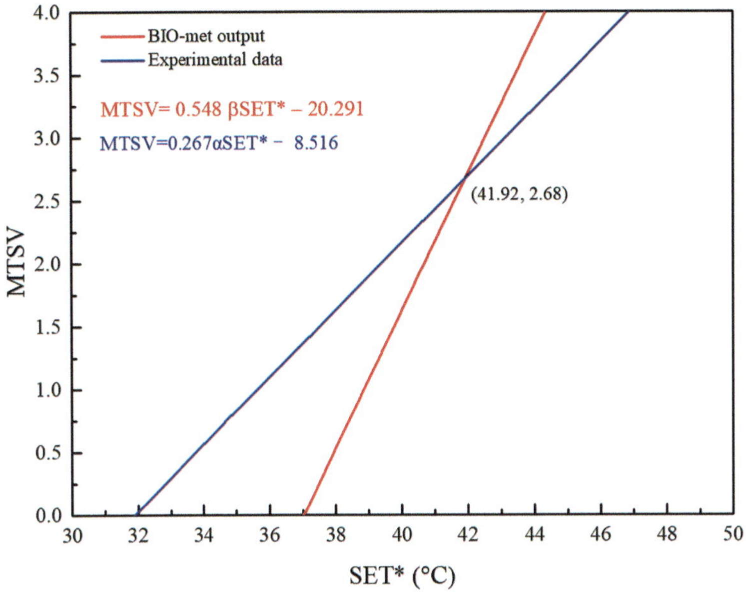


**Fig. 18** Comparison of SET* values between BIO-met output and experimental data

Although systematic deviations exist between BIO-met outputs and experimental thermal comfort models, certain consistent patterns can be identified. These differences may stem from the software's inability to account for thermal sensation compensation caused by environmental variables—such as wind variability during dynamic activities. Furthermore, the model does not fully incorporate the effect of fluctuating metabolic rates during movement. As established in earlier research, such omissions can lead to inaccuracies in thermal index evaluation [59].

To improve the accuracy of thermal index models generated via BIO-met, it is necessary to develop enhanced dynamic computation strategies grounded in empirical field data.

By equating Eqs. (6)–(8) and (11), the relationship between the software-calculated thermal index models (β PET, β-UTCI, and β SET *) and the experimentally improved thermal models (α-PET, α-UTCI, and α-SET *) can be obtained:

$$\alpha\, PET = 0.615\, \beta\, PET + 12.257 \tag{12}$$

$$\alpha\, UTCI = 0.703\, \beta\, UTCI + 12.533, \tag{13}$$

$$\alpha\, SET^{*} = 2.053\, \beta\, SET^{*} - 44.121. \tag{14}$$

For clarity of comparison, the refinements made to the three thermal index models are collectively illustrated in Fig. 19. In general, the α-thermal index models, grounded

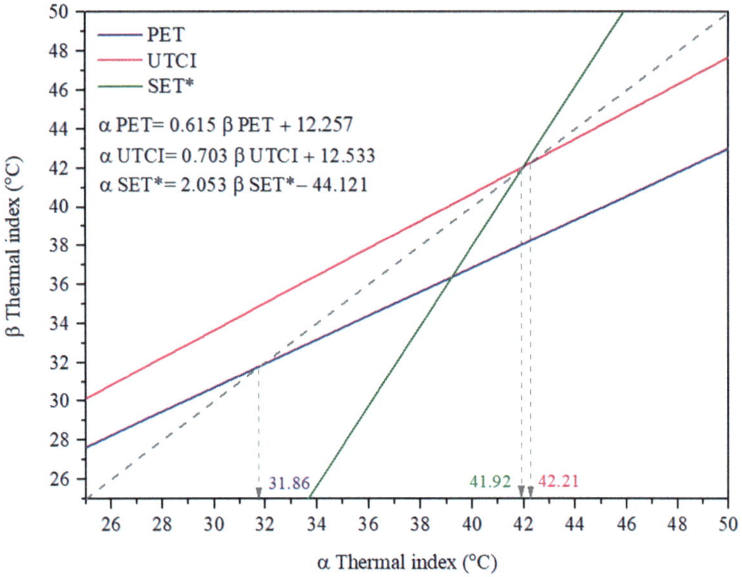


**Fig. 19** Improved BIO-met output models of PET, UTCI, and SET*

in on-site studies, account for both the variations in metabolic rate during dynamic walking and the corresponding changes in mean thermal sensation vote (MTSV) under such conditions. Their reliability in assessing dynamic thermal perception for individuals walking at consistent speeds across different metabolic levels within transitional spaces in Guangzhou has been validated [5, 59].

In contrast, the BIO-met thermal index model employs an algorithm predicated on a relatively stable urban microclimate over time. It incorporates averaged human metabolic values over a given period and does not reflect dynamic metabolic fluctuations. Moreover, the software's thermal comfort modeling is limited to hourly average thermal index outputs and cannot provide minute-accurate calculations. As a result, the thermal index values generated by the software only offer a reference under the assumption of a mean metabolic level across one hour.

Transforming Eq. (12) into Eq. (14) reveals a clear linear relationship between the experimentally enhanced α-thermal index model and the β-thermal index model derived from ENVI-met software. The output values from the software can be refined through straightforward equation-based adjustments, thereby optimizing the hourly averaged thermal index estimates.

This conversion process, while still delivering only an hourly average thermal index value, significantly enhances the accuracy of the thermal comfort model within that interval. These linear conversion formulas offer an optimized reference for improving thermal comfort model algorithms when using ENVI-met software to evaluate transitional spaces in the Guangzhou region during summer.

# 4 Discussion

## *4.1 Accuracy Analysis of the ENVI-Met Simulation Results*

Based on the analysis of simulation outcomes detailed in Sect. 3, discrepancies were identified between the ENVI-met software predictions and actual microenvironment measurements, encompassing temperature distribution, wind speed patterns, and thermal index modeling. Even with model parameters carefully calibrated to mirror real conditions, deviations in simulated outputs suggest inherent limitations in the software's computational algorithms. Some researchers have pointed out that the atmospheric numerical modeling in ENVI-met lacks rigor, with additional concerns regarding its simulation of heat fluxes and localized air temperatures [73–75]. In 2018, Stella et al. systematically summarized constraints within ENVI-met, covering shortcomings in its soil, vegetation, and ancillary modules [63].

In the current study, ENVI-met simulations diverged from empirical measurements in both air temperature and wind speed: temperatures were generally overestimated, while wind speeds were underestimated.

The accuracy of temperature simulations in ENVI-met remains a contested issue. Several studies report that the software tends to overestimate ambient air temperature [50, 76, 77], while others note underestimation relative to real conditions [75, 78, 79]. Some comparative analyses of daytime and nighttime simulations reveal that temperatures are underestimated during the day but overestimated at night [80]. A study conducted on a U.S. university campus indicated that ENVI-met could simulate regional temperature patterns with reasonable accuracy, though precision was higher in the center of the domain compared to the edges, where influences from adjacent areas are challenging to fully incorporate [67]. Another study, set in a courtyard setting during summer, identified significant deviations of 2–4 °C between ENVI-met predictions and measured values [81]. Similarly, a summer simulation of mixed-use and commercial neighborhood microclimates in the Denver metropolitan area also showed consistent overestimation, with simulated temperatures averaging 1–2 °C higher than observed—a result aligned with the findings of this study [82].

Furthermore, in their investigation of wind-speed behaviour, Song and co-workers noted that ENVI-met has a pronounced tendency to inflate the true wind velocity [83]. Krüger et al. carried out a field trial in 2011 [84] in which the meteorological-station reading was fed into ENVI-met as the incoming velocity and the outcome was then contrasted with on-site recordings taken 2.1 m above a downtown canyon. They discovered that whenever the imposed breeze stayed beneath 2 m/s the modelled and observed figures matched, yet once the forcing speed climbed above 2 m/s the software systematically overshot the real magnitude. In the present campaign, more than 70% of the monitored gusts remained under the 2 m/s threshold. Nevertheless, because open-air airflow could not be restrained, sporadic surges appeared at specific moments throughout the observation window; such spikes were absent from the ENVI-met outputs and are regarded as the primary cause of the observed discrepancies.

## 4.2 Analysis of the Accuracy of the BIO-Met Output Results

There have been limited review studies on the constraints of the BIO-met program; however, numerous studies have conducted comparative assessments between PET/ UTCI simulations and empirical data. In an investigation on pedestrian thermal comfort in residential neighborhoods, [76] demonstrated that the discrepancy between ENVI-met derived PET and field-measured PET was 2.8 K when the experimental PET was 25 °C, and this gap widened to 3.6 K when the experimental value reached 40 °C [76]. Yin et al. observed that the simulated PET notably underestimated real conditions during low-radiation intervals, with the peak deviation attaining 3.7 °C [85]. A separate study carried out in three representative public spaces in Athens revealed that the calculated root-mean-square errors (RMSE %) for UTCI and PET were 4% and 7%, respectively, reflecting satisfactory accuracy of these predictive models [86].

Nonetheless, examination of dynamic conditions remains scarce in current literature. To enhance the precision of urban planning, further refinement of the BIO-met software is essential. Although the present study's efforts to refine the algorithm of the BIO-met program are not thoroughly comprehensive, they offer valuable insights for future enhancements.

# 5 Conclusions

This chapter examined the accuracy of ENVI-met and its BIO-met module in simulating outdoor thermal comfort indices, using an on-site walking experiment in a transitional space as a validation benchmark. The findings indicate that ENVI-met can approximate local microclimates, but with systematic deviations: simulated air temperature was consistently overestimated by up to 2 °C, and wind dynamics were insufficiently reproduced. Despite these discrepancies, comparisons between simulation-derived β-model indices and empirically adjusted α-models revealed convergent trends, with systematic overestimation or underestimation depending on index values. SET* displayed an opposite trend relative to PET and UTCI, underscoring the complexity of index behavior across models. By integrating field observations with simulation outputs, a preliminary calibration algorithm was developed to refine BIO-met predictions, offering improved accuracy in representing outdoor thermal comfort conditions. These results underscore the importance of combining computational modeling with empirical validation to enhance the reliability of thermal comfort assessment, and provide valuable methodological guidance for applying ENVI-met in urban planning, particularly in transitional spaces where microclimatic conditions are highly dynamic.

**Acknowledgements** This study was supported by the National Natural Science Foundation of China [Project No. 51978180] and GuangDong Basic and Applied Basic Research Foundation [grant number 2021A1515011671].

# References


1. L. Sheng, X. Tang, H. You, Q. Gu, H. Hu, Comparison of the urban heat island intensity quantified by using air temperature and Landsat land surface temperature in Hangzhou, China. Ecol. Indic. **72**, 738–746 (2017)
2. D. Zhou, J. Xiao, S. Bonafoni, C. Berger, K. Deilami, Y. Zhou, S. Frolking, R. Yao, Z. Qiao, J. Sobrino, Satellite remote sensing of surface urban heat Islands: progress, challenges, and perspectives. Remote. Sens. **11**(1), 48 (2018)
3. C.H. Lim, J. Ryu, Y. Choi, S.W. Jeon, W.-K. Lee, Understanding global PM2.5 concentrations and their drivers in recent decades (1998–2016). Environ. Int. **144**, 106011 (2020)
4. L. Tian, J. Cao, R. Li, X. Pu, L. Shi, R. Wang, H. Liu, C. Xu, Z. Tong, S.S. Zhou, Landscape grain effect in Yancheng coastal wetland and its response to landscape changes. Int. J. Environ. Res. Public Health **16**(12), 2225 (2019)
5. Z. Fang, T. Tang, Z. Zheng, X. Zhou, W. Liu, Y. Zhang, Thermal responses of workers during summer: an outdoor investigation of construction sites in South China. Sustain. Cities Soc. **66**, 102705 (2021)
6. H. Johnson, K.R. Sari, G. McGregor, J. Stedman, M. Gibbs, H. Walton, L. Cook, E. Black, The impact of the 2003 heat wave on mortality and hospital admission in England. Health Stat. Quart. **25**, 6–11 (2003)
7. S. Muthers, G. Laschewski, A. Matzarakis, The summer 2003 and 2015 in South-West Germany: heat waves and heat-related mortality in the context of climate change. Atmosphere **8**, 224 (2017)
8. European Commission Buildings; Energy Information Administration (EIA) How much energy is consumed in residential and commercial buildings in the United States? (2018)
9. W.J. Fisk, A.H. Rosenfeld, Estimates of improved productivity and health from better indoor environments. Indoor Air **7**, 158–172 (1997)
10. P. MacNaughton, J. Pegues, U. Satish, S. Santanam, J. Spengler, J. Allen, Economic, environmental and health implications of enhanced ventilation in office buildings. Int. J. Environ. Res. Public Health **12**, 14709–14722 (2015)
11. E. Landsberg, *The Urban Climate* (Academic Press, London, 1981), pp. 5–10
12. R. Oke, City size and the urban heat island. Atmos. Environ. **7**, 69–79 (1973)
13. F. Fezer, Wieweit verbessern Griinfl/ichen das Siedlungsklima. Heidelb. Univ. J. **57**, 77–79 (1976)
14. Finke, Zuordnung und Mischung von bebauten und begrunten Flachen. Schriftenr. Stadtebau Forsch. 03.044, BMRBS, Bonn, 60–90 (1976)
15. M. Yıldırım, Shading in the outdoor environments of climate-friendly hot and dry historical streets: the passageways of Sanliurfa, Turkey. Environ. Impact Assess. Rev. **80** (2020)
16. S. Mazouz, M.S. Zerouala, Shading as a modulator for the design of urban layouts based on vernacular experiences. Energy Build. **29**, 11–15 (1998)
17. D. Shi, J. Song, J. Huang, C. Zhuang, R. Guo, Y. Gao, Synergistic cooling effects (SCEs) of urban green-blue spaces on local thermal environment: a case study in Chongqing, China. Sustain. Cities Soc. **55** (2020)
18. Z.N. Wang, *Architectural Space Art and Technology* (1987)
19. H. Zhang, M. Tang, W. Zhao, Investigation of the urban walking space sun-shading, in *Proceedings of the National Conference on Building Environment and Building Energy Conservation*, (2007)
20. J. Liu, X. Zhang, J. Niu, K.T. Tse, Pedestrian-level wind and gust around buildings with a 'lift-up' design: assessment of influence from surrounding buildings by adopting LES. Build. Simul. **12**(6), 1107–1118 (2019)
21. J. Liu, J. Niu, Delayed detached eddy simulation of pedestrian-level wind around a building array—the potential to save computing resources. Build. Environ. **152**, 28–38 (2019)

22. M. Bruse, H. Fleer, Simulating surface–plant–air interactions inside urban environments with a three dimensional numerical model. Environ. Model. Softw. **13**(3–4) 373–384 (1998)
23. L. Shashua-Bar, M.E. Hoffman, The Green CTTC model for predicting the air temperature in small urban wooded sites. Build. Environ. **37**(12), 1279–1288 (2002)
24. A. Matzarakis, F. Rutz, H. Mayer, Modelling radiation fluxes in simple and complex environemnts-application of the RayMan model. Int. J. Biometeorol. **51**, 323–334 (2007)
25. H. Swaid, H. M.E, Prediction of urban air temperature variations using the analytical CTTC model. Energy Build. **14**(4), 313–324 (1990)
26. J. Eichhorn, *Entwicklung und Anwendung eines dreidimensionalen mikroskaligen Stadtklima-Modells (in German)* (Johannes Gutenberg-Universität in Mainz, Mainz, 1989)
27. Y. Yang, E. Gatto, Z. Gao, R. Buccolieri, T.E. Morakinyo, H. Lan, The "plant evaluation model" for the assessment of the impact of vegetation on outdoor microclimate in the urban environment. Build. Environ. **159**, 106151 (2019)
28. J. Wang, C.D. Lu, L.W. Niu, F.F. Zhang, Simulation and verification of the influence of road vegetation structure on the diffusion of atmospheric inhalable particulate matter. Trans. Chin. Soc. Agric. Eng. **34**(20), 225–232 (2018)
29. Y. Wang, U. Berardi, H. Akbari, Comparing the effects of urban heat island mitigation strategies for Toronto, Canada. Energy Build., 2–19 (2016)
30. U. Berardi, J. Graham, Investigation of the impacts of microclimate on PV energy efficiency and outdoor thermal comfort. Sustain. Cities Soc. **62**, 102402 (2020)
31. U. Berardi, Z. Jandaghian, J. Graham, Effects of greenery enhancements for the resilience to heat waves: a comparison of analysis performed through mesoscale (WRF) and microscale (ENVI-met) modeling. Sci. Total. Environ. (2020)
32. K.M. Wai, C. Yuan, A. Lai, K.N. Peter, Relationship between pedestrian-level outdoor thermal comfort and building morphology in a high-density city. Sci. Total. Environ. **708**, 134516 (2020)
33. D. Lai, W. Liu, T. Gan, K. Liu, Q. Chen, A review of mitigating strategies to improve the thermal environment and thermal comfort in urban outdoor spaces. Sci. Total. Environ. **661**, 337–353 (2019)
34. J. Huang, D. Shi, Z. Fang, Y. Gao, C. Zhuang, J. Zhai, Impact of short-term thermal experience on thermal sensation: a case study of Chongqinng, China. Build. Environ. **179** (2020)
35. P.O. Fanger, Thermal comfort: analysis and applications in environmental engineering. Therm. Comf. Anal. Appl. Environ. Eng. (1972)
36. H.J. Cho, J.W. Jeong, Evaluation of thermal comfort in an office building served by a liquid desiccant-assisted evaporative cooling air-conditioning system. Energy Build. **171**, 361–370 (2018)
37. C. Chun, A. Tamura, Thermal comfort in urban transitional spaces. Build. Environ. **40**, 633–639 (2005)
38. S.A. Damiati, S.A. Zaki, H.B. Rijal, S. Wonorahardjo, Field study on adaptive thermal comfort in office buildings in Malaysia, Indonesia, Singapore, and Japan during hot and humid season. Build. Environ. **109**, 208–223 (2016)
39. Z. Fang, X. Feng, J. Liu, Z. Lin, C.M. Mak, J. Niu, K.T. Tse, X. Xu, Investigation into the differences among several outdoor thermal comfort indices against field survey in subtropics. Sustain. Cities Soc. **44**, 676–690 (2019)
40. T. Xi, Q. Li, A. Mochida, Q. Meng, Study on the outdoor thermal environment and thermal comfort around campus clusters in subtropical urban areas. Build. Environ. **52**, 162–170 (2012)
41. A. Gulyas, J. Unger, A. Matzarakis, Assessment of the microclimatic and human comfort conditions in a complex urban environment: modelling and measurements. Build. Environ. **41**, 1713–1722 (2006)
42. S. Thorsson, M. Lindqvist, S. Lindqvist, Thermal bioclimatic conditions and patterns of behaviour in an urban park in Goteborg, Sweden. Int. J. Biometeorol. **4**, 149–156 (2004)

43. A. Matzarakis, E. Rudel, M. Zygmuntowski, E. Koch, Bioclimatic maps for tourism purposes. Phys. Chem. Earth **35**, 57–62 (2010)
44. H.H. Ali, R. Al-Hashlamun, Assessment of indoor thermal environment in different prototypical school buildings in Jordan. Alex. Eng. J. **58**, 699–711 (2019)
45. P. Weihs, H. Staiger, B. Tinz, E. Batchvarova, H. Rieder, L. Vuilleumier, M. Maturilli, G. Jendritzky, The uncertainty of UTCI due to uncertainties in the determination of radiation fluxes derived from measured and observed meteorological data. Int. J. Biometeorol. **56**(3), 537–555 (2012)
46. Z. Fang, Z. Lin, C.M. Mak, J. Niu, K.T. Tse, Investigation into sensitivities of factors in outdoor thermal comfort indices. Build. Environ. **128**, 129–142 (2018)
47. ENVI-met Applications Homepage: Quickstart_biomet (envi-met.info) (13 Oct 2021). https://envi-met.info/doku.php?id=apps:biomet
48. Y. Su Q. Zhao, N. Zhou Improvement Strategies for Thermal Comfort of a City Block Based on PET Simulation—a case study of Dalian, a cold-region city in China, (2021)
49. M.U. Cilek, C. Uslu, Modeling the relationship between the geometric characteristics of urban green spaces and thermal comfort: the case of Adana city. Sustain. Cities Soc. **79**, 103748 (2022)
50. F. Salata, I. Golasi, R. de Lieto Vollaro, A. de Lieto Vollaro, Urban microclimate and outdoor thermal comfort. A proper procedure to fit ENVI-met simulation outputs to experimental data. Sustain. Cities Soc. **26**, 318–343 (2016)
51. Z. Liu, W. Cheng, C.Y. Jim, T.E. Morakinyo, Y. Shi, E. Ng, Heat mitigation benefits of urban green and blue infrastructures: a systematic review of modeling techniques, validation and scenario simulation in ENVI-met V4. Build. Environ. **200**(455), 107939 (2021)
52. M. Karimimoshaver, M.S. Shahrak, The effect of height and orientation of buildings on thermal comfort. Sustain. Cities Soc. **79**, 103720 (2022)
53. M. Binarti, D. Koerniawan, S. Triyadi, A. Matzarakis, The predicted effectiveness of thermal condition mitigation strategies for a climate-resilient archaeological park. Sustain. Cities Soc. **76**, 103457 (2022)
54. H. Farhadi, M. Faizi, H. Sanaieian, Mitigating the urban heat island in a residential area in Tehran: investigating the role of vegetation, materials, and orientation of buildings. Sustain. Cities Soc. **46**, 101448 (2019)
55. D. Elgheznawy, S. Eltarabily, The impact of sun sail-shading strategy on the thermal comfort in school courtyards. Build. Environ. **202**(1), 108046 (2021)
56. D. Fiala, G. Havenith, P. Bröde, B. Kampmann, G. Jendritzky, UTCI-Fiala multi-node model of human heat transfer and temperature regulation. Int. J. Biometeorol. **56**(3), 429–441 (2012)
57. ISO 7730, 2005 Ergonomics of the thermal environment—instruments for measuring physical quantities (2005)
58. ANSI/ASHRAE Standard 55-2020: Thermal Environmental Conditions for Human Occupancy American Society of Heating Refrigerating and Air-Conditioning Engineers (ASHRAE) (Atlanta, Georgia, 2020)
59. Y. Zhang, J. Liu, Z. Zheng, Z. Fang, X. Zhang, Y. Gao, Y. Xie, Analysis of thermal comfort during movement in a semi-open transition space. Energy Build. **225**(3), 110312 (2020)
60. J. Yang, X. Hu, H. Feng, S. Marvin, Verifying an ENVI-met simulation of the thermal environment of Yanzhong Square Park in Shanghai. Urban For. Urban Green. **66**, 127384 (2021)
61. A. Forouzandeh, Prediction of surface temperature of building surrounding envelopes using holistic microclimate ENVI-met model. Sustain. Cities Soc. **70**,102878 (2021)
62. Z. Wang, M. Ni, S. Hu, B.X. Chen, A practical approach of urban green infrastructure planning to mitigate urban overheating: a case study of Guangzhou. J. Clean. Prod. **287**, 124995 (2021)

63. S. Tsoka, A. Tsikaloudaki, T. Theodosiou, Analyzing the ENVI-met microclimate model's performance and assessing cool materials and urban vegetation applications–a review. Sustain. Cities Soc. **43**, 55–76 (2018)
64. L. Qiu, H. Zhang, W. Zhang, D. Lai, R. Li, Effect of existing residential renovation strategies on building cooling load: cases in three Chinese cities. Energy Build. **253**, 111548 (2021)
65. S.R. Yang, T.P. Lin, An integrated outdoor spaces design procedure to relieve heat stress in hot and humid regions. Build. Environ. **99**, 149–160 (2016)
66. Li Applicability of thermal indices in urban outdoor space in hot and humid zone. SCUT (Doctoral dissertation) (2017)
67. R.L. Chow, C.A. Pope, A.J.B. Martin, Observing and modeling the nocturnal park cool island of an arid city: horizontal and vertical impact. Theor. Appl. Climatol. **103**(1–2), 197–211 (2011)
68. J.A. Cruz, A.C. Blanco, J.J. Garcia, J.A. Santos, A.D. Moscoso, Evaluation of the cooling effect of green and blue spaces on urban microclimate through numerical simulation: a case study of Iloilo River Esplanade, Philippines. Sustain. Cities Soc. **74**(3), 103184 (2021)
69. H. Huang, *Human and Thermal Environment* (Science Press, Beijing, 2011)
70. H. Matzarakis, M.G.I. Mayer, Applications of a universal thermal index: physiological equivalent temperature. Int. J. Biometeorol. **43**(2), 76–84 (1999)
71. D. Bröde, K. Fiala, I.H. Błażejczyk, G. Jendritzky, B. Kampmann, B. Tinz, G. Havenith, Deriving the operational procedure for the Universal Thermal Climate Index (UTCI). Int. J. Biometeorol. **56**(3), 481–494 (2012)
72. Z. Fang, Y. Zhang, Y. Yang, X. Zhou, Z. Zheng, Experimental investigation of standard effective temperature (SET*) adapted in human walking in both indoor and transition thermal environment. Sci. Total. Environ. **793**(1), 148421 (2021)
73. G. Maggiotto, R. Buccolieri, M.A. Santo, L.S. Leo, S. Di Sabatino, Validation of temperature-perturbation and CFD-based modelling for the prediction of the thermal urban environment: the Lecce (IT) case study. Environ. Model. Softw. **60**, 69–83 (2014)
74. R. Buccolieri, M. Maggiotto, S. Di Sabatino, Evaluation of mitigation strategies to improve pedestrian comfort in a typical Mediterranean city, in *ICUC9-9th International Conference on Urban Climate Jointly with 12th Symposium on the Urban Environment*, (2015)
75. J.A. Acero, J. Arrizabalaga, Evaluating the performance of ENVI-met model in diurnal cycles for different meteorological conditions. Theor. Appl. Climatol., 1–15 (2016)
76. H. Lee, H. Mayer, L. Chen, Contribution of trees and grasslands to the mitigation of human heat stress in a residential district of Freiburg, Southwest Germany. Landsc. Urban Plan. **148**, 37–50 (2016)
77. A. Qaid, H.B. Lamit, D.R. Ossen, R.N.R. Shahminan, Urban heat island and thermal comfort conditions at micro-climate scale in a tropical planned city. Energy Build. **133**, 577–595 (2016)
78. T.E. Morakinyo, L. Kong, K.K.-L. Lau, C. Yuan, E. Ng, A study on the impact of shadow-cast and tree species on in-canyon and neighborhood's thermal comfort. Energy Build. **115**, 1–17 (2017)
79. M.B. Sosa, E.N. Correa, M.A. Cantón, Neighborhood designs for low-density social housing energy efficiency: case study of an arid city in Argentina. Energy Build. (2018)
80. W.T.L. Chow, A.J. Brazel, Assessing xeriscaping as a sustainable heat island mitigation approach for a desert city. Energy Build. **47**, 170–181 (2012)
81. V.P. López-Cabeza, C. Galán-Marín, C. Rivera-Gómez, J. Roa-Fernández, Courtyard microclimate ENVI-met outputs deviation from the experimental data. Energy Build. **144**, 129–141 (2018)
82. M.P. Heris, A. Middel, B. Muller, Impacts of form and design policies on urban microclimate: assessment of zoning and design guideline choices in urban redevelopment projects. Landsc. Urban Plan. **202**, 103870 (2020)
83. B.G. Song, K.H. Park, S.G. Jung, Validation of ENVI-met model with in situ measurements considering spatial characteristics of land use types. J. Korean Assoc. Geogr. Inf. Stud. **17**, 156–172 (2014)
84. L. Krüger, F.O. Minella, F. Rasia, Impact of urban geometry on outdoor thermal comfort and air quality from field measurements in Curitiba, Brazil. Build. Environ., 621–634 (2011)

85. S. Yin, W. Lang, Y. Xiao, Z. Xu, Correlative impact of shading strategies and configurations design on pedestrian-level thermal comfort in traditional shophouse neighbourhoods, Southern China. Sustainability, 1355 (2019)
86. A. Tseliou, I.X. Tsiros, Modeling urban microclimate to ameliorate thermal sensation conditions in outdoor areas in Athens (Greece). in *Building Simulation*, vol. 9, no. 3. (Tsinghua University Press, 2016)

# Exploring Thermal Comfort and Safety in Various Outdoor Scenarios

# Thermal Comfort and Safety During Commuting Activities in Semi-open Transitional Space

**Abstract** With rapid urbanization, interest in outdoor thermal comfort has increased, particularly in semi-open transitional spaces such as areas beneath overhead shelters that are frequently used during commuting. This chapter presents an experimental study conducted at Guangzhou University to investigate transient thermal comfort during walking and resting activities in such spaces. Thirty participants performed 20 min walking trials at different speeds, followed by a 10 min rest, while both subjective thermal responses and objective microclimatic data were collected. The analysis focused on the influence of metabolic rate variation and the relationship between mean thermal sensation votes (MTSV) and widely used indices, including PET and UTCI. Results show that the time required for metabolic rate stabilization is significantly longer in semi-open spaces compared with indoor settings, and that turbulence intensity has a pronounced effect on convective heat transfer. Neglecting this factor leads to an overestimation of body heat storage by about 20%. Based on these findings, a refined predictive model was developed that effectively captures thermal comfort across both dynamic and stationary conditions, offering practical implications for designing safe and comfortable commuting environments.

## 1 Introduction

Large urban centers are facing acute urban heat island challenges [1], associated with the continual growth of constructed areas [2]. Structures are responsible for around 40% of worldwide energy use, profoundly affecting non-renewable resource depletion and climate change [3, 4]. Furthermore, individuals spend over 90% of their time inside buildings [5]. A substantial portion of energy in both residential and commercial structures—amounting to 48%—is consumed by heating, ventilation, and air conditioning (HVAC) systems [6]. While numerous studies have explored climate change mitigation through modifications in building energy usage patterns [7–9], their outcomes remain constrained. Hence, it is important to motivate people to participate in more outdoor activities and reduce duration in climate-controlled settings, which may help alleviate the urban heat island phenomenon. Accordingly,

 
Z. Fang et al., *Outdoor Thermal Comfort and Heat Stress Mitigation*, Environmental Science and Engineering, https://doi.org/10.1007/978-981-95-8899-2_12

enhancing the quality and appeal of outdoor thermal conditions is crucial [10–12]. In addition, well-designed and agreeable transitional zones in urban settings significantly promote outdoor engagement.

Transition spaces comprise between 10 and 40% of the total volume across different building classifications [13]. These areas are generally classified into two types: those linking structures without immediate contact to the exterior, and buffer zones that act as intermediaries between inside and outside environments. The latter, often referred to as partially enclosed transition areas, typically include sheltered features like overhead structures and eaves [14]. In contrast to completely open areas, the thermal conditions in transition spaces are more manageable due to more consistent radiation levels. Thermal comfort in these zones remains highly susceptible to short-wave radiation and external influences. During hot summer conditions, such spaces yield superior thermal comfort compared to unsheltered outdoor settings, since shading mitigates direct sun exposure. Prior studies indicate that surface temperatures in shaded regions can be up to 5.2 °C lower than in sun-exposed areas [15]. Reduced solar radiation can markedly improve thermal comfort. Thus, integrating shaded areas is a widespread strategy in urban design [16–18], with semi-open spaces often developed to provide shading.

Wind represents another key element influencing outdoor thermal comfort [19–21]. Typically, fluctuations in thermal conditions are highly variable and transient in open-air settings [22]. Within contemporary high-density urban layouts, significant research emphasis has been placed on pedestrian comfort within wind microclimates [23–26]. One study conducted in a humid climate indicated that enhancing wind conditions offers limited relief from thermal stress caused by solar radiation during intense summer heat [27]. Transition spaces can synergize shading and wind effects to improve outdoor comfort. They can be planned to offer partial protection from solar radiation while allowing penetration of outdoor breezes [28, 29]. Therefore, semi-open transition spaces deserve increased focus for fostering comfortable outdoor thermal environments.

Historically, expectations for the thermal conditions in transition spaces were relatively low. The variable nature of these spaces also complicates the analysis of their thermal parameters. As a result, numerous studies have primarily focused on thermally stable interior environments [30–32]. Only a limited number of researches have addressed thermal comfort in semi-open transition areas [33–43]. Numerous thermal indices have been established to assess different thermal conditions, including those in transition spaces. In 2006, Epstein and colleagues [44] suggested 40 indicators for assessing thermal environments. Commonly used metrics, including the predicted mean vote (PMV), standard effective temperature (SET), physiologically equivalent temperature (PET), and universal thermal climate index (UTCI), have been employed in earlier research to evaluate thermal comfort [45, 46]. For assessing indoor thermal comfort, PMV [47] and SET [48] are acknowledged as conforming to international standards. PET and UTCI are more commonly applied to outdoor settings. How well these indices apply to evaluating thermal comfort in transition spaces remains uncertain.

Walking is dynamic in nature. Both the external thermal conditions and the physiological state of an individual, such as metabolic rate, play important roles in influencing thermal comfort. The average metabolic rates of building occupants have been continuously tracked using portable instruments [49]. In assessing thermal environments, Havenith et al. [50] explored various personal factors and concluded that the metabolic rates listed in the ISO 8996 standard [51] were not entirely precise. To understand the impacts on thermal comfort in transition spaces, studies have looked into metabolic rate during physical activity. Through a field investigation, Larsen and Colton [52] studied metabolic responses to activity in office settings, demonstrating that variations in metabolic rate during exercise considerably influence thermal sensation. Schnorr et al. [53] also examined metabolic adjustment during recovery. Additionally, Fabbri [54] employed questionnaires for data collection and confirmed the feasibility of estimating metabolic rate from individuals' instantaneous heart rates; this technique was applied to determine participants' heart rates throughout the survey.

This chapter aimed to evaluate the thermal comfort of a semi-open transition space. Thirty participants were asked to walk through the transition space at varying speeds. Throughout the experiment, thermal environment parameters and physiological indicators were recorded simultaneously, and subjective thermal feedback was gathered. The effects of wind on thermal comfort and changes in participants' metabolic rates were examined. Additionally, this chapter explored the suitability of thermal comfort indices for evaluating transition spaces in both stationary and moving conditions.

# 2 Methods

## *2.1 Experimental Design*

### 2.1.1 Experimental Conditions

The field measurements were taken from June 2019 to July 2019. Table 1 summarizes the environmental parameters of the field conditions. The air temperature ranged at 34.0 ± 2.5 °C, and the relative humidity ranged at 85.0 ± 10.0%.

On-site data collection was conducted in the overhead shelter of a university building, as illustrated in Fig. 1. The architectural characteristic of this shelter is that it features an overhead cover without enclosed sides. As a result, partial sunlight penetration is allowed, and wind can flow freely from all directions. Two thermal environment monitors were positioned at both the starting and ending locations of the walking path to measure climate parameters. The instruments were installed at a height of 1.1 m adjacent to each participant [55]. All data were collected in compliance with ASHRAE Standard 55-2017 [56].

**Table 1** Detailed information of the test conditions

| Classification | Air temperature, $T_a$ (°C) | Relative humidity, RH (%) | Black globe temperature, $T_g$ (°C) | Air velocity, $V_a$ (m/s) |
|---|---|---|---|---|
| Max. value | 35.7 | 95.0 | 36.1 | 7.7 |
| Min. value | 31.6 | 80.2 | 31.8 | 0.0 |
| Average value | 33.9 | 85.9 | 34.1 | 1.9 |
| Standard deviation | 1.1 | 3.8 | 1.2 | 1.2 |

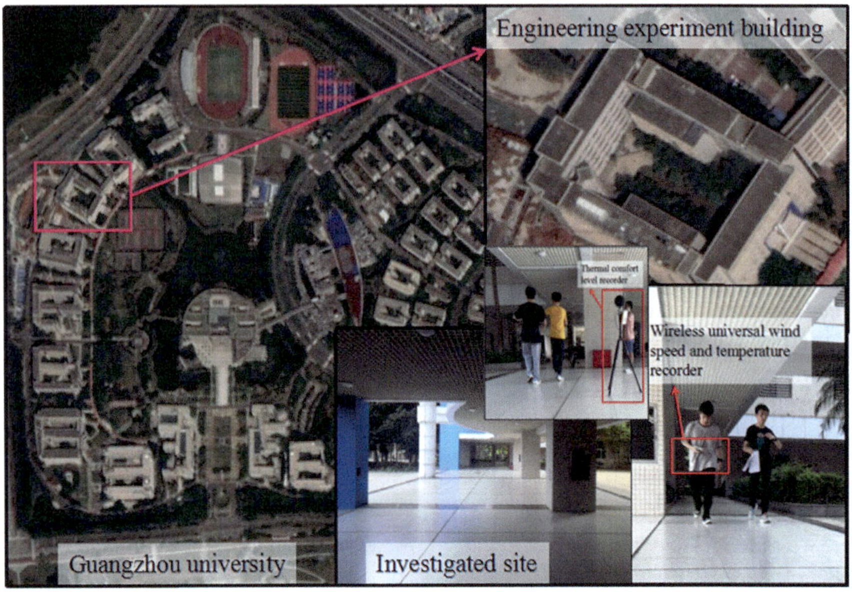


**Fig. 1** Investigated site in the college building

### 2.1.2 Experimental Protocol

The experiment was designed to examine dynamic variations in physiological indicators, including heart rate, auditory canal temperature, and metabolic rate, as well as body heat storage, across different walking speeds. The study also accounted for the effects of environmental factors—such as air temperature, relative humidity, wind speed, and mean radiant temperature—on dynamic thermal comfort.

Participants were instructed to walk within an overhead layer at specified speeds: 1.2, 1.4, 1.6, and 2.0 m/s. A metronome was used to maintain consistent pacing across trials. Speeds of 1.2 and 1.4 m/s were categorized as slow walking, while 1.6 and 2.0 m/s were classified as fast walking. Throughout the experiment, wireless combined anemometer-thermometer instruments were positioned at navel height, adjusted individually based on subject stature, ranging on average from 80.0 to 115.0 cm. This placement minimized excessive shaking caused by walking motion and avoided elevation near the breathing zone.

Each subject's evaluation was structured into three stages. During the first stage (control condition, also a preparation phase), subjects remained seated for 30 min. In the second stage (exercise phase), conducted immediately after the first, subjects walked continuously for 20 min. The third stage (recovery phase) required subjects to sit and rest for 10 min. The assessment procedure is illustrated in Fig. 2.

### 2.1.3 Testing Instruments

During the assessment, environmental parameters within the transition space—including air temperature ($T_a$), relative humidity (RH), globe temperature ($T_g$), and air velocity ($V_a$)—were recorded concurrently. Two physiological indicators were also monitored: heart rate and auditory canal temperature. Heart rates were collected using smart bracelets, while auditory canal temperatures were measured with an infrared tympanic thermometer. Although multiple studies have indicated that rectal temperature [57, 58] most accurately reflects core body temperature, this method is difficult to implement in assessments of this kind. Hence, auditory canal temperature was adopted for This chapter. The equipment utilized throughout the evaluations is listed in detail in Table 2.

## 2.2 *Subjective Survey and Measurement*

A group of 30 healthy university students were enlisted as participants, made up of 18 males and 12 females (specific details are presented in Table 3). Their mean age was 22.3 years, with an average body mass index (BMI) of 19.2 kg/m$^2$. The basal metabolic rate (BMR) [59] was found to be lower among female participants compared to males. Each subject was required to participate in all three phases across four different walking speeds. Their anthropometric measurements (height, weight, BMI, and BMR) were assessed using the same instrument. Throughout the study, the insulation value of clothing (Icl) was determined based on questionnaire responses. These calculations referenced standard insulation values provided in the ASHRAE Handbook [60]. During the experimental sessions, participants were provided with standardized attire totaling approximately 0.57 clo, which included short-sleeved shirts, lightweight pants, socks, and sneakers.

The questionnaire [61] used during the study was divided into two sections. In the first phase, TSV and MTSV under Phase 1 conditions were documented. During the second phase, in addition to TSV and MTSV, physiological indicators were also gathered. The thermal response voting scales are summarized in Table 4. The scales employed in this chapter adhere to the specifications outlined in ASHRAE Standard 55 [56] and ISO 7730 [61].

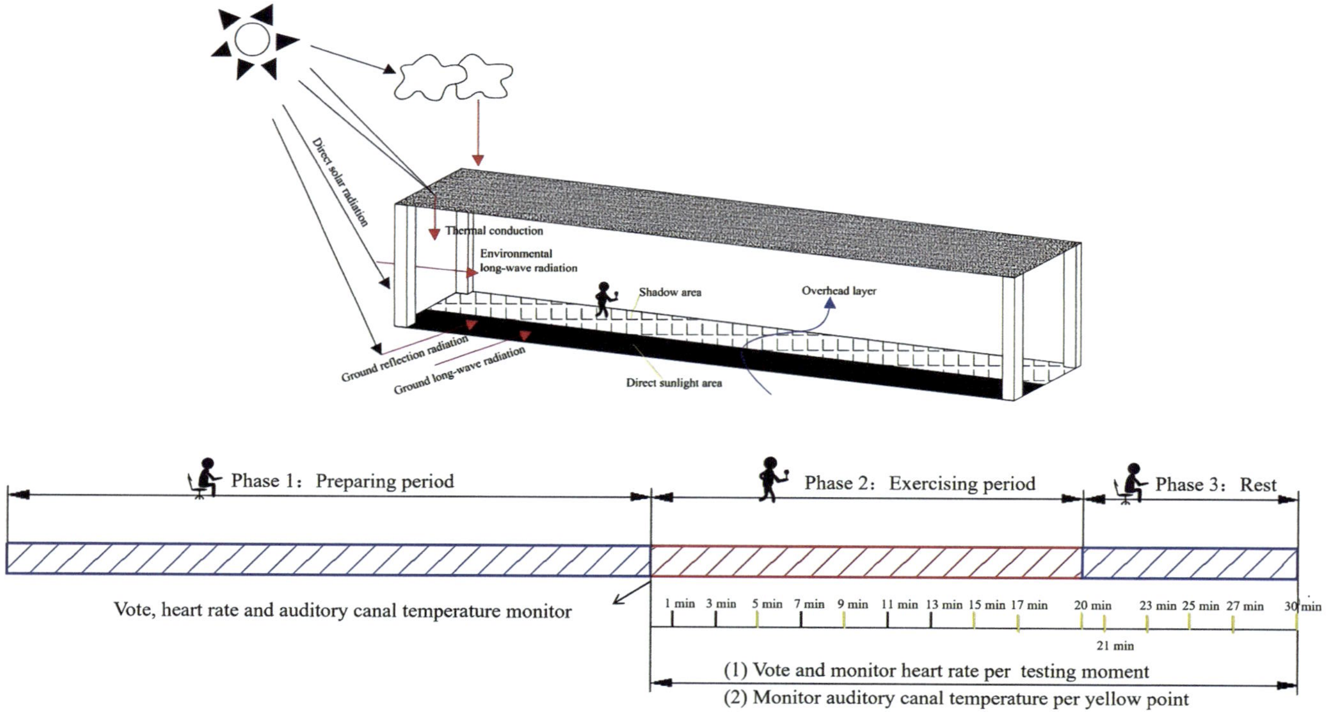


**Fig. 2** Assessment process

**Table 2** Details of instruments used in this assessment

| Instrument | Type | Parameter | Measuring range | Accuracy | Sampling rate (s) |
|---|---|---|---|---|---|
| Thermal comfort level recorder | SSDZY-1 | $T_a$ (°C) | − 20.0 °C ~ + 80.0 °C | ± 0.3 °C | 2 |
| | | RH (%) | 0.01–99.9%RH | ± 2.0%RH (10.0–90.0%RH) | 2 |
| | | $T_g$ (°C) | − 20.0–80.0 °C | ± 0.3 °C | 2 |
| Wireless universal wind speed and temperature recorder | WFWZY-1 | $V_a$ (m/s) | 0.05–5.00 m/s | 5% ± 0.05 m/s | 2 |
| Smart bracelet | 3XMSH05HM | HR (bpm) | 30–250 bpm | ± 3 bpm | 2 |
| Infrared tympanic thermometer | TB-300 | T(°C) | 32.0–42.2 °C | ± 0.2 °C | 1 |

**Table 3** Detailed information on the subjects

| Gender | Sample size | Age (years) | Height (cm) | Weight (kg) | BMI (kg/m$^2$) | BMR (kcal) |
|---|---|---|---|---|---|---|
| Male | 18.0 | 22.5 ± 1.5 | 175.3 ± 5 | 66.0 ± 8.0 | 19.3 ± 1.5 | 1479.1 ± 150.0 |
| Female | 12.0 | 21.5 ± 2.5 | 161.0 ± 3 | 50.2 ± 1.5 | 19.0 ± 1.3 | 1197.0 ± 20.0 |
| Average | – | 22.3 | 171.6 | 57.8 | 19.2 | 1337.4 |

**Table 4** Summary of thermal response vote scale

| Thermal sensation | Air movement sensation | Thermal preference | Humidity preference | Acceptability |
|---|---|---|---|---|
| − 3 cold<br>− 2 cool<br>− 1 slightly cool<br>0 neutral<br>+ 1 slightly warm<br>+ 2 warm<br>+ 3 hot | − 3 much too strong<br>− 2 too strong<br>− 1 slightly strong<br>0 just right<br>+ 1 slightly weak<br>+ 2 too weak<br>+ 3 much too weak | − 1 cooler<br>0 no change<br>+ 1 warmer | − 1 lower<br>0 no change<br>+ 1 higher | − 1 unacceptable<br>1 acceptable |

## 2.3 Data Analysis

In this chapter, data on MTSV, mean air movement sensation vote (MMSV), auditory canal temperature, heart rate, and metabolic rate were collected. Their relationships were established using linear regression. PET was computed with the RayMan software [62]. UTCI, which is founded on the multi-node model of human thermal regulation [63, 64], can be accessed via the project's official website (www.utci.org).

### 2.3.1 Determining Metabolic Rate

The metabolic rate of the human body was estimated based on the mass of inhaled oxygen ($O_2$) and exhaled carbon dioxide ($CO_2$) [65]. Alternatively, it can be evaluated using heart rate [66], a method considered more practical than the respiratory approach, which necessitates measuring the volumes of inhaled $O_2$ and exhaled $CO_2$. Therefore, in this evaluation, the heart rates of participants were recorded to establish the metabolic rate. The resulting metabolic rate was computed as indicated in Eq. (1) [66]:

$$\mathrm{M} = \frac{HR - HR_0}{RM} + M_0 \tag{1}$$

### 2.3.2 Determining Heat Storage

The heat storage (S) of the human body was calculated [67, 68] according to Eqs. (2–6), as follows:

$$\mathrm{S} = \mathrm{M} - \mathrm{W} - \mathrm{R} - \mathrm{C} - \mathrm{E}. \tag{2}$$

The mechanical work (W) done by the human body depends on the intensity of activity. The mechanical efficiency (η) is usually used to indicate the situation in which the human body performs mechanical work. The mechanical efficiency is defined as follows:

$$\eta = \frac{\mathrm{W}}{M} \tag{3}$$

Even when people are doing effective work, the value of η is very low, usually in the range of 0–20%. In most cases, including walking activities on the level surface as in this experiment, η is regarded as zero [69].

The values of radiant heat exchange (R), convective heat exchange (C), and evaporating heat exchange (E) can be determined using the following formulas [70, 71]:

$$\mathrm{R} = f_{cl} h_r (\mathrm{T}_{cl} - \mathrm{T}_{mrt}) \tag{4}$$

$$\mathrm{f}_{cl} = \frac{1}{(1 + 0.155 \times (h_c + h_r) \times I_{cl})} \tag{4.1}$$

$$\begin{aligned} \mathrm{h_r} =& \mathrm{f_{rad}} \varepsilon [(\mathrm{T_{cl}} + 273.15)^2 + (\mathrm{T_{mrt}} + 273.15)^2] \\ & \times [(T_{\mathrm{cl}} + 273.15) + (T_{\mathrm{mrt}} + 273.15)] \times 5.67 \times 10^{-8} \end{aligned} \tag{4.2}$$

$$\begin{aligned} \mathrm{T_{cl}} =& 35.7 - 0.032\mathrm{M} - 0.18 I_{\mathrm{cl}} \{-3.4 \times 10^{-8} f_{\mathrm{cl}} [(\mathrm{T_{cl}} \\ & + 273.15)^4 - (\mathrm{T_{mrt}} + 273.15)^4)] + \mathrm{f_{cl}} \mathrm{h_c} (\mathrm{T_{cl}} - \mathrm{T_a})\} \end{aligned} \tag{4.3}$$

$$\mathrm{C} = \mathrm{f_{cl}} h_c (\mathrm{T}_{cl} - \mathrm{T}_a) \tag{5}$$

$$\mathrm{h_c} = 9.93 \mathrm{u}^{0.54} \times \left(1 + 1.03 TI \times u^{0.54}\right) \tag{5.1}$$

$$\mathrm{TI} = \frac{\sigma}{\mathrm{u}} \tag{5.2}$$

$$\mathrm{E} = \mathrm{E_{sk}} + (C_{res} + E_{res}) \tag{6}$$

$$\mathrm{E_{sk}} = \omega (p_{sk} - p_a) R_{ecl} h_e \tag{6.1}$$

$$\mathrm{R_{ecl}} = \frac{1}{(1 + 0.143 \times h_c \times I_{cl})} \tag{6.2}$$

$$\mathrm{LR} = \frac{\mathrm{h_e}}{h_c} \tag{6.3}$$

$$\mathrm{T_{sk}} = 33.876 - 0.641 M \tag{6.4}$$

$$\mathrm{C_{res}} + E_{res} = 0.0014 M (34 - \mathrm{T}_a) + 0.0173 M (5.87 - p_a) \tag{6.5}$$

# 3 Results

## 3.1 Analysis of Thermal Parameters

Unlike indoor environments, parameters such as air temperature ($T_a$), mean radiant temperature ($T_{mrt}$), relative humidity (RH), and surrounding air velocity ($V_s$) exerted a more immediate influence on human thermal perception within the transition space. By computing the mean thermal sensation vote (MTSV) for every 0.5 °C increment in $T_a$ and $T_{mrt}$, for each 1% change in RH, and for every 0.1 m/s shift in $V_s$, the MTSV corresponding to each interval of $T_a$, $T_{mrt}$, RH, and $V_s$ was derived, as illustrated in Fig. 3. The impacts of $T_a$, $T_{mrt}$, and RH on MTSV were not pronounced during the experiment. In contrast, the coefficient of determination ($R^2$) reflecting the linear relationship between $V_s$ and MTSV was the highest, indicating a pronounced influence of surrounding air velocity on thermal sensation. Furthermore, fluctuations in surrounding air velocity were more substantial compared to those of other thermal parameters, as presented in Table 5. It can be inferred that metabolic rate is not the sole factor influencing thermal sensation during walking; the distributed air velocity around individuals also plays a critical role. Hence, it is essential to analyze the correlation between distributed air velocity and thermal sensation at each walking speed.

Figure 4 illustrates the changes in the surrounding air velocity observed during the experimental period. Natural wind variation was relatively moderate: the mean velocity measured 1.5 m/s, fluctuating within a range of ± 0.5 m/s. As walking speed rose, the air velocity around the subjects also increased, surpassing that of the ambient wind. Concurrently, fluctuations in $V_s$ became more pronounced. The main cause was that natural wind within the transitional space displayed certain irregular variations. The influence of movement-induced disturbance on surrounding airflow was considerable. At lower walking speeds, the disturbance effect on $V_s$ remained

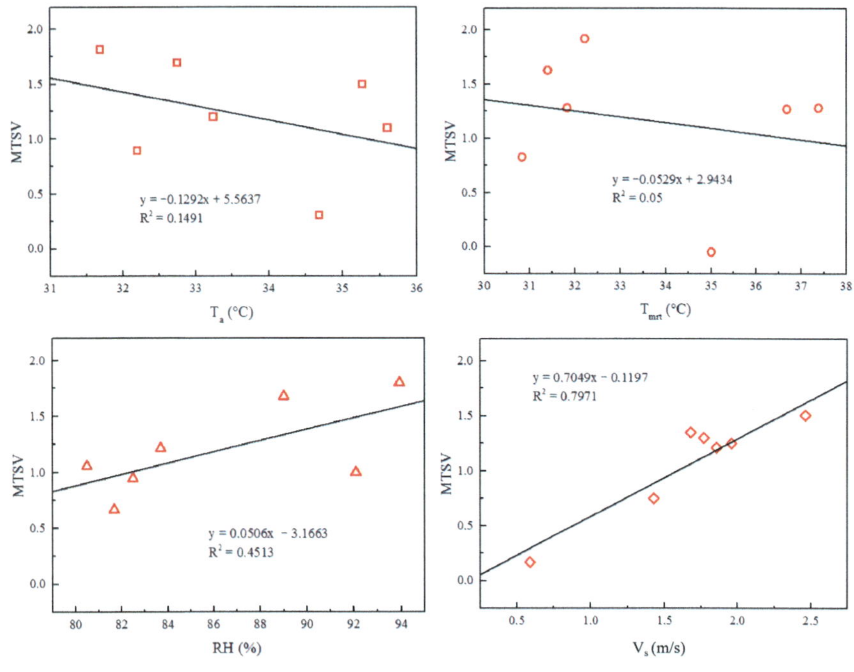


**Fig. 3** Relationship between MTSV and thermal parameters ($T_a$, $T_{mrt}$, RH, and $V_s$)

**Table 5** Variation of different primary environmental parameters

| Classification | | $T_a$ (°C) | | | RH (%) | | | $T_g$ (°C) | | | $V_s$ (m/s) | | |
|---|---|---|---|---|---|---|---|---|---|---|---|---|---|
| | | Max. | Min. | Average | Max. | Min. | Average | Max. | Min. | Average | Max. | Min. | Average |
| Stable state | | 34.1 | 31.6 | 33.8 | 95.0 | 82.8 | 85.9 | 34.5 | 31.8 | 34.1 | 1.8 | 1.2 | 1.5 |
| Walking speed | 1.2 m/s | 35.7 | 34.7 | 34.2 | 91.3 | 88.8 | 90.1 | 36.1 | 34.9 | 34.5 | 3.3 | 1.1 | 1.9 |
| | 1.4 m/s | 33.5 | 32.2 | 32.4 | 89.6 | 80.2 | 85.2 | 33.9 | 32.3 | 32.6 | 2.4 | 1.1 | 1.8 |
| | 1.6 m/s | 34.1 | 32.5 | 33.3 | 88.6 | 86.1 | 87.3 | 34.3 | 32.8 | 33.6 | 3.4 | 0.9 | 2.0 |
| | 2.0 m/s | 34.1 | 32.6 | 33.2 | 92.1 | 86.7 | 90.3 | 34.3 | 31.8 | 33.0 | 3.6 | 1.5 | 2.3 |

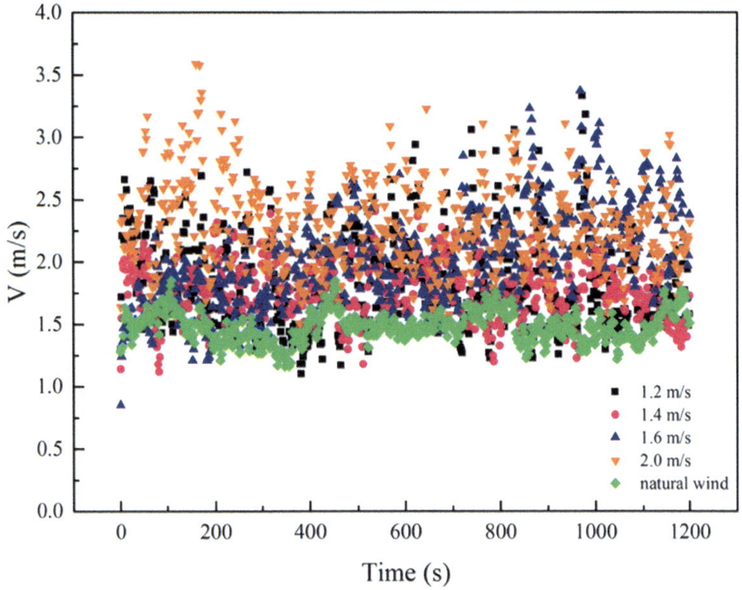


**Fig. 4** Variation of air velocity surrounding subjects at different walking speeds

consistent, averaging 1.8 m/s. In contrast, faster walking produced a more substantial disturbance in airflow, resulting in an average $V_s$ value of 2.16 m/s.

## 3.2 *Analysis of Thermal Response*

The distribution of participant votes is presented in Fig. 5, which illustrates the distinct thresholds for each walking speed. Significant variations in thermal perception were observed across different walking speeds. Nearly 60% of participants reported experiencing a neutral sensation (0) during slow walking. In contrast, close to 70% indicated a neutral sensation during fast walking. The proportion of participants reporting a neutral experience during slow walking fell to one-third of the total. Conversely, votes on the warm-biased side represented merely 35% of the total in the slow-walking condition.

Figure 6 illustrates the variation in MTSV relative to walking speed. All MTSV slopes showed a marked increase as walking speeds rose from 1.2 to 2.0 m/s. Furthermore, MTSV rose progressively across different speeds over the duration of walking and approached stabilization between 15 and 20 min. The higher vote recorded in the final steady phase may be attributed to faster walking speeds. The values returned to those observed in Phase 1 within 10 min after walking ended.

The changes in MMSV over time are displayed in Fig. 7. MMSV remained low at the start of exercise but rose sharply within the first minute. Between 1 and 20 min,

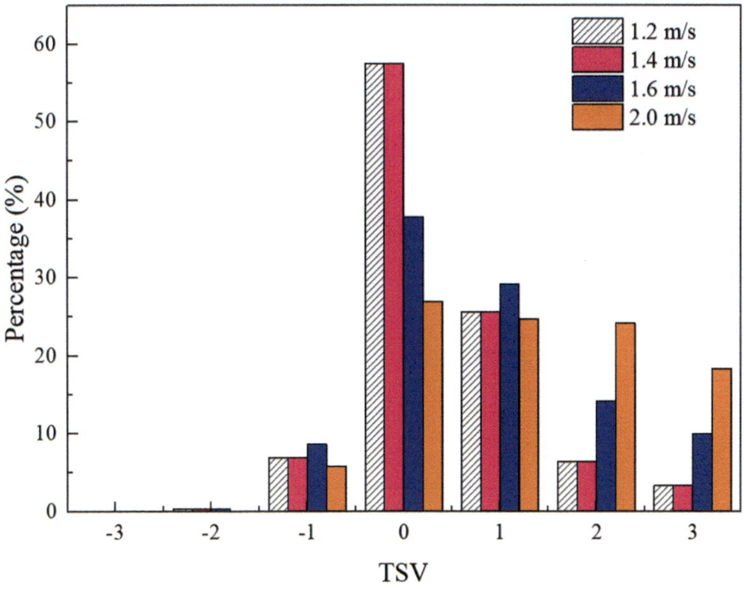


**Fig. 5** Distribution of voting values according to different walking speeds

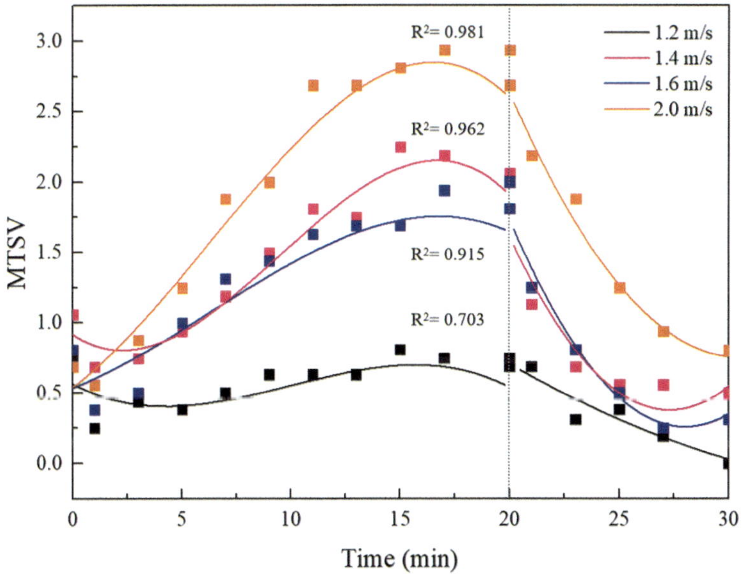


**Fig. 6** Variations in MTSV within a time frame. *Note* The dotted line at 20 min represents the boundary between phase 2 and phase 3, both in Figs. 6 and 7

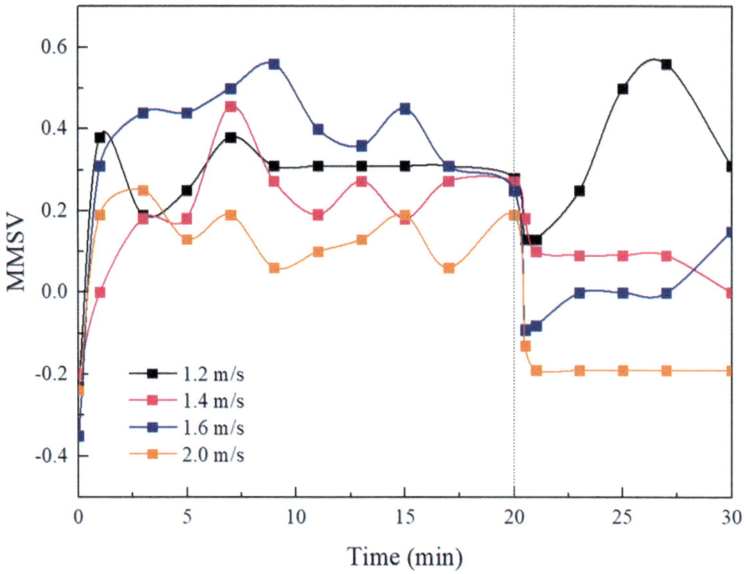


**Fig. 7** Variations in MMSV within a timeframe

noticeable fluctuations in MMSV occurred, with the maximum variation reaching approximately 0.3. By the end of the exercise, MMSV was lower during fast walking compared to slow walking, with the lowest value occurring at a speed of 2.0 m/s. During Phase 3, when walking was maintained at 1.2 m/s, subjects' MMSV continued to increase for 7 min before declining. Under other speed conditions, it decreased immediately and then stabilized. The most pronounced reduction in MMSV was observed at a speed of 2.0 m/s.

## *3.3 Analysis of Physiological Indices*

Prior research has established positive relationships between heart rate and both exercise intensity and energy metabolism. Specifically, as physical exertion rises, the demand for energy metabolism increases, resulting in a corresponding elevation in heart rate [65]. Consequently, heart rate can serve as a reliable indicator for assessing both exercise intensity and actual metabolic rate. Additionally, a positive correlation has been observed between auditory canal temperature and metabolic activity [68]. Based on this evidence, both heart rate and auditory canal temperature were monitored concurrently to estimate the actual metabolic rate.

Analysis of the measurements showed that the subjects' heart rates varied between 81 and 142 bpm, while auditory canal temperatures ranged from 36.1 to 37.2 °C.

Under normal conditions, during low-to-medium intensity exercise, heart rate fluctuations in healthy individuals generally remain below 100 bpm, and changes in auditory canal temperature typically do not exceed about 1 °C [72, 73]. The results obtained in this experiment fall within these expected ranges, supporting further analysis of variations in both heart rate and auditory canal temperature.

Changes in auditory canal temperature are presented in Fig. 8. In Fig. 8a, $\Delta T$ is defined as T(time(i)) −T(time(i = 0)), whereas in Fig. 8b, $\Delta T$ represents T(time(i)) −T(time(i−1)). By comparing values at a given time (i) with those at the previous interval (i−1) within an acceptable range, the dynamic fluctuation process can be considered relatively stable. Therefore, each value was also compared with the preceding time point.

As illustrated in Fig. 8a, when walking at 2.0 m/s, the average auditory canal temperature exhibited a decreasing trend during the initial 3 min, whereas an increase was observed at other speeds. After 7 min, the average auditory canal temperature reached a dynamic equilibrium. Some laboratory-based exercise studies conducted under stable environmental conditions have confirmed a rise in body temperature during physical activity [65, 66, 68]. However, initiating walking at 2.0 m/s immediately induced considerable airflow disturbance. This airflow enhanced convective heat transfer, influencing the thermal exchange process across the subjects' skin surfaces. The improved heat transfer contributed to a reduction in skin surface temperature.

A comparable study performed in a badminton gym [74] reported a consistent decrease in average auditory canal temperature throughout the exercise period, independent of walking speed. The discrepancy between the two studies can be attributed to differences in location. Radiation intensity in the transition space was significantly higher than in the badminton gym, leading to increased bodily heat production. Under identical exercise intensity, this resulted in greater heat storage, which was reflected in the elevated auditory canal temperatures observed.

Figure 8b illustrates that between the 11th and 20th minutes, irrespective of whether the subjects adopted a brisk or a leisurely gait, the instantaneous change $\Delta T(T_i - T_{i-1})$ stayed inside the narrow band of ± 0.075 °C, indicating that the average body temperature was essentially stable. Throughout this recovery interval the auditory-canal temperature rebounded swiftly. When the volunteers moved at 1.2 m/s, the auditory-canal temperature displayed a pronounced drop within the first minute; this was ascribed to the mild air-current disturbance generated by the walking motion itself. Once exercise stopped, the metabolic rate fell promptly. Over the remainder of the pause the auditory-canal temperature oscillated within ± 0.05 °C. The same initial fluctuation pattern was observed at every other walking velocity.

Figure 9 presents the heart-rate disparities. In panel (a) ΔHR is defined as the value at instant i minus the value at the initial instant i = 0, whereas in panel (b) it is computed as the difference between the reading at i and the immediately preceding one at i−1. Panel (a) reveals that the cardiac frequency surged conspicuously during the opening minute of exertion, after which its ascent decelerated. Panel (b) demonstrates that, from the third to the 20th minute, the heart rate stayed inside a steady band of ± 10 bpm, signifying that its fluctuation had become virtually constant.

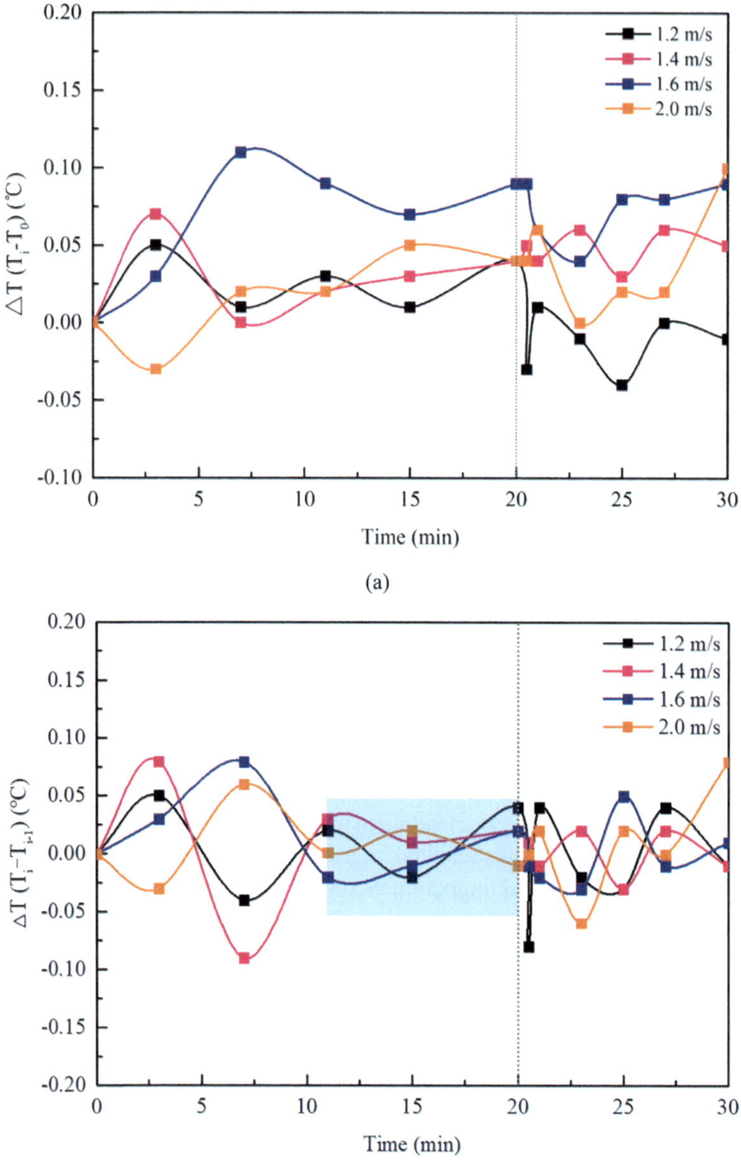


**Fig. 8** Differences in auditory canal temperature: **a** compared with the initial value, **b** compared with the final value

Measurements taken in the badminton hall mirror the trends depicted here [74]. Once the recovery interval began, the heart rate plunged dramatically within the first minute and thereafter tapered off gradually. Ten minutes into the rest phase the ΔHeart-rate ($T_i - T_{i-1}$) had fallen below 5 bpm, returning to a level comparable with that recorded during the seated baseline.

Figure 10 illustrates changes in the average metabolic rate throughout the experimental duration. ASHRAE Standard 55-2017 provides reference values for energy metabolism rates of adult males at various activity levels. According to the standard, walking at 1.2 m/s corresponds to a metabolic rate of 2.6 met, and at 1.8 m/s, the value is 3.8 met. These estimates pertain to walking on a level surface. Using linear interpolation, the metabolic rates at 1.4 m/s, 1.6 m/s, and 2.0 m/s were calculated as 3.0 met, 3.4 met, and 4.0 met, respectively. The results indicate that the mean metabolic rate was lower than these estimated values during the first 5 min of walking. It approached the reference value by the fifth minute under slower walking conditions. During faster walking, the actual metabolic rate slightly exceeded the estimated level. A possible explanation is that increased convection due to faster walking enhances heat dissipation from the body surface. This activates the body's thermal regulation mechanism, accelerating cardiac output and consequently elevating metabolic rate. As shown in Fig. 11, variations in heat storage over time synchronize with changes in metabolic rate.

# 4 Establishment of the Thermal Index Model

## 4.1 Correlation Between Thermal Sensation and Thermal Indices

To explore the impact of further factors on occupants' thermal comfort in transitional areas, the MTSV linked to every single-degree variation in the two selected thermal metrics—PET and UTCI—was derived; the dataset spans both the exercise and rest intervals (Fig. 12).

As shown in Fig. 12, UTCI and PET both demonstrated a strong linear correlation with MTSV throughout the rest period, with the coefficient of determination $R^2$ exceeding 0.7. In contrast, during walking, the relationship between PET and MTSV was weak, and no significant linear association was observed between UTCI and MTSV.

According to Fig. 10, metabolic rates across various walking speeds increased steadily from an initial low during rest to a stable plateau, a transition lasting approximately 5–10 min. The walking interval was categorized into two distinct phases based on metabolic trends: 0–9 min was designated as the metabolic increase phase, and 11–20 min as the metabolically stable phase. Figure 13 displays the correlations between both thermal indices and MTSV across these phases; subfigures (a) and (b) illustrate MTSV and UTCI, while (c) and (d) depict MTSV and PET. The

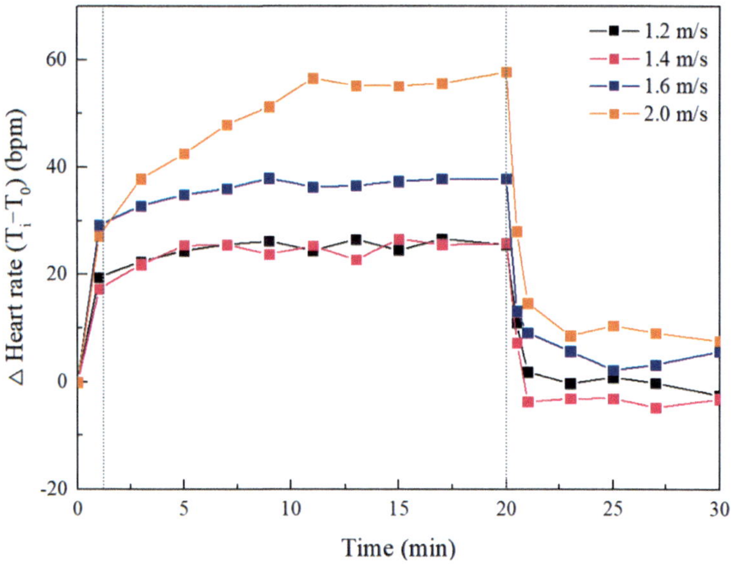


(a)

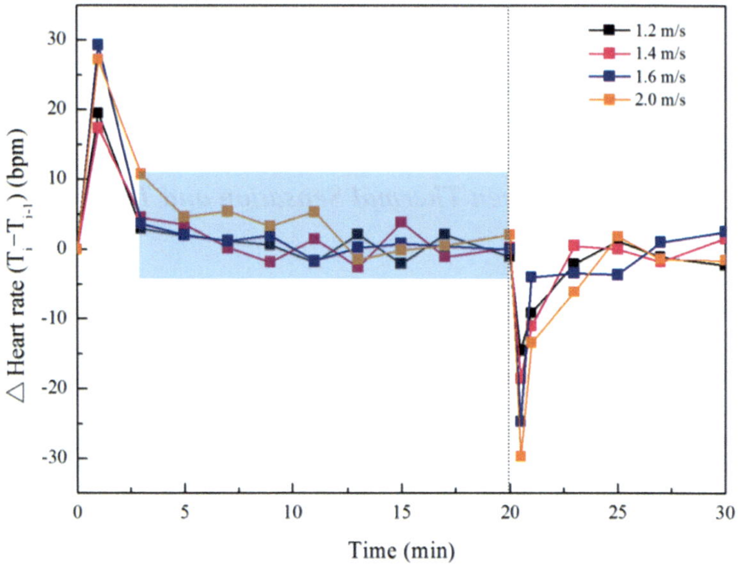


(b)

**Fig. 9** Differences in heart rate: **a** compared with the initial value, **b** compared with the final value

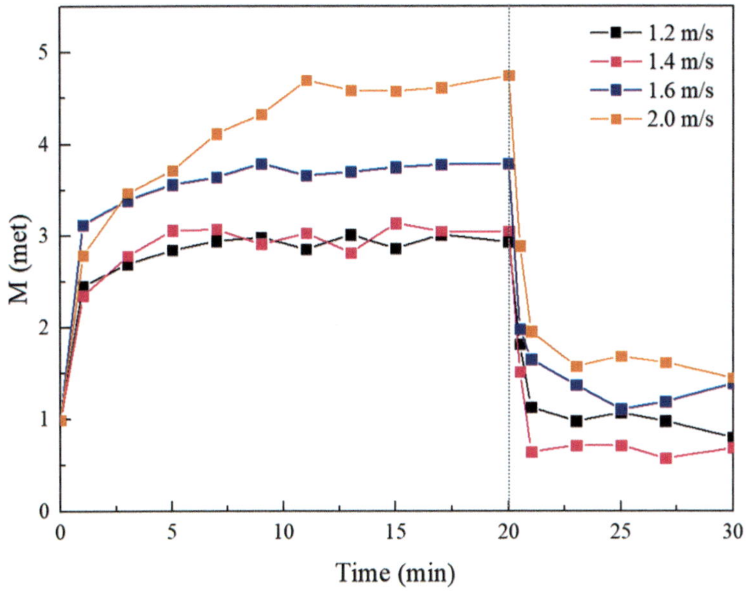


**Fig. 10** Variations in the mean metabolic rate over time

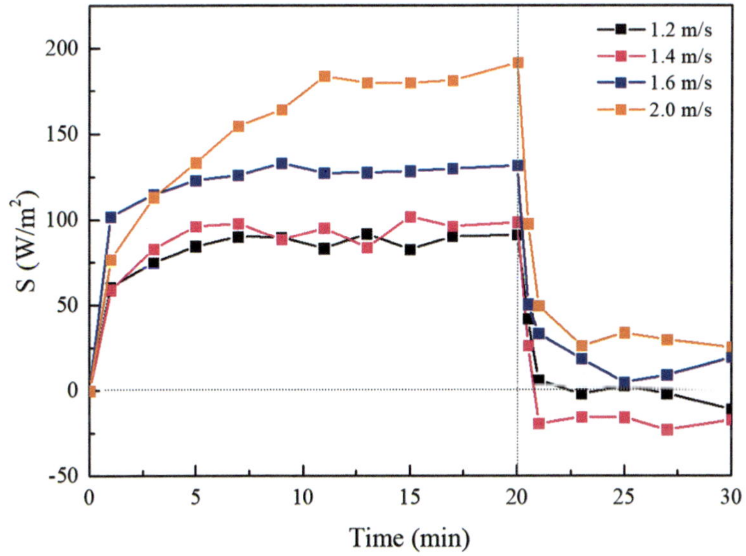


**Fig. 11** Variations in heat storage over time

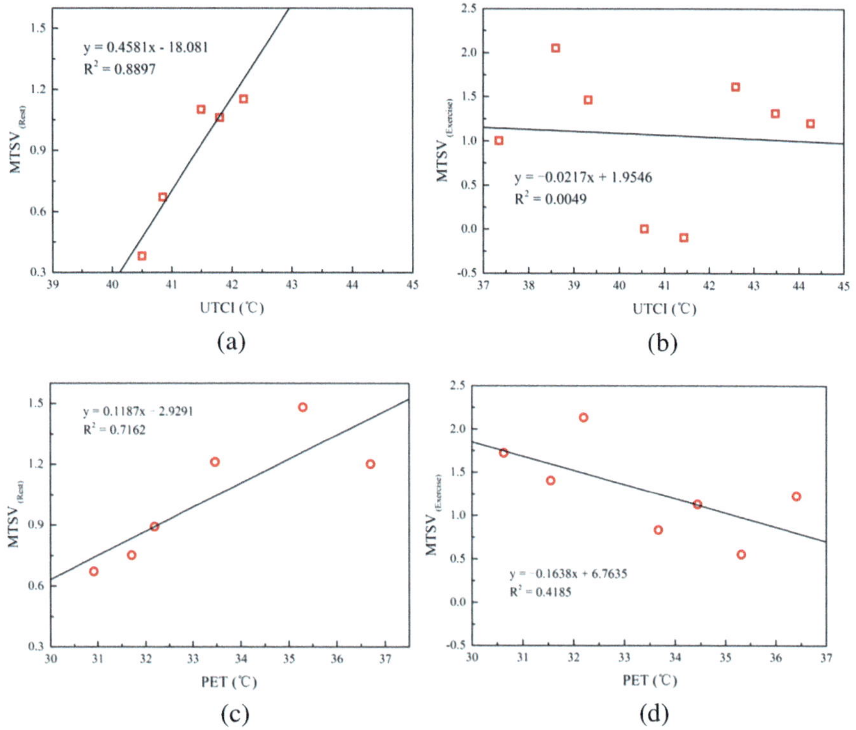


**Fig. 12** Correlations between thermal indices and MTSV: **a, c** rest period, **b, d** exercise period

linear association between MTSV and both indices was markedly stronger during the metabolic increase phase compared to the full exercise period. Specifically, the $R^2$ value between MTSV and PET surpassed 0.7, indicating a robust correlation.

Numerous outdoor thermal comfort investigations have established that under metabolically steady states—such as during seated rest [75–81]—a strong linear relationship exists between UTCI and MTSV. This finding was corroborated in the present study, as evidenced in Fig. 13. However, unlike results from previous static comfort research, a negative correlation emerged between MTSV and UTCI during the 11–20 min period. This phenomenon occurred because elevated UTCI values coincided with lower metabolic rates, leading to reduced MTSV ratings. Conversely, even under low UTCI conditions, higher metabolic activity resulted in elevated MTSV. These patterns suggest that metabolic rate exerts considerable influence on thermal sensation.

Furthermore, PET was derived using the RayMan software by incorporating environmental parameters and average metabolic rate. It is important to note that during actual walking, metabolic rate (M) fluctuated dynamically. Hence, the observed linear relationship between PET and MTSV during metabolically stable periods remains consistent and interpretable.

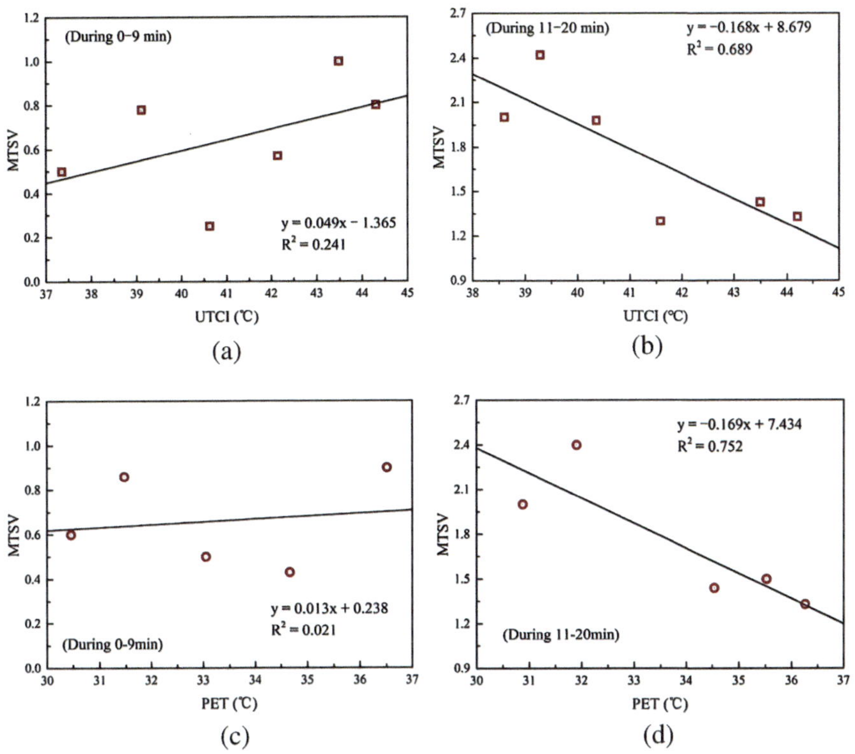


**Fig. 13** Relationship between MTSV and UTCI during different periods of exercise

## *4.2 Establishment of the Relationships' Process*

Based on the results presented above, we hypothesize that the nonlinear relationship between the thermal index and MTSV arose from the gradual variation in metabolic rate. Accordingly, we examined the relationship between ΔMTSV (defined as the difference between MTSV during the exercise period and MTSV in phase 1) and ΔM (the difference between the current metabolic rate and the initial metabolic rate). As illustrated in Fig. 14, thermal sensation shifted consistently as metabolism increased. A strong linear correlation was observed between these variables, supporting the validity of our hypothesis. Thus, the primary factor behind the lack of a strong alignment between thermal indices and MTSV was the continuous variation in metabolic rate during walking.

Concerning the nonlinear association between PET and MTSV during walking, it was observed that slight variations in the input metabolic rate did not lead to clearly discernible changes in the resulting PET value. This may be attributed to inherent constraints of the RayMan software [43]. Such a limitation also partly accounts for the weak correlation observed between PET and MTSV. The UTCI values were

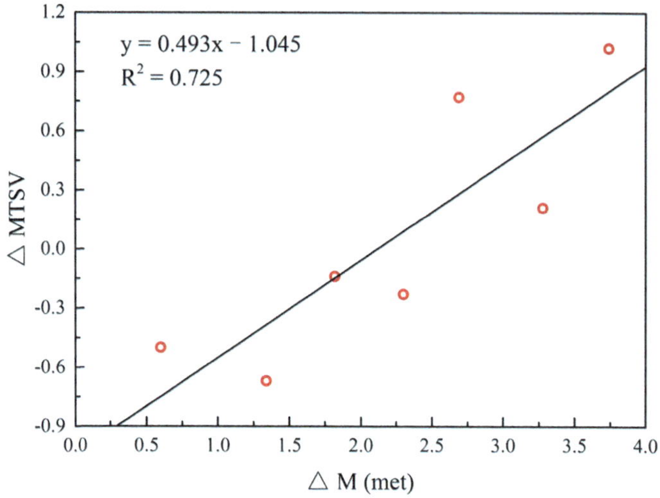


**Fig. 14** Relationship between ΔMTSV and ΔM

computed using the web-based platform (www.utci.org), which incorporates only the influences of environmental parameters—$T_a$, $T_{mrt}$, RH, and $V_a$—without accounting for variations in metabolic rate (M) throughout dynamic conditions.

Since the dominant influencing factor was the dynamic variation in metabolic rate, an enhanced model was employed to capture its progressive alterations during physical activity. Using the stepwise regression module in SPSS 20, regression analysis was performed with ΔM and UTCI (or PET) throughout the exercise period as independent variables, and MTSV as the dependent variable. The significance test yielded a result of Sig. < 0.01, indicating that the independent variables possess strong predictive power for the dependent variable. Accordingly, the following formulas were derived:

$$\mathrm{MTSV} = -0.235\mathrm{UTCI} + 1.225\Delta\mathrm{M} + 7.745 \tag{7}$$

$$\mathrm{MTSV} = -0.235\mathrm{PET} + 0.626\Delta\mathrm{M} - 18.858 \tag{8}$$

Next, we transformed Eqs. (7) and (8) based on the correlation between the MTSV and the thermal index during the rest phase (Eqs. 9 and 10 below), so that both the rest period and the exercise period are represented by the same formula:

$$\mathrm{MTSV} = 0.4581(-0.51\mathrm{UTCI} + 2.67\Delta\mathrm{M} + 56.38) - 18.081 \tag{9}$$

$$\mathrm{MTSV} = 0.1187(-1.98\mathrm{PET} + 5.27\Delta\mathrm{M} + 87.89) - 2.9291 \tag{10}$$

In the above formulas, the contents in red and in brackets are defined as α UTCI (Eq. 11) and α PET (Eq. 12), representing the improved thermal indices, respectively.

$$\alpha\ \text{UTCI} = -0.51\text{UTCI} + 2.67\Delta\text{M} + 56.38 \tag{11}$$

$$\alpha\ \text{PET} = -1.98\text{PET} + 5.27\Delta\text{M} + 87.89 \tag{12}$$

By this approach, refined versions of the thermal indices UTCI and PET were produced, enabling the derived values to furnish a more accurate forecast of participants' MTSV, both in the resting state and while walking.

These enhanced formulations are displayed in Figs. 15 and 16; the α-UTCI and α-PET figures yielded by the updated models exhibit a strong linear correspondence with MTSV, registering $R^2$ coefficients of 0.9 and 0.8, respectively, under either resting or walking condition.

Consequently, the adjusted model offers a practical instrument for assessing human thermal comfort inside transitional zones.

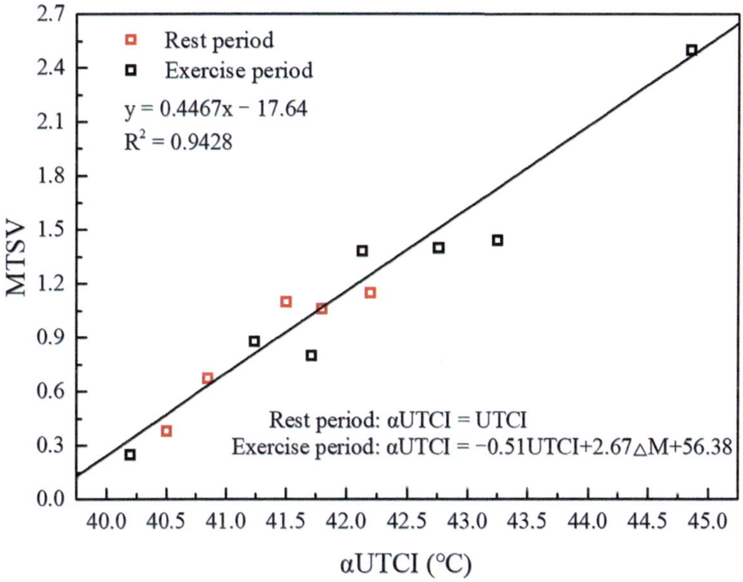


**Fig. 15** The modified model of UTCI

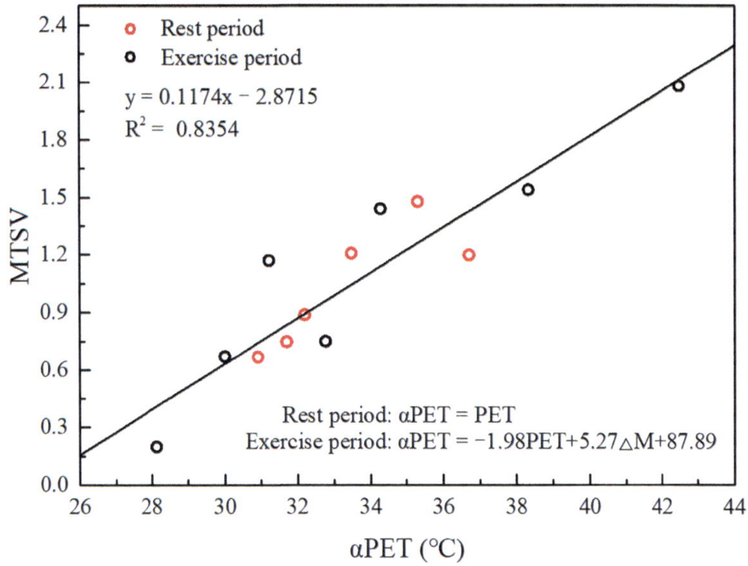


**Fig. 16** The modified model of PET

# 5 Discussion

## *5.1 Variation of MTSV with Walking Speed*

During this test, the air temperature in the overhead layer measured 34.0 ± 2.5 °C. As shown in Fig. 17, the subjects' MTSV values across different speeds were above the comfort zone, suggesting that most participants experienced varying levels of discomfort while walking in the transition space. Only at a maintained speed of 1.2 m/s did the MTSV approach 0.5, placing it within the acceptable range. Some individual TSV values did fall inside the comfort zone. By extrapolating the linear regression line (transition space), which intersects the comfort zone at two points, the walking speed predicted to lie within the comfort zone under these conditions was approximately 0.40–0.98 m/s.

In comparison, the comfort zone speed range observed in a naturally ventilated indoor space (badminton gym) was 0.67–1.40 m/s, with MTSV values remaining within the comfort range during slow walking [74]. A comparison of the two experiments reveals that both the comfortable walking speed range and the maximum comfortable speed were lower in the transition space. This difference is likely due to the higher air temperature during the overhead layer test. Besides the influence of wind speed and individual metabolism, thermal sensation in the transition space may also be affected by combined ground radiation and environmental long-wave radiation. Therefore, to improve thermal comfort when moving through an overhead layer, it is advisable to limit walking speed to 1.3 m/s or lower.

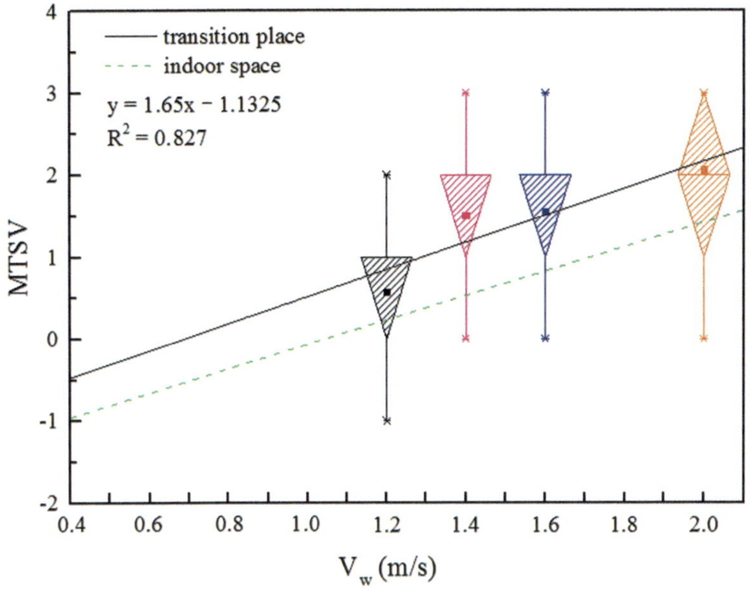


**Fig. 17** Variation of MTSV against walking velocity ($V_W$). *Note* The data of indoor space (green line) is taken from previous research [74]

Since the predicted comfort zone speed range for the transition space falls outside the experimentally tested speeds, this range is used here solely for comparative purposes within this section, and its precise boundaries require further validation.

## *5.2 Analysis of Disturbance Air Velocity on Heat Storage*

This chapter was conducted in a transition space under an overhead layer, which was open to wind from all directions. Participants performing the walking tests were significantly influenced by turbulent airflow. Data presented in Table 5 show that the mean ambient air speed across all walking speeds ranged between 1.8 and 2.3 m/s. In time-frequency analysis, wind is characterized not only by average velocity, but also by turbulence intensity, integral length scale, and power spectral density. Among these, turbulence intensity affects thermal comfort for two primary reasons [82]. The first, suggested by Meyer [83], is that increased turbulence enhances convective heat loss, leading to reduced skin temperature. The second explanation, proposed by Fanger [84, 85], argues that higher turbulence does not decrease mean skin temperature, but causes stronger temperature fluctuations. This amplifies the activity of thermal receptors in the skin compared to a steady airflow at the same average speed. Consequently, under conditions of strong wind speed variation in transition spaces, disregarding the impact of dynamic airflow would introduce errors. Moreover, heat exchange between wind and the human body occurs mainly through convective heat transfer (C), the key influencing factor of which is the convective heat transfer coefficient ($h_c$).

The formula for $h_c$ ($h_c = 8.6u^{0.6}$) proposed by Mitchell in 1974 [86] is widely regarded as the most accurate averaging method under steady wind conditions. De Dear et al. [87] were among the first to measure convective heat transfer coefficients separately for different parts of the body. The wind speeds used in their experiments encompassed most ranges encountered in daily outdoor environments. Research by Ono et al. [88] demonstrated that under turbulent wind conditions, $h_c$ is a function of both wind speed and turbulence intensity. Based on simulation experiments, Yu et al. [71] computed convective heat transfer coefficients in turbulent scenarios and established a correction rule for calculating the convective heat transfer coefficient of the human body in unsteady flow ($h_c = 9.93u^{0.54} \times (1 + 1.03TI \times u^{0.54})$). This correction accounts for the effect of wind turbulence on convective heat exchange between the human body and the external environment.

During this experiment, participants were exposed to turbulent winds of varying intensities, resulting from a combination of natural airflow and disturbance wind generated by their own movement. The correction formula for the convective heat transfer coefficient ($h_c$) introduced by Yu et al. [71] was applied to compute convective heat exchange, and the body heat storage curve was derived using Eq. 5 (Fig. 11). In a similar walking test conducted in a badminton gym [74], where indoor natural wind was stable, the disturbance wind induced by walking exhibited low average speed and minimal fluctuation, resulting in low turbulence. For simplicity, the steady-state $h_c$ formula was used to estimate convective heat transfer indoors. In contrast, the present study was carried out in a transition space. Analysis of collected data revealed considerable fluctuation in wind speed. Neglecting the effect of turbulent wind under these conditions could lead to significant experimental inaccuracies.

To evaluate the impact of turbulent wind intensity around subjects walking in a transition space, we computed $h_c$ using both dynamic and steady-state approaches, yielding two heat storage curves as depicted in Fig. 18. The heat storage level obtained with the modified $h_c$ was generally lower than that from the unadjusted $h_c$, indicating that disregarding the turbulence intensity of the surrounding wind may lead to an underestimation of convective heat transfer and consequently an overestimation of body heat storage. Based on the data in Fig. 18, the maximum stored heat was overestimated by approximately 20%. Experiments by Yu et al. [71] also confirmed that ignoring turbulence intensity at wind speeds between 2 and 3 m/s may cause $h_c$ to be underestimated by about 30%, and could reduce the sensitivity of thermal comfort models to variations in airflow.

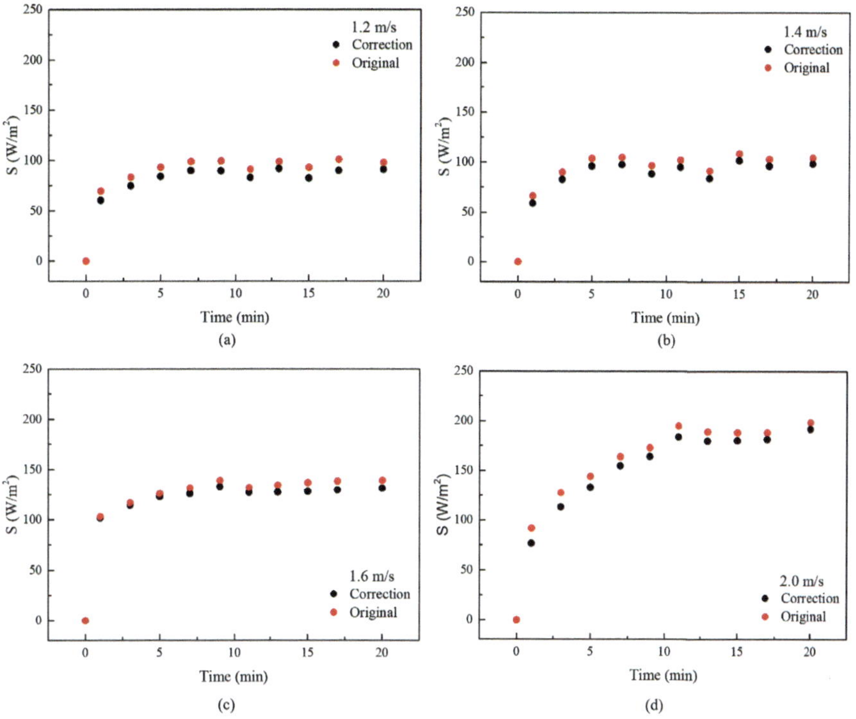


**Fig. 18** Heat storage calculated by different methods

# 6 Conclusions

This chapter investigated thermal comfort during commuting activities in a semi-open transitional space, combining field experiments with subjective and objective monitoring to capture dynamic pedestrian responses. The results confirm that ambient air velocity exerts the strongest influence on thermal perception in such environments, not only by altering convective heat transfer but also by affecting the stabilization of metabolic rate during walking. Compared with indoor conditions, semi-open spaces prolong the time needed for metabolic adjustment, with stabilization typically occurring after 9–11 min of walking, and returning to resting levels within 3–5 min after stopping. Once metabolic equilibrium is reached, thermal sensation shows a strong linear correlation with PET and UTCI, enabling reliable prediction of thermal comfort through these indices. Importantly, the study highlights that overlooking the role of turbulent airflow leads to a substantial overestimation of body heat storage, by around 20%, thus underrepresenting the cooling effect of convective processes. The refined thermal index model developed here demonstrates improved predictive accuracy for both dynamic and stationary phases, underscoring the necessity of integrating physiological adaptation and airflow dynamics into outdoor thermal comfort assessments.

These findings provide actionable insights for enhancing comfort and safety in semi-open transitional spaces, particularly those designed for commuting activities in hot and humid urban environments.

**Acknowledgments** This work was supported by the National Natural Science Foundation of China (Project No. 51978180), the Guangzhou Science and Technology Plan Project (No. 201804010252), and Fundamental Research Funds for the Central Universities (Project No. 2020CDJQY-A009). The authors thank all the workers who participated in the surveys.

## References

1. H. Radhi, S. Sharples, Quantifying the domestic electricity consumption for air-conditioning due to urban heat islands in hot arid regions. Appl. Energy **112**, 371–380 (2013)
2. A. Ghaffarianhoseini, U. Berardi, A. Ghaffarianhoseini, K. Al-Obaidi, Analyzing the thermal comfort conditions of outdoor spaces in a university campus in Kuala Lumpur, Malaysia. Sci Total. Environ **666**, 1327–1345 (2019)
3. European Commission (2018) Buildings
4. Energy Information Administration (EIA) How much energy is consumed in residential and commercial buildings in the United States?
5. N.E. Klepeis, W.C. Nelson, W.R. Ott, J.P. Robinson, A.M. Tsang, J.P. Switzer, W.H. Engelmann, The national human activity pattern survey (NHAPS): a resource for assessing exposure to environmental pollutants. J. Expo. Anal. Environ. Epidemiol. **11**(3), 231 (2001)
6. Z. Ren, D. Chen, Modelling study of the impact of thermal comfort criteria on housing energy use in Australia. Appl. Energy **210**, 152–166 (2018)
7. R. Hwang, W. Shih, T. Lin, K. Huang, Simplification and adjustment of the energy consumption indices of office building envelopes in response to climate change. Appl. Energy **230**, 460–470 (2018)
8. M. Santamouris, Analyzing the heat island magnitude and characteristics inone hundred Asian and Australian cities and regions. Sci. Total. Environ. **512–513**, 582–598 (2015)
9. C. Fabiani, A.L. Pisello, E. Bou-Zeid, J. Yang, F. Cotana, Adaptive measures for mitigating urban heat islands: the potential of thermochromic materials to control roofing energy balance. Appl. Energy **247**, 155–170 (2019)
10. K.M. Wai, C. Yuan, A. Lai, P.K.N. Yu, Relationship between pedestrian-level outdoor thermal comfort and building morphology in a high-density city. Sci. Total. Environ. **708**, 134516 (2020)
11. D. Lai, W. Liu, T. Gan, K. Liu, C. Chen, A review of mitigating strategies to improve the thermal environment and thermal comfort in urban outdoor spaces. Sci. Total. Environ. **661**, 337–353 (2019)
12. J. Huang, D. Shi, Z. Fang, Y. Gao, C. Zhuang, J. Zhai, Impact of short-term thermal experience on thermal sensation: a case study of Chongqinng, China. Build. Environ. **179** (2020)
13. A. Pitts, J.B. Saleh, Transition spaces and thermal comfort—opportunities for optimising energy use, in *PLEA2006—The 23rd Conference on Passive and Low Energy Architecture*, (Geneva, 2006)
14. Z.N. Wang in *Architectural Space Art and Technology* (1987)
15. H. Zhang, M. Tang, W. Zhao, Investigation of the urban walking space sun-shading, in *Proceedings of the National Conference on Building Environment and Building Energy Conservation*, (2007)
16. M. Yıldırım, Shading in the outdoor environments of climate-friendly hot and dry historical streets: the passageways of Sanliurfa, Turkey. Environ. Impact Assess. Rev. **80** (2020)

17. S. Mazouz, M.S. Zerouala, Shading as a modulator for the design of urban layouts based on vernacular experiences. Energy Build. **29**, 11–15 (1998)
18. D. Shi, J. Song, J. Huang, C. Zhuang, R. Guo, Y. Gao, Synergistic cooling effects (SCEs) of urban green-blue spaces on local thermal environment: a case study in Chongqing, China. Sustain. Cities Soc. **55** (2020)
19. Y. Zhang, D. Shi, R. Guo, C. zhuang, K. Zhao, Single image modeling (SIM) for predicting the temperature and airflows of outdoor air zones in regional planning. Sustain. Cities Soc. **53** (2020)
20. B. Hong, B. Lin, Numerical studies of the outdoor wind environment and thermal comfort at pedestrian level in housing blocks with different building layout patterns and trees arrangement. Renew. Energy **73**, 18–27 (2015)
21. Y. Du, C.M. Mak, Y. Li, A multi-stage optimization of pedestrian level wind environment and thermal comfort with lift-up design in ideal urban canyons. Sustain. Cities Soc. **46** (2019)
22. R.D. Dear, J. Kim, Thermal comfort inside and outside building. Adv. Environ. Wind. Eng. (2016)
23. X. Zhang, A.U. Weerasuriya, B. Lu, K.T. Tse, C.H. Liu, Y. Tamura, Pedestrian-level wind environment near a super-tall building with unconventional configurations in a regular urban area. Build. Simulation. (2019). https://doi.org/10.1007/s12273-019-0588-3
24. X. Zhang, K.T. Tse, A.U. Weerasuriya, K.C.S. Kwok, J. Niu, Z. Lin, C.M. Mak, Pedestrian-level wind conditions in the space underneath lift-up buildings. J. Wind. Eng. Ind. Aerodyn. **179**, 58–69 (2018)
25. X. Zhang, K.T. Tse, A.U. Weerasuriya, S.W. Li, K.C.S. Kwok, C.M. Mak, J. Niu, Z. Lin, Evaluation of pedestrian wind comfort near 'lift-up' buildings with different aspect ratios and central core modifications. Build. Environ. **124**, 245–257 (2017)
26. K.T. Tse, X. Zhang, A.U. Weerasuriya, S.W. Li, K.C.S. Kwok, C.M. Mak, J. Niu, Adopting "lift-up" building design to improve the surrounding pedestrian-level wind environment. Build. Environ. **117**, 154–165 (2017)
27. Y. Xie, J. Liu, T. Huang, J. Li, J. Niu, C.M. Mak, T. Lee, Outdoor thermal sensation and logistic regression analysis of comfort range of meteorological parameters in Hong Kong. Build. Environ. **155**, 175–186 (2019)
28. J. Liu, X. Zhang, J. Niu, K.T. Tse, Pedestrian-level wind and gust around buildings with a 'lift-up' design: assessment of influence from surrounding buildings by adopting LES. Build. Simul. **12**(6), 1107–1118 (2019)
29. J. Liu, J. Niu, Delayed detached eddy simulation of pedestrian-level wind around a building array—the potential to save computing resources. Build. Environ. **152**, 28–38 (2019)
30. H.J. Cho, J.W. Jeong, Evaluation of thermal comfort in an office building served by a liquid desiccant-assisted evaporative cooling air-conditioning system. Energy Build. **171**, 361–370 (2018)
31. C. Chun, A. Tamura, Thermal comfort in urban transitional spaces. Build. Environ. **40**, 633–639 (2005)
32. S.A. Damiati, S.A. Zaki, H.B. Rijal, S. Wonorahardjo, Field study on adaptive thermal comfort in office buildings in Malaysia, Indonesia, Singapore, and Japan during hot and humid season. Build. Environ. **109**, 208–223 (2016)
33. A.H. Abdullah, Q. Meng, L. Zhao, F. Wang, Field study on indoor thermal environment in an atrium in tropical climates. Build. Environ. **44**, 431–436 (2009)
34. S. Hussain, P.H. Oosthuizen, A. Kalendar, Evaluation of various turbulence models for the prediction of the airflow and temperature distributions in atria. Energy Build. **48**, 18–28 (2012)
35. H. Liu, Z. Lian, Z. Gong, Y. Wang, G. Yu, Thermal comfort, vibration, and noise in Chinese ship cabin environment in winter time. Build. Environ. **135**, 104–111 (2018)
36. F. Wang, K. Pichatwatana, S. Roaf, L. Zhao, Z. Zhu, J. Li, Developing a weather responsive internal shading system for atrium spaces of a commercial building in tropical climates. Build. Environ. **71**, 259–274 (2014)
37. M.J. Holmes, J.N. Hacker, Climate change, thermal comfort and energy: meeting the design challenges of the 21st century. Energy Build. **39**, 802–817 (2007)

38. Y.H. Yau, A preliminary thermal comfort study in tropical buildings located in Malaysia Int. J. Mech. Mater. Eng. **3**, 119–126 (2008)
39. J. Han, G. Zhang, Q. Zhang, J. Zhang, J. Liu, L. Tian, et al., Field study on occupants' thermal comfort and residential thermal environment in a hot-humid climate of chin. Build. Environ. **42**, 4043–4050 (2007)
40. H.B. Chen, Y.X. Zhu, Energy saving and thermal comfort problems of residential buildings which adopt concentrated HVAC system. J. Harbin Inst. Technol. **14**, 176–178 (2007)
41. H. Yoshino, Y. Yoshino, Q. Zhang, A. Mochida, N. Li, Z. Li, et al., Indoor thermal environment and energy saving for urban residential buildings in China. Energy Build. **38**, 1308–1319 (2006)
42. S.H. Hong, J. Gilbertson, T. Oreszczyn, G. Green, I. Ridley, A field study of thermal comfort in low-income dwellings in England before and after energy efficient refurbishment. Build. Environ. **44**, 1228–1236 (2009)
43. T. Huang, J. Li, Y. Xie, J. Niu, C.M. Mak, Simultaneous environmental parameter monitoring and human subject survey regarding outdoor thermal comfort and its modelling. Build. Environ. **125**, 502–514 (2017)
44. Y. Epstein, D.S. Moran, Thermal comfort and the heat stress indices. Ind. Health. **44**, 388–398 (2006)
45. H.H. Ali, R. Al-Hashlamun, Assessment of indoor thermal environment in different prototypical school buildings in Jordan. Alex. Eng. J. **58**, 699–711 (2019)
46. Z. Fang, Z. Lin, C.M. Mak, J. Niu, K.-T. Tse, Investigation into sensitivities of factors in outdoor thermal comfort indices. Build. Environ. **128**, 129–142 (2018)
47. P.O. Fanger, Thermal comfort: analysis and applications in environmental engineering. Therm. Comf. Anal. Appl. Environ. Eng. (1972)
48. A.P. Gagge, Rational temperature indices of man's thermal environment and their use with a 2-node model of his temperature regulation. Fed. Proc. **32** (1973)
49. M.H. Hasan, F. Alsaleem, M. Rafaie, Sensitivity study for the PMV thermal comfort model and the use of wearable devices biometric data for metabolic rate estimation. Build. Environ. **110**, 173–183 (2016)
50. G. Havenith, I. Holmér, K. Parsons, Personal factors in thermal comfort assessment: clothing properties and metabolic heat production. Energy Build. **34**(6), 581–591 (2002)
51. ISO 8996, *Ergonomics of Thermal Environments—Determination of Metabolic Heat Production*, (ISO, Geneva, 1989)
52. T.F. Colton, E. Larsen, Exercise physiology: the response of metabolic rate to physical activity, in *Proceedings of the 23rd Workshop/Conference of the Association for Biology Laboratory Education (ABLE)*, vol. 23 (2002), pp. 235–254
53. D. Schnorr, G. Dörner, F. Stahl, W. Rohde, H.M. Guddat, Human heat balance during post exercise recovery: separating metabolic and nonthermal effects. Am. J. Physiol. **294**(2), 1586–1592 (2008)
54. K. Fabbri, *Indoor Thermal Comfort Perception: A Questionnaire Approach Focusing on Children* (Springer, New York City, NY, USA, 2015)
55. ISO 7726, *Ergonomics of the Thermal Environment—Instruments for Measuring Physical Quantities* (1998)
56. ANSI/ASHRAE Standard 55–2017, *Thermal Environmental Conditions for Human Occupancy American Society of Heating Refrigerating and Air-Conditioning Engineers (ASHRAE)* (Atlanta, Georgia, 2017)
57. J.E. Greenleaf, B.L. Castle, External auditory canal temperature as an estimate of core temperature. J. Appl. Physiol. **32**(2), 194–198 (1972)
58. J.G. Modell, C.R. Katholi, S.M. Kumaramangalam, E.C. Hudson, D. Graham, Unreliability of the infrared tympanic thermometer in clinical practice: a comparative study with oral mercury and oral electronic thermometers. South. Med. J. **91**(7), 649–654 (1998)
59. J.E. Hayter, C.J.K. Henry, A re-examination of basal metabolic rate predictive equations: the importance of geographic origin of subjects in sample selection. Eur. J. Clin. Nutr. **48**(10), 702–707 (1994)
60. R.A. Engineers, *ASHRAE Handbook: Fundamentals* (2017)

61. ISO, International Standard 7730, Ergonomics of the Thermal Environment—Analytical Determination and Interpretation of Thermal Comfort Using Calculation of the PMV and PPD Indices and Local Thermal Comfort Criteria International Standard Organization, Geneva, (2005)
62. A. Matzarakis, F. Rutz, H. Mayer, Modelling radiation fluxes in simple and complex environemnts—application of the RayMan model. Int. J. Biometeorology. **51**, 323–334 (2007)
63. P. Weihs, H. Staiger, B. Tinz, E. Batchvarovam, H. Rieder, L. Vuilleumier, M. Maturilli, G. Jendritzky, The uncertainty of UTCI due to uncertainties in the determination of radiation fluxes derived from measured and observed meteorological data. Int. J. Biometeorol. **56**, 537–555 (2012)
64. D. Fiala, G. Havenith, P. Bröde, B. Kampmann, G. Jendritzky, UTCI-Fiala multi-node model of human heat transfer and temperature regulation. Int. J. Biometeorol. **56**(3), 429–441 (2012)
65. W. Li, Y. Zhang, L. Zhang, Z. Li, The research of the sport burden grade standard corresponding to the heart rate. Nat. Sci. J. Harbin Normal Univ. **24**(5), 99–102 (2008)
66. P.O. Fanger, *Thermal Comfort: Analysis and Application in Environmental Engineering* (Danish Technology Press, Copenhagen, 1970)
67. Y. Nishi, Measurement of thermal balance of man. Stud. Environ. Sci. **10**, 29–39 (1981)
68. T. Sakoi, T. Mochida, Y. Kurazumi, K. Kuwabara, Y. Horiba, S. Sawada, Heat balance model for a human body in the form of wet bulb globe temperature indices. J. Therm. Biol. **71**, 1–9 (2018)
69. D.A. Mcintyre, *Indoor Climate* (Applied Science Publisher, London, 1980)
70. H. Wang, S. Hu, Experimental study on comfortable skin temperature under sweating condition. Heat. Vent. Air Cond. **5**, 110–113 (2014)
71. Y. Yu, J. Liu, K. Chauhan, R.D. Dear, J. Niu, Experimental study on convective heat transfer coefficients for the human body exposed to turbulent wind conditions. Build. Environ. **169**, 106533 (2020)
72. Z. Fang, H. Liu, B. Li, X. Du, A. Baldwin, Investigation of the effects of temperature for supplied air from a personal nozzle system on thermal comfort of air travelers. Build. Environ **12**, 82–97 (2018)
73. W.L. Kenney, J. Wilmore, D. Costill, *Physiology of Sport and Exercise*, 6th edn. (Human Kinetics, Champaign, IL, 2015)
74. Y. Zhang, X. Zhou, Z. Zhen, M.O. Oladokun, Z. Fang, Experimental investigation into the effects of different metabolic rates of body movement on thermal comfort. Build. Environ. **168** (2020)
75. M.W. Yahia, E. Johansson, Evaluation the behavior of different thermal indices by investigating various outdoor urban environments in the hot dry city of Damascus, Syria. Build. Environ. **57**, 615–630 (2013)
76. B. Yang, T. Olofsson, G. Nair, A. Kabansi, Outdoor thermal comfort under subarctic climate of north Sweden—a pilot study in Umea. Sustain. Cities Soc. **28**, 387–397 (2017)
77. Z. Fang, X. Feng, X. Xu, X. Zhou, Z. Lin, Y. Ji, Investigation into outdoor thermal comfort conditions by different seasonal field surveys in China Guangzhou. Int. J. Biometeorol. **63**(10), 1357–1368 (2019)
78. H. Zhang, *Human Thermal Sensation and Comfort in Transient and Non-uniform Thermal Environments*. Ph.D. Thesis, University of California Berkeley (2003)
79. Y. Zhang, H. Chen, Q. Meng, Thermal comfort in buildings with split airconditioners in hot-humid area of China. Build. Environ. **64**, 213–224 (2013)
80. L. Zhao, X. Zhou, L. Li, S. He, R. Chen, Study on outdoor thermal comfort on a campus in a subtropical urban area in summer. Sustain. Cities Soc. **22**, 164–170 (2016)
81. Y. Zhu, J. Liu, A. Hagishima, J. Tanimoto, Y. Yao, Z. Ma, Evaluation of coupled outdoor and indoor thermal comfort environment and anthropogenic heat. Build. Environ. **2**, 1018–1025 (2007)
82. Q. Jia, R. Zhao, Impact of wind spectra on human thermal comfort. J. Tsinghua Univ (SCI. & Tech.) . **41** (2001)
83. E. Mayer. *Physical Causes for Draft: Some New Findings* (1987)

84. P.O. Fanger, A.K. Melikov, H. Hanzawa, J. Ring, Air turbulence and sensation of draught. Energy Build. **12**, 21–39 (1988)
85. P.O. Fanger, N.K. Christensen, Perception of draught in ventilated spaces. Ergonomics **29**, 215–235 (1986)
86. D.A. McIntyre, *Indoor Climate* (Applied Science Publisher, London, 1981), pp. 152–154
87. R.J. De Dear, E. Arens, Z. Hui, M. Oguro, Convective and radiative heat transfer coefficients for individual human body segments. Int. J. Biometeorol. **40**, 141–156 (1997)
88. T. Ono, S. Murakami, R. Ooka, T. Omori, Numerical and experimental study on convective heat transfer of the human body in the outdoor environment. J. Wind Eng. Ind. Aerodyn. **96**, 1719–1732 (2008)

# Thermal Comfort and Safety for High-Intensity Outdoor Laborers: An Outdoor Investigation of Construction Sites in South China

**Abstract** High-intensity outdoor labor in hot climates poses significant risks to worker health, safety, and productivity, making reliable evaluation of thermal stress a priority in occupational environments. This chapter presents findings from a field study conducted at construction sites in Southern China during the summer of 2019, where both microclimatic conditions and physiological indicators were monitored to assess thermal safety. Relationships between environmental parameters (air temperature, mean radiant temperature, air velocity, and relative humidity), physiological responses (heart rate and ear canal temperature), and cognitive performance were examined alongside workers' subjective feedback. Results indicate that once air temperature exceeds 34 °C, reductions in work intensity become necessary to prevent discomfort and physiological overload. Elevated heart rate and ear canal temperature were closely linked to heat exposure, while cognitive performance declined as thermal stress intensified, especially among older workers. Overall, nearly 40% of participants reported discomfort, and operational efficiency fell by approximately 8% for every marked increase in heat load. These findings underscore the importance of integrating physiological and psychological monitoring into occupational safety frameworks, offering valuable insights for the development of adaptive strategies to protect outdoor laborers under extreme thermal conditions.

## 1 Introduction

Climate change is intensifying global warming, resulting in elevated outdoor temperatures and a strengthened urban heat island effect, which adversely affects both urban sustainability and public health [1]. Simultaneously, worldwide urbanization is advancing, bringing about a consistent increase in energy use by buildings, thereby further aggravating climate change [2, 3]. Projections suggest that 68% of the global population will be living in urban areas by 2050 [4, 5]. This swift urban spread is predicted to induce further warming in cities. Over the next hundred years, the average global temperature is anticipated to increase by 3 °C [6], and extreme heat events are likely to become more common, extended, and intense compared

 
Z. Fang et al., *Outdoor Thermal Comfort and Heat Stress Mitigation*, Environmental Science and Engineering, https://doi.org/10.1007/978-981-95-8899-2_13

to pre-industrial times across various global regions [7]. From 2000 to 2013, China witnessed a notable rise in the frequency, strength, and duration of extreme heat occurrences [8], representing a substantial public health risk. Manual laborers are more vulnerable to heatstroke and other heat-induced conditions due to contact with heatwaves. Conversely, pedestrians may suffer from physical unease and persistent health issues from extended time outdoors. Hence, there is expanding focus on adapting urban planning and architecture to improve thermal comfort under hot weather [1]. Multiple prior researches have investigated approaches for accurately measuring outdoor thermal comfort [9, 10] and for boosting the quality of outdoor settings through design and policy measures [11–13]. Yet, only a small number of studies have concentrated on the thermal comfort of construction workers operating under high temperatures. Media accounts from Hong Kong noted 43 heat—and stress-associated incidents on construction sites from 2007 to 2011, which led to 11 fatalities [14]. In China, deaths are recorded every year as a result of working in extremely hot conditions, and because of its large number of manual workers, the country is expected to be among those most severely affected in coming years [15].

With industrialization and metropolitan growth accelerating, greater attention is being given to threats like heat-triggered health issues—such as heat stress and heat exhaustion—facing workers in severely hot occupational environments. According to Chinese standards, a high-temperature workplace is characterized by a dry-bulb temperature reading of 35 °C or higher, often in the presence of strong thermal radiation, high humidity, and other environmental elements, either individually or together [16]. Studies concerning heat stress [16, 17] specify that high-temperature settings include living spaces over 35 °C and workplaces above 32 °C. Earlier examinations have indicated that warm climates elevate the occurrence of heat-linked health problems and sicknesses, including skin cancer, allergic reactions, heart and blood circulation diseases, and kidney malfunction [18–21]. Such research has found a strongly positive connection between elevated temperatures and rates of job-related accidents. Intense heat can cause undesirable consequences like exhaustion, decreased work capability, attention deficits, major health threats, and reduced output [22]. Worldwide climate projections indicate that periods of extreme heat will grow more frequent during the twenty-first century. This will increase the danger of heat-related sickness and fatalities for construction workers who perform demanding physical tasks outside [23]. Evaluations based on international datasets estimate that ability to work in the warmest periods will drop by 20% by 2050 and by 60% by 2200 because of global warming [24]. Increases in air temperature are associated with an estimated 5.3% reduction in efficiency for open-air manual labor across the world [25], emphasizing that exposure to extreme heat presents a major hazard and difficulty for outdoor workers.

Numerous studies [26–31] indicate that heat-wave events are linked to the well-being of city dwellers. Significant correlations have been observed between rising temperatures and elevated mortality rates [26, 27]. Extreme heat can lead to serious health issues such as respiratory, cardiovascular, and cerebrovascular complications, and in some cases, fatalities [29, 30]. Earlier research has emphasized pathological dimensions, including identifying temperature thresholds that pose health risks during extreme heat, and exploring connections between the incidence and mortality of heat-related illnesses and factors such as age, gender, pre-existing health conditions, and socioeconomic status [32, 33]. An investigation in Hong Kong showed that hot weather heightened heat stress among construction workers and even resulted in fatalities [34]. To clarify the connection between key outdoor thermal parameters and thermal comfort, researchers have carried out measurements and on-site surveys in various locations, gathering information to examine how subjective human thermal perception outdoors correlates with primary thermal environment indicators [18, 35–39]. Zare et al. [18] examined the relationship between heat stress indices and physiological metrics among workers in Iran. Their findings revealed a strong association between the Universal Thermal Climate Index (UTCI) and the Wet Bulb Globe Temperature (WBGT). Through subjective surveys, Nag et al. [39] found that the maximum tolerable duration of thermal exposure was around 80–85 min when core body temperature was between 38 and 38.2 °C, decreasing to about 40–45 min at 39 °C. Thus, thermal comfort is influenced by (i) the thermal equilibrium between the human body and environmental conditions, and (ii) human physiological and psychological responses [18, 36–40].

Furthermore, several studies suggest that thermal conditions considerably influence work performance [41–43]. Tsutsumia et al. [41] noted that in low-humidity settings, most individuals reported a comfortable state without notable differences in subjective thermal sensation. Although work efficiency did not vary significantly across humidity levels, fatigue onset was quicker. Ismaila et al. [44] identified temperature as the primary factor affecting work output, followed by lighting and relative humidity. Consequently, some experts have advised consuming cool menthol beverages to enhance the productivity of laborers in hot and humid conditions [45]. However, in the construction sector, tasks vary widely [15], making actual work efficiency difficult to assess. Multiple studies indicate that reduced cognitive function is a key factor behind declines in work capacity. A strong relationship exists between cognitive skills and adaptive performance [46]. Higher cognitive abilities support better adaptation, which can preserve lives, improve academic outcomes, and benefit collective entities (e.g., teams or organizations) by sustaining efficiency [46]. Therefore, it is important to study how hot environments affect cognitive functions such as perception, memory, attention, reasoning, and creativity [46–48].

Existing research [11–14] has consistently indicated that the thermal comfort experienced in outdoor settings is shaped by a range of meteorological variables, such as ambient temperature ($T_a$), relative humidity (RH), wind speed ($V_a$), and mean radiant temperature ($T_{mrt}$). Nevertheless, studies examining how these thermal conditions affect both comfort and work efficiency in hot outdoor occupational settings have notable gaps, calling for more holistic approaches that incorporate the following aspects.

Much of the prior work [46–49] on thermal conditions and cognitive function in construction settings has omitted real-world, on-site evaluation. The construction sector is known for its labor-intensive nature and high physical demands, with workers frequently operating outdoors under intense solar radiation and elevated temperatures [49]. Yet, a significant number of earlier studies have employed simulated conditions or laboratory-based experiments. Outcomes from such simulations, theoretical models, or tightly regulated experimental environments often fail to accurately reflect genuine outdoor working conditions [16].

According to an earlier report [15, 45], analyses of thermal influences on heat stress and cognitive performance have often overlooked temporal fluctuations, relying instead on aggregated daily averages. This approach limits the ability to capture precise thermal conditions and variations in cognitive functioning across the full workday. Hence, there is a clear need for research conducted in actual construction environments that accounts for different times within a worker's shift.

The current study examines how high-temperature conditions at outdoor construction sites affect heat stress and productivity across various times of day, through simultaneous monitoring of environmental parameters and physiological responses. This chapter distinguishes itself from earlier efforts by incorporating real-site measurements of heat exposure, including the collection of physiological data and subjective feedback from workers. Age was also taken into account as a variable during field data collection. Participants were asked to complete survey questionnaires, while physiological markers—specifically, heart rate (HR) and auditory canal temperature (ACT)—were measured. By identifying key factors in the outdoor thermal environment that influence health, this chapter analyzes subjective thermal assessments, physiological indicators, and efficiency metrics. The outcomes advance knowledge regarding the impacts of high-temperature outdoor settings on construction workers' well-being and performance, offering a basis for establishing safety guidelines tailored to this workforce in China.

## 2 Methods

Between July 18 and August 9, 2019, measurements of thermal environmental parameters, subjective thermal assessments, and physiological parameters of workers were carried out at four outdoor construction sites characterized by high-temperature conditions (Fig. 1). An automated on-site monitoring device was used to continuously record outdoor thermal environment data. Data collection for both physiological parameters and subjective thermal evaluations took place from 8:00 AM to 6:00 PM throughout the field study period (refer to Sect. 2.2).

### *2.1 Survey Area*

Guangzhou is situated in the Pearl River Delta in southern China, with geographical coordinates spanning 112 °E–114.2 °E and 22.3 °N–24.1 °N (Fig. 1). The area experiences a typical subtropical climate characterized by mild springs and autumns, as well as extended hot and humid summers. July and August are the hottest months,

**Fig. 1** Survey area in Guangzhou and locations of the four construction sites (A–D)

during which the maximum outdoor air temperature frequently surpasses 36 °C. Annual average relative humidity (RH) shows little variation, generally ranging from 70 to 90%, and at times exceeding 90% [50, 51]. Meteorological data collected between 2014 and 2018 were used to compute monthly averages and the distribution ranges of daily RH and air temperature ($T_a$). Throughout July and August of those years, daily RH varied from 45 to 95%, while Ta fluctuated between 24 °C and 36 °C (Fig. 2). The summertime climate in Guangzhou is often marked by high temperatures and elevated humidity levels, leading to widespread outdoor discomfort among residents.

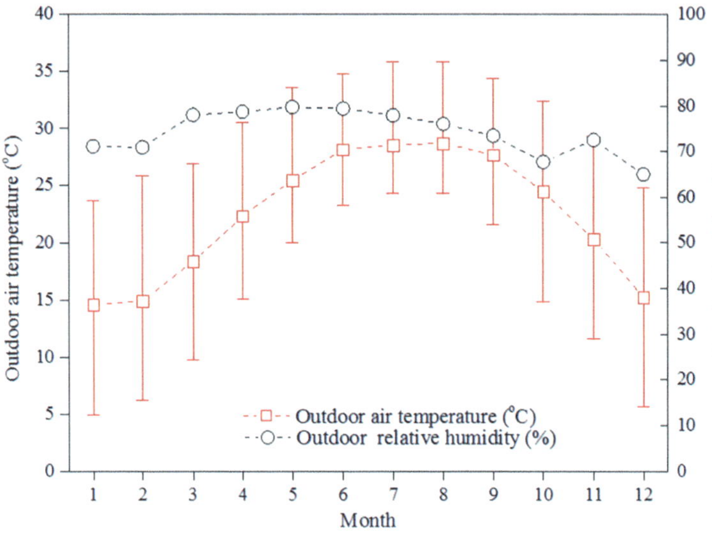


**Fig. 2** Monthly range and mean of the daily outdoor relative humidity and air temperature from 2014 to 2018

## 2.2 Subjects

A total of 1063 male construction workers, aged 22–56 years, from four chosen construction sites took part in the on-site study, which included both questionnaire administration and physiological measurements. Images captured during the field investigation are provided in Fig. 3.

**Fig. 3** Questionnaires and physiological tests conducted at four different construction sites (**a**–**d**)

## 2.3 Measured Parameters and Instruments

Environmental parameters at the selected construction sites were monitored with an SSDZY—1 thermal comfort meter (Table 1). In accordance with ISO 7726 [53], measurements were taken at a height of 1.1 m in proximity to the subjects. A standard matte black globe thermometer, 0.15 m in diameter (D), was employed to measure the globe temperature ($T_g$), with a globe emissivity ($\varepsilon_g$) of 0.95. The mean radiant temperature ($T_{mrt}$) was derived using Eq. (1) [54]:

$$T_{\text{mrt}} = \left\{ (T_g + 273)^4 + \left[ \frac{(1.1 \times 10^8 \times V_a^{0.6})}{(\varepsilon_g \times D^{0.4})} \right] \times (T_g - T_a) \right\}^{1/4} - 273 \quad (1)$$

where $T_a$ is the ambient temperature, $T_g$ is the globe temperature, and $V_a$ is the wind speed.

Workers' physiological indicators, including heart rate (HR), blood pressure (BP), and auditory canal temperature (ACT), were measured using high-precision test watches and sensitive ear thermometers. Data collection was performed by the investigator using these instruments. Each physiological assessment had a duration of approximately 30 s.

Table 1 summarizes the detailed instrument information for measuring $T_a$, RH, $T_g$, $V_a$, HR, and BP.

**Table 1** Detailed information of instruments used in this chapter

| Equipment | Model | Parameter | Measuring range | Accuracy | Sampling rate (s) |
|---|---|---|---|---|---|
| (1) Thermal comfort instrument | SSDZY-1 | $T_a$ (°C) | – 20 °C to + 80 °C | ± 0.3 °C | 30 |
| | | RH (%) | 0.01–99.9% RH | ± 2% RH (10–90%) | 30 |
| | | $T_g$ (°C) | – 20–80 °C | ± 0.3 °C | 30 |
| (2) Universal wind speed recorder | WFWZY-1 | $V_a$ (m/s) | 0.05–5.00 m/s | 5% ± 0.05 m/s | 30 |
| (3) Infrared auditory canal thermometer | TB-300 | $T$ (°C) | 32.0–42.2 °C | ± 0.2 °C | 1 |
| (4) ADZ | AV5 | HR (beats/min) | 40–195 beats/min | ± 5% | 2 |
| | | BP (mmHg) | 0–300 mmHg | ± 3 mmHg | 2 |

## 2.4 Survey Questionnaire

The survey questionnaire consisted of two sections: one addressing thermal comfort and the other evaluating cognitive ability. The thermal comfort section collected information on gender, age, height, weight, clothing specifics, activity levels, thermal symptoms, thermal sensations, and thermal acceptability. Ten potential heat-related symptoms were listed: weakness, dizziness, nausea, profuse sweating, chest tightness, pallor, tachycardia, stomach discomfort, irritability, and headache. Thermal sensation was assessed using scales defined in ASHRAE Standard 55 [55] and ISO 7730 [56]. The conventional ASHRAE seven-point thermal sensation scale ranges from − 3 (cold) to 3 (hot). Given that workers are frequently subjected to hot and humid outdoor conditions, a nine-point thermal sensation scale was applied for more precise evaluation: − 4 (very cold) to 4 (very hot) [50, 56]. Thermal acceptability was recorded using a two-point scale: 0 (acceptable) and 1 (unacceptable).

The cognitive ability questionnaire measured workers' responsiveness and task efficiency. Multiple earlier studies have examined the relationship between task performance and thermal conditions [57–62]. Findings from memory span and image recognition tests indicated that participants' memory performance was better in cooler environments and diminished under warmer conditions [62]. Consequently, the accuracy of tasks requiring substantial working memory also declined at elevated temperatures. During the on-site study, workers were asked to connect numbers in ascending order within 40 s. Each worker received a unique set of numbers for the test.

## 2.5 Procedure

Environmental parameters were logged continuously from the start to the end of each workday at one-minute intervals, with instruments placed adjacent to work areas. Throughout the work period, physiological metrics—such as heart rate (HR), blood pressure (BP), and auditory canal temperature (ACT)—along with subjective feedback and cognitive ability questionnaires, were gathered every 30 min. A detailed outline of the field survey process is provided in Fig. 4. The entire procedure, which extended over 10 h and 10 min, comprised five phases: preparation, pre-work, morning work period, rest break, and afternoon work period. Throughout the data collection, average values were computed for each thermal environmental parameter. It is important to note that all physiological measurements and subjective responses were administered while workers were pausing to rest. Completing the questionnaires—covering both cognitive performance and thermal perception—required approximately 4 min.

Throughout the entire process, cognitive ability testing was employed to evaluate the relative state of the workers. Such tests are widely utilized to measure attention, reaction time, working memory, and visual memory in subjects [63]. Numerous prior

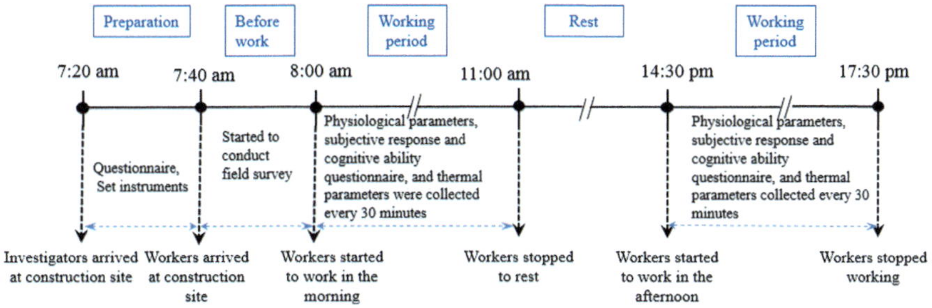


**Fig. 4** Schematic of the field survey procedure

studies have indicated that high temperatures significantly affect cognitive performance following physical exertion [64–68]. At outdoor construction sites, the variety of tasks performed is considerable [15], making it challenging to gauge actual work efficiency directly. As a result, many earlier studies have adopted cognitive ability tests to evaluate the performance of construction workers [63, 69, 70].

Originally developed to measure productivity changes among office workers under varying environmental conditions, cognitive assessments examine factors such as typing speed, comprehension, coordination, memory, and pattern recognition across different work settings [57–62]. Subsequent research extended these methods to evaluate the work efficiency of manual laborers [64–68]. The neurobehavioral approach, through psychometric testing, has been validated as a rational method for assessing productivity impacts. This is because the central nervous system is particularly sensitive to environmental disruptions, more so than peripheral physiological responses. Psychometric tests quantitatively measure the accuracy and speed of four key neurobehavioral functions, enabling precise evaluation [62]. Consequently, the neurobehavioral method offers a systematic and quantitative means to assess how indoor environmental quality influences the productivity of office workers. The cognitive ability test questionnaire applied in this chapter has also been used in previous research [63, 71].

In this field survey, workers were asked to connect numbers in ascending order within 40 s. The questionnaire layout is illustrated in Fig. 5. Starting from "START: 00", participants connected points based on numerical sequence. The task was halted if the time exceeded 40 s. Each worker received a unique version of the questionnaire with different number combinations. The accuracy of the connections was used to assess the cognitive ability of the participants.

## 2.6 *Statistical Analysis*

The required sample size was determined using the population size, margin of error, confidence level, and expected response distribution. In this chapter, the mean thermal sensation vote (MTSV) was computed for each 1 °C interval of air temperature ($T_a$)

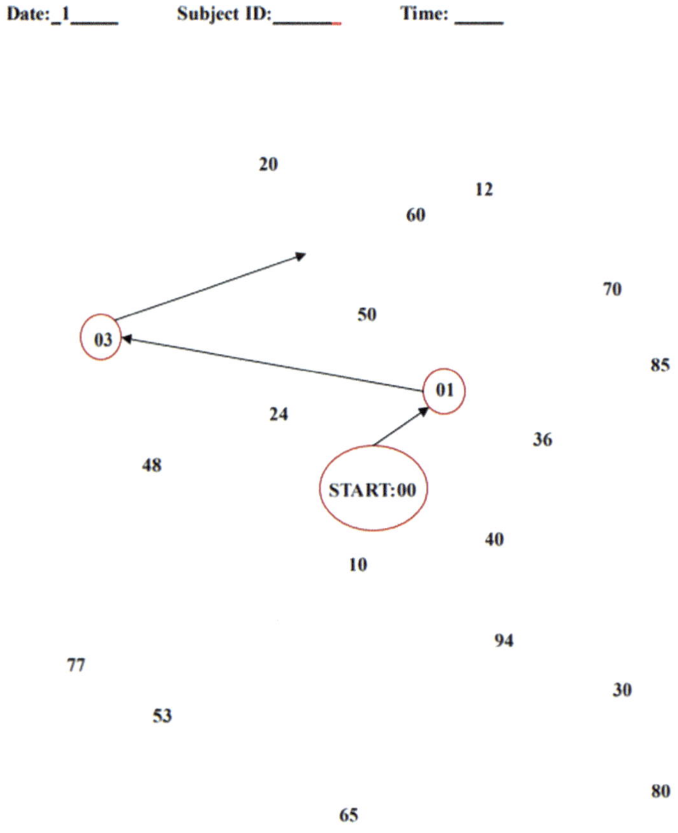


**Fig. 5** Cognitive ability questionnaire

and mean radiant temperature ($T_{mrt}$). Similarly, the mean sunshine sensation vote (MSSV) was calculated per 1 °C interval of $T_{mrt}$.

Linear regression, as the most commonly applied method for evaluating thermal indices, was used to investigate the influence of varying temperatures ($T_a$ and $T_{mrt}$) on subjective responses, physiological indicators, and cognitive test outcomes [72–75]. Additionally, curvilinear regression was utilized to examine associations between physiological parameters and thermal variables. All data were analyzed using SPSS v20.0 (SPSS Inc., Chicago, IL, USA), with statistical significance defined at $p = 0.05$ (one-tailed).

# 3 Results

## 3.1 Thermal and Personal Parameters

The anthropometric data obtained from the 1063 questionnaires completed by workers are summarized in Table 2.

The recorded outdoor thermal parameters—including air temperature ($T_a$), globe temperature ($T_g$), relative humidity (RH), air velocity ($V_a$), mean radiant temperature ($T_{mrt}$), and clothing insulation ($I_{cl}$)—are presented in Table 3. The average Ta was 38.84 °C, with a maximum of 47.19 °C, while the mean RH reached 67.4%, peaking at 93.7%. Throughout the study, all participants wore short-sleeved shirts, thin trousers or jeans, socks, and sneakers, resulting in a mean $I_{cl}$ value of 0.5 clo.

**Table 2** Anthropometric data of subjects (SD: standard deviation)

| Sex | Number | Age in years (SD) | Height in m (SD) | Weight in kg (SD) | Body surface area in $m^2$ (SD)[a] | Positive index in kg 1/3.$m^{-1}$ (SD)[b] |
|---|---|---|---|---|---|---|
| Male | 1063 | 38.1 (9.1) | 1.72 (4.16) | 67.9 (4.8) | 1.79 (0.07) | 2.38 (0.06) |
| Parameters | | Minimum | | Maximum | | Mean |
| Age (years) | | 22 | | 56 | | 38.1 |
| Height (m) | | 1.65 | | 1.81 | | 1.72 |
| Weight (kg) | | 60.0 | | 77.0 | | 67.9 |

*Note* [a] Body surface area (A) was determined through the DuBois area: $A = 0.202\ W^{0.424}\ h^{0.725}$ [76], where W is weight and h is height. [b] Ponderal index = $W^{1/3}/h$

**Table 3** Outdoor thermal and total clothing insulation data

| Parameters | Abbreviation | Minimum | Maximum | Mean | Standard deviation |
|---|---|---|---|---|---|
| Ambient temperature | $T_a$ (°C) | 26.55 | 47.19 | 38.84 | 3.74 |
| Globe temperature | $T_g$ (°C) | 26.93 | 57.90 | 45.52 | 6.74 |
| Relative humidity | RH (%) | 45.8 | 93.7 | 67.4 | 11.5 |
| Air velocity | $V_a$ (m/s) | 0.21 | 2.43 | 1.44 | 0.47 |
| Mean radiant temperature | $T_{mrt}$ (°C) | 27.21 | 63.93 | 49.03 | 8.38 |
| Natural wet-bulb temperature | $T_w$ (°C) | 24.68 | 44.66 | 36.42 | 3.68 |
| Total clothing insulation | $I_{cl}$ (clo) | 0.43 | 0.58 | 0.50 | 0.08 |

## 3.2 *Thermal Response Analysis*

### 3.2.1 Distribution of Heat Symptoms

As indicated in Fig. 6a, over 40% of the surveyed workers reported various symptoms. The predominant discomfort symptoms were rapid heartbeat and heavy sweating, representing 30% and 40% of responses, respectively. Other reported symptoms—fatigue, headache, irritability, and chest tightness—accounted for 10%, 7%, 10%, and 5%, respectively (Fig. 6b). More than 80% of workers with prolonged outdoor exposure experienced excessive sweating (classified as "a lot of sweat" and "sweat profusely"), each constituting 40% of the total (Fig. 6c). Increased dehydration may impair the body's ability to tolerate heat. Typically, workers perspire less prior to starting work. When cumulative stress becomes excessive, individuals may experience weakness, dizziness, distress, or discomfort, and continued work under high temperatures raises the risk of heat stroke [50]. Thus, outdoor labor in hot environments entails health risks, and extended exposure to such conditions should be minimized.

### 3.2.2 Distribution of Thermal Sensation Vote

Figure 7 displays the distribution of thermal sensation votes (TSV), indicating that over 95% of workers reported feeling "hot" (40.08%) or "very hot" (56.11%) at the construction sites during work periods. Fluctuations in $T_a$, $T_g$, RH, and $T_{mrt}$ confirm that the outdoor thermal conditions were extremely hot. Before responding to the survey, workers had been engaged in strenuous activity, and their high metabolic rates led to significant sweating. As a result, the majority indicated that they felt hot.

Clear linear correlations are illustrated in Fig. 8 between MTSV and both $T_a$ (Fig. 8a; $R^2 = 0.8463$) and $T_{mrt}$ (Fig. 8b; $R^2 = 0.7085$), which correlate with Guangzhou's hot and humid summer climate. Air temperature and solar radiation substantially influence thermal comfort in open environments, and shading can help mitigate solar exposure. As observed in Fig. 8a, once $T_a$ surpassed 34.3 °C, MTSV exceeded 3, reflecting that most workers perceived conditions as hot and uncomfortable. After extended work under such temperatures, rest should be considered to avoid potential heat-related injuries. Similarly, as shown in Fig. 8b, MTSV also rose above 3 once $T_{mrt}$ exceeded 38 °C.

### 3.2.3 Assessment of the Environment

Based on the survey responses from workers regarding sunlight intensity, airflow intensity, and relative humidity (RH), the distributions of "acceptable" and "unacceptable" ratings for the thermal environment are illustrated in Fig. 9. Over 70% of workers found sunlight intensity unacceptable due to intense solar radiation, while

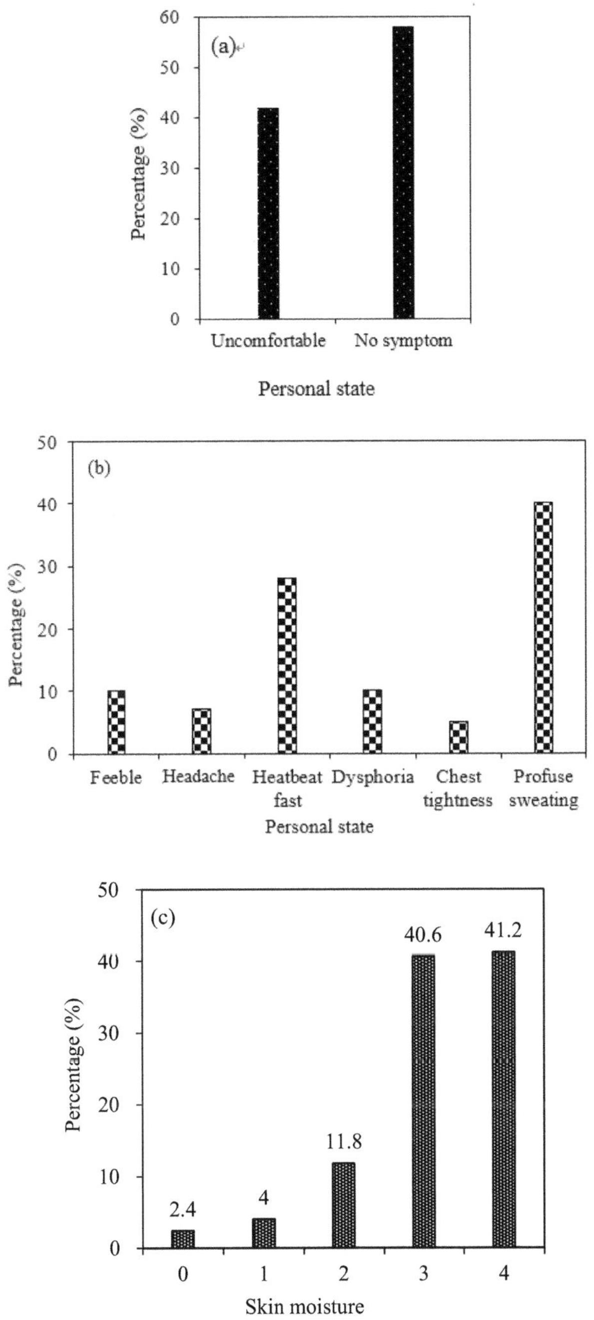


**Fig. 6** Distribution of **a** uncomfortable symptoms, **b** various symptoms, and **c** skin moisture for the surveyed workers. In **c**, 0: no sweat; 1: a little sweat; 2: obvious sweat; 3: a lot of sweat; 4: sweating profusely

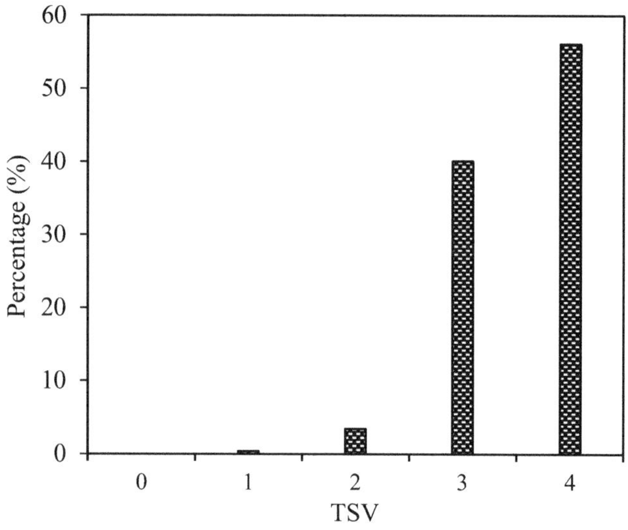


**Fig. 7** Distribution of thermal sensation votes (TSV). 0: neutral; 1: slightly warm; 2: warm; 3: hot; 4: very hot

more than 80% regarded airflow intensity (associated with air velocity, $V_a$) and RH as acceptable. Consequently, close to half of the workers considered the overall thermal conditions at the construction sites to be acceptable.

### 3.2.4 Evaluation of Solar Radiation

Figure 10 displays the distribution of sunshine sensation votes (SSV), showing that over 35% of workers described conditions as "too sunny" and nearly 40% as "very sunny" at their construction sites. Fewer than 5% perceived the sunlight as neutral or slightly sunny. The mean sunshine sensation vote (MSSV) was computed using the bin method with 1 °C intervals for $T_{mrt}$. Regression analysis revealed a strong relationship between MSSV and $T_{mrt}$ (Fig. 11; $R^2 = 0.945$), with a *p*-value below 0.05 confirming statistical significance. Throughout the field study, $T_{mrt}$ consistently measured above 30 °C. The MSSV derived from responses exceeded 2, underscoring the substantial influence of solar radiation intensity on SSV.

### 3.2.5 Evaluation of Wind Speed and RH

On-site measurements at the construction sites indicated that 80% of air velocity ($V_a$) readings fell between 0.9 and 2.1 m/s, with fewer than 15% below 1 m/s, while 60% of relative humidity (RH) values were between 45 and 70%, and over 15% exceeded 80% RH. Survey results showed that 75% of workers perceived air movement as

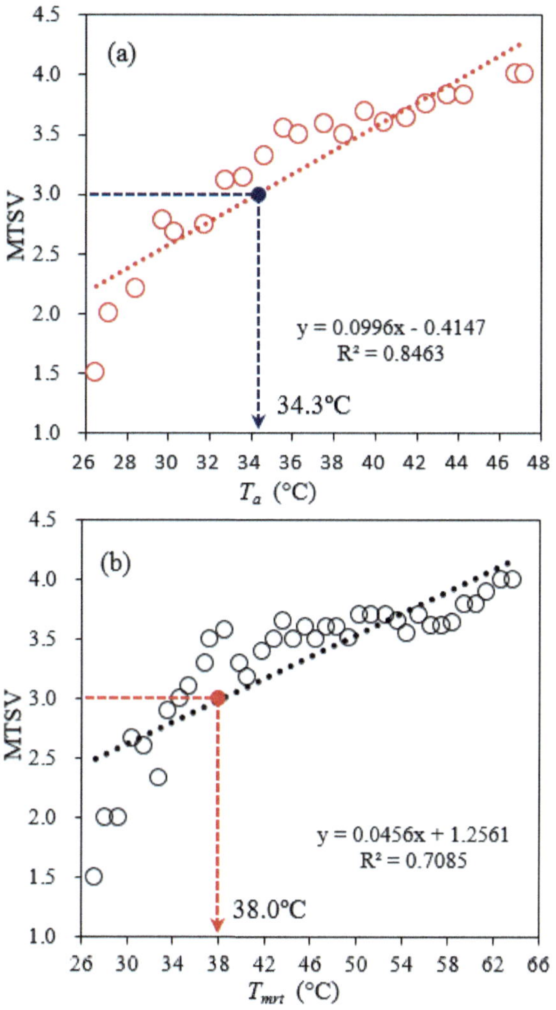


**Fig. 8** Mean thermal sensation vote (MTSV) with respect to **a** ambient temperature ($T_a$) and **b** mean radiant temperature ($T_{mrt}$)

slightly low or neutral, and more than 70% considered the outdoor RH conditions neutral or slightly humid (Fig. 12). Consequently, a majority of workers deemed these environmental conditions acceptable.

## 3.3 Analysis of Physiological Indices

Heart rate (HR) serves as a key physiological indicator of the thermal strain imposed by ambient temperature ($T_a$) and work intensity. Consequently, HR variability is

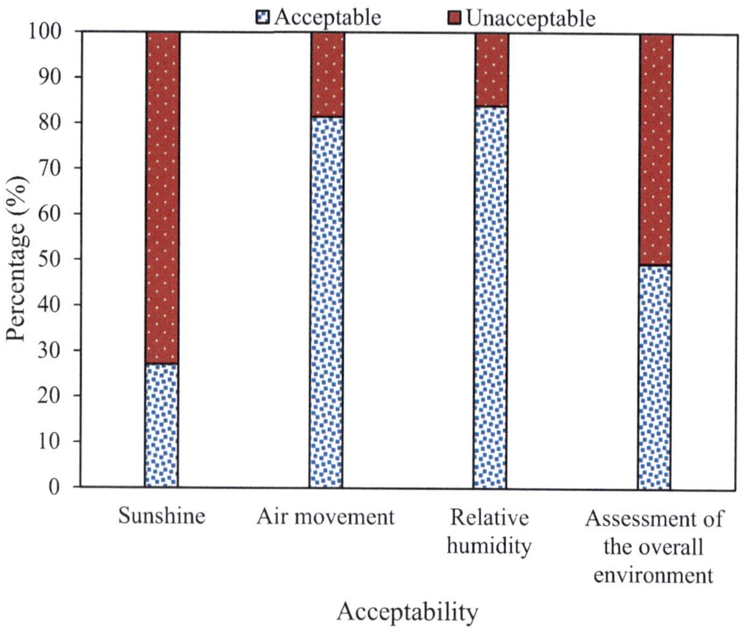


**Fig. 9** Percentage of acceptability of environmental factors from the workers' survey

**Fig. 10** Distribution of sunshine sensation votes (SSV) from the workers' survey. 0: neutral; 1: slightly sunny; 2: sunny; 3: very sunny; 4: too sunny

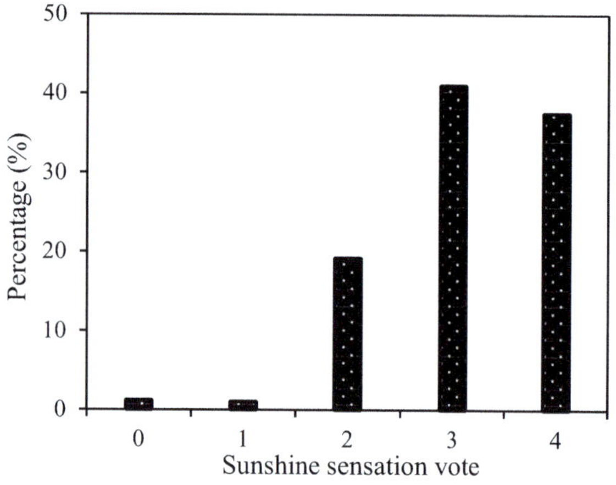


widely used to assess health effects associated with occupational exertion or environmental heat exposure [77–79]. Auditory canal temperature (ACT), measured in the ear near the cochlea, offers a practical approximation of core body temperature [75, 80, 81]. Key benefits of ACT include its non-invasive nature, user-friendly application, and responsiveness to dynamic changes in thermal conditions [82].

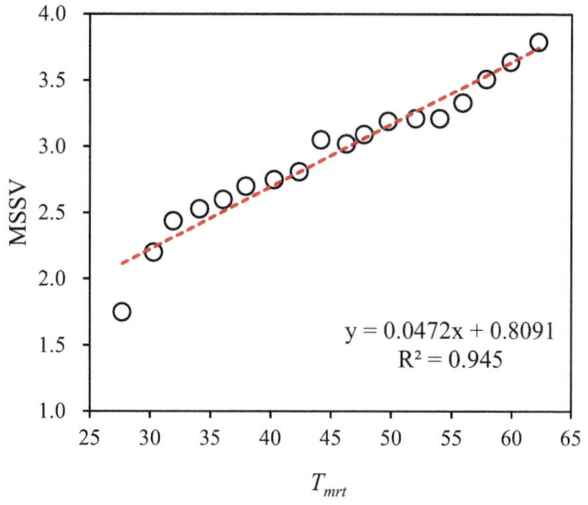


**Fig. 11** Mean sunshine sensation vote (MSSV) with respect to the mean radiant temperature ($T_{mrt}$)

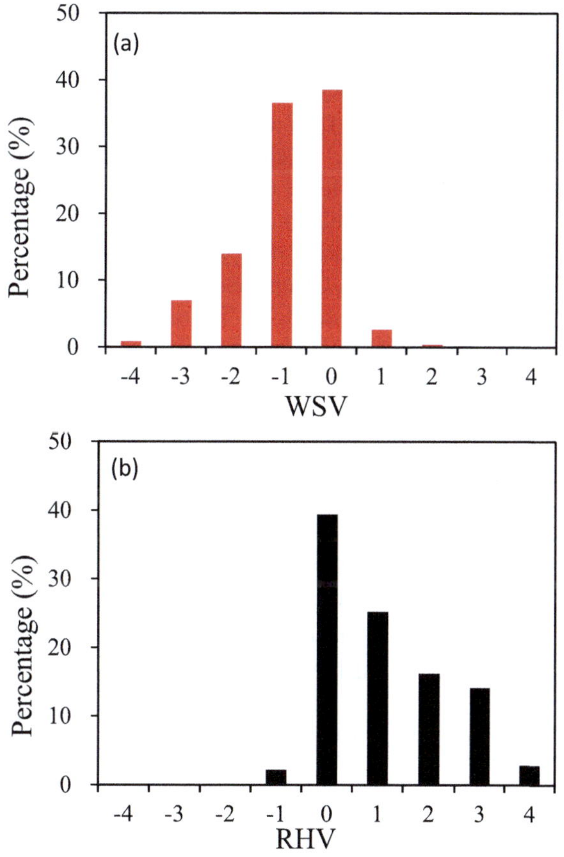


**Fig. 12** Distribution of **a** wind speed votes (WSV) (− 4: too low; − 3: very low; − 2: low; − 1: slightly low; 0: neutral; 1: slightly strong; 2: strong; 3: very strong; 4: too strong) and **b** relative humidity votes (RHV) (− 4, too dry; − 3, very dry; − 2, dry; − 1, slightly dry; 0, neutral; 1, slightly wet; 2, wet; 3, very wet; 4, too wet)

As summarized in Table 4, the ranges and mean values are provided for HR, systolic blood pressure (SBP), diastolic blood pressure (DBP), and ACT. While mean values—except for HR—fall within normal bounds, maximum readings surpassed recommended thresholds, largely due to high work demands in a hot outdoor setting. Previous studies indicate that the human body approaches its tolerance limit when the rate of water loss reaches 0.02–0.03% per unit time [16]. When evaporative cooling is insufficient to offset metabolic and environmental heat gain, internal heat accumulation occurs, raising body temperature. This triggers increased blood flow and sweat secretion, elevating cardiovascular strain and consequently raising blood pressure [50].

Figure 13 illustrates changes in HR over the work period. Initially, at 8:00 AM, HR was relatively low (around 85 beats/min). As work commenced, HR rose with prolonged activity and stabilized at a high level after 9:30 AM. A decline began after 11:00 AM, mainly due to the lunch break. In the afternoon, following rest, HR again increased when work resumed and then plateaued. Differences in HR between morning and afternoon sessions were not significant (Fig. 13). As shown in Fig. 14, nearly 50% of HR values fell between 90 and 100 beats/min, and over 35% exceeded 110 beats/min. These findings indicate that workers consistently exhibited elevated heart rates while working in hot conditions.

The associations between HR and air temperature ($T_a$), as well as HR and mean radiant temperature ($T_{mrt}$), were evaluated using a 2 °C bin interval. As illustrated in Fig. 15, a pronounced quadratic polynomial regression was observed (Fig. 15a: $R^2 = 0.7335$; Fig. 15b: $R^2 = 0.6933$), demonstrating that HR rose with increasing Ta and $T_{mrt}$. These correlations are statistically significant, with *p*-values below 0.05. The majority of recorded HR values exceeded 90 beats/min, and over 30% were above 100 beats/min. Given these HR measurements, further investigation is needed to establish safe thresholds for $T_a$ and $T_{mrt}$ in outdoor work environments.

As a key physiological parameter, mean auditory canal temperature (MACT) also exhibited significant relationships with both $T_a$ and $T_{mrt}$ (Fig. 16), with $p$-values < 0.05. The coefficient of determination ($R^2$) was 0.6031 for MACT relative to $T_a$ and 0.6044 relative to $T_{mrt}$, reflecting reasonably strong dependencies. Similar to HR, MACT increased alongside rising $T_a$ and $T_{mrt}$ for outdoor construction workers.

**Table 4** Descriptive statistics for physiological parameters

| Parameters | Abbreviation (units) | Minimum | Maximum | Mean | Standard deviation |
|---|---|---|---|---|---|
| Heart rate | HR (beats/min) | 70.0 | 134.0 | 95.5 | 9.4 |
| Systolic blood pressure | SBP (mmHg) | 108.0 | 139.0 | 124.9 | 6.7 |
| Diastolic blood pressure | DBP (mmHg) | 72.0 | 96.0 | 82.6 | 32.9 |
| Auditory canal temperature | ACT (°C) | 35.2 | 37.2 | 36.6 | 4.6 |

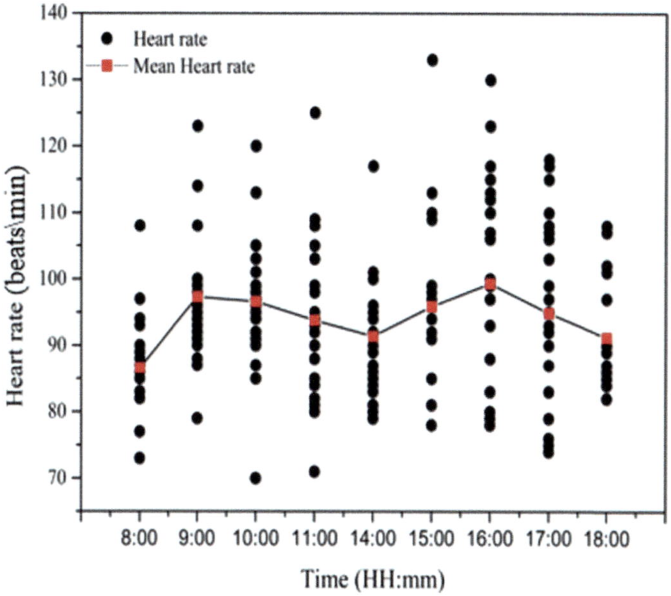


**Fig. 13** Variation in HR with respect to working time

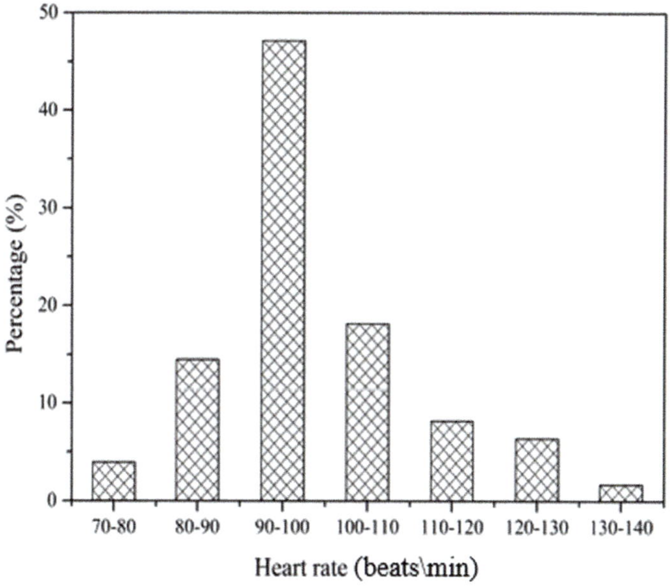


**Fig. 14** Distribution of HR

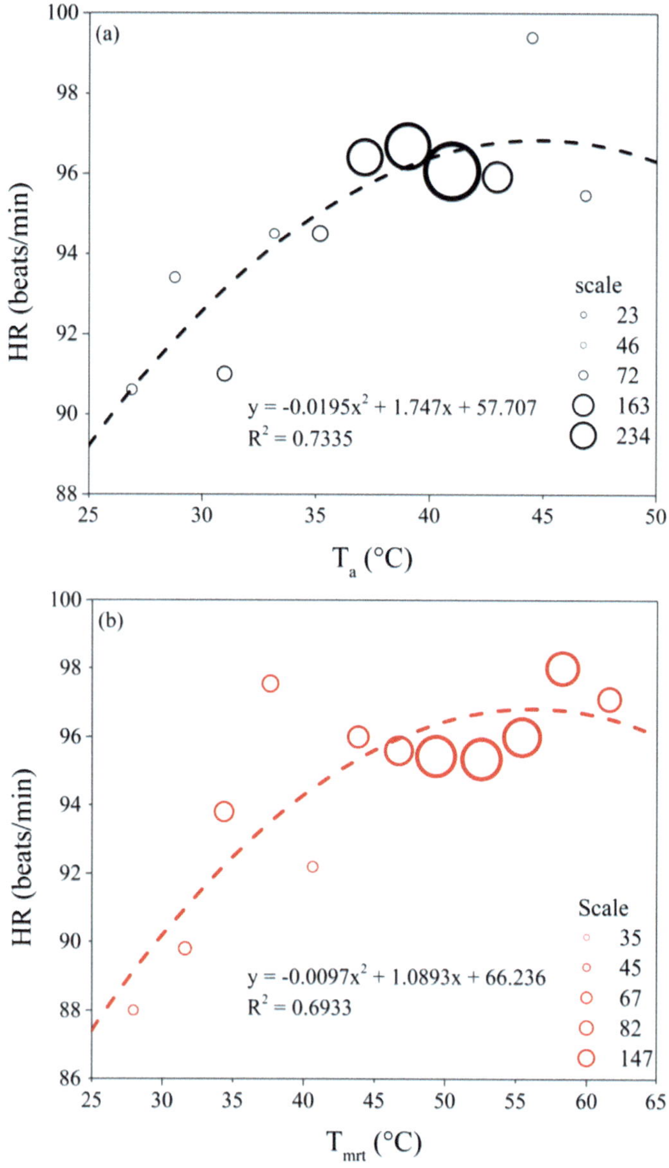


**Fig. 15** Heart rate (HR) with respect to **a** ambient temperature ($T_a$) and **b** mean radiant temperature ($T_{mrt}$). *Note* The size of the circle refers to the proportion of the amount of data

During the field study, most MACT readings were above 36.5 °C, indicating consistently elevated ACT levels attributable to the hot environment and high metabolic activity. This suggests sustained high heat exchange between the workers and their thermal surroundings.

The linear regressions of MTSV, HR, and ACT revealed that only HR and MTSV had a significant linear relationship (Fig. 17a: $R^2 = 0.8275$; Fig. 16b: $R^2 = 0.2213$), with $p$-values $< 0.05$. Correspondingly, the workers' TSV was also affected by HR.

## 3.4 Cognitive Performance Assessment

### 3.4.1 Effects of Working Time and Age

Based on the age information provided in the questionnaires, workers were categorized into three groups: under 30 years old, 30–40 years old, and over 40 years old. Figure 18 displays the number of workers and their corresponding efficiency levels across these age groups throughout the workday. For all age categories, working efficiency declined between 8:00 AM and 11:00 AM. After the lunch break, accuracy rates were higher compared to the morning session, but decreased again as work continued in the afternoon. This indicates that the lunch interval had a restorative effect on performance accuracy. Figure 17 further shows that cognitive performance was lower in older age groups, highlighting a considerable influence of age on cognitive function.

### 3.4.2 Effects of Thermal Parameters

Figure 19 illustrates the relationship between task accuracy and both air temperature ($T_a$) and mean radiant temperature ($T_{mrt}$), revealing strong negative linear correlations ($T_a$: $R^2 = 0.78$; $T_{mrt}$: $R^2 = 0.87$) and a general decline in accuracy as $T_a$ and $T_{mrt}$ rise. These associations are statistically significant, with $p$-values below 0.05. With prolonged working hours, outdoor temperatures steadily increased, leading to a greater heat burden on the workers. High temperatures and strenuous labor contribute to fatigue, markedly reducing cognitive sharpness and accuracy. As human responsiveness diminishes and error rates climb, the risk of life-threatening accidents—especially in hazardous settings—becomes more pronounced. It is therefore essential to establish safety thresholds for $T_a$ and $T_{mrt}$ at construction sites to protect workers from heat-related harm.

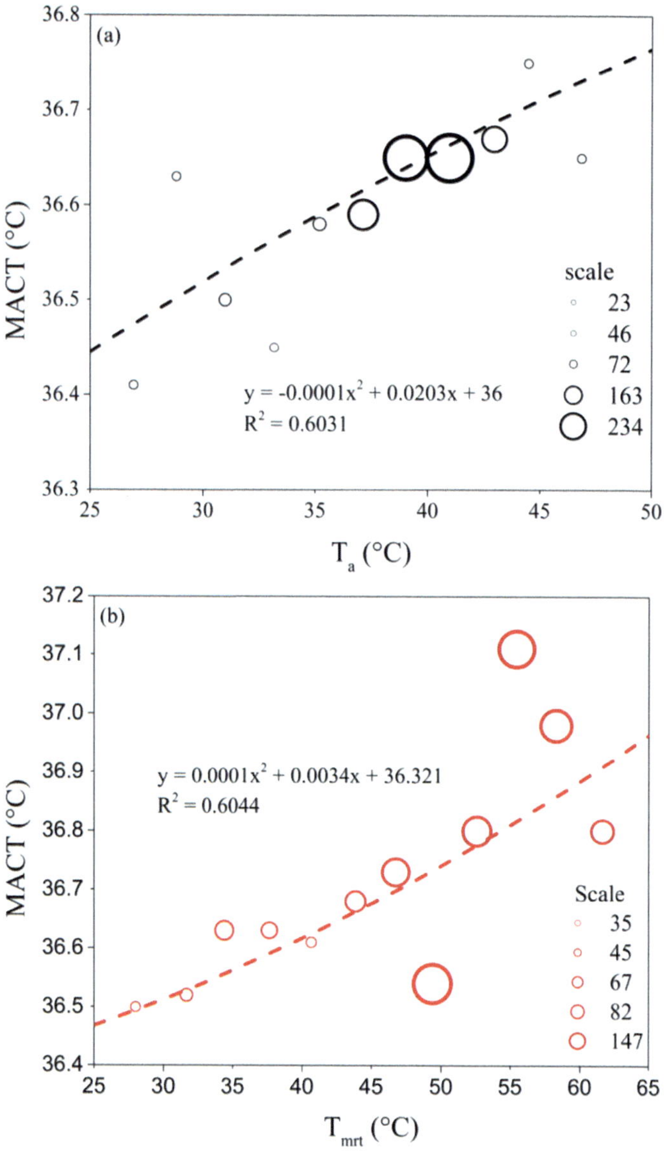


**Fig. 16** Mean auditory canal temperature (MACT) with respect to **a** ambient temperature ($T_a$) and **b** mean radiant temperature ($T_{mrt}$). *Note* The size of the circle refers to the proportion of the amount of data

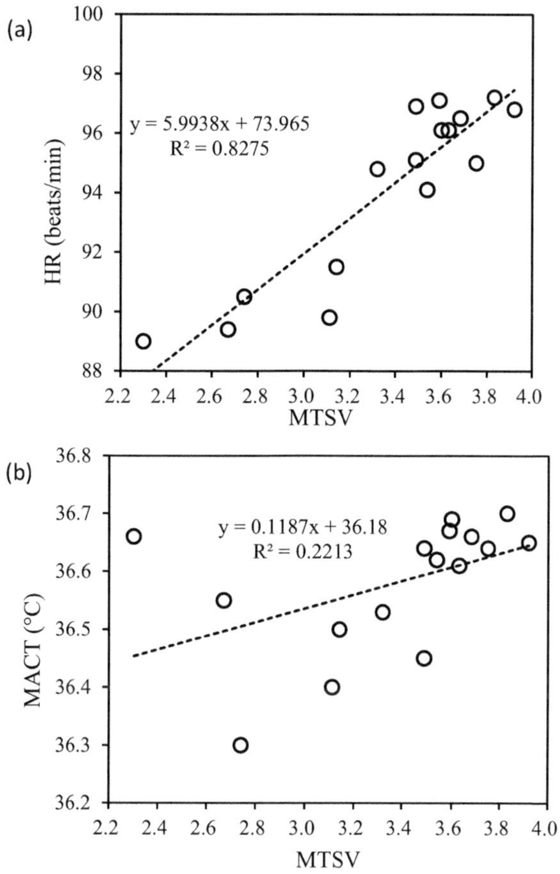


**Fig. 17** MTSV with respect to **a** HR and **b** MACT

# 4 Discussion

## *4.1 Effects of Temperature*

In the context of heat exposure, projected temperature increases due to global warming are expected to pose serious threats to worker health and safety across many nations [83–85]. This includes elevated risks of occupational injuries from hot weather in Guangzhou, China [86]. Outdoor workers generally exhibit greater heat resistance than indoor workers owing to enhanced thermal adaptation [35], and some individuals may demonstrate higher personal tolerance to heat. In this chapter, 50% of workers rated their overall thermal environment as acceptable (Fig. 9); however, 68% reported symptoms such as rapid heart rate and heavy sweating, and a small number were unable to work due to heat-related illness. Recorded air temperatures

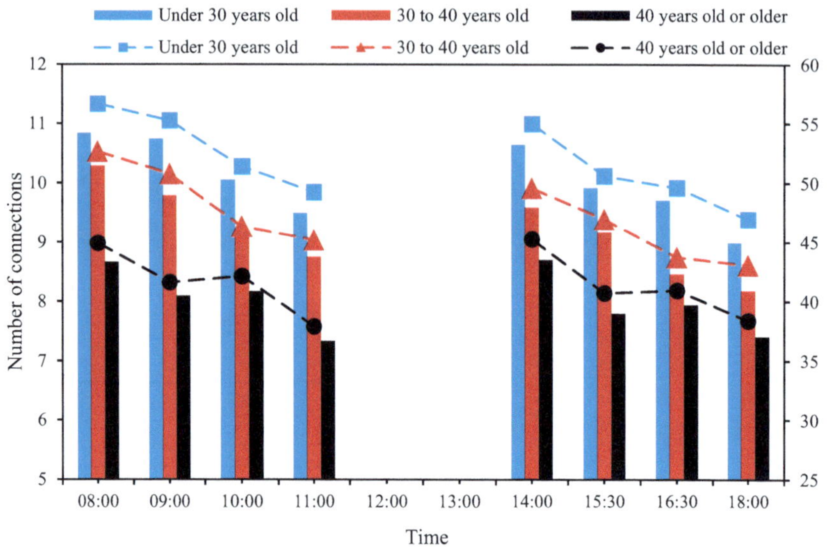


**Fig. 18** Number of cognitive connections achieved by workers in the cognitive assessment and the working efficiency of the three different age groups

($T_a$) varied from 26.55 to 47.19 °C (Table 3), indicating persistently severe microclimates at the construction sites. Previous studies typically reported $T_a$ between 20 and 40 °C [29, 39, 87–91]. The likelihood of airflow variations and intense solar radiation was high, both influencing $T_a$. Nonetheless, nearly 70% of workers considered air velocity ($V_a$) and relative humidity (RH) conditions acceptable.

In the sites studied, $T_a$ and mean radiant temperature ($T_{mrt}$) were the most influential factors for thermal comfort, aligning with earlier research [37, 39, 92–94]. Fang et al. [39] found that the maximum acceptable $T_a$ for students during military training was 32.5 °C, with a thermal unacceptability rate of 60%. Furthermore, as shown in Fig. 6, when $T_a$ surpassed 34.6 °C, the mean thermal sensation vote (MTSV) exceeded 3 ("hot"). This is consistent with prior standards defining extreme heat thresholds in China: above 35 °C for living environments and above 32 °C for working conditions [16, 17]. Sharma et al. [64] noted a significant decline in mental performance beyond 32.2 °C (basic effective temperature) in humid heat, and above 33 °C in dry heat. Based on field studies, Zare et al. [18] reported mean values of UTCI, WBGT, WBDT, and TSI as 34.59 ± 1.51 °C, 24.59 ± 0.61 °C, 28.37 ± 0.82 °C, and 31.51 ± 0.7 °C, respectively. To prevent thermal discomfort and injury when $T_a$ exceeds 34 °C, it is necessary to reduce work intensity at construction sites.

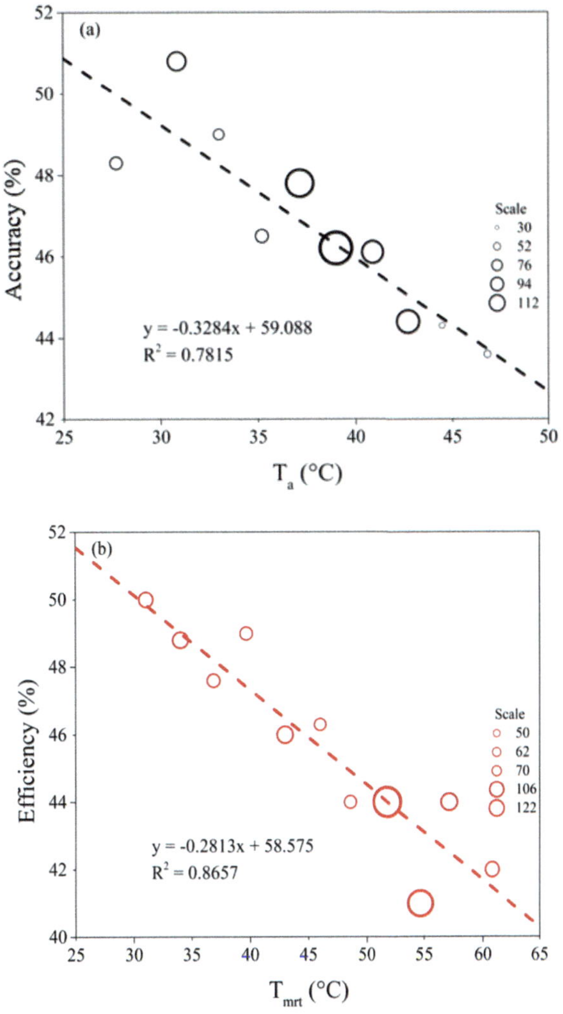


**Fig. 19** Working efficiency with respect to **a** ambient temperature ($T_a$) and **b** mean radiant temperature ($T_{mrt}$). *Note* The size of the circle refers to the proportion of the amount of data

## 4.2 Effects on Heart Rate

Physiological measures often require continuous or repeated monitoring, raising practical challenges for daily use in occupational settings. Some studies [14, 95, 96] reported no strong correlation between physiological indicators and thermal conditions, largely due to brief assessment periods and limited sample sizes. Therefore, extended field data collection is essential to accurately evaluate how thermal environments affect physiological responses [95].

Internationally standardized electrocardiogram guidelines define the normal resting heart rate (HR) for adults as 60–100 beats/min [97], and HR has been established as a reliable marker of physical adaptation and endurance [98, 99]. Prolonged elevation in HR increases metabolic rate and internal heat storage, elevating cardiac strain and raising the risk of health issues such as chest tightness and myocardial infarction [100–102]. The World Health Organization (WHO) recommends an upper limit of 110 beats/min for resting HR. High work intensity further raises the likelihood of physical harm [103] and reduces productivity by 9% for moderate and 25% for heavy workloads, respectively [104].

In this chapter, heat exposure rose steadily with extended work hours, leading to marked increases in physiological measures, with HR consistently exceeding 90 beats/min. As shown in Fig. 13, nearly 50% of workers had an HR above 100 beats/min, indicating sustained cardiovascular strain in hot conditions. According to the GRACE risk-prediction model [105], each 30-beat/min increase in HR elevates event risk by 30%. Previous research also notes significant HR fluctuations during work and recovery phases [100, 106], with strong correlations between HR, metabolic rate, and air temperature ($T_a$) [99, 107]. Regulating both thermal exposure duration and work intensity is critical to preventing heat-related injuries.

This chapter used binomial regression to model the relationship between HR and $T_a$ (as well as $T_{mrt}$). When Ta surpassed 34 °C, HR exceeded 94 beats/min. Similarly, when the mean thermal sensation vote (MTSV) was greater than 3, HR was above 92 beats/min. The average minimum HR among all participants was 84.6 beats/min, and the safety risk increased by 11.9% beyond threshold values [105]. To minimize discomfort, it is advisable to maintain HR around 94 beats/min and below 100 beats/min [87].

## 4.3 *Cognitive Performance with Respect to Age*

Human decision-making capacity is recognized to influence both work efficiency and accident rates in high-temperature settings [41, 108–110]. Therefore, evaluating the sensitivity of factors that shape cognitive performance and impact productivity is critically important. Several studies suggest that the skill level demanded by the task, worker proficiency, and task complexity significantly moderate the relationship between employee age and job performance [111–114]. In roles with lower skill requirements, performance tends to improve rapidly with experience and typically peaks at an earlier stage [114]. Using productivity models, Li et al. [14] reported that the proportion of direct work time declined by 0.72% for each one-year increase in a worker's age. Other research has identified an inverted U-shaped relationship between age and performance, with optimal performance occurring around 40 years old [115, 116].

In this chapter, cognitive performance across all three age groups was influenced by both working duration and thermal conditions (Fig. 19). As thermal environment parameters rose, accuracy decreased by approximately 8%. Li et al. [14] also observed that a 1 °C increase in wet-bulb globe temperature led to a 0.57% reduction in direct work time and a 0.74% rise in idle time. Additionally, a 5% accuracy gap was noted among different age groups, indicating that individuals in physically demanding occupations under hot conditions are particularly susceptible to cognitive decline [117]. These findings can assist managers in identifying high-risk workers and appropriately modifying their workloads.

### *4.4 Limitations*

In this chapter, no detailed classification of worker activity levels was performed. Variations in work intensity may result in differences in physiological responses. The influence of differing work intensities on thermal comfort should be addressed in future research. During the field survey, the majority of workers were between 30 and 45 years old. Differences in physiological indicators across age groups were not statistically significant. Future studies should include a broader age range to better evaluate age-related effects.

Moreover, notable discrepancies exist between earlier chamber experiments and field-based studies. It is challenging to replicate high-temperature conditions in chamber settings that fully mirror real-world environments. In field studies, parameters tend to be more complex and variable. Additionally, self-reported data are inherently subject to response bias, which is a widely recognized limitation in survey-based research [118, 119]. However, due to its random and unpredictable nature, quantifying this bias is difficult.

To mitigate the impact of response bias, the field survey was conducted from July 18 to August 9, 2019, with a total of 1063 questionnaires collected. For a more precise assessment of response bias, future work should include controlled chamber experiments to compare differences between field and laboratory results. In chamber settings, thermal conditions can be precisely created and maintained constant, allowing isolation of self-response effects from other variables. This approach would help quantify response bias, and it is recommended that subsequent investigations integrate both chamber experiments and field surveys.

## 5 Conclusions

This chapter examined the thermal comfort and safety of construction workers performing high-intensity outdoor labor in Guangzhou, integrating on-site microclimatic monitoring with physiological and cognitive assessments. The results show that air temperature and mean radiant temperature were the dominant factors shaping thermal comfort and safety, with both heart rate and ear canal temperature rising consistently with prolonged exposure. Although no incidents of heatstroke or collapse occurred, over 40% of participants experienced discomfort, and more than one-third of measured heart rates exceeded 110 beats per minute. The study further highlights that once air temperature surpasses 34 °C and mean radiant temperature exceeds 38 °C, reducing work intensity becomes necessary to limit physiological strain.

Cognitive function was also found to deteriorate under extended heat exposure, particularly among older workers, suggesting that age-related vulnerability should be considered in occupational safety practices. Taken together, these findings demonstrate that thermal stress in outdoor labor contexts involves complex interactions between environmental heat, physiological status, and mental performance, and they provide a scientific basis for refining guidelines aimed at safeguarding health and maintaining productivity in hot and humid climates.

**Acknowledgements** This work was supported by the National Natural Science Foundation of China (Project No. 51978180), Research Project of Guangzhou University (No. YG2020003). The authors thank all workers who participated in the surveys.

## References

1. L. Zefeng, L. Binyiu, D.B. Robert, Exploring the suitable assessment method and best performance of human energy budget models for outdoor thermal comfort in hot and humid climate area. Sustain. Cities Soc. **63**, 102323 (2020)
2. D. Bienvenido-Huertas, J.A. Pulido-Arcas, C. Rubio-Bellido, A. Pérez-Fargallo, Influence of future climate changers scenarios on the feasibility of the adaptive comfort model in Japan. Sustain. Cities Soc. **61**, 102303 (2020)
3. C. Zhou, Z. Fang, X. Xu, X. Zhang, Y. Ding, X. Jiang, Y. Ji, Using long short-term memory networks to predict energy consumption of air-conditioning systems. Sustain. Cities Soc. **55**, 102000 (2020)
4. F. Jeff, K. Elias, G. Stephane, A. Luc, Impact of urban heat island mitigation measures on microclimate and pedestrian comfort in a dense urban district of Lebanon. Sustain. Cities Soc. **61**, 102375 (2020)
5. United Nations Department of Economic and social affairs, 2018 revision of world urbanization prospects (2018). Retrieved from https://www.un.org/development/desa/publications/2018-revision-of-world-urbanization-prospects.html
6. T. Kjellstrom, I. Holmer, B. Lemke, Workplace heat stress, health and productivity- an increasing challenge for low and middle-income countries during climate change. Glob. Health Action **2**, 1–6 (2009)
7. F.L. Silvana, F. Celina, Energy efficiency, thermal resilience, and health during extreme heat events in low-income housing in Argentina. Energy and Buildings **231**, 110576 (2020)
8. L. Long, Z. Yong, Population exposure to extreme heat in China: frequency, intensity, duration and temporal trends. Sustain. Cities Soc. **60**, 102282 (2020)
9. M.B. Modest, J. Erik, W.Y. Moohammed, Urbanites' outdoor thermal comfort in the informal urban fabric of warmhumid Dar es Salaam Tanzania. Sustain. Cities Soc. **62**, 102380 (2020)
10. S. Zare, N. Hasheminezhad, K. Sarebanzadeh, F. Zolala, R. Hemmatjo, D. Hassanvan, Assessing thermal comfort in tourist attractions through objective and subjective procedures based on ISO 7730 standard: a field study. Urban Clim. **26**, 1–9 (2018)
11. D. Manob, D. Arijit, M. Sukanya, Outdoor thermal comfort in different settings of a tropical planning region: a study on Sriniketan-Santiniketan planning Area (SSPA) Eastern India. Sustain. Cities Soc. **63**, 102433 (2020)
12. S.Y. Chan, C.K. Chau, On the study of the effects of microclimate and park and surrounding building configuration on thermal comfort in urban parks. Sustain. Cities Soc. **64**, 102512 (2021)
13. S. Atwaa, M.G. Ibrahim, R. Murata, Evaluation of plantation design methodology to improve the human thermal comfort in hot-arid climatic responsive open spaces. Sustain. Cities Soc. **59**, 102198 (2020)

14. L. Xiaodong, H.C. Kwan, Z. Yimin, L. Ying, Evaluating the impacts of high-temperature outdoor working environments on construction labor productivity in China: a case study of rebar workers. Build. Environ. **95**, 42–52 (2016)
15. Y. Sadia, L. Hong, W. Yuxin, L. Baizhan, Physiological responses of acclimatized construction workers during different work patterns in a hot and humid subtropical area of China. J. Build. Eng. **30**, 101281 (2020)
16. J. Zhao, N. Zhu, S. Lu, Productivity model in hot and humid environment based on heat tolerance time analysis. Build. Environ. **44**, 2202–2207 (2009)
17. X. Shi, N. Zhu, G. Zheng, The combined effect of temperature, relative humidity and work intensity on human strain in hot and humid environments. Build. Environ. **69**, 72–80 (2013)
18. S. Zare, H.E. Shirvan, R. Hemmatjo, F. Nadri, Y. Jahani, K. Jamshidzadeh, P. Paydar, A comparison of the correlation between heat stress indices (UTCI, WBGT, WBDT, TSI) and physiological parameters of workers in Iran. Weather Clim. Extremes. **26**, 100213 (2019)
19. F.B. Nerbass, R. Pecoits-Filho, W.F. Clark, J.M. Sontrop, C.W. McIntyre, L. Moist, Occupational heat stress and kidney health: from farms to factories. Kidney Int. Rep. **2**, 998–1008 (2017)
20. M. Szewczyk, J. Pazdrowski, P. Golusinski, A. Danczak-Pazdrowska, Ł. Łuczewski, S. Marszałek, E. Majchrzak, W. Golusinski, Basal cell carcinoma in farmers: an occupation group at high risk. Int. Arch. Occup. Environ. Health **89**, 497–501 (2016)
21. J.Y. Takakura, S. Fujimori, K. Takahashi, Y. Hijioka, T. Hasegawa, Y. Honda, T. Masui, Cost of preventing workplace heat-related illness through worker breaks and the benefit of climate-change mitigation. Environ. Res. Lett. **12**(6), 064010 (2017)
22. B.M. Varghese, A. Hansen, P. Bi, D. Pisaniello, Are workers at risk of occupational injuries due to heat exposure? A comprehensive literature review. Saf. Sci. **110**, 380–392 (2018)
23. IPCC, Fifth Assessment Report on Climate Change, Intergovernmental Panel on Climate Change, World Meteorological Organization (2014)
24. J.P. Dunne, R.J. Stouffer, J.G. John, Reductions in labour capacity from heat stress under climate warming. Nat. Clim. Change **3**(6), 563–566 (2013)
25. N. Watts, M. Amann, S. Ayeb-Karlsson et al., The Lancet Countdown on health and climate change: from 25 years of inaction to a global transformation for public health. Lancet **391**, 581–630 (2018)
26. K. Ward, S. Lauf, B. Kleinschmit, W. Endlicher, Heat waves and urban heat islands in Europe: a review of relevant drivers. Sci. Total. Environ. **569–570**, 527–539 (2016)
27. W. Zeng, X. Lao, S. Rutherford, Y. Xu, X. Xu, H. Lin et al., The effect of heat waves on mortality and effect modifiers in four communities of Guangdong Province, China. Sci. Total. Environ. **482–483**, 214–221 (2014)
28. L.P. Wong, H. Alias, N. Aghamohammadi, S. Aghazadeh, N.M. Sulaiman, Urban heat island experience, control measures and health impact: a survey among working community in the city of Kuala Lumpur. Sustain. Cities Soc. **35**, 660–668 (2017)
29. G. Li, M. Zhou, Y. Cai, Y. Zhang, X. Pan, Does temperature enhance acute mortality effects of ambient particle pollution in Tianjin City, China. Sci. Total. Environ. **409**, 1811–1817 (2011)
30. Q. Yin, J. Wang, The association between consecutive days' heat wave and cardiovascular disease mortality in Beijing China. BMC Publ. Health. **17**(1), 223 (2017)
31. C.S. Fong, N. Aghamohammadi, L. Ramakreshnan, N.M. Sulaiman, P. Mohammadi, Holistic recommendations for future outdoor thermal comfort assessment in tropical Southeast Asia: a critical appraisal. Sustain. Cities Soc. **46**, 101428 (2019)
32. Z.F. Lian, B.Y. Liu, R.D. Brown, Exploring the suitable assessment method and best performance of human energy budget models for outdoor thermal comfort in hot and humid climate area. Sustain. Cities Soc. **63**, 102423 (2020)
33. S. Hajat, B. Armstrong, M. Baccini et al., Impact of high temperatures on mortality: is there an added heat wave effect. Epidemiology **17**(6), 632–638 (2006)
34. Y. Xu, A.P.C. Chan, *An investigation of heat related incidents on construction workers in Hong Kong*, in the 19th Cross-Strait Conference on Occupational Safety and Health, Hong Kong, 3rd November (2011)

35. C. Du, B. Li, Modification of the Predicted Heat Strain (PHS) model in predicting human thermal responses for Chinese workers in hot environments. Build. Environ. **165**, 106349 (2019)
36. Z. Fang, X. Feng, J. Liu, Z. Lin, C.M. Mak, J. Niu, K.T. Tse, X. Xu, Investigation into the differences among several outdoor thermal comfort indices against field survey in subtropics. Sustain. Cities Soc. **44**, 676–690 (2019)
37. P. Kumar, A. Sharma, Study on importance, procedure, and scope of outdoor thermal comfort—A review. Sustain. Cities Soc. **63**, 102423 (2020)
38. Z. Fang, X. Feng, X. Xu, X. Zhou, Z. Lin, Y. Ji, Investigation into outdoor thermal comfort conditions by different seasonal field surveys in China Guangzhou. Int. J. Biometeorol. **63**(10), 1357–1368 (2019)
39. P.K. Nag, S.P. Ashtekar, A. Nag et al., Human heat tolerance in simulated environment. Indian J. Med. Res. **105**, 226–234 (1997)
40. Z. Zamanian, Z. Sedaghat, M. Hemehrezaee, F. Khajehnasiri, Evaluation of environmental heat stress on physiological parameter. J. Environ. Heal. Sci. Eng. **15**, 24 (2017)
41. H. Tsutsumia, S. Tanabea, J. Harigaya, Effect of humidity on human comfort and productivity after step changes from warm and humid. Build. Environ. **42**(12), 4034–4042 (2007)
42. A.R. Ismaila, C.M. Nizamb, M. Hanifiah et al., The impact of workers productivity under simulated environmental factor by taguchi analysis. APCBEE Proc. **10**, 263–268 (2014)
43. D.G. Mook, *Energy, Arousal and Action* (WW Norton & Co Publishers, New York, 1987)
44. K. Srinavin, S. Mohamed, Thermal environment and construction workers' productivity: some evidence from Thailand. Build. Environ. **38**(2), 339–345 (2003)
45. T. Tran Trong, F. Riera, K. Rinaldi, W. Briki, O. Hue, Ingestion of a cold temperature/menthol beverage increases outdoor exercise performance in a hot, humid environment. PLoS ONE **9**(4), 161–171 (2015)
46. S. Lukasz, How important is cognitive ability when adapting to changes? A meta-analysis of the performance adaptation literature. Pers. Individ. Differ. **166**, 110178 (2020)
47. R. Wheeler, Adaptive performance, cognitive ability and the moderating effect of task characteristics. Doctoral dissertation. Retrieved from. University of Oklahoma (2012)
48. B. Morgan, S. D'Mello, R. Abbott, G. Radvansky, M. Haass, A. Tamplin, Individual differences in multitasking ability and adaptability. Hum. Factors **55**(4), 776–788 (2013). https://doi.org/10.1177/0018720812470842
49. M. Gillen, J.L. Gittleman, Path forward: emerging issues and challenges. J. Saf. Res. **41**(3), 301–306 (2010)
50. Z. Fang, X. Xu, X. Zhou, S. Deng, H. Wu, J. Liu, Z. Lin, Investigation into the thermal comfort of university students conducting outdoor training. Build. Environ. **149**, 26–38 (2019)
51. Y. Zhang, H. Chen, Q. Meng, Thermal comfort in buildings with split air-conditioners in hot-humid area of China. Build. Environ. **64**, 213–224 (2013)
52. R.A. Engineers, ASHRAE Handbook: Fundamentals (2017)
53. ISO 7726 Ergonomics of the Thermal Environment-Instruments for Measuring Physical Quantities (1998)
54. ANSI/ASHRAE, Standard 55–2017: Thermal Environmental Conditions for Human Occupancy, American Society of Heating Refrigerating and Air-Conditioning Engineers (ASHRAE), Atlanta, Georgia (2017)
55. ISO, International, Standard 7730, Ergonomics of the Thermal Environment-Analytical Determination and Interpretation of Thermal Comfort Using Calculation of the PMV and PPD Indices and Local Thermal Comfort Criteria, International Standard Organization, Geneva (2005)
56. Z. Fang, Z. Lin, C.M. Mak, J. Niu, K.T. Tse, Investigation into sensitivities of factors in outdoor thermal comfort indices. Build. Environ. **128**, 129–142 (2018)
57. O.A. Seppänen, W. Fisk, Some quantitative relations between indoor environmental quality and work performance or health. HVAC R. Res. **12**, 957–973 (2006)
58. L. Lan, P. Wargocki, Z. Lian, Quantitative measurement of productivity loss due to thermal discomfort. Energy Build. **43**, 1057–1062 (2011)

59. L. Lan, Z. Lian, L. Pan, Q. Ye, Neurobehavioral approach for evaluation of office workers' productivity: the effects of room temperature. Build. Environ. **44**, 1578–1588 (2009)
60. Y. Geng, W. Ji, B. Lin, Y. Zhu, The impact of thermal environment on occupant IEQ perception and productivity. Build. Environ. **121**, 158–167 (2017)
61. J. Jiang, D. Wang, Y. Liu, Y. Xu, J. Liu, A study on pupils' learning performance and thermal comfort of primary schools in China. Build. Environ. **134**, 102–113 (2018)
62. L. Lan, Z. Lian, Use of neurobehavioral tests to evaluate the effects of indoor environment quality on productivity. Build. Environ. **44**, 2208–2217 (2009)
63. C. Yuyan, T. Meihui, L. Weiwei, High temperature impairs cognitive performance during a moderate intensity activity. Build. Environ. **186**, 107372 (2020)
64. V.M. Sharma, K. Sridharan, G. Pichan, M.R. Panwar, Influence of heat-stress induced dehydration on mental functions. Ergonomics **29**(6), 791–799 (1986)
65. P.M. Gopinathan, G. Pichan, V.M. Sharma, Role of dehydration in heat stress—induced variations in mental performance. Arch. Environ. Health **43**(1), 15–17 (1988)
66. S.S. Radakovic, J. Maric, M. Surbatovic et al., Effects of acclimation on cognitive performance in soldiers during exertional heat stress. Mil. Med. **172**(2), 133–136 (2007)
67. S. Racinais, N. Gaoua, J. Grantham, Hyperthermia impairs short-term memory and peripheral motor drive transmission. J. Physiol. **586**(19), 4751–4762 (2008)
68. S.J. Robinson, J. Leach, P.J. Owen-Lynch, S.I. Sünram-Lea, Stress reactivity and cognitive performance in a simulated firefighting emergency. Aviat. Space Environ. Med. **84**(6), 592–599 (2013)
69. H. Yu, Z. Xu, C. He, N. Jiang, An initial research on cognitive ability and their estimate stardand of seafaring student. J. Zhejiang Vocat. Tech. Inst. Transp. **5**, 30–33 (2004)
70. H. Yu, L. Li, W. Zhu, X. Chang, Perceptibility test and analysis in personnel of a certain unit. Chin. J. Naut. Med. Hyperbar. Med. **15**, 295–298 (2008)
71. X. Tian, Z. Fang, W. Liu, Decreased humidity improves cognitive performance at extreme high indoor temperature. Indoor Air **10**, 608–627 (2020)
72. J.E. Partington, R.G. Leiter, Partington's pathways test. Psychol. Serv. Center J. **1**, 11–20 (1949)
73. Y. Zhang, J. Liu, Z. Zheng, Z. Fang, X. Zhang, Y. Gao, Y. Xie, Analysis of thermal comfort during movement in a semi-open transition space. Energy Build. **225**, 110312 (2020)
74. C. Fu, Z. Zheng, C.M. Mak, Z. Fang, M.O. Oladokun, Y. Zhang, T. Tang, Thermal comfort study in prefab construction site office in subtropical China. Energy Build. **217**, 1099958 (2020)
75. Y. Zhang, X. Zhou, Z. Zheng, M.O. Oladokun, Z. Fang, Experimental investigation into the effects of different metabolic rates of body movement on thermal comfort. Build. Environ. **168**, 106489 (2020)
76. Z. Fang, H. Liu, B. Li, M. Tan, O.M. Olaide, Experimental investigation on thermal comfort model between local thermal sensation and overall thermal sensation. Energy Build. **158**, 1286–1295 (2018)
77. J.F. Thayer, S.S. Yamamoto, J.F. Brosschot, The relationship of autonomic imbalance, heart rate variability and cardiovascular disease risk factors. Int. J. Cardiol. **141**(2), 122–131 (2010)
78. F. Togo, M. Takahashi, Heart rate variability in occupational health a systematic review. Health **47**(6), 589–602 (2009)
79. F.L. Chang, Y.M. Sun, K.H. Chuang, D.J. Hsu, Work fatigue and physiological symptoms in different occupations of high-elevation construction workers. Appl. Ergon. **40**, 591–596 (2009)
80. K. Cooper, W. Cranston, E. Snell, Temperature in the external auditory meatus as an index of central temperature changes. J. Appl. Physiol. **19**, 1032–1034 (1964)
81. R. Hansen, T. Olds, D. Richards, C. Richards, B. Leelarthaepin, Infrared thermometry in the diagnosis and treatment of heat exhaustion. Int. J. Sports Med. **17**, 66–70 (1996)
82. J.Y. Lee, K. Nakao, N. Takahashi, S.Y. Son, I. Bakri, Y. Tochihara, Validity of infrared tympanic temperature for the evaluation of heat strain while wearing impermeable protective clothing in hot environments. Ind. Health **49**, 714–725 (2011)

83. O.M. Hyatt, B. Lemke, T. Kjellstrom, Regional maps of occupational heat exposure: past, present, and potential future. Glob. Health Action **3**(1), 5715 (2010)
84. T. Kjellsreom, B. Lemke, M. Otto, Mapping occupational heat exposure and effects in south-east Asia: ongoing time trends 1980–2011 and future estimates to 2050. Ind. Health **51**, 56–67 (2013)
85. C. Roelofs, D. Wegman, Workers: the climate canaries. Am. J. Publ. Health **104**(10), 1799–1801 (2014)
86. R. Sheng, C. Li, Q. Wang, L. Yang, J. Bao, K. Wang et al., Does hot weather affect work-related injury? A case-crossover study in Guangzhou. China. Int. J. Hyg. Environ. Health **221**(3), 423–428 (2018)
87. C.H. Sprague, P.E. McNall, The effects of fluctuating temperature and relative humidity on the thermal sensation (thermal comfort) of sedentary subjects. ASHRAE Trans. **76**(1), 146–156 (1970)
88. R.G. Nevins, R.R. Gonzalez, Y. Nishi et al., Effect of changes in ambient temperature and level of humidity on comfort and thermal sensations. ASHRAE Trans. **81**(2), 169–182 (1975)
89. L.G. Berglund, R.R. Gonzales, Application of acceptable temperature drifts to build environments as a mode of energy conservation. ASHRAE Trans. **84**(1), 110–121 (1978)
90. F.H. Rohles, J.E. Laviana, W. Runbai et al., The human response to temperature drifts in a simulated office environment. ASHRAE Trans. **91**(1), 116–123 (1985)
91. Q. Deng, J. Zhao, W. Liu, Y. Li, Heatstroke at home: prediction by thermoregulation modeling. Build. Environ. **137**, 147–156 (2018)
92. W. Liu, Y. Zhang, Q. Deng, The effects of urban microclimate on outdoor thermal sensation and neutral temperature in hot-summer and cold-winter climate. Energy Build. **128**, 190–197 (2016)
93. M. Nikolopoulou, S. Lykoudis, Thermal comfort in outdoor urban spaces: analysis across different European countries. Build. Environ. **41**(11), 1455–1470 (2006)
94. M.A. Ruiz, E.N. Correa, Adaptive model for outdoor thermal comfort assessment in an Oasis city of arid climate. Build. Environ. **85**, 40–51 (2015)
95. J. Bowen, A. Hinze, C. Griffiths, Investigating real-time monitoring of fatigue indicators of New Zealand forestry workers. Accident Anal. Prev. **126**, 1222–2141 (2019)
96. J.D. Runkle, C. Cui, C. Fuhrmann, S. Stevens, J. Del Pinal, M.M. Sugg, Evaluation of wearable sensors for physiologic monitoring of individually experienced temperatures in outdoor workers in southeastern U.S. Environ. Int. **129**, 229–238 (2019)
97. L.G. Horan, The quest for optimal electrocardiography. Am. J. Cardiol. **41**, 126–129 (1978)
98. R. McCraty, F. Shaffer, Heart rate variability: new perspectives on physiological mechanisms, assessment of self-regulatory capacity, and health risk. Glob. Adv. Health Med. **4**(1), 46–61 (2015)
99. K.C. Parsons, International standards for the assessment of the risk of thermal strain on clothed workers in hot environments. Ann. Occup. Hyg. **43**(5), 297–308 (1999)
100. L. Brouha, Physiologic effect of work on the heart. Heart Ind. **1**, 47–104 (1960)
101. G.D. Giannoglou, Y.S. Chatzizisis, C. Zamboulis, G.E. Parcharidis, D.P. Mikhailidis, G.E. Louridas, Elevated heart rate and atherosclerosis: an overview of the pathogenetic mechanisms. Int. J. Cardiol. **126**, 302–312 (2008)
102. M. Hara, M. Nishino, M. Taniike, N. Makino, H. Kato, Y. Egami, R. Shutta, H. Yamaguchi, J. Tanouchi, Y. Yamada, High incidence of thrombus formation at 18 months after paclitaxel-eluting stent implantation: angioscopic comparison with sirolimus-eluting stent. Am. Heart J. **159**, 905–910 (2010)
103. J. Xiang, P. Bi, D. Pisaniello, A. Hansen, Health impacts of workplace heat exposure: an epidemiological review. Ind. Health **52**, 91–101 (2014)
104. S.S. Cheung, J.K.W. Lee, J. Oksa, Employment standards **1**(164),148–164 (2016)
105. C.B. Granger, R.J. Goldberg, O. Dabbous, K.S. Pieper, K.A. Eagle, C.P. Cannon et al., Fox KA predictors of hospital mortality in the global registry of acute coronary events. Arch. Intern. Med. **163**, 2345 (2003)

106. P.J. Cole, L.M. LeMura, T.A. Klinger, K. Strohecker, T.R. McConnell, Measuring energy expenditure in cardiac patients using the body media (TM) armband versus indirect calorimetry: a validation study. J. Sports Med. Phys. Fitness **44**(3), 262–271 (2004)
107. L. Brouha, M.E. Maxfield, P.E. Smith Jr., G.J. Stopps, Discrepancy between heart rate and oxygen consumption during work in the warmth. J. Appl. Physiol. **18**, 1095–1098 (1963)
108. S.M. Alavinia, D. Molenaar, A. Burdorf, Productivity loss in the workforce: associations with health, work demands, and individual characteristics. Am. J. Ind. Med. **52**(1), 49–56 (2009)
109. W.J. Meerding, W. Ijzelenberg, M.A. Koopmanschap, J.L. Severens, A. Burdorf, Health problems lead to considerable productivity loss at work among workers with high physical load jobs. J. Clin. Epidemiol. **58**(5), 517–523 (2005)
110. J.A. Alvarez, E. Emory, Executive function and the frontal lobes: a meta-analytic review. Neuropsychol. Rev. **16**(1), 17–42 (2006)
111. A. Clark, A. Oswald, P. Warr, Is job satisfaction U-shaped in age? J. Occup. Organ. Psychol. **69**(1), 57–81 (1996)
112. E. Katz, M. Coleman, The Influence of research on career development at academic colleges of education in Israel. J. Educ. Teach. **28**(1), 45–61 (2002)
113. R. Kanfer, P.I. Aekerman, Aging adult development and motivation. Acad. Manage. Rev. **29**(3), 440–458 (2004)
114. P.R. Sparrow, D.R. Davis, Effects of age, tenure, training, and job complexity on technical performance. Psychol. Aging **3**(1), 307–314 (1988)
115. S.R. Rhodes, Age-related differences in work attitudes and behavior: a review and conceptual analysis. Psychol. Bull. **93**(1), 328–367 (1983)
116. E.A. Bosman, Age-related differences in motoric aspects of transcription typing skill. Psychol. Aging **8**(2), 87–102 (1993)
117. E.K. O'Neal, P. Bishop, Effects of work in a hot environment on repeated performances of multiple types of simple mental tasks. Int. J. Ind. Ergon. **40**(1), 77–81 (2010)
118. R. Robert, T. Vidhura, G.H. Laura, Measuring bias in self-reported data. Int. J. Behav. Healthc. Res. **2**(4), 320–332 (2011)
119. G.S. Howard, Response-shift bias: a problem in evaluating interventions with pre/post self-reports. Eval. Rev. **41**, 93–106 (1980)

# Thermal Comfort and Safety for Outdoor High-Intensity Student Military Training

**Abstract** Heat stress poses a serious threat to health and performance during strenuous outdoor training, increasing the risk of heat-related illnesses such as heat cramps, heat exhaustion, and heat stroke. Despite its importance, limited research has focused on thermal comfort and safety during high-intensity student military training conducted in hot environments. This chapter investigates the relationships between environmental conditions, thermal indices—including Wet Bulb Globe Temperature (WBGT), Physiological Equivalent Temperature (PET), and Universal Thermal Climate Index (UTCI)—and subjective heat perceptions during a university military training program in Southern China. Freshmen students completed questionnaires on their thermal experiences, while simultaneous microclimate measurements were collected during training sessions. Results showed that some participants experienced heat syncope, and more than 30% reported profuse sweating, with air temperature exerting the strongest influence on thermal comfort. Analysis of regression models suggested the need to adjust stress categories for WBGT, PET, and UTCI to better reflect the vulnerability of students engaged in high-intensity outdoor training. These findings highlight the necessity of refining thermal safety thresholds to ensure both the well-being and training effectiveness of young populations under extreme heat conditions.

## 1 Introduction

Global climate change is constantly raising concerns regarding the public's occupational health due to rising temperatures [1], which increases the environmental temperature for outdoor activities—including work, training, and entertainment. Some previous studies have demonstrated that human activities are influenced by weather conditions [2]. Exposure to high-temperature environments may cause people to suffer severe consequences, such as sudden death or damage to major body organs and physiological functions. It can also indirectly increase injuries, disease risks, and the likelihood of work- and training-related accidents due to heat exposure [3–6], especially in extremely hot environments—namely, temperatures

 
Z. Fang et al., *Outdoor Thermal Comfort and Heat Stress Mitigation*, Environmental Science and Engineering, https://doi.org/10.1007/978-981-95-8899-2_14

above 35 °C for daily living or above 32 °C for work or training [7–9]. A previous study [10] reported that a 1 °C increase in maximum temperature is linked to a 0.2% increase in injury claims up to 37 °C, which is similar to the results reported by Adam-Poupart et al. [11]. When engaging in activities in extremely hot environments, humans face a significant risk of changes in various physiological parameters—such as increased heart rate, body temperature, blood pressure, and sweat production [12–15]—as well as heat-related disorders like heat stroke, heat cramps, and heat exhaustion [16, 17]; these factors raise health risks in extremely hot environments [18, 19]. To reduce health risks in such environments, people should undergo heat acclimatization training under scientific guidance to protect their health and safety. At the same time, hot outdoor thermal environments need to be accurately evaluated for outdoor activities, including work, training, and entertainment.

With the increase in thermal health incidents, more attention is being paid to this serious thermal issue, which has driven the rapid development of thermal stress indices for evaluating high-temperature environments. Since 1950, over 60 heat stress indices have been proposed to assess high-temperature environments and predict the likelihood of human heat strain [1]. The wet-bulb globe temperature (WBGT), one of the most well-known heat stress indices, has been widely used as a measure to evaluate the risk of heat disorders. Initially, WBGT was adopted to assess the thermal environment for military training. During the 1940s and 1950s, heat illness was a severe problem in the US Army, Navy, and Marine Corps. Between 1942 and 1944—a three-year period—198 soldiers died of heat stroke in US Army training camps, an incidence four times higher than that during active service in tropical, temperate, and desert climate zones [20, 21]. Later, Yaglou and Minard [22] proposed WBGT to prevent heat-related casualties at military training centers. Since then, it has been incorporated into standards established by the National Institute for Occupational Safety and Health [23] and the International Organization for Standardization [24], as well as into the threshold limit values defined by the American Conference of Governmental Industrial Hygienists [25]. The American College of Sports Medicine has also used WBGT to develop general guidelines for exercising in hot conditions. Coris et al. [26] concluded that strenuous exercise in conditions with WBGT > 28 °C is considered "high risk," while exercise is advised to stop when WBGT > 32 °C [27], and cooling devices are recommended when WBGT > 28 °C. For broader application, in some previous studies [28], WBGT was defined to account for potential variations in clothing types. In other earlier studies [29, 30], the heat balance equation between the surrounding environment and the human body was considered to clarify the calculation of WBGT. Meanwhile, several researchers have calculated this index for outdoor spaces based on meteorological data.

Physiological Equivalent Temperature (PET) also ranks among the most commonly used indices for measuring heat stress in outdoor spaces. It is the output of the Munich Energy Balance Model for Individuals (MEMI) [31], and is described as the equivalent air temperature in a typical indoor environment—one that produces approximately the same human body thermal perceptions as an outdoor environment [32]. Meanwhile, independent physiological parameters such as height, age, and activity level were taken into account when calculating metabolic and sweat rates in the PET index. Matzarakis et al. [33] provided PET values corresponding to different grades of thermal perception in Europe, while Lin and Matzarakis [34] reported distinct thermal perception grades in a subtropical region; this highlights the impact of adaptive comfort on human thermal perception across various climatic settings. Several studies have also documented PET's applications in evaluating complex outdoor environments across different climatic regions [35–39].

The Universal Thermal Climate Index (UTCI), one of the latest indices for assessing human thermal comfort, can be defined as the air temperature of a hypothetical environment that elicits a physiological response similar to that of a real environment [40]. It derives from an approach proposed over a decade ago by the International Society of Biometeorology Commission, which was later reinforced by the European Cooperation in Science and Technology Action 730 [41]. UTCI has been widely applied to evaluate outdoor thermal environments [42–48], particularly extremely hot ones.

Several studies have been conducted to calibrate these thermal indices using empirical thermal sensation data. The results have shown strong relationships between thermal indices (including WBGT, PET, and UTCI) and mean thermal sensation votes across different climate zones [49–52]. Based on survey data, the neutral temperature and its ranges for different climates, countries, or regions have been identified. However, all these studies only involved subjects engaged in outdoor entertainment activities; few previous investigations have focused on the effects of outdoor training (or similar military training) on thermal sensation and thermal health. In China, college freshmen are required to participate in outdoor training at the start of the academic year—typically in September, at the beginning of the autumn semester. September weather is often unfavorable, especially in southern China, which has a hot and humid climate. Therefore, the impact of weather on the thermal sensation and thermal health of college freshmen needs to be analyzed.

This chapter aimed to analyze evidence on the relationships between high-intensity outdoor training and both thermal comfort and thermal health, as well as between thermal indices (WBGT, PET, and UTCI) and environmental parameters. The field survey was conducted in the subtropical climate of Guangzhou, where college freshmen from Guangzhou University were invited as volunteers to complete a questionnaire after each training session, with simultaneous recording of microclimate parameters. Additionally, this chapter attempted to evaluate the performance of the WBGT, PET, and UTCI indices in quantifying thermal sensation for outdoor training in this climate, and to determine their heat stress categories for designing outdoor training schedules for freshmen.

# 2 Methods

## 2.1 *Climate Characteristics of the Site*

Guangzhou is situated in the Pearl River Delta in southern China, between longitudes of 112 °E and 114.2 °E and latitudes of 22.3 °N and 24.1 °N. It is one of China's largest economic centers and trading ports. Guangzhou features a typical subtropical climate, characterized as mild in spring and autumn, hot and dry in summer, and cool in winter. Based on monthly temperature statistics from 2011 to 2017, the average annual temperature reaches 22.8 °C, ranging from 8.05 °C in January to 34.5 °C in July [53]. The coldest months of the year are January and February, while July and August are the hottest. The maximum daily temperature in July often exceeds 36 °C, and the minimum daily temperature in January is usually higher than 10 °C. However, the average relative humidity (RH) varies slightly throughout the year, between 70 and 90% [54].

## 2.2 *Protocol of Field Data*

The field survey was conducted on the campus of Guangzhou University in Guangzhou. Figure 1 shows the layout and location of the surveyed site. The investigation was conducted over 7 days in July 2017, between the eighth and 20th. A description of the survey process is presented in Fig. 2. During the training procedure, the students were always instructed to train for approximately 45 min. The training programs included standing at attention, marching, quick marching, and practicing boxing. Few studies have specified the metabolic rates for standing at attention, marching, quick marching, and practicing boxing. However, according to the ASHRAE Handbook [55], these training programs may correspond to standing (1.2 met), walking at 0.9 m/s (2.0 met), walking at 1.2 m/s (2.6 met), and calisthenics/exercise (4.0 met). After each training segment, the drillmaster would pause for a rest period, which could last 20–25 min. Since outdoor training could not be interrupted, the students—serving as volunteers—were invited to complete the questionnaire during the rest sessions, with microclimate measurements taken simultaneously. Thus, this process lasted for roughly 20 min. As the field survey was conducted during the rest sessions, the metabolic rate was determined based on actual conditions, including standing or sitting postures.

## 2.3 *Micrometeorological Measurements*

During the field survey, the thermal environment parameters were also measured at the same time. The main microclimatic parameters were gathered in line with some

**Fig. 1** Overview of the locations studied

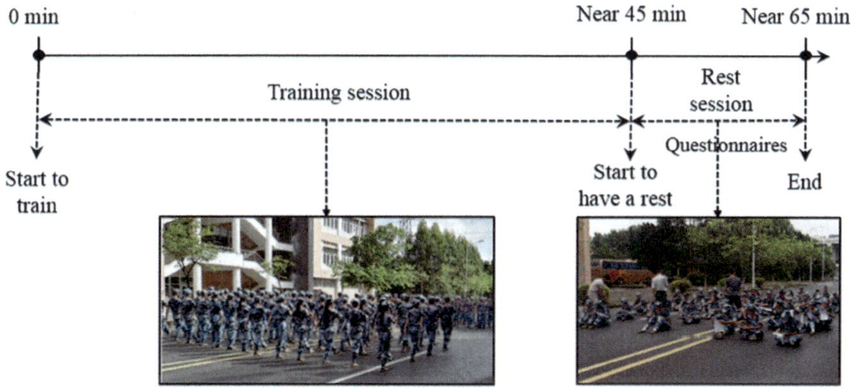


**Fig. 2** Survey procedure

Standards [56, 57], including air temperature ($T_a$), relative humidity (RH), globe bulb temperature ($T_g$), and wind speed ($V_a$). The measurements were carried out at a height of 0.6 m (with the subjects seated) and at a height of 1.1 m near the subjects, in accordance with ISO 7726 [58] During the measurement process, Ta and RH were recorded by the temperature and humidity sensor ZDR-20 respectively, while $T_g$ was recorded by the globe thermometer JTR10. The measurement range

**Table 1** Detailed information of the instruments

| Sensor | Meteorological parameters | Measuring range | Accuracy |
|---|---|---|---|
| ZDR-20 | $T_a$ | – 40–100 °C | ± 0.5 °C; |
| | $RH$ | 0–100% | ± 3% |
| JTR10 | $T_g$ | 5–120 °C | ± 0.2 °C |
| Kanomax model KA22 | $V_a$ | 0–50 m/s | ± 2% |

was 5–120 °C, which covers the microclimate conditions of Guangzhou, and the accuracy was ± 0.2 °C. Table 1 presents the range, accuracy, and sensor types of the devices. In this chapter, standard globe thermometers were used to measure the globe bulb temperature; these thermometers have a diameter of 150 mm (D = 0.15 m) and are painted matt black ($\varepsilon_g = 0.95$), and they were used to calculate the mean radiant temperature ($T_{mrt}$) based on Eq. (1) [58].

$$T_{mrt} = \left\{ \left(T_g + 273\right)^4 + \left[ \frac{\left(1.1 \times 10^8 \times V_a^{0.6}\right)}{\left(\varepsilon_g \times D^{0.4}\right)} \right] \times \left(T_g - T_a\right) \right\}^{1/4} - 273 \quad (1)$$

To enable the registration of sudden variations in $T_{mrt}$ with a globe thermometer, the globe must have a sufficiently short response time. Thorsson et al. [59] compared the difference in $T_{mrt}$ between internal radiation measurements and a 38 mm flat gray globe thermometer. The results showed that most of the effects of changes in radiation, temperature, and wind speed were averaged out by using 5 min mean values. Thus, the 38 mm flat gray globe thermometer is a simple and accurate method for estimating $T_{mrt}$ in an outdoor urban environment. For this reason, some previous studies have proposed using a gray globe thermometer with an approximate diameter of 38 mm for outdoor applications [59–62]. Other studies have adopted a 70 mm-diameter gray globe thermometer to evaluate the outdoor thermal environment [37, 63]. Although the standard 150 mm-diameter copper globe thermometer may require up to 20 min to reach equilibrium [64], it has been used in numerous field survey studies [44, 51, 52, 65]. In this investigation, the field survey was conducted during rest sessions, lasting approximately 20 min. To reduce the impact of fluctuations, the instruments were set to record thermal parameters every 10 s. Subsequently, the 20 min average value of the thermal parameters was calculated to evaluate thermal comfort. Therefore, the average wind speed was utilized.

## 2.4 Survey Questionnaire

During each survey session, the volunteers' personal information—including gender, age, height, weight, detailed clothing, and activity—and thermal response data (including heat symptoms, thermal sensation, and thermal acceptability) were documented. Before conducting this investigation, to ensure the questionnaire's accuracy, the authors carried out a pre-survey based on the questionnaire draft to finalize the questionnaire content. In hot environments, most people felt uncomfortable during high-intensity training, and most heat symptoms focused on 10 possible conditions: weakness, dizziness, nausea, profuse sweating, chest tightness, pale face, rapid heartbeat, stomach discomfort, dysphoria, and headache. These symptoms were therefore included in the questionnaire.

The scope of thermal sensation was in line with the thermal environment standards specified in ASHRAE Standard 55 [56] and ISO 7730 [57] The traditional ASHRAE 7-point thermal sensation vote scale is as follows: – 3: cold, – 2: cool, – 1: slightly cool, 0: neutral, 1: slightly warm, 2: warm, and 3: hot. Some studies have used a 9-point thermal sensation scale to evaluate indoor thermal environments (– 4: very cold, – 3: cold, – 2: cool, – 1: slightly cool, 0: neutral, 1: slightly warm, 2: warm, 3: hot, and 4: very hot) [43, 54] Given the students' exposure to the hot and humid outdoor thermal environment, the ASHRAE 7-point scale was expanded to an 11-point scale to suit the current study's questionnaire. The 11-point thermal sensation scale was thus as follows: – 5: insufferably cold, – 4: very cold, – 3: cold, – 2: cool, – 1: slightly cool, 0: neutral, 1: slightly warm, 2: warm, 3: hot, 4: very hot, and 5: insufferably hot. In summer and after training, the – 5 and – 4 scales were not applicable, so these two scales were omitted. Thermal acceptability used a 2-point scale: 0: acceptable; 1: unacceptable. The data were analyzed statistically using SPSS.

For each interviewee, the clothing insulation ($I_{clo}$) was also calculated. In line with the method proposed in the ASHRAE Handbook [55], the questionnaire included a list of clothing ensembles, allowing interviewees to select the option most similar to their actual attire. A list of clothing insulation values for common individual garments can be obtained from the ASHRAE Handbook [55]. A simpler and nearly accurate summation formula [66] is as follows

$$I_{cl} = \sum_{i} I_{clu,i} \quad (2)$$

In this survey, the total clothing insulation was computed using Eq. (2). Detailed information regarding clothing insulation was collected during the field survey.

## 2.5 Evaluation Indices

In each questionnaire, the values of $T_a$, RH, $T_g$, and $V_a$ were linked. As a result, the votes provided by interviewees regarding their thermal perception and acceptability were related to the corresponding values of several thermal comfort indices, including WBGT, PET, and UTCI.

WBGT, an outdoor thermal comfort index, was proposed by Yaglou and Minard [22] to prevent heat-related casualties at military training centers, and has been used as a measure of heat stress in certain standards [24] and institutions [23, 25]. Taking the effects of solar load into account, WBGT was calculated using two equations. When there are no effects of solar load, WBGT was computed using the natural wet bulb temperature ($T_w$) and $T_g$, as shown in Eq. (3). In an outdoor thermal environment with solar load, WBGT is expressed using $T_w$, $T_g$, and $T_a$ in Eq. (4).

The weighted averages of $T_w$, $T_g$, and $T_a$ are shown in both equations.

$$\text{WBGT} = 0.7T_w + 0.3T_g \tag{3}$$

$$\text{WBGT} = 0.7T_w + 0.2T_g + 0.1T_a \tag{4}$$

The most obvious difference between WBGT and PET is that WBGT does not integrate the human energy balance, focusing only on environmental conditions. PET is described by the concept of equivalent temperature: it refers to the air temperature of a typical indoor room that produces the same core and skin temperatures as the actual complex outdoor conditions. In this typical room, there is no radiation ($T_{mrt} = T_a$), the air speed is below 0.1 m/s, and the vapor pressure is 1200 Pa (equivalent to 50% RH at 20 °C). Meanwhile, the metabolic rate of a standardized person was set to 80 W for light activity, with a clothing heat resistance of 0.9 clo. In this chapter, PET was calculated using RayMan software [33]. Developed by Rutz, Matzarakis, and Mayer at the University of Freiburg [33], RayMan has been widely used to evaluate the outdoor thermal environment. Thus, PET values were obtained based on the survey data.

Based on the Fiala multi-node model [40], UTCI is defined as the equivalent ambient temperature of a reference environment that elicits the same physiological response in a reference person as the actual environment. In this chapter, UTCI was calculated using the "UTCI calculator" software available on the utci.org website. Calculating UTCI requires $T_a$, RH, $T_{mrt}$, and $V_a$ measured at a height of 10 m above the ground. The wind speed was actually measured at a height of 1.1 m above the ground, and then converted to the required input $V_i$ in accordance with Eq. (5):

$$V_i = V_a Log(10/0.01)/Log(x/0.01) \tag{5}$$

## 2.6 Statistical Analysis

Using the data of thermal environment parameters ($T_a$, RH, $V_a$, and $T_{mrt}$) and personal parameters (age, sex, weight, height, activity, and clothing insulation), the non-parametric Kruskal-Wallis H test method was employed to identify the distribution status of the measurement data. The measurement data from the two locations may have significant differences, which influences the results. Therefore, the deviation between the measurement values and the mean value was calculated using Eq. (6) [67].

$$D(x)_{dj} = x_{dji} - \frac{1}{m}\sum_{m}^{j=1} x_{dji} \tag{6}$$

In Eq. (6), x represents the value of thermal parameters, d denotes the survey days, and j indicates the survey sites.

The non-parametric Spearman correlation test was used to identify any significance among the mean thermal sensation vote (MTSV), $T_a$, RH, $V_a$, and $T_{mrt}$. Based on the standards, thermal acceptability is indirectly derived from the PMV model, ranging from negative (cool) to neutral to positive (warm) [68]. In ASHRAE Standard 55 [56], the range of acceptable temperature was proposed by determining 80% acceptability. Thermal acceptability has been used to assess thermal conditions in many previous studies [69, 70]. In this chapter, regarding thermal acceptability, subjects were directly asked in the questionnaire whether their thermal conditions were "acceptable" or "unacceptable."

The percentages of actual unacceptable votes relative to MTSV, WBGT, PET, and UTCI were fitted. Meanwhile, the MTSV within a 1 °C interval of WBGT, PET, and UTCI was calculated, and then the linear regressions of MTSV versus WBGT, PET, and UTCI were fitted. Additionally, based on the regression models, the heat categories of the MTSV scale—i.e., "slightly warm," "warm," "hot," "very hot," and "insufferably hot"—and the different temperature intervals were determined. This method has been adopted to identify the heat stress categories of different thermal comfort indices (WBGT, PET, and UTCI) in some previous studies [37, 51, 52, 63, 71] In this investigation, all analyses were based on a 95% confidence interval at a significance level of 0.05.

# 3 Results

## *3.1 Statistical Summary of Thermal Environment Parameter and Personal Parameters*

In total, 1100 completed questionnaires—filled out by 835 (76%) male volunteers and 265 (24%) female volunteers—were gathered for analysis. All volunteers were students, with ages ranging from 17 to 25 years and a mean age of 19.18 years. Their height and weight varied from 150 to 195 cm and 39 to 91 kg, respectively, with mean values of 168 cm and 57.61 kg. The body surface area and Ponderal index are summarized in Table 2.

**Table 2** Anthropometric data of subjects

| Sex | Number | Age (years) | Height (m) | Weight (kg) | Body surface area ($m^2$)[b] | Ponderal index (kg1/3 $m^{-1}$)[c] |
|---|---|---|---|---|---|---|
| Male | 835 | 19.19 (0.84)[a] | 1.71 (0.06) | 60.14 (8.78) | 1.69 (0.15) | 2.27 (0.15) |
| Female | 265 | 19.15 (0.7) | 1.59 (0.06) | 49.04 (6.22) | 1.48 (0.11) | 2.29 (0.08) |
| Male + female | 1100 | 19.18 (0.8) | 1.68 (0.08) | 57.61 (9.5) | 1.64 (0.17) | 2.28 (0.13) |

*Note* [a]Standard deviation. [b]Body surface area was determined through the DuBois area, $A = 0.202\ W^{0.424}\ h^{0.725}$ [72]. [c]Ponderal index = $W^{1/3}/h$

**Table 3** Statistical results of outdoor microclimate parameters

| Parameters | Minimum | Maximum | Mean | Standard deviations |
|---|---|---|---|---|
| $T_a$ (°C) | 28.30 | 33.40 | 31.15 | 1.43 |
| $T_g$ (°C) | 29.40 | 45.60 | 34.79 | 4.33 |
| *RH* (%) | 59.90 | 88.60 | 76.23 | 9.16 |
| $V_a$ (m/s) | 0.150 | 1.10 | 0.55 | 0.29 |
| $I_{clo}$ (clo) | 0.36 | 0.54 | 0.50 | 0.08 |
| $T_{mrt}$ (°C) WBGT | 30.18 | 59.70 | 39.92 | 8.50 |
| (°C) | 27.30 | 31.42 | 29.32 | 1.03 |
| PET (°C) | 29.70 | 47.60 | 35.55 | 4.97 |
| UTCI (°C) | 32.70 | 42.00 | 36.80 | 2.39 |

The ranges, means, and standard deviations of the measured outdoor thermal parameters (including $T_a$, $T_g$, RH, $V_a$, and $T_{mrt}$), a personal parameter ($I_{clo}$), and several thermal comfort indices (including WBGT, PET, and UTCI) are presented in Table 3. The mean $T_a$ and $T_g$ were 31.15 °C and 34.79 °C, respectively, while the mean $T_{mrt}$ was 39.94 °C; the maximum Tmrt was approximately 59.70 °C, and $T_{mrt}$ was significantly higher than $T_a$. The mean RH attained 76.23%, indicating high humidity during the field survey. The mean wind speed was 0.55 m/s, and the mean $I_{clo}$ was 0.5 clo. The main reason for this was that the students were required to wear uniform clothing for training. The ranges of WBGT, PET, and UTCI were 27.3 to 31.42 °C, 29.7 to 47.60 °C, and 32.7 to 42.00 °C, respectively, and their mean values are summarized in Table 3.

## 3.2 *Distribution of Heat Symptoms*

During the training, the number of students who experienced syncope was documented, and these students were required to stop training and rest in the designated area. The other students continued with their training. Figure 3a shows the percentages of students with and without syncope. During the training, 4.2% of the students had syncope due to exposure to the hot outdoor thermal environment, while 95.8% were able to endure the training—of this group, 79.38% felt uncomfortable. Only 20.62% of students had no symptoms at all, as shown in Fig. 3b.

Among the 79.38% who felt uncomfortable, most students (30.36%) reported "profuse sweating." The percentages of students experiencing "weakness" and "dysphoria" were 16.99% and 12.28%, respectively. Additionally, 4.35% and 3.11% of students reported a headache and pale face, respectively—these data indicate that a hot and humid outdoor thermal environment poses potential risks for training.

The main reasons for this are that humans are homeothermic, meaning their internal body temperature varies only slightly within a very narrow range around

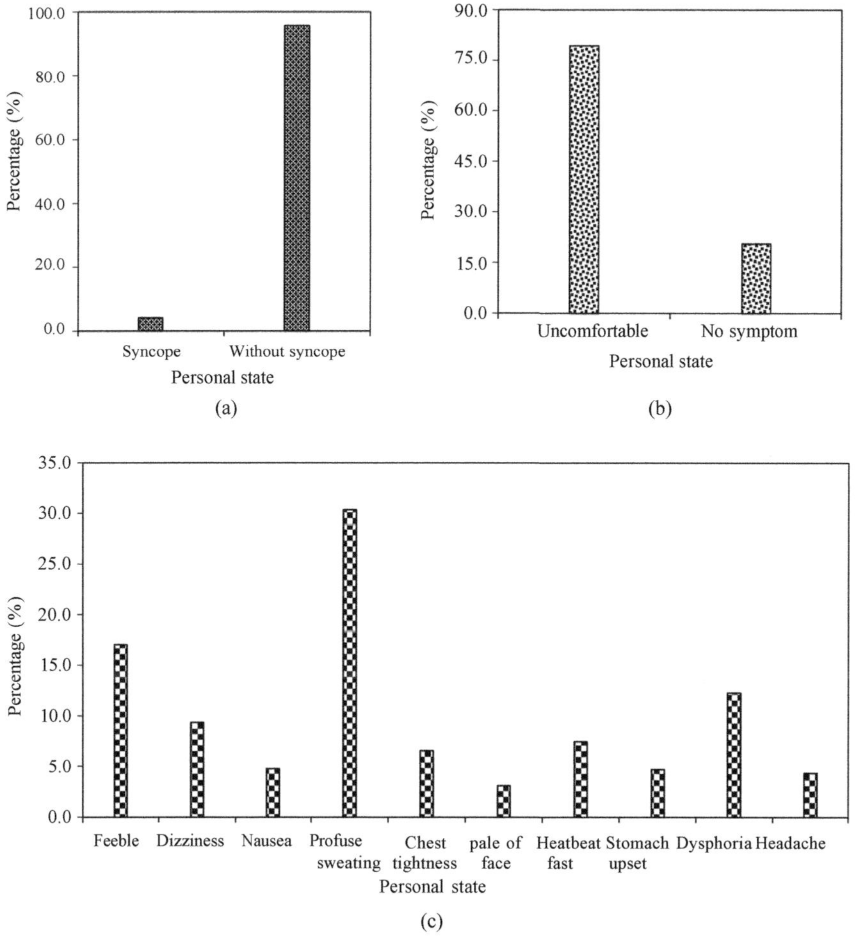


**Fig. 3** Distribution of personal state: **a** ill symptom, **b** uncomfortable symptom, **c** various symptoms

the 37 °C "set point." Under normal conditions, fluctuations in body temperature are usually no more than approximately 1 °C [73–75]. In a hot and humid environment during training, the human body sweats heavily to maintain body heat balance; sweat evaporation helps regulate body temperature. When the water loss rate per unit time reaches 0.02–0.03%, the human body reaches its tolerance limit [76]. However, if evaporation is too limited to dissipate the environmental and metabolic heat loads, heat accumulates in the body and body temperature rises. Cardiovascular strain increases as more blood is pumped to the skin, and additional sweat is secreted. Accelerated dehydration may approach the limit of human heat tolerance. When the circulatory system is overloaded by the combined stresses of heat, training, upright posture, and possibly dehydration, humans may feel weak, dizzy, uncoordinated, or unwell. If they continue exercising in the heat, they may develop heat exhaustion or heat stroke [77]. Thus, it is essential to accurately evaluate the outdoor thermal environment for training to minimize the risks associated with prolonged exposure.

## 3.3 *Distribution of Thermal Sensation Vote*

The proportions of thermal sensation votes are displayed in Fig. 4. Over the entire course of this field survey, a large majority of participants reported feeling "hot" (36.04%), "very hot" (16.42%), and "insufferably hot" (11.04%). The combined share of these groups reached 63.5%, meaning more than half of the volunteers. Only a small number of respondents selected "cold," "cool," and "slightly cool," with these three categories adding up to just 2.29%. Changes in thermal parameters—including air temperature ($T_a$), globe temperature ($T_g$), relative humidity (RH), and mean radiant temperature ($T_{mrt}$)—confirmed that the outdoor thermal environment during this survey was typical of a hot and humid climate. Meanwhile, prior to completing the questionnaires, the volunteers had engaged in training that resulted in a high metabolic rate. Their main activity during training was quick stepping (as shown in Fig. 5), which led to noticeable sweating. For these reasons, most of the participants felt hot throughout the survey.

## 3.4 *Variations in MTSV Against the Thermal Parameters*

For analyzing the sensitivity of thermal parameters to thermal sensation in the hot and humid subtropical region of southern China under training conditions, the MTSV of respondents in each interval group (1 °C for $T_a$ and $T_{mrt}$, 0.1 m/s for $V_a$, and 5% for RH) was computed. The relationships between MTSV and key thermal parameters—including $T_a$, $T_{mrt}$, $V_a$, and RH—were plotted, as shown in Fig. 6.

Figure 6a and b shows the relationships between MTSV and $T_a$, and MTSV and $T_{mrt}$, respectively. The strong linear regression models were statistically significant. The results indicated that based on the survey data, $T_a$ and $T_{mrt}$ exhibited stronger

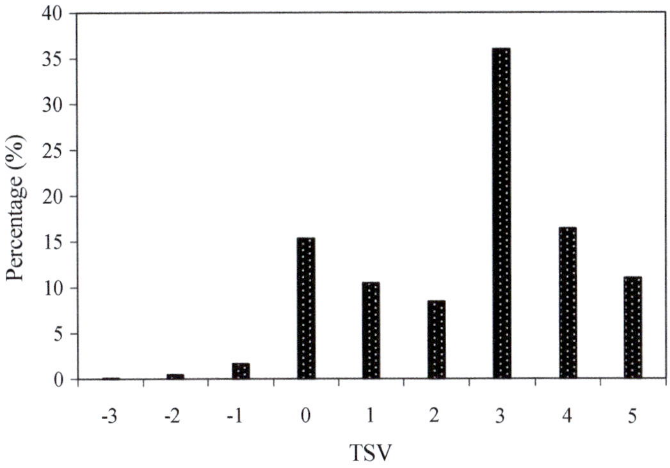


**Fig. 4** Distribution of thermal sensation votes: – 3: cold, – 2: cool, – 1: slightly cool, 0: neutrality, 1: slightly warm, 2: warm, 3: hot and 4: very hot, 5: insufferably hot

**Fig. 5** Students' outdoor training session

correlations with MTSV ($R^2 = 0.7425$ and $R^2 = 0.8989$, respectively). The slopes of the regression lines represent the volunteers' sensitivity to changes in the indices; MTSV increased as $T_a$ and $T_{mrt}$ rose. When comparing the differences between the slopes of the $T_a$ and $T_{mrt}$ lines, the slope of $T_a$ was 0.494—much higher than that of $T_{mrt}$ (0.073). This implies a 2.03 °C change in $T_a$ and a 13.5 °C change in $T_{mrt}$ per

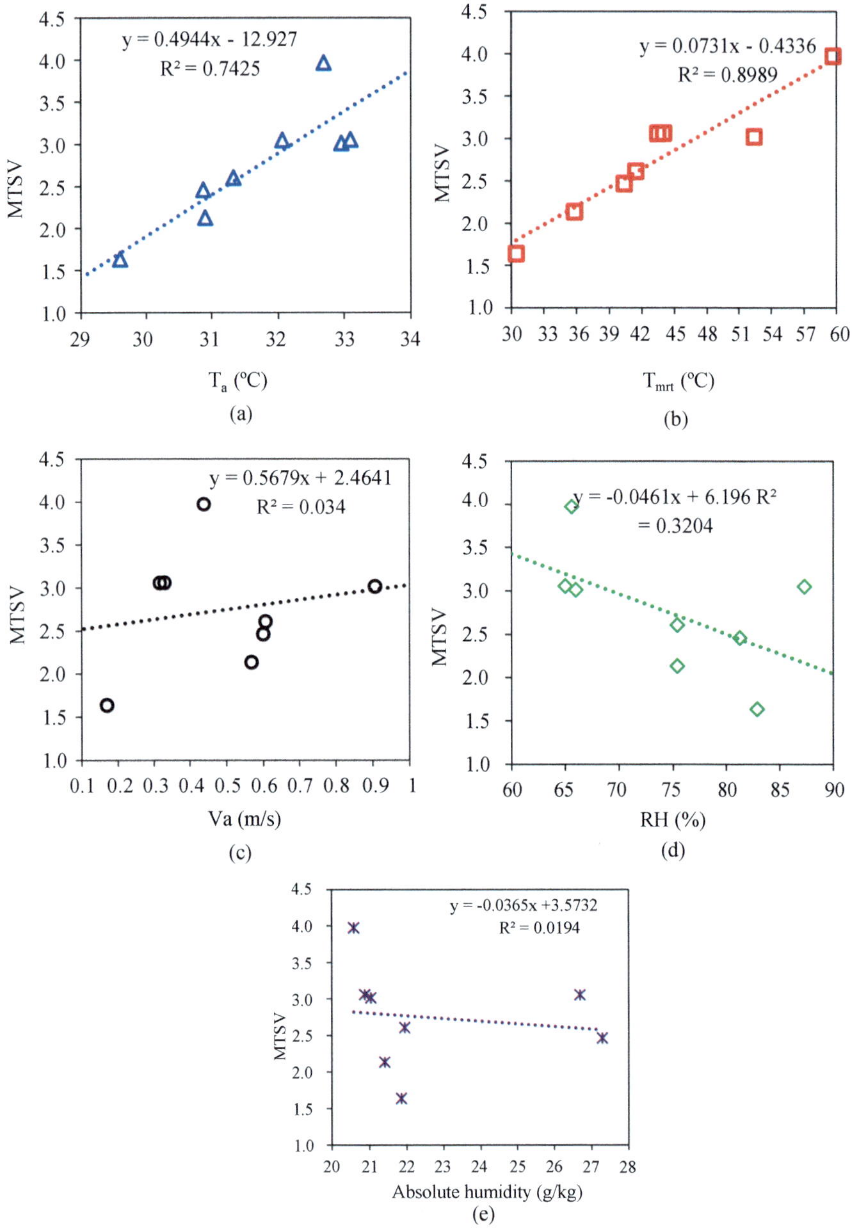


**Fig. 6** Variations in MTSV against the thermal parameters: **a** $T_a$; **b** $T_{mrt}$; **c** $V_a$; **d** $RH$; **e** absolute humidity

MTSV unit. Thus, the sensitivity of $T_a$ to MTSV was far more significant than that of $T_{mrt}$.

Figure 6c and d shows the correlations between MTSV and $V_a$, and MTSV and RH, respectively. These two figures clearly illustrate the effects of $V_a$ and RH on MTSV. Both $R^2$ values were lower than 0.4; in particular, the $R^2$ for $V_a$ was only 0.034, meaning statistical variation in $V_a$ was unnecessary. However, the effects of RH on thermal sensation were unclear, as shown in Fig. 6d. The main reason for this was that during the survey, the students were often dripping with sweat regardless of whether RH was high or low, which weakened the effects of RH. This result differed from previous indoor thermal experiment results—some prior studies have reported that MTSV increased with RH [78] Most of these previous studies were conducted in indoor climate chambers, where air temperature was set below 32 °C and there was no influence of sunlight. To further analyze the effects of humidity, the variation in MTSV relative to absolute humidity can also be considered, as shown in Fig. 6e. Additionally, there was no strong linear relationship between absolute humidity and MTSV, which was similar to the case with RH. Thus, the effects of RH on MTSV in outdoor thermal environments need to be analyzed in future research.

## 3.5 Thermal Sensation with Thermal Indices

By determining the MTSV for 1 °C intervals of thermal indices (WBGT, PET, and UTCI) throughout the study duration, the corresponding MTSV values for each index bin were derived, as illustrated in Fig. 7. Simple linear regression analyses were conducted between MTSV and these thermophysiological indicators. The results demonstrated strong linear relationships, with all $R^2$ values surpassing 0.75. Positive regression slopes indicated that MTSV increased progressively with rising values of WBGT, PET, and UTCI. The slope for WBGT (0.586) was the steepest, while PET showed the most gradual slope (0.104), and UTCI registered an intermediate value (0.238), revealing WBGT's predominant influence on MTSV variations.

Due to the prevailing hot and humid conditions, the thermal comfort indices predominantly clustered within higher ranges (Fig. 7). Concurrently, the training regimen induced elevated metabolic rates that persisted briefly post-exercise. Consequently, the synergistic effect of environmental heat stress and heightened metabolism drove MTSV beyond 1.5 units, as depicted in Fig. 7.

Based on the regression models in Fig. 7, thermal stress categories for each index were established and compiled in Table 4. All tabulated MTSV values exceeded unity, rendering lower stress categories practically irrelevant. Cross-comparison of stress categories revealed that most participants reported "insufferably hot" sensations when WBGT surpassed 32.7 °C, PET exceeded 53.9 °C, or UTCI climbed above 44.9 °C. These thresholds warranted training termination and mandatory rest periods. Conversely, thermal environments remained acceptable within the "slightly

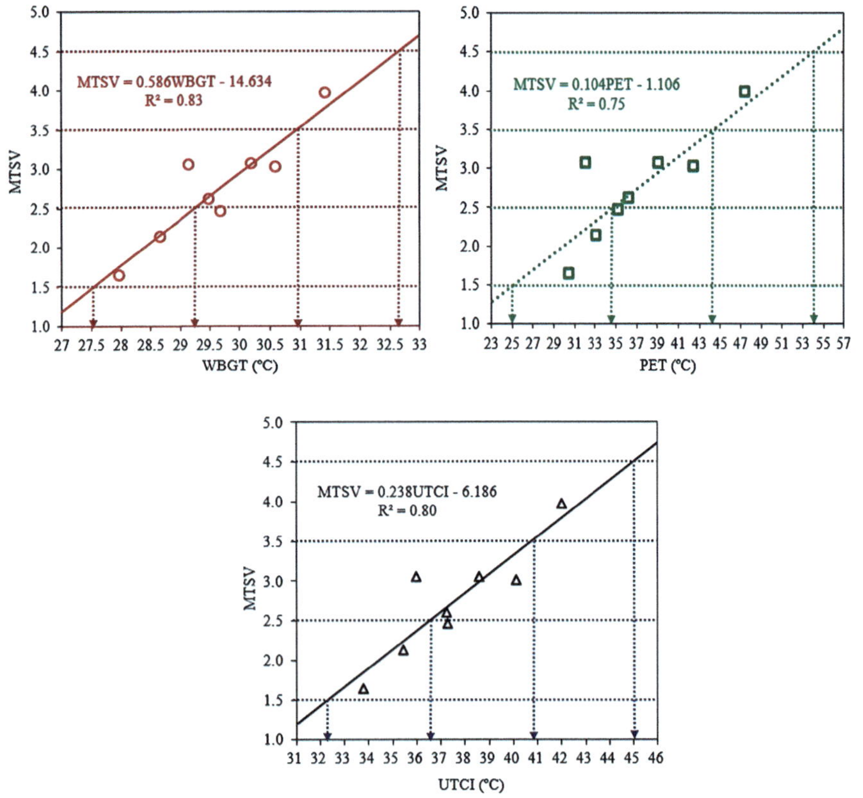


**Fig. 7** Correlations between MTSV and thermal comfort indices: **a** WBGT; **b** PET; **c** UTCI

cool" to "slightly warm" sensation range. The maximum tolerable limits were established at 27.5 °C for WBGT, 25.1 °C for PET, and 32.3 °C for UTCI, providing valuable criteria for evaluating outdoor activity conditions.

Notably, regional disparities emerged in PET and UTCI heat stress categorizations. Within the MTSV range of 1.5–2.5, the PET spectrum in this chapter (25.1–34.7 °C) registered lower than Tianjin [63] (31–36 °C), Taichung [34] (34–38 °C), and Freiburg [79] (29–35 °C), but higher than Umeå [52] (24–32 °C). This variability primarily stems from physiological acclimatization to local climates. However, the broader PET range observed herein (25.1–34.7 °C) reflects the inclusion of heat acclimation training, contrasting with recreational outdoor activities surveyed in other regions. Parallel patterns emerged for UTCI comparisons, with distinct regional stress categorizations. Students reported insufferable heat at UTCI > 44.9 °C, lower than Tianjin's threshold [63] (> 47 °C) but exceeding those documented in Athens [49] (> 37.7 °C) and Umeå [52] (> 40 °C).

**Table 4** Assumed thermal sensations for different stress categories

| Thermal sensation | MTSV | WBGT | PET (°C) | | | | | UTCI (°C) | | | | Heat stress |
|---|---|---|---|---|---|---|---|---|---|---|---|---|
| | | | This study | Tianjin | Taichung | Freiburg | Umeå | This study | Tianjin | Athens | Umeå | |
| Neutral | − 0.5–0.5 | – | – | 11–24 | 26–30 | 18–23 | 10–17 | – | 12–25 | 17.4–24.5 | 12–17 | No thermal stress |
| Slightly warm | 0.5–1.5 | – | – | 24–31 | 30–34 | 23–29 | 17–24 | – | 25–33 | 24.5–29.1 | 17–29 | Moderate heat stress |
| Warm | 1.5–2.5 | 27.5–29.2 | 25.1–34.7 | 31–36 | 34–38 | 29–35 | 24–32 | 32.3–36.5 | | | | |
| Hot | 2.5–3.5 | 29.2–31.0 | 34.7–44.3 | 36–46 | 38–42 | 35–41 | 32–39 | 36.5–40.7 | 33–39 | 29.1–34.1 | 29–34 | Strong heat stress |
| Very hot | 3.5–4.5 | 31.0–32.7 | 44.3–53.9 | > 46 | > 42 | > 41 | > 39 | 40.7–44.9 | 39–47 | 34.1–37.7 | 34–40 | Very strong heat stress |
| Insufferably hot | > 4.5 | > 32.7 | >53.9 | | | | | > 44.9 | > 47 | > 37.7 | > 40 | Extreme heat stress |

## 3.6 Thermal Sensation and Thermal Acceptability

The relationships between the percentage of unacceptability and each thermal index are presented in Fig. 8a–d. E functions were applied to the regression models. The $R^2$ values for MTSV, WBGT, PET, and UTCI were 0.68, 0.603, 0.32, and 0.34, respectively, and all P values were lower than 0.05. Thus, the regression models were statistically significant.

Based on these models, the 20% unacceptability threshold for each thermal comfort index was computed. The MTSV value corresponding to 20% unacceptability was 1.99, which is higher than the requirement specified in ASHRAE 55. This indicates that in hot and humid thermal environments, humans have the adaptability to adjust to thermal sensations. In this field survey, the training conditions expanded the acceptable range of thermal indices. Therefore, as shown in Fig. 8, when the percentage of thermal acceptability equals 20%, the upper limits of the thermal

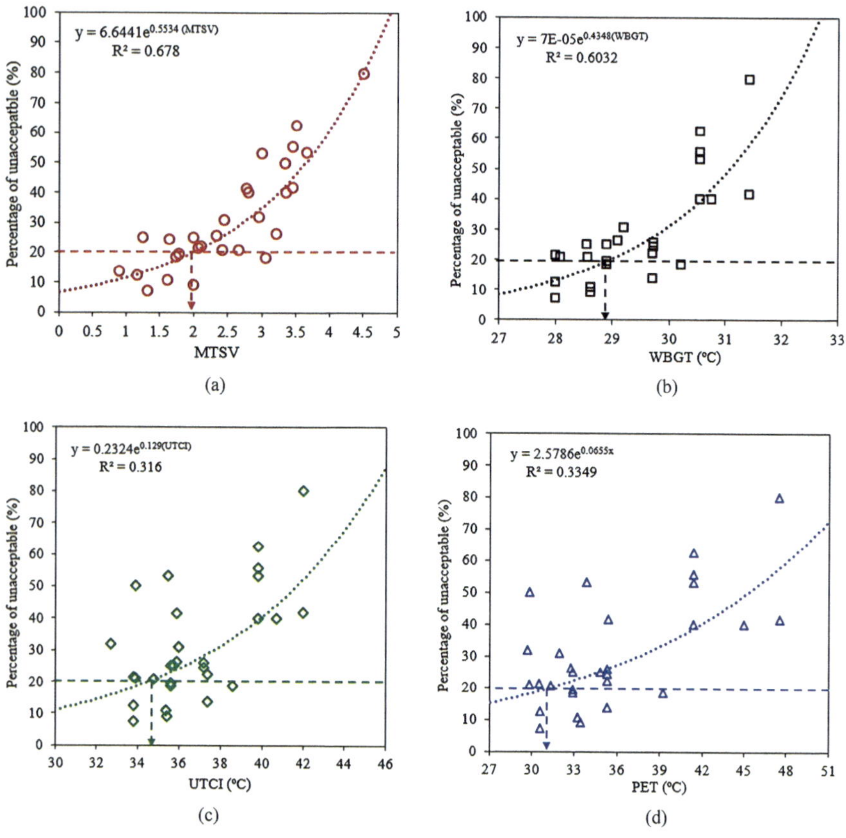


**Fig. 8** Relations between the percentage of unacceptability and each thermal index: **a** MTSV, **b** WBGT, **c** PET, **d** UTCI

indices were determined: WBGT (28.9 °C), PET (34.5 °C), and UTCI (31.3 °C). This finding is beneficial for evaluating the thermal environment during training.

When comparing the stress categories in Table 4, most MTSVs fell within the range of 1.5–2.5, which shows that the subjects demonstrated thermal adaptability in the outdoor thermal environment. Thus, the typical stress categories need to be revised for this region.

# 4 Discussion

## 4.1 Sensitivity of Thermal Parameters to Thermal Comfort Indices

Many previous studies have demonstrated that human thermal comfort is a complex function of atmospheric conditions and personal factors [35–38]. The effects of microclimate on outdoor thermal comfort are more complex than those on indoor comfort, due to the random variations in outdoor thermal parameters. Many researchers have emphasized that human comfort levels also depend on meteorological parameters (such as $T_a$, RH, $V_a$, and solar radiation) and subjective factors [80, 81]. Correlations between microclimate variables and MTSV were also analyzed to assess the sensitivity of thermal parameters. However, the sensitivity of different thermal parameters varies.

Based on the outdoor thermal comfort assessment of an arid climate in an oasis city, Nikolopoulou and Lykoudis [80] reported that MTSV showed a stronger correlation with globe temperature than with air temperature. Middel et al. [82] found that globe temperature was the only significant predictor of TSV in the regression of observed meteorological variables, highlighting the importance of the radiative environment for outdoor thermal comfort in hot and dry climates. These results were consistent with those of previous studies that found thermal sensation had a better correlation with globe temperature than with air temperature in Europe [80] Ali and Patnaik [83] also found that among all four independent variables tested, $T_g$ significantly affected people's thermal sensation votes. There is a strong relationship between $T_g$ and $T_{mrt}$; therefore, $T_{mrt}$, as one of the influencing factors, should also be focused on.

Makaremi et al. [84] reported that the effect of $T_{mrt}$ on thermal comfort was stronger than that of $T_a$ in Malaysia. Other researchers also identified $T_{mrt}$ as the most important variable for outdoor thermal comfort [85–87] However, Matje et al. [88], whose test location was Birmingham, UK, claimed that solar radiation and RH did not show a strong relationship. Ruiz and Correa [89] reported that the highest correlation was found with $T_a$; solar radiation did not have a high correlation, and wind speed also did not show a significant linear correlation with MTSV. In summer, RH had a positive correlation; in winter, the linear relationship between RH and MTSV was not significant. Thus, regardless of whether humidity has a good correlation with comfort, different seasonal climates need to be considered. However, "damp" or "dry" conditions are associated with increased discomfort [80].

Regarding wind speed, Makaremi et al. [84] pointed out that higher $V_a$ can improve thermal comfort, but lower $V_a$ does not affect people's thermal sensation in outdoor spaces [84] Based on the above analysis, it is found that their results differed even though they applied the same methodology to similar fields. The current study found that there were strong linear relationships between MTSV and $T_a$, and between MTSV and $T_{mrt}$—especially with $T_a$. The effect of air temperature on MTSV was the most significant: an approximate 2 °C variation in $T_a$ caused a 1-unit change in MTSV. This result was similar to that of Ruiz and Correa [89] However, wind speed and RH did not show a strong relationship. From Fig. 6c, wind speed data were concentrated in the range of 0.1–1 m/s, and the effect of wind speed was slight—especially in the hot outdoor thermal environment.

During and after training, most students sweated. However, during the survey, the effect of enhancing heat loss due to low wind speed was insignificant, which was similar to the results of Ali et al. [83] In addition, the effect of RH on MTSV can be ignored. The main reason for this was that RH was concentrated at a high level in this subtropical climate zone, always exceeding 70%. Thus, the training period should be shortened on sunny days when temperatures are high and solar radiation is intense.

## 4.2 Different Stress Categories of Thermal Comfort Indices

To evaluate the outdoor thermal environment, based on extensive experimental and survey data, the WBGT, PET, and UTCI values corresponding to various levels of thermal perception and physiological stress were identified and are presented in Table 5 [1, 90–93]. Since the investigations were conducted in a hot summer, all MTSVs were above zero; thus, Table 5 only includes five levels of thermal perception and physiological stress (ranging from − 0.5 to 4.5). From Fig. 7, MTSVs were concentrated in the range of 1.5–4.5, so four levels of thermal perception and physiological stress were compared in Table 5.

When comparing the original heat stress values of thermal indices (for normal activity) and the modified thermal indices (for outdoor training), within the same thermal sensation range, the values of the modified thermal indices were significantly higher than those of the original ones. For instance, when examining WBGT, it is easy to observe that the range for "very hot or very strong heat stress" in the present study (30.95–32.65 °C) is much higher than the required range (28–30 °C), which is due to physiological adaptation to the hot and humid subtropical climate [37] The other thermal comfort indices—including PET and UTCI—showed a similar trend to WBGT. The PET range for "hot" (34.7–44.3 °C) was also much higher than the range reported by Matzarakis et al. (35–41 °C), and the UTCI range for "hot" (36.5–40.7 °C) was higher than that in the EU, a finding that has been demonstrated in many previous studies [37, 52].

**Table 5** Modified thermal indices

| Thermal perception | Stress category | MTSV | Recommended of WBGT | Original WBGT (°C) | Modified WBGT (°C) | Original PET (°C) | PET (°C) (Guangzhou normal activity) | Modified PET (°C) | Original UTCI (°C) | UTCI (°C) (Guangzhou normal | Modified UTCI (°C) |
|---|---|---|---|---|---|---|---|---|---|---|---|
| Comfortable | No thermal stress | – 0.5–0.5 | No danger | < 18 | – | 18–23 | | – | 9–26 | | – |
| Slightly warm | Moderate heat stress | 0.5–1.5 | | | – | 23–29 | | – | 26–32 | | – |
| Warm | | 1.5–2.5 | Caution | 18–23 | 27.5–29.2 | 29–35 | 31.3–39.9 | 25.1–34.7 | 26–32 | 31.6–38.8 | 32.29–36.50 |
| Hot | Strong heat stress | 2.5–3.5 | Extreme caution | 23–28 | 29.2–31.0 | 35–41 | 39.9–50.9 | 34.7–44.3 | 32–38 | 38.8–47.8 | 36.50–40.70 |
| Very hot | Very strong heat stress | 3.5–4.5 | Danger | 28–30 | 31.0–32.7 | > 41 | > 50.9 | 44.3–53.9 | 38–46 | > 47.8 | 40.70–44.90 |
| | Extreme heat stress | > 4.5 | Extreme danger | ≥ 30 | ≥ 32.7 | | | ≥ 53.9 | ≥ 46 | | ≥ 44.90 |

In addition, normal entertainment activities are more or less voluntary, while outdoor training is a mandatory activity—this difference is significant. The results comparing normal activities and outdoor training in Guangzhou are summarized in Table 5. Within the same thermal sensation range, the differences between the thermal indices for normal activities and outdoor training were significant: the thermal indices for normal activities were higher than those for outdoor training. When MTSV fell in the range of 1.5–2.5, the PET range for normal activities was 31.3 to 39.9 °C, which is approximately 5 °C higher than the PET range for outdoor training (25.1 to 34.7 °C). The UTCI results were similar to the PET results, mainly because outdoor training is a mandatory activity. Students must follow regulations with high concentration, which may boost blood circulation and metabolic rate, increasing sweating and raising MTSV. Thus, in the same MTSV-based thermal environment, the thermal indices for normal activities were higher than those for outdoor training.

Based on the current survey data, the ranges for "extreme heat stress or extreme danger" were analyzed with consideration for thermal health. During training, human metabolism is much higher than during regular walking, so the impact of heat exposure during training in hot outdoor thermal environments on human health is significant. Many previous studies [87, 88] have focused on entertainment and similar activities, with few examining military training. However, it is essential to define the ranges for "extreme heat stress or extreme risk." When thermal comfort indices exceed a certain threshold—e.g., WBGT $\geq$ 32.7 °C, PET $\geq$ 53.9 °C, or UTCI $\geq$ 44.9 °C—it is recommended to reduce outdoor activities as a precaution.

## 5 Conclusion

This chapter presented a field survey conducted during a university's summer military training program, combining on-site microclimatic monitoring with questionnaire data to evaluate student thermal comfort and safety. The findings revealed that heat stress during training significantly affected students' physical and subjective responses, with 4.2% experiencing syncope, nearly 80% reporting discomfort, and over 30% suffering from profuse sweating. Air temperature was identified as the dominant factor shaping mean thermal sensation votes, followed by mean radiant temperature, underscoring the risks of prolonged exposure without adequate shading or cooling measures. Regression analyses further demonstrated that current stress classifications for WBGT, PET, and UTCI require downward adjustment to better represent the thermal tolerance of freshmen students, with revised thresholds indicating critical limits beyond which training conditions become insufferably hot and potentially hazardous. These results stress the importance of developing tailored heat safety guidelines for student military training, as their metabolic demands differ from leisure activities and may amplify heat vulnerability. Looking ahead, the integration of portable cardiopulmonary monitoring devices into training studies will enable precise measurement of metabolic rates across different drills, allowing for a deeper understanding of how training duration and intensity influence thermal tolerance. Together, these insights contribute to safer and more sustainable approaches to outdoor high-intensity training in hot climates.

**Acknowledgments** This chapter was fully supported by a grant from the Research Grants Council of the Science technology project of Guangzhou, China (Project No. 201804010252), and the Opening Funds of State Key Laboratory of Building Safety and Built Environment and National Engineering Research Center of Building Technology (Project No. BSBE2018-03). The authors thank all students who participated in the surveys.

## References

1. S. Zare, N. Hasheminejad, H.E. Shirvan, R. Hemmatjo, K. Sarebanzadeh, S. Ahmadi, Comparing universal thermal climate index (UTCI) with selected thermal indices/environmental parameters during 12 months of the year. Weather. Clim. Extrem. **19**, 49–57 (2018)
2. I. Burton, K.L. Ebi, G. McGregor, Biometeorology for adaptation to climate variability and change, Spring, Part of the Biometeorology book series, vol. 1 (2009), pp. 1–5
3. Y. Zhao, A. Ducharne, B. Sultan, P. Braconnot, R. Vautard, Estimating heat stress from climate-based indicators: Present-day biases and future spreads in the CMIP5 global climate model ensemble. Environ. Res. Lett. **10**(8) (2016)
4. L. Harduar Morano, T.L. Bunn, M. Lackovic, A. Lavender, G.T. Dang, J.J. Chalmers, Occupational heat-related illness emergency department visits and inpatient hospitalizations in the southeast region, 2007-2011. Am. J. Ind. Med. **58**(10), 1114–1125 (2015)
5. L. Harduar Morano, S. Watkins, K. Kintziger, A comprehensive evaluation of the burden of heat-related illness and death within the Florida population. Int. J. Environ. Res. Public Health **13**(6), 551 (2016)
6. The National Institute for Occupational Safety and Health, *Heat Stress* (CDC, 2015). Available from. http://www.cdc.gov/niosh/topics/heatstress/. Updated 19 July 2016; cited 22 Jul 2016
7. X.Q. He, Safety Engineering, first ed., China University of Mining and Technology Press, Beijing, 2000
8. J. Zhao, N. Zhu, S. Lu, Productivity model in hot and humid environment based on heat tolerance time analysis. Build. Environ. **44**(11), 2202–2207 (2009)
9. X. Shi, N. Zhu, G. Zheng, The combined effect of temperature, relative humidity and work intensity on human strain in hot and humid environments. Build. Environ. **69**, 72–80 (2013)
10. J. Xiang, P. Bi, D. Pisaniello, A. Hansen, T. Sullivan, Association between high temperature and work-related injuries in Adelaide, South Australia, 2001–2010. Occup. Environ. Med. **71**(4), 246–252 (2014)
11. A. Adam-Poupart, A. Smargiassi, M.A. Busque, P. Duguay, M. Fournier, J. Zayed, F. Labrèche, Summer outdoor temperature and occupational heat-related illnesses in Quebec (Canada). Environ. Res. **134**, 339–344 (2014)
12. S.L. Lu, N. Zhu, Experimental research on physiological index at the heat tolerance limits in China. Build. Environ. **42**(12), 4016–4021 (2007)
13. J.R. Brotherhood, Heat stress and strain in exercise and sport. J. Sci. Med. Sport **11**(1), 6–19 (2008)
14. R.T. Gun, G.M. Budd, Effects of thermal, personal and behavioural factors on the physiological strain, thermal comfort and productivity of Australian shearers in hot weather. Ergonomics **38**(7), 1368–1384 (1995)

15. K. Pantavou, G. Theoharatos, A. Mavrakis, M. Santamouris, Evaluating thermal comfort conditions and health responses during an extremely hot summer in Athens. Build. Environ. **46**(2), 339–344 (2011)
16. G. Zheng, N. Zhu, Z. Tian, Y. Chen, B. Sun, Application of a trapezoidal fuzzy AHP method for work safety evaluation and early warning rating of hot and humid environments. Saf. Sci. **50**(2), 228–239 (2012)
17. T.P. Warren, A. Mitsuru, Y.T. Frank, J.P. Christopher, Temperature and air pollution as risk factors for heat stroke in Tokyo, July and August 1980–1995. Environ. Health Perspect. **107**(11), 911–916 (1999)
18. A.P.C. Chan, W. Yi, D.P. Wong, M.C.H. Yam, D.W.M. Chan, Determining an optimal recovery time for construction rebar workers after working to exhaustion in a hot and humid environment. Build. Environ. **58**, 163–171 (2012)
19. D.E. Hancher, H.A. Abd-Elkhalek, The effect of hot weather on construction labor productivity and costs. Cost Eng. **40**(3), 32–36 (1998)
20. E. Schickele, Environment and fatal heat stroke-an analysis of 157 cases occurring in the army in the U.S. during World War II. Mil. Surg. **100**, 235–256 (1947)
21. E.L. Cook, Epidemiological approach to heat trauma. Mil. Med. **116**, 317–322 (1955)
22. C.P. Yaglou, D. Minard, Control of heat casualties at military training centers. Am. Med. Assoc. Arch. Ind. Health **16**(4), 302–316 (1957)
23. NIOSH, Criteria for a recommended standard: occupational exposure to heat and hot environments, in *U.S. Department of Health and Human Services, Centers for Disease Control and Prevention, National Institute for Occupational Safety and Health. DHHS (NIOSH)* ed. by B. Jacklitsch, W.J. Williams, K. Musolin, coca, a., J.-H. Kim, N. Turner, O.H. Cincinnati (Publication, 2016–2106, 2016)
24. ISO, *Assessment of Heat Stress Using the WBGT (Wet Bulb Globe Temperature) Index* (International Organization for Standardization, Geneva, Switzerland, 2017)
25. ACGIH, Documentation of the threshold limit values for chemical substances and physical agents and biological exposure indices, seventh ed., 2015 supplement, in *Proceedings of American Conference of Governmental Industrial Hygienists*, (Cincinnati, Ohio, 2015). ISBN: 978-1-607260-82-0
26. E.E. Coris, A.M. Ramirez, D.J. Van Durme, Heat illness in athletes. Sports Med. **34**(1), 9–16 (2004)
27. M.T. Smith, M. Reid, S. Kovalchik, T.O. Woods, R. Duffield, Heat stress incident prevalence and tennis match play performance at the Australian open. J. Sci. Med. Sport **21**, 467–472 (2018)
28. T.E. Bernard, C.D. Ashley, Short-term heat stress exposure limits based on wet bulb globe temperature adjusted for clothing and metabolic rate. J. Occup. Environ. Hyg. **6**(10), 632–638 (2009)
29. T. Sakoi, T. Mochida, An interpretation of WBGT for outdoor environment based on heat transfer engineering and its adjustment factors. Jpn. J. Biometeorol **52**(1), 17–28 (2015) (in Japanese with English abstract)
30. T. Sakoi, T. Mochida, Y. Kurazumi, K. Kuwabara, Y. Horiba, S. Sawada, Heat balance model for a human body in the form of wet bulb globe temperature indices. J. Therm. Biol. **71**, 1–9 (2018)
31. M.H. Elnabawi, N. Hamza, S. Dudek, Thermal perception of outdoor urban spaces in the hot arid region of Cairo, Egypt. Sustain. Cities Soc. **22**, 136–145 (2016)
32. P. Höppe, The physiological equivalent temperature—a universal index for the biometeorological assessment of the thermal environment. Int. J. Biometeorol. **43**, 71–75 (1999)
33. A. Matzarakis, F. Rutz, H. Mayer, Modelling radiation fluxes in simple and complex environemnts—application of the RayMan model. Int. J. Biometeorol. **51**, 323–334 (2007)
34. T.P. Lin, A. Matzarakis, Tourism climate and thermal comfort in Sun moon Lake, Taiwan. Int. J. Biometeorol. **52**, 281–290 (2008)
35. H. Mayer, J. Holst, P. Dostal, F. Imbery, D. SchindlerHuman, Thermal comfort in summer within an urban street canyon in Central Europe. Meteorologische Zeitschrif **17**, 241–250 (2008)

36. J. Niu, J. Liu, T. Lee, Z. Lin, C.M. Mak, K.T. Tse, B.S. Tang, K.C.S. Kwok, A new method to assess spatial variations of outdoor thermal comfort: onsite monitoring results and implications for precinct planning. Build. Environ. **91**, 263–270 (2015)
37. B. Yang, T. Olofsson, G. Nair, A. Kabansi, Outdoor thermal comfort under subarctic climate of North Sweden—a pilot study in Umea. Sustain. Cities Soc. **28**, 387–397 (2017)
38. S. Amindeldar, S. Heidari, M. Khalili, The effect of personal and microclimatic variables on outdoor thermal comfort: a field study in Tehran in cold season. Sustain. Cities Soc. **32**, 153–159 (2017)
39. K. Pantavou, S. Lykoudis, M. Nikolopoulou, L.X. Tsiros, Thermal sensation and climate: a comparison of UTCI and PET thresholds in different climates. Int. J. Biometeorol. **7**, 1–14 (2018)
40. K. Błażejczyk, P. Broede, D. Fiala, G. Havenith, I. Holmér, G. Jendritzky, A. Kunert, Principles of the new universal thermal climate index (UTCI) and its application to bioclimatic research in European scale. Miscellanea Geographica **14**, 91–102 (2010)
41. F. Pappenberger, G. Jendritzky, H. Staiger, E. Dutra, F. Di Giuseppe, D.S. Richardson, H.L. Cloke, Global forecasting of thermal health hazards: The skill of probabilistic predictions of the universal thermal climate index (UTCI). Int. J. Biometeorol. **59**, 311–323 (2015)
42. P.K. Cheung, C.Y. Jim, Determination and application of outdoor thermal benchmarks. Build. Environ. **123**, 333–350 (2017)
43. Z. Fang, Z. Lin, C.M. Mak, J. Niu, K.T. Tse, Investigation into sensitivities of factors in outdoor thermal comfort indices. Build. Environ. **128**, 129–142 (2018)
44. I. Golasi, F. Salata, E.de Lieto Vollaro, M.Coppi, Complying with the demand of standardization in outdoor thermal comfort: a first approach to the global outdoor comfort index (GOCI), Build. Environ. 130(2018) 104–119
45. J. Huang, C. Zhou, Y. Zhuo, L. Xu, Y. Jiang, Outdoor thermal environments and activities in open space: an experiment study in humid subtropical climates. Build. Environ. **103**, 238–249 (2016)
46. T. Huang, J. Li, Y. Xie, J. Niu, C.M. Mak, Simultaneous environmental parameter monitoring and human subject survey regarding outdoor thermal comfort and its modelling. Build. Environ. **125**, 502–514 (2017)
47. E.L. Kruger, P. Drach, Identifying potential effects from anthropometric variables on outdoor thermal comfort. Build. Environ. **117**, 230–237 (2017)
48. E.L. Krüger, C.A. Tamura, P. Bröde, M. Schweiker, A. Wagner, Short- and long-term acclimatization in outdoor spaces: exposure time, seasonal and heat wave adaptation effects. Build. Environ. **116**, 17–29 (2017)
49. K. Pantavou, G. Theoharatos, M. Santamouris, D. Asimakopoulos, Outdoor thermal sensation of pedestrians in a Mediterranean climate and a comparison with UTCI. Build. Environ. **66**, 82–95 (2013)
50. Y. Wang, Z. Nia, Y. Peng, B. Xia, Local variation of outdoor thermal comfort in different urban green spaces in Guangzhou, a subtropical city in South China. Urban For. Urban Green. **32**, 99–112 (2018)
51. F. Salata, I. Golasi, R.D.L. Vollaro, A.D.L. Vollaro, Outdoor thermal comfort in the Mediterranean area. A transversal study in Rome, Italy. Build. Environ. **96**, 46–61 (2016)
52. M. Hadianpour, M. Mahdavinejad, M. Bemanian, F. Nasrollahi, Seasonal differences of subjective thermal sensation and neutral temperature in an outdoor shaded space in Tehran, Iran. Sustain. Cities Soc. **39**, 751–764 (2018)
53. 2345 Weather Report. http://tianqi.2345.com/wea_history/59287.htm
54. Y. Zhang, H. Chen, Q. Meng, Thermal comfort in buildings with split air-conditioners in hot-humid area of China. Build. Environ. **64**, 213–224 (2013)
55. R.A. Engineers, *ASHRAE Handbook: Fundamentals*. 2017
56. ANSI/ASHRAE, *Standard 55–2017: Thermal Environmental Conditions for Human Occupancy*. Atlanta, Georgia: American Society of Heating Refrigerating and Air-Conditioning Engineers (ASHRAE); 2017

57. ISO, International Standard *7730, Ergonomics of the Thermal Environment–Analytical Determination and Interpretation of Thermal Comfort Using Calculation of the PMV and PPD Indices and Local Thermal Comfort Criteria* (International Standard Organization, Geneva, 2005)
58. ISO 7726, *Ergonomics of the Thermal Environment—Instruments for Measuring Physical Quantities* (1998)
59. S. Thorsson, F. Lindberg, I. Eliasson, B. Holmer, Different methods for estimating the mean radiant temperature in an outdoor urban setting. Int. J. Climatol. **27**, 1983–1993 (2007)
60. M.W. Yahia, E. Johansson, Evaluating the behaviour of different thermal indices by investigating various outdoor urban environments in the hot dry city of Damascus, Syria. Int. J. Biometeorol. **57**, 615–630 (2013)
61. E. Johansson, S. Thorsson, R. Emmanuel, E. Krüger, Instruments and methods in outdoor thermal comfort studies—the need for standardization. Urban Clim. **10**, 346–366 (2014)
62. T. Sharmin, K. Steemers, A. Matzarakis, Analysis of microclimatic diversity and outdoor thermal comfort perceptions in the tropical megacity Dhaka, Bangladesh. Build. Environ. **94**, 734–750 (2015)
63. D. Lai, D. Guo, Y. Hou, C. Lin, Q. Chen, Studies of outdoor thermal comfort in northern China. Build. Environ. **77**, 110–118 (2014)
64. D.A. McIntyre, *Indoor Climate* (Architectural Publication Series, London, 1980)
65. W. Liu, Y. Zhang, Q. Deng, The effects of urban microclimate on outdoor thermal sensation and neutral temperature in hot-summer and cold-winter climate. Energy Build. **128**, 190–197 (2016)
66. B.W. Olesen, A new and simpler method for estimating the thermal insulation of a clothing ensemble. ASHRAE Trans. **91**(2), 478–492 (1985)
67. Y. Wang, Z. Ni, Y. Peng, B. Xia, Local variation of outdoor thermal comfort in different urban green spaces in Guangzhou, a subtropical city in South China. Urban For. Urban Green. **32**, 99–112 (2018)
68. O. Fanger, *Thermal Comfort. Analysis and Applications in Environmental Engineering* (Danish Technical Press, Copenhagen, 1970)
69. Z. Zhang, Y. Zhang, E. Ding, Acceptable temperature steps in the hot—humid area of China. Build. Environ. **121**, 190–199 (2017)
70. M. Xu, B. Hong, J. Mi, S. Yan, Outdoor thermal comfort in an urban park during winter in cold regions of China. Sustain. Cities Soc. **43**, 208–220 (2018)
71. C. Piselli, V.L. Castaldo, I. Pigliautile, A.L. Pisello, F. Cotana, Outdoor comfort conditions in urban areas: On citizens' perspective about microclimate mitigation of urban transit areas. Sustain. Cities Soc. **39**, 16–36 (2018)
72. Z. Fang, H. Liu, B. Li, M. Tan, O.M. Olaide, Experimental investigation on thermal comfort model between local thermal sensation and overall thermal sensation. Energ. Build. **158**, 1286–1295 (2018)
73. K. Parsons, *Human Thermal Environments: The Effects of Hot, Moderate, and Cold Environments on Human Health, Comfort, and Performance*, third ed. (CRC Press, 2014)
74. F.L. Boca Raton, L. Sherwood, *Human Physiology: From Cells to Systems*, ninth ed. (Cengage learning, Boston, MA, 2015)
75. W.L. Kenney, J. Wilmore, D. Costill, *Physiology of Sport and Exercise*, sixth ed. (Human Kinetics, Champaign, IL, 2015)
76. J. Zhao, N. Zhum, S. Lu, Productivity model in hot and humid environment based on heat tolerance time analysis. Build. Environ. **44**, 2202–2207 (2009)
77. G.M. Budd, Wet-bulb globe temperature (WBGT)—its history and its limitations. J. Sci. Med. Sport **11**, 20–32 (2008)
78. M Lan, *Effects of Air Humidity and Velocity on Human Thermal Sensation in Summer*, Ph.D, (Chongqing University, 2012)
79. A. Matzarakis, H. Mayer, Another kind of environmental stress: thermalstress. WHO News **18**, 7–10 (1996)

80. M. Nikolopoulou, S. Lykoudis, Thermal comfort in outdoor urban spaces: analysis across different European countries. Build. Environ. **41**, 1455–1470 (2006)
81. J. Yin, Y. Zheng, R. Wu, J. Tan, D. Ye, W. Wang, An analysis of influential factors on outdoor thermal comfort in summer. Int. J. Biometeorol. **56**, 941–948 (2012)
82. A. Middel, N. Selover, B. Hagen, N. Chhetri, Impact of shade on outdoor thermal comfort—a seasonal field study in Tempe, Arizona. Int. J. Biometeorol. **60**, 1849–1861 (2016)
83. S.B. Ali, S. Patnaik, Thermal comfort in urban open spaces: objective assessment and subjective perception study in tropical city of Bhopal, India. Urban Climate **24**, 954–967 (2018)
84. N. Marialena, E. Salleh, M.Z. Jaafar, A.H. GhaffarianHoseini, A thermal comfort conditions of shaded outdoor spaces in hot and humid climate of Malaysia. Build. Environ. **48**, 7–14 (2012)
85. F. Ali-Toudert, H. Mayer, Effects of asymmetry, galleries, overhanging façades and vegetation on thermal comfort in urban street canyons. Sol. Energy **81**(6), 742–754 (2007)
86. H. Mayer, J. Holst, P. Dostal, F. Imbery, D. Schindler, Human thermal comfort in summer within an urban street canyon in Central Europe. Meteorol. Z. **17**, 241–250 (2008)
87. H. Lee, H. Mayer, D. Schindler, Importance of 3-D radiant flux densities for outdoor human thermal comfort on clear-sky summer days in Freiburg, Southwest Germany. Meteorol. Z. **23**, 315–330 (2014)
88. N. Metje, M. Sterling, C.J. Baker, Pedestrian comfort using clothing values and body temperatures. J. Wind Eng. Ind. Aerodyn. **96**(4), 412–435 (2008)
89. M.A. Ruiz, E.N. Correa, Adaptive model for outdoor thermal comfort assessment in an oasis city of arid climate. Build. Environ. **85**, 40–51 (2015)
90. A. Matzarakis, H. Mayer, M.G. Iziomon, Applications of a universal thermal index: physiological equivalent temperature. Int. J. Biometeorol. **43**, 76–84 (1999)
91. K. Błażejczyk, UTCI-nowy wskaźnik oceny obciążeń cieplnych człowieka. UTCI new index for assessment of heat stress in man. Przeglad Geogr. **82**, 49–71 (2010)
92. K. Błażejczyk, *BioKlima.* (Universal Tool for Bioclimatic and Thermophysiological Studies, 2017). https://www.igipz.pan.pl/Bioklima-zgik.html. Accessed 4 Nov 2017
93. A. Young, *Universal Thermal Climate Index.* http://www.utci.org/utci_doku.php. Accessed 15 April 2017

# Design Strategies and Implications

# The Mitigating Effect of Shaded Spaces (Including Solar-Induced Shadows from Trees and Buildings) on Thermal Discomfort in Urban Construction

**Abstract** Ground radiation is a critical factor influencing outdoor mean radiant temperature (MRT), yet its variation across different materials and shading conditions remains insufficiently explored. This chapter investigates the dynamic impacts of ground radiation on MRT in shaded and unshaded environments, focusing on four representative urban surfaces: slabstone, asphalt, tile, and lawn. Field measurements conducted from morning to evening revealed that in unshaded areas, ground radiation generally followed the pattern of incident solar radiation, while in shaded areas inconsistencies occasionally emerged. Longwave radiation remained relatively stable across both settings. Across all surfaces, about 50% of ground radiation ($R_{str,\ g}$) was absorbed by the human body. The mean maximum hourly MRT change caused by $R_{str,\ g}$ was 6.4 °C in unshaded conditions and within 2 °C in shaded ones. These results highlight the importance of shaded spaces, whether from trees or buildings, in mitigating thermal discomfort and provide practical guidance for optimizing surface design to improve urban microclimates.

## 1 Introduction

A large body of research has concentrated on how the thermal environment influences individuals' outdoor thermal comfort [1–4]. These works have consistently identified that several physical variables, such as air temperature, relative humidity, wind velocity, and mean radiant temperature (MRT) [5], play important roles, with air temperature and MRT being especially influential [6, 7]. In recent years, investigations into outdoor MRT have gained increasing attention [8–10], as MRT not only strongly shapes human thermal perception outdoors, but can also be substantially adjusted through microclimate interventions, including solar shading strategies [11–14]. Importantly, MRT outdoors is determined by more than just direct solar radiation, which changes over time and distributes unevenly as diffuse radiation; it also incorporates longwave radiation emitted by the surrounding built and natural surfaces that interact with the human body [7, 15]. Since people are generally in close contact

 
Z. Fang et al., *Outdoor Thermal Comfort and Heat Stress Mitigation*, Environmental Science and Engineering, https://doi.org/10.1007/978-981-95-8899-2_15

with the ground, the contribution of ground radiation becomes critical. Previous findings have demonstrated that ground radiation can significantly alter MRT [16, 17], although the degree of influence varies with environmental conditions and ground material types [18].

In outdoor settings, ground materials vary considerably, and the surface temperature distribution is often heterogeneous [8, 19, 20]. The radiation from the ground consists of both shortwave and longwave components, part of which is received and absorbed by the human body [21, 22]. Ground-reflected shortwave radiation results from the reflection of incoming shortwave flux at the surface, making it dependent on the albedo of the material. In contrast, the behavior of longwave radiation near the ground is more complicated [19]. It arises from both the reflection and the emission of longwave energy by the surface; consequently, it is influenced by the ground's emissivity and reflectivity as well as the surface temperature, which itself is controlled by the ground heat balance [23, 24].

Based on radiative transfer principles and the ground heat balance [25–27], both shortwave and longwave radiation from the ground are strongly influenced by boundary conditions as well as the properties of surface materials. Moreover, because incoming radiation changes markedly throughout the day, the intensity of ground radiation also fluctuates, thereby altering its influence on MRT [28, 29]. Under shading, the amount of shortwave radiation reaching the ground is greatly reduced, which in turn modifies both the shortwave and longwave components of ground radiation [11, 30]. On one side, less solar energy is reflected upward from the ground; on the other, the thermal loading of shortwave radiation on the surface is weakened, lowering the ground temperature and consequently decreasing the emitted longwave radiation. As a result, the total radiative energy absorbed by the human body in contact with the ground declines, thereby lowering MRT, which is defined as the uniform blackbody temperature that produces the same radiative heat exchange with a person as the actual environment [31]. In contrast, under direct solar exposure, ground surfaces can heat up considerably, which leads to an increase in longwave emission from the ground [12, 32]. Hence, ground radiation patterns differ substantially between shaded and unshaded contexts. Furthermore, since longwave emission is connected to the thermophysical properties of ground materials, the reflected longwave radiation may vary across surfaces even when boundary conditions are identical [33, 34]. Nevertheless, the diurnal variation characteristics of these ground radiation components remain scarcely explored.

The portion of ground radiation ($R_{str,\ g}$, W/m$^2$) that is absorbed by the human body can be regarded as a parameterized indicator of how ground radiation influences MRT. This absorption depends on the overall intensity of ground radiation, which may vary considerably over time as a result of dynamic interactions between the ground and the surrounding environment. Furthermore, $R_{str,\ g}$ is influenced by the relative contributions of shortwave and longwave components, since the human body exhibits different absorptivity values for each type of radiation. Typically, the absorptivities for shortwave and longwave ($\alpha_k$ and $\alpha_l$, respectively) radiation are about 0.7 and 0.97, respectively [21]. Beyond the magnitude of $R_{str,\ g}$, the impact of $R_{str,\ g}$ on MRT is also associated with the total radiation ($S_{str}$, W/m$^2$) absorbed by the body. Despite the clear importance of ground radiation for MRT, systematic and continuous quantitative evidence remains insufficient.

Therefore, the objective of this research was to evaluate how ground radiation influences outdoor mean radiant temperature by carrying out continuous monitoring of various ground radiation components. The study examined the temporal patterns of radiation variation from morning through evening for different ground surface types. Moreover, the relationship between ground radiation and MRT was further assessed in terms of the changes in $R_{str,\,g}$ and $S_{str}$.

## 2 Materials and Methods

### 2.1 Field Measurements Sites and Method

This investigation was carried out in Guangzhou City, China, a region known for its hot summers and mild winters. The field campaign took place on the campus of Guangzhou University [35] in the Panyu District. As illustrated in Fig. 1 and summarized in Table 1, eight observation sites (P1–P8) were established, covering four common pavement types: slabstone, asphalt, tile, and lawn. For each pavement category, one location was exposed to direct solar radiation (unshaded: P1, P3, P5, and P7), while the paired site was positioned in shaded conditions (P2, P4, P6, and P8). The monitoring was carried out during the winter season, spanning from December 24, 2021, to January 8, 2022. Each site was observed for a full day, with measurements conducted between approximately 08:00 and 17:00 local time.

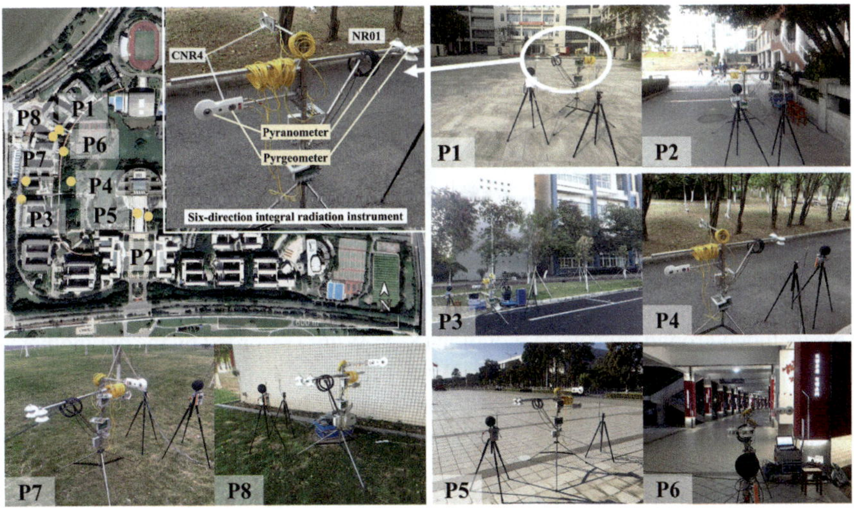


**Fig. 1** Measurement points and the six-direction integral radiation instrument used in this chapter

**Table 1** Measurement date and configuration

| Site no | Measurement date | Presence of shade | Air temperature (°C) | Pavement materials |
|---|---|---|---|---|
| P1 | 2021/12/24 | No | 15–24 | Slabstone |
| P2 | 2022/1/7 | Yes | 7–11 | |
| P3 | 2022/1/11 | No | 12–17 | Asphalt |
| P4 | 2022/1/14 | Yes | 3–7 | |
| P5 | 2022/1/18 | No | 13–17 | Tile |
| P6 | 2022/1/12 | Yes | 3–9 | |
| P7 | 2022/1/16 | No | 13–15 | Lawn |
| P8 | 2022/1/8 | Yes | 7–9 | |

**Table 2** Instrument specifications

| Instrument | | Spectral area | Measurement range | Response time | Field of view | Measurement direction |
|---|---|---|---|---|---|---|
| NR01 | Pyranometer | 0.285–3 μm | 0–2000 W/m$^2$ | 18 s | 180° | Vertical |
| | Pyrgeometer | 4.5–42 μm | – | 18 s | 150° | |
| CNR4 | Pyranometer | 0.3–2.8 μm | 0–2000 W/m$^2$ | 18 s | 180° | Horizontal |
| | Pyrgeometer | 4.5–42 μm | – | 18 s | 180° | |

Radiation measurements were obtained using a six-directional integral radiation system (Fig. 1), which incorporated six pyranometers and six pyrgeometers sourced from NR01 (Hukseflux) and CNR4 (KIPP&ZONEN), as detailed in Table 2. These sensors were oriented vertically (upward and downward) as well as horizontally (front, back, left, and right) to capture shortwave and longwave radiation arriving from each direction. The device was mounted at a height of about 1.1 m, representing the typical center of gravity of the human body, and radiation data were collected at intervals of every 5 s.

## 2.2 *Ground Radiation and MRT*

At each observation site, the readings obtained from the downward-facing sensors of the six-direction radiation system were defined as the ground radiation ($R_g$, W/m$^2$). This included ground shortwave radiation (K $\uparrow$, W/m$^2$) recorded by the pyranometer and ground longwave radiation (L $\uparrow$, W/m$^2$) recorded by the pyrgeometer. In the same way, the signals captured by the upward-facing sensors represented the radiation incident upon the ground surface, consisting of shortwave (K $\downarrow$, W/m$^2$) and longwave (L $\downarrow$, W/m$^2$) components.

A portion of the ground radiation is absorbed by the human body, and $R_{str,\,g}$ depends on $R_g$ together with the view factor ($F_{g-p}$) between the ground and the human body. It was determined through Eq. (1), where $K_{str,\,g}$ ($W/m^2$) and $L_{str,g}$ ($W/m^2$) represent the shortwave and longwave components of ground radiation absorbed by the body, respectively. Prior research has indicated that, for large open ground surfaces, the $F_{g-p}$ value can be taken as 0.5 [19, 36].

$$R_{str,\,g} = K_{str,\,g} + L_{str,\,g} = \alpha_k F_{g-p} K \uparrow + \alpha_l F_{g-p} L \uparrow \tag{1}$$

In addition to $R_{str,\,g}$, the human body also receives radiation ($R_{str,\,o}$, $W/m^2$) from the surrounding environment. Thus, $S_{str}$ can be obtained using Eq. (2). Following the approach proposed by Thorsson [21], MRT is derived through Eq. (3), where σ denotes the Stefan–Boltzmann constant, $\sigma = 5.67 \times 10^{-8}$; $\varepsilon_p$ represents the emissivity of the human body, taken as 0.97; and $K_{hor-i}$ ($W/m^2$) and $L_{hor-i}$ ($W/m^2$) correspond to the shortwave and longwave radiation fluxes from the horizontal plane, respectively, accounting for contributions from the four cardinal directions.

The absolute and relative contributions of $R_{str,\,g}$ to MRT were assessed through Eqs. (4) and (5), respectively. In Eq. (4), $E_D$ (°C) represents the variation in MRT induced by $R_{str,\,g}$, whereas in Eq. (5), $E_R$ (%) denotes the proportion of $R_{str,\,g}$ change within the overall variation of absorbed radiation. For both equations, the variables with subscripts i or i + 1 indicate hourly mean values; for instance, $S_{str,\,8}$ refers to the average of $S_{str}$ calculated from the 5 s interval data collected between 8:00:00 and 8:59:55.

$$S_{str} = R_{str,\,g} + R_{str,\,o} \tag{2}$$

$$\begin{aligned} MRT &= \sqrt[4]{\frac{S_{str}}{\varepsilon_p \cdot \sigma}} - 273.15 \\ &= \sqrt[4]{\frac{0.06(a_k K\downarrow + a_l L\downarrow) + 0.06(a_k K\uparrow + a_l L\uparrow) + 0.22\left(a_k \sum_1^4 K_{hor-i} + a_l \sum_1^4 L_{hor-i}\right)}{\varepsilon_p \cdot \sigma}} \\ &\quad - 273.15 \end{aligned} \tag{3}$$

$$E_D = \left(\sqrt[4]{\frac{S_{str,\,i} + \left(R_{str,\,g,\,i+1} - R_{str,\,g,\,i}\right)}{\varepsilon_p \cdot \sigma}} - 273.15\right) - MRT_i \tag{4}$$

$$E_R = \frac{\left|R_{str,\,g,\,i+1} - R_{str,\,g,\,i}\right|}{\left|R_{str,\,g,\,i+1} - R_{str,\,g,\,i}\right| + \left|R_{str,\,o,\,i+1} - R_{str,\,o,\,i}\right|} \times 100\% \tag{5}$$

# 3 Results

## 3.1 *Variations in Ground Radiation*

### 3.1.1 Unshaded Environment

As illustrated in Fig. 2, K ↓ exhibited pronounced variability caused by intermittent cloud cover, with particularly strong fluctuations observed at P1, P5, and P7. These variations were more evident around midday compared with morning or late-afternoon periods. Even though the measurements were taken in winter, the magnitude of K ↓ remained high, reaching peak values of over 900 $W/m^2$. The temporal patterns of K ↑ and K ↓ were almost the same; however, their magnitudes were markedly lower than those of K ↓, since ground shortwave radiation arises solely from the reflection of incoming shortwave flux. In addition, at P1, P3, P5, and P7, the average values K ↑ were below 71 $W/m^2$, while the maximum K ↑ did not exceed 356 $W/m^2$.

As indicated in Fig. 2, the variations of L ↓ and L ↑ were much smaller compared with those of K ↓ and K ↑. The magnitude of L ↑ was governed by the ground surface temperature and was linked to K ↓ + L ↓. Even though K ↓ showed considerable temporal variation, the corresponding L ↑ fluctuations remained minor. At P1, P5, and P7, L ↑ exhibited only slight changes during the observation period, with standard deviations of 15.1 $W/m^2$, 13.1 $W/m^2$, and 25.8 $W/m^2$, respectively. In contrast, from 11:00 to 17:00, L ↑ rose notably without a comparable increase in L ↓. This occurred because the asphalt surface temperature rose substantially under the influence of high K ↓ values (Fig. 2c). Since L ↑ is primarily dictated by surface temperature, and most pavements possess high emissivity but low reflectivity [37, 38], direct solar exposure can lead to elevated ground surface temperature and consequently large L ↑ values. As shown in Fig. 2, all L ↑ values exceeded those of L ↓, with L ↑/L ↓ ratios ranging between 104.5% and 150.8%.

As depicted in Fig. 3, the pattern of K ↑ + L ↑ closely followed the trend of incident radiation (K ↑ + L ↑), reaching its peak value at midday. During the early hours, ground radiation was nearly equivalent to the incident radiation, whereas in the late afternoon it became greater than the incident flux because of the decline in K ↓. By contrast, it was lower than the incident radiation after the morning period and before dusk. The largest discrepancy between ground and incident radiation occurred around noon, primarily K ↓ since reached its maximum while K ↑ stayed comparatively small owing to the low ground albedo.

### 3.1.2 Shaded Environment

As illustrated in Fig. 4, both the magnitudes and fluctuations of K ↓ and K ↑ were markedly lower in shaded settings compared with unshaded ones. With the exception of P2, the mean K ↓ values at all other sites were below 16 $W/m^2$, while the mean

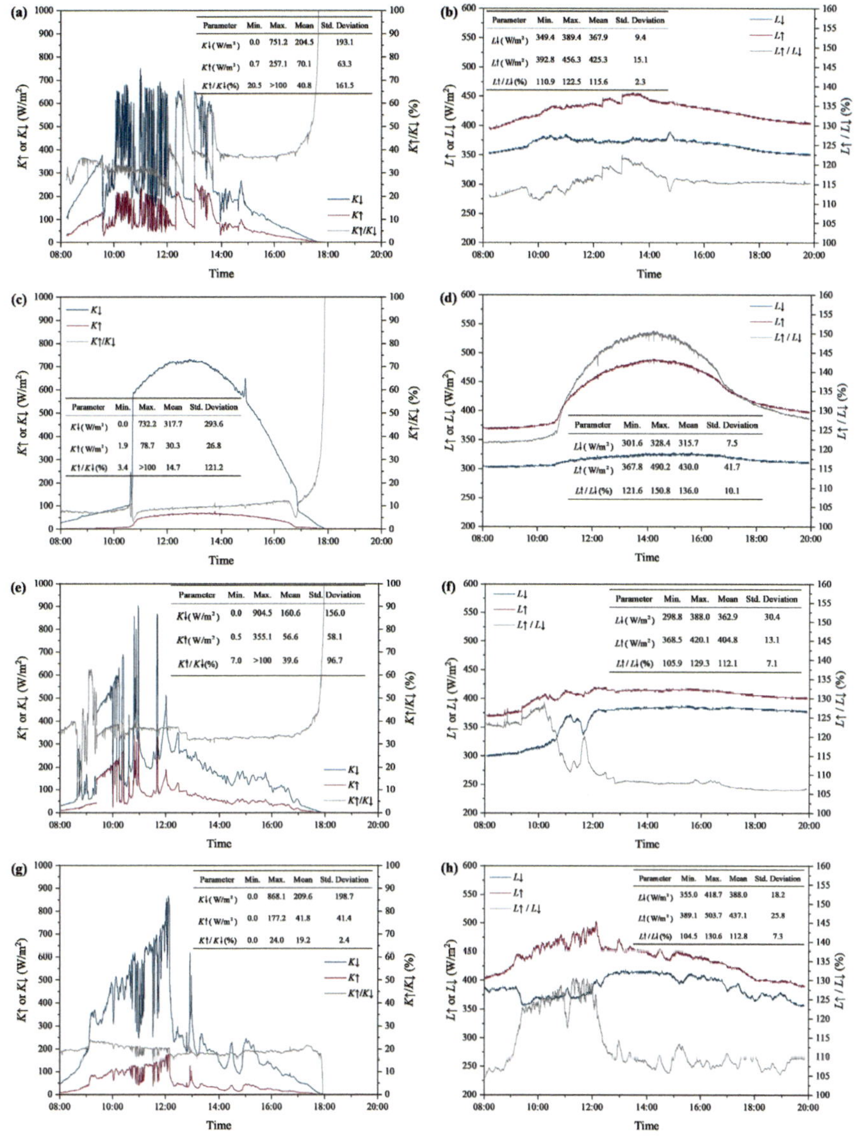


**Fig. 2** Variations in the shortwave and longwave radiations for different ground materials in the unshaded environment: **a** P1 shortwave radiation (Slabstone), **b** P1 longwave radiation (Slabstone), **c** P3 shortwave radiation (Asphalt), **d** P3 longwave radiation (Asphalt), **e** P5 shortwave radiation (Tile), **f** P5 longwave radiation (Tile), **g** P7 shortwave radiation (Lawn), **h** P7 longwave radiation (Lawn)

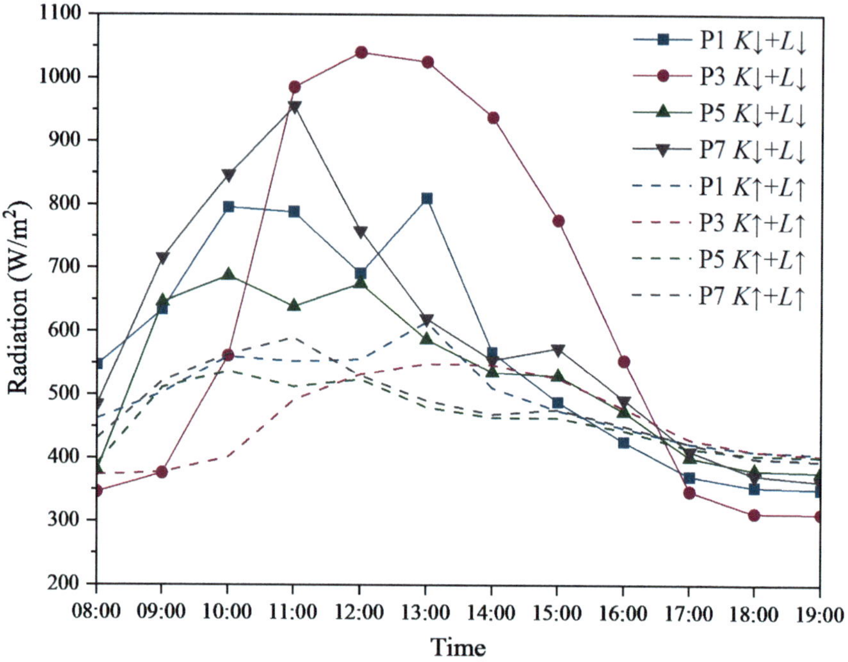


**Fig. 3** Incident ground radiation and ground longwave radiation in the unshaded environment

K ↑ values were under 7 W/m$^2$. The relatively higher levels and variability of K ↓ and K ↑ at P2 and P4, compared with P6 and P8, resulted from partial radiation penetration beneath the tree canopy. In contrast, although direct shortwave input was blocked by adjacent buildings, K ↓ at P6 and P8 mainly originated from lateral directions and represented reflected flux. As evident in Fig. 4, even though K ↓ and K ↑ remained small under shade, L ↓ and L ↑ were comparatively higher. Moreover, the temporal variations of L ↑ aligned well with those of L ↓. Because L ↑ was closer to L ↓, the L ↑/L ↓ ratio in shaded conditions was lower than in unshaded ones, with mean values between 100.2% and 103.4%. Figure 5 further shows that the dynamics of $R_g$ also corresponded to those of K ↓ + L ↓. Additionally, $R_g$ approached K ↓ + L ↓, while in some cases such as P6, $R_g$ even exceeded K ↓ + L ↓. Notably, the maximum $R_g$ did not necessarily appear at midday, likely due to shading effects on K ↓ + L ↓.

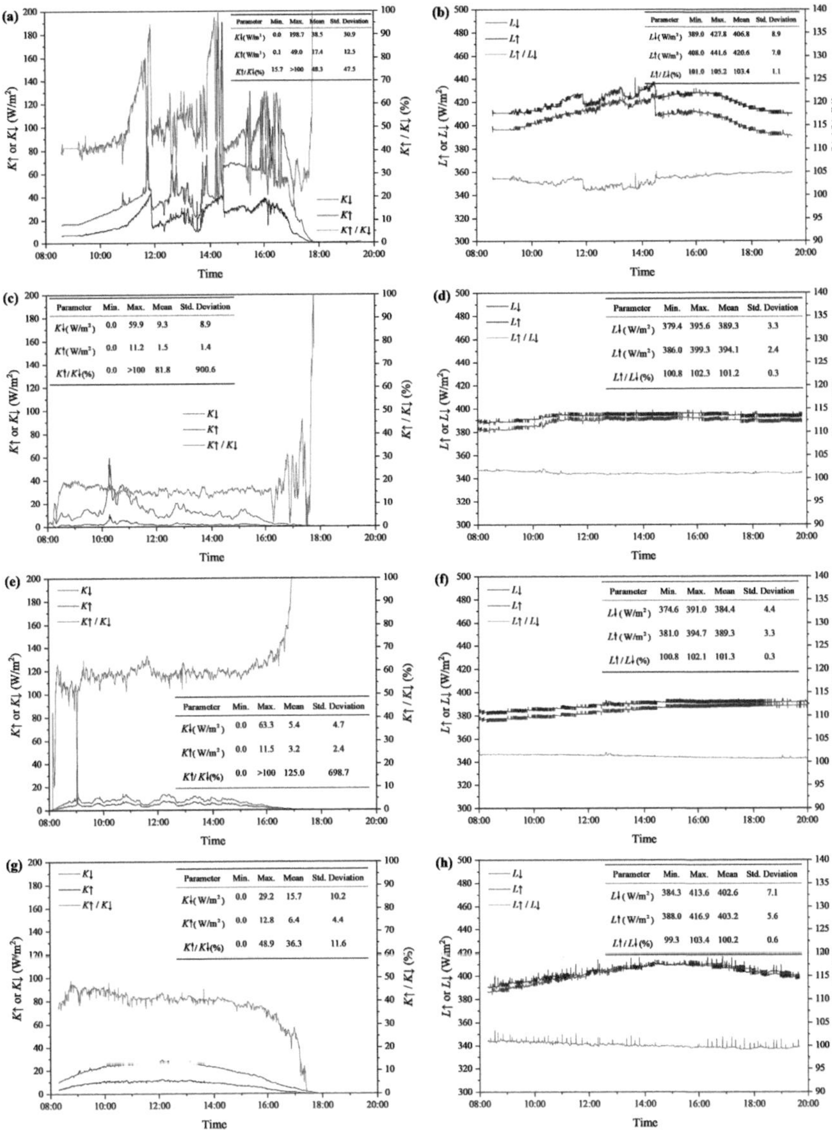


**Fig. 4** Variations in the shortwave and longwave radiation for different ground materials in the shaded environment: **a** P2 shortwave radiation (Slabstone), **b** P2 longwave radiation (Slabstone), **c** P4 shortwave radiation (Asphalt), **d** P4 longwave radiation (Asphalt), **e** P6 shortwave radiation (Tile), **f** P6 longwave radiation (Tile), **g** P8 shortwave radiation (Lawn), **h** P8 longwave radiation (Lawn)

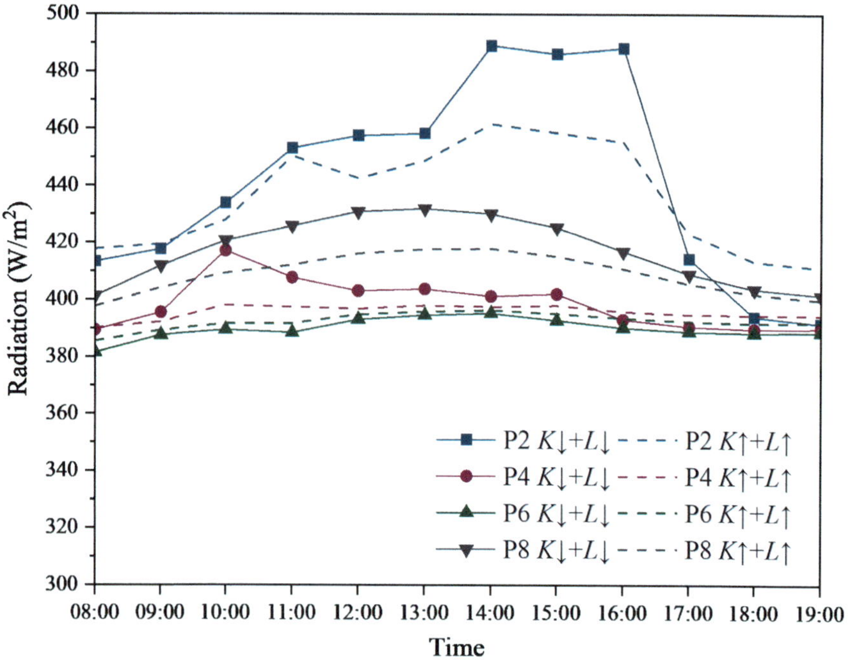


**Fig. 5** Incident ground radiation and ground longwave radiation in the shaded environment

## *3.2 Effect of Different Ground Materials on MRT*

### 3.2.1 Absorbed Ground Radiation in Incident Radiation

$R_{str,\,g}$ was directly proportional to $R_g$, thus linking $K\downarrow + L\downarrow$ with the type of ground surface material. As presented in Fig. 6a, the variation patterns of $R_{str,\,g}/(K\downarrow + L\downarrow)$ for unshaded pavements composed of different materials were alike, showing higher values in the morning and evening but lower ones around noon. Throughout the day, the $R_{str,\,g}/(K\downarrow + L\downarrow)$ values at P1, P5, and P7 were nearly consistent, falling between 28.3% and 56.2%. Compared with the $R_{str,\,g}/(K\downarrow + L\downarrow)$ value of P2, where ranged from 23.4% to 63.5%, the values were relatively smaller during early and late hours but became larger around midday. The maximum $R_{str,\,g}/(K\downarrow + L\downarrow)$ discrepancy in appeared at 14:00 between P1 and P2, with a difference of 14.5%.

The variations in $R_{str,\,g}$ could stem from the absorbed shortwave ($K_{str,\,g}$, W/m$^2$) or longwave ($L_{str,\,g}$, W/m$^2$) components, and the proportions of $K_{str,g}$ or $L_{str,g}$ to $R_{str,\,g}$ may vary among different surface materials. According to Fig. 6a, the ratios of $L_{str,\,g}$ to $R_{str,\,g}$ ($L_{str,\,g}/R_{str,\,g}$, %) across various grounds and times were relatively similar, lying within the range of 79.2%–100%. Moreover, the highest mean $L_{str,g}/R_{str,\,g}$ was recorded at P3 (95.6%), followed in sequence by P7 (94.1%), P5 (91.2%), and P1 (90.2%).

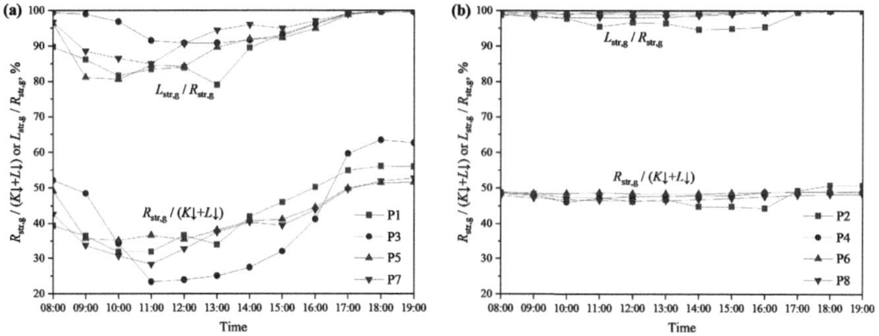


**Fig. 6** Percentage of ground radiation absorbed by the human body to incident ground radiation in the unshaded and shaded environments: **a** unshaded environment, **b** shaded environment

As depicted in Fig. 6b, variations of $R_{str,\,g}/(K\downarrow + L\downarrow)$ at P2, P4, P6, and P8 under shaded conditions were relatively limited, ranging between 44.5% and 50.9%. At P2, however, $R_{str,\,g}/(K\downarrow + L\downarrow)$ showed notable afternoon fluctuations because the tree canopy created shading, yet the upward sensors were intermittently exposed to direct solar beams. This caused higher $K\downarrow$ values (see Fig. 4b) and consequently reduced $R_{str,\,g}/(K\downarrow + L\downarrow)$ values. In addition, $R_{str,\,g}/(K\downarrow + L\downarrow)$ rose again in the evening at P2 owing to the relatively large $L\uparrow$. The greatest $R_{str,\,g}/(K\downarrow + L\downarrow)$ difference among P2, P4, P6, and P8 was 4.4%, observed between P2 and P6 at 16:00. Furthermore, as illustrated in Fig. 6b, the ratios of $L_{str,\,g}/R_{str,\,g}$ (94.7–100%) across different surfaces under shade were close to those under unshaded conditions, likely because $K\downarrow$ was smaller in shaded than in unshaded scenarios.

### 3.2.2 Absorbed Ground Radiation in Total Absorbed Radiation

The influence of $R_{str,\,g}$ on $S_{str}$ or MRT varied across different pavement materials. Under unshaded conditions, the temporal changes of $R_{str,\,g}$ followed the same trend as $S_{str}$, with larger values observed around midday and in the afternoon, as illustrated in. The ranges of in $R_{str,\,g}$ in $S_{str}$ ($R_{str,\,g}/S_{str}$, %) were comparable among P1, P3, P5, and P7, being 43.0–51.6%, 40.5–55.0%, 42.7–50.8%, and 43.3–50.2%, respectively. The corresponding mean $R_{str,\,g}/S_{str}$ values were 48.0%, 46.3%, 48.4%, and 47.2%. In addition, all $R_{str,\,g}/S_{str}$ ratios became higher during the evening because of the relatively stronger reduction in $S_{str}$.

Based on Eqs. (4) and (5), Fig. 8 illustrates the hourly influence of $R_{str,\,g}$ on MRT during the observation period. For the unshaded cases (P1, P3, P5, and P7), the $E_D$ of all four surface materials varied considerably throughout the daytime but showed only slight fluctuations in the evening. This pattern occurred because $R_{str,\,g}$ underwent larger changes in response to variations in $K\downarrow$ during the day. The maximum absolute $E_D$ values reached 5.4 °C, 6.4 °C, 7.9 °C, and 6.0 °C at P1, P3, P5, and P7, respectively. Likewise, $E_R$ also displayed marked variations, with peak $E_R$

values of 86.2%, 84.6%, 64.2%, and 44.8% at P1, P3, P5, and P7, and corresponding mean $E_R$ values of 48.2%, 32%, 34.4%, and 35.8%. From and Fig. 8, it is evident that $E_R$ was generally smaller than $R_{str,\,g}/S_{str}$, suggesting that while $R_{str,\,g}$ has a notable impact on $S_{str}$, $\Delta R_{str,\,g}$ is not the dominant source of $\Delta S_{str}$, since L ↑ changes more slowly over time as a result of ground thermal inertia.

As illustrated in Fig. 7, both $S_{str}$ and $R_{str,\,g}$ showed smaller variations in shaded settings (P2, P4, P6, and P8) compared with unshaded ones. Because $S_{str}$ and $R_{str,\,g}$ values remained relatively stable, the fluctuations of $R_{str,\,g}/S_{str}$ were also limited under shade. At P2, P4, P6, and P8, the $R_{str,\,g}/S_{str}$ values ranged between 44.5–50.5%, 49.0–49.9%, 49.5–49.9%, and 47.8–49.2%, respectively. Nevertheless, the average $R_{str,\,g}/S_{str}$ in shaded conditions was nearly the same as that under unshaded conditions. The mean/values for P2, P4, P6, and P8 were 48.0%, 49.6%, 49.7%, and 48.5%, with the largest difference of 3.2% observed between P3 and P4. As shown in Fig. 8, $E_D$ exhibited only minor changes (− 2.0–1.5 °C) across the four shaded ground materials. Furthermore, since $R_{str,\,o}$ changed only slightly, even small changes $R_{str,\,g}$ could lead to relatively large $E_R$ values. The maximum $E_R$ values at P2, P4, P6, and P8 were 99.1%, 91.1%, 89.2%, and 92.3%, while the corresponding mean values were 33.8%, 26.5%, 48.9%, and 36.4%. From Figs. 7 and 8, it can also be seen that $E_R$ was generally smaller than $R_{str,\,g}/S_{str}$ at most times.

# 4 Discussion

## 4.1 *Characteristics of Ground Radiation*

In unshaded settings, the variation of $R_g$ closely corresponded to K ↓ + L ↓ for the different surface materials. In contrast, under shaded conditions, $R_g$ did not follow the pattern of K ↓ + L ↓. Throughout the day, K ↑ changed considerably, whereas L ↑ remained steady. Moreover, although K ↓ + L ↓ dropped sharply in the evening for both shaded and unshaded cases, L ↑ did not decline significantly because of the influence of ground thermal inertia [26]. The observed trends of K ↑ and L ↑ agree with findings reported in earlier studies [16, 19, 39]. Hence, the primary driver of $R_g$ variation was K ↑, and its behavior shifted with K ↓ changes. Under shaded conditions, as radiation K ↓ was blocked by buildings or tree canopies, $R_g$ showed only slight daytime variation. Compared with unshaded conditions, both K ↑ and L ↑ were reduced under shade, particularly K ↑. This indicates that shading can effectively lower human heat stress [40, 41]. Such radiation variation patterns are valuable for guiding microclimate optimization during the day. For instance, shading can reduce both solar and ground radiation simultaneously. Furthermore, the contribution of ground longwave radiation to human thermal load in the evening deserves additional consideration.

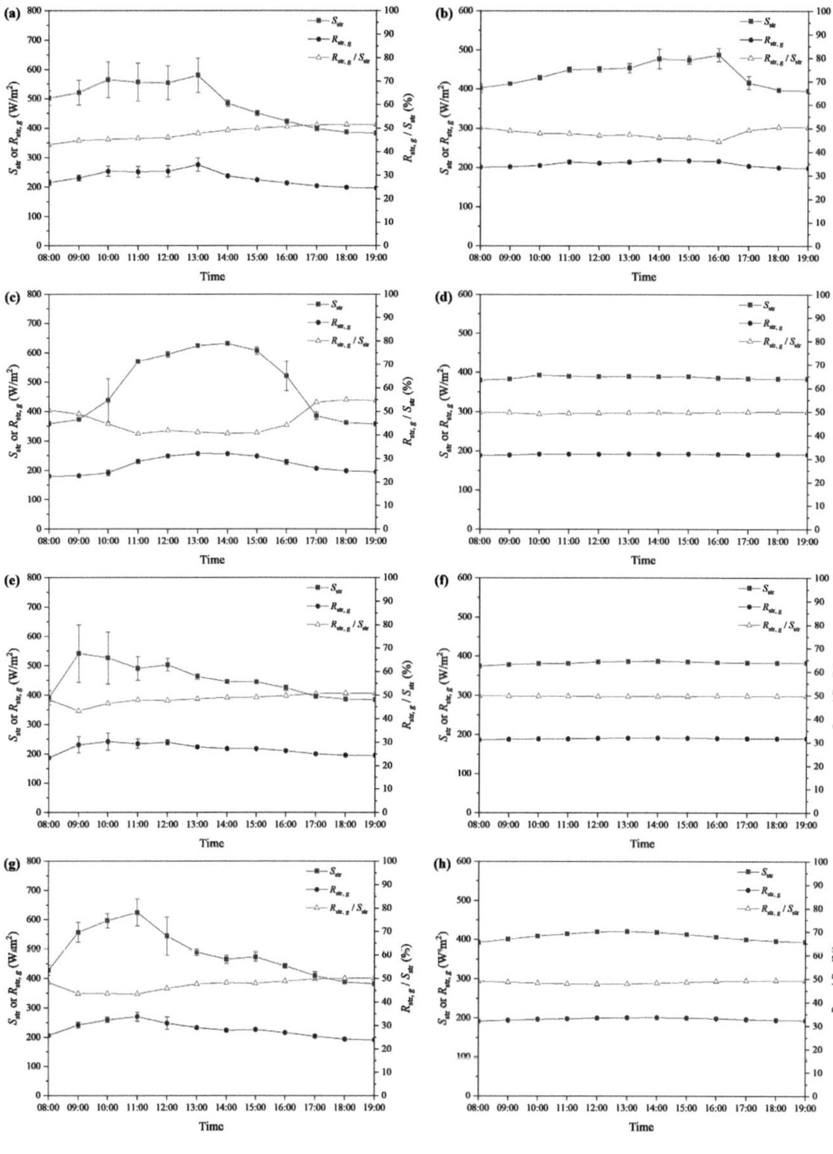


**Fig. 7** Total and ground radiation absorbed by the human body in the unshaded and shaded environments: **a** P1 (Unshaded slabstone), **b** P2 (Shaded slabstone), **c** P3 (Unshaded asphalt), **d** P4 (Shaded asphalt), **e** P5 (Unshaded tile), **f** P6 (Shaded tile), **g** P7 (Unshaded lawn), **h** P8 (Shaded lawn)

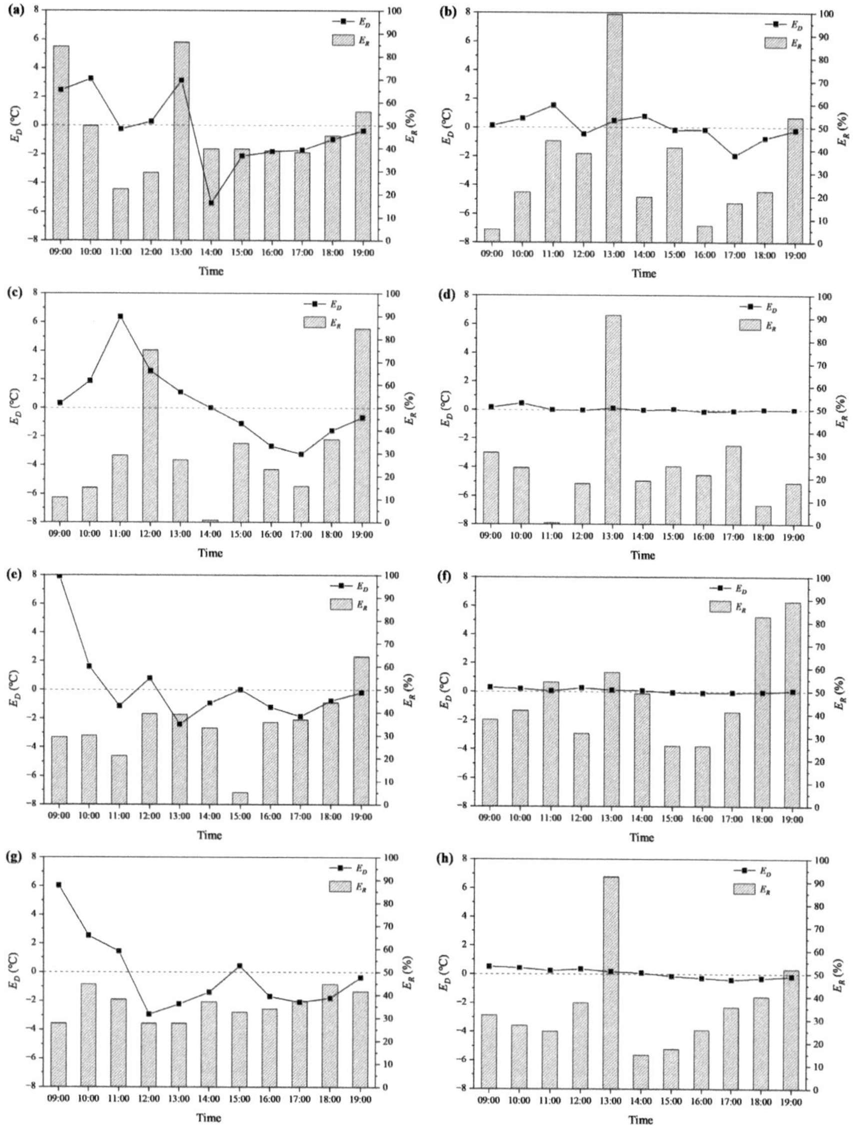


**Fig. 8** Effects of absorbed ground radiation on MRT in the unshaded and shaded environments: **a** P1 (Unshaded slabstone), **b** P2 (Shaded slabstone), **c** P3 (Unshaded asphalt), **d** P4 (Shaded asphalt), **e** P5 (Unshaded tile), **f** P6 (Shaded tile), **g** P7 (Unshaded lawn), **h** P8 (Shaded lawn)

## 4.2 Absorbed Ground Radiation

The findings demonstrated that $R_{str,\,g}$ exhibited comparable characteristics. Despite examining different pavement materials under varying weather conditions, the variation patterns of $R_{str,\,g}/(K\downarrow + L\downarrow)$ remained alike. Because of the relatively smaller $\alpha_k$, $K\downarrow$ had little influence on $K_{str,g}$ or $R_{str,\,g}$, yet it could notably alter $R_{str,\,g}/(K\downarrow + L\downarrow)$. In contrast, due to the larger $L\uparrow$ and $\alpha_l$, $R_{str,\,g}$ was predominantly controlled by $L\uparrow$, with $L_{str,g}/R_{str,\,g}$ values exceeding 79% and 94% in unshaded and shaded cases, respectively. These outcomes are consistent with earlier research [19, 36]. Therefore, lowering $R_{str,\,g}$ through the reduction of $L\uparrow$ and $\alpha_l$ may represent an effective approach. Moreover, $L\uparrow$ was largely dependent on surface temperature of the ground, which can be reduced by selecting alternative materials, while $\alpha_l$ could be mitigated by adjusting human clothing properties.

The variation in $R_{str,\,g}$ closely followed the $S_{str}$ trends in both shaded and unshaded conditions, and $R_{str,\,g}/S_{str}$ remained high at nearly 50%, suggesting that $R_{str,\,g}$ had a strong influence on MRT. Furthermore, fluctuations in $R_{str,\,g}$ could substantially affect $S_{str}$. In unshaded areas, this produced considerable MRT differences, with the four ground materials showing an average maximum hourly change of 6.4 °C. In contrast, the effect was weaker in shaded areas because $R_{str,\,g}$ varied only slightly. Under shade, both $S_{str}$ and $R_{str,\,g}$ stayed relatively steady from morning to evening. These findings indicate that ground radiation, modulated by surface materials, plays an important role in shaping MRT. Nevertheless, meteorological conditions remain the primary driver of MRT. As illustrated in Fig. 7, $S_{str}$ (which determines MRT) shifted throughout the day with weather changes such as air temperature and incident radiation, while the relative MRT levels among different ground materials were also influenced by these weather variations.

## 4.3 Limitations and Future Research Directives

This chapter examined the characteristics of ground radiation variations and their influence on MRT, yet several limitations remain for future improvement. First, the ground radiation estimation simplified the surrounding effects by assuming $F_{g-p} = 0.5$. In reality, radiation measured at a point may also include contributions from nearby façades, meaning that $K\uparrow$ and $L\uparrow$ are not exclusively from the ground surface and $F_{g-p}$ may deviate from 0.5. Although $K\uparrow$ and $L\uparrow$ actual ground radiation still dominates and because of the large $F_{g-p}$ [19], the role of adjacent surfaces should be incorporated in subsequent analyses. Second, the radiation performance of different materials was assessed under varying weather conditions, which makes it difficult to separate their quantitative impacts on MRT. Ground temperature is controlled not only by incident, absorbed, reflected, and emitted radiation but also by meteorological factors such as wind and atmospheric humidity. Thus, future work should evaluate different ground materials under uniform boundary conditions, or perform simultaneous radiation measurements for direct comparison. This would also allow development of targeted optimization strategies for material selection. Finally, seasonal variability should be considered, since both the demand for ground radiation and its influence on MRT change with seasonal climatic conditions.

# 5 Conclusions

Radiation measurements were performed for four ground materials in unshaded and shaded environments. The changes in $R_g$, including K ↑ and L ↑, were analyzed at each study site from 8:00 to 20:00 (UTC + 8). $R_{str,\,g}/(K\downarrow + L\downarrow)$ was calculated to analyze the characteristic and variable mechanism of $R_{str,\,g}$. Subsequently, the direct and relative effects of $R_{str,\,g}$ on MRT were analyzed using $E_D$ and $E_R$, respectively. The main findings were as follows:

(1) In the unshaded environment, the changes in $R_g$ and K ↓ + L ↓ were consistent, but were occasionally inconsistent in the shaded environment. The changes in L ↑ were relatively stable in both unshaded and shaded environments, and did not decrease significantly due to the reductions in K ↓ + L ↓.
(2) When the incident radiation was the same, the changes in $R_{str,\,g}$ for different ground materials were similar. $R_{str,\,g}$ mainly originated from L ↑, with $L_{str,g}/R_{str,\,g}$ being > 79% and 94% in the unshaded and shaded environments, respectively. Moreover, $R_{str,\,g}$ considerably affected MRT, with $R_{str,\,g}/S_{str}$ being approximately 50% across different ground surfaces. Lastly, the mean maximum MRT change per hour caused by $R_{str,\,g}$ was 6.4 °C in the unshaded environment, but it was within 2 °C in the shaded environment.

**Acknowledgements** This work was supported by the National Natural Science Foundation of China (Project No. 51978180), Guang Dong Basic and Applied Basic Research Foundation (2021A1515011671).

# References

1. Z. Fang, X. Feng, J. Liu, Z. Lin, C.M. Mak, J. Niu, K.-T. Tse, X. Xu, Investigation into the differences among several outdoor thermal comfort indices against field survey in subtropics. Sustain. Cities Soc. **44**, 676–690 (2019)
2. Z. Fang, X. Feng, X. Xu, X. Zhou, Z. Lin, Y. Ji, Investigation into outdoor thermal comfort conditions by different seasonal field surveys in China, Guangzhou. Int. J. Biometeorol. **63**(10), 1357–1368 (2019)
3. B. Sayad, D. Alkama, H. Ahmad, J. Baili, N.H. Aljahdaly, Y. Menni, Nature-based solutions to improve the summer thermal comfort outdoors. Case Stud. Therm. Eng. **28**, 101399 (2021)
4. R. Zhu, X. Zhang, L. Yang, Y. Liu, Y. Cong, W. Gao, Correlation analysis of thermal comfort and physiological responses under different microclimates of urban park. Case Stud. Therm. Eng. **34**, 102044 (2022)

5. Z. Zheng, Y. Zhang, Y. Mao, Y. Yang, C. Fu, Z. Fang, Analysis of SET* and PMV to evaluate thermal comfort in prefab construction site offices: case study in South China. Case Stud. Therm. Eng. **26**, 101137 (2021)
6. X. Feng, Z. Zheng, Y. Yang, Z. Fang, Quantitative seasonal outdoor thermal sensitivity in Guangzhou, China. Urban Clim. **39**, 100938 (2021)
7. N. Kántor, J. Unger, The most problematic variable in the course of human-biometeorological comfort assessment—the mean radiant temperature. Cent. Eur. J. Geosci. **3**(1), 90–100 (2011)
8. H. Guo, D. Aviv, M. Loyola, E. Teitelbaum, N. Houchois, F. Meggers, On the understanding of the mean radiant temperature within both the indoor and outdoor environment, a critical review. Renew. Sustain. Energy Rev. **117**, 109207 (2020)
9. C.V. Gál, N. Kántor, Modeling mean radiant temperature in outdoor spaces, a comparative numerical simulation and validation study. Urban Clim. **32**, 100571 (2020)
10. H. Lee, S. Jo, S. Park, A simple technique for the traditional method to estimate mean radiant temperature. Int. J. Biometeorol. **66**(3), 521–533 (2022)
11. A. Speak, L. Montagnani, C. Wellstein, S. Zerbe, The influence of tree traits on urban ground surface shade cooling. Landsc. Urban Plan. **197**, 103748 (2020)
12. M.A. Rahman, V. Dervishi, A. Moser-Reischl, F. Ludwig, H. Pretzsch, T. Rötzer, S. Pauleit, Comparative analysis of shade and underlying surfaces on cooling effect. Urban For. Urban Green. **63**, 127223 (2021)
13. S.E.O. Oliveira, C.C. de Melo Costa, M. Chiquitelli Neto, F.A. Dalla Costa, A.S.C. Maia, Effects of shade location and protection from direct solar radiation on the behavior of Holstein cows. Int. J. Biometeorol. **63**(11), 1465–1474 (2019)
14. N. Karvatte, E.S. Klosowski, R.G. de Almeida, E.E. Mesquita, C.C. de Oliveira, F.V. Alves, Shading effect on microclimate and thermal comfort indexes in integrated crop-livestock-forest systems in the Brazilian Midwest. Int. J. Biometeorol. **60**(12), 1933–1941 (2016)
15. W. Yao, Z. Li, X. Wang, Q. Zhao, Z. Zhang, L. Lin, A simplified model exploration research of new anisotropic diffuse radiation model. Energy Convers. Manage. **126**, 724–735 (2016)
16. F. Lindberg, S. Onomura, C. Grimmond, Influence of ground surface characteristics on the mean radiant temperature in urban areas. Int. J. Biometeorol. **60**(9), 1439–1452 (2016)
17. M. Taleghani, The impact of increasing urban surface albedo on outdoor summer thermal comfort within a university campus. Urban Clim. **24**, 175–184 (2018)
18. H. Zhao, G. Xu, Y. Shi, J. Li, Y. Zhang, The characteristics of dynamic and non-uniform thermal radiation experienced by pedestrians in a street canyon. Build. Environ. **222**, 109361 (2022)
19. Z. Li, X. Feng, W. Chen, Z. Fang, Quantifying the effect of ground view factor and ground temperature on outdoor mean radiant temperature. Sustain. Cities Soc. **84**, 104030 (2022)
20. H. Zhao, Y. Zhang, Method of spherical triangle-division for solving view factors in built environment and its application for non-uniform thermal radiation environments. Build. Environ. **222**, 109360 (2022)
21. S. Thorsson, F. Lindberg, I. Eliasson, B. Holmer, Different methods for estimating the mean radiant temperature in an outdoor urban setting. Int. J. Climatol.: J. R. Meteorol. Soc. **27**(14), 1983–1993 (2007)
22. Z. Li, X. Feng, X. Fan, J. Sun, Z. Fang, Effect of direct solar projected area factor on outdoor thermal comfort evaluation: a case study in Shanghai, China. Urban Clim. **41**, 101033 (2022)
23. W. Morrison, S. Kotthaus, S. Grimmond, Urban surface temperature observations from ground-based thermography: intra-and inter-facet variability. Urban Clim. **35**, 100748 (2021)
24. A. Zajch, W.A. Gough, Seasonal sensitivity to atmospheric and ground surface temperature changes of an open earth-air heat exchanger in Canadian climates. Geothermics **89**, 101914 (2021)
25. J.R. Howell, M.P. Mengüç, K. Daun, R. Siegel, Thermal Radiation Heat Transfer, Seventh ed. (CRC Press, 2020)
26. Y. Qin, J.E. Hiller, Understanding pavement-surface energy balance and its implications on cool pavement development. Energy Build. **85**, 389–399 (2014)
27. Y. Xu, Z. Li, H. Liu, M. Jia, Q. Wang, M. Zhang, Y. Xu, Modeling of the dynamic behaviors of heat transfer during the construction of roadway using moving mesh. Case Stud. Thermal Eng. **26**, 100958 (2021)

28. A. Middel, E.S. Krayenhoff, Micrometeorological determinants of pedestrian thermal exposure during record-breaking heat in Tempe, Arizona: introducing the MaRTy observational platform. Sci. Total. Environ. **687**, 137–151 (2019)
29. J. Du, C. Sun, Q. Xiao, X. Chen, J. Liu, Field assessment of winter outdoor 3-D radiant environment and its impact on thermal comfort in a severely cold region. Sci. Total. Environ. **709**, 136175 (2020)
30. N. Kántor, A. Kovács, Á. Takács, Small-scale human-biometeorological impacts of shading by a large tree. Open Geosci. **8**(1), 231–245 (2016)
31. ASHRAE, Thermal Environmental Conditions for Human Occupancy (ANSI/ASHRAE Standard 55, 2020)
32. M. Napoli, L. Massetti, G. Brandani, M. Petralli, S. Orlandini, Modeling tree shade effect on urban ground surface temperature. J. Environ. Qual. **45**(1), 146–156 (2016)
33. Q. Zeng, J. Cheng, L. Dong, Assessment of the long-term high-spatial-resolution global land surface satellite (GLASS) surface longwave radiation product using ground measurements. IEEE J. Sel. Top. Appl. Earth Obs. Remote. Sens. **13**, 2032–2055 (2020)
34. G. Yan, Z.-H. Jiao, T. Wang, X. Mu, Modeling surface longwave radiation over high-relief terrain. Remote Sens. Environ. **237**, 111556 (2020)
35. P. Tang, Q. Li, Evaluation of the observation methods of outdoor mean radiant temperature in a subtropical city. Build. Environ. **207**, 108462 (2022)
36. Z. Li, X. Feng, Z. Fang, A modified method to measure outdoor mean radiant temperature: comparison between two-hemisphere method and six-direction integral method. Build. Environ. **221**, 109292 (2022)
37. J. Coakley, Reflectance and Albedo, Surface, Encyclopedia of Atmospheric Sciences (2003), pp. 1914–1923
38. H. Li, J. Harvey, A. Kendall, Field measurement of albedo for different land cover materials and effects on thermal performance. Build. Environ. **59**, 536–546 (2013)
39. N. Kántor, C.V. Gál, Á. Gulyás, J. Unger, The impact of façade orientation and woody vegetation on summertime heat stress patterns in a central European square: comparison of radiation measurements and simulations. Adv. Meteorol. (2018)
40. D. Lai, W. Liu, T. Gan, K. Liu, Q. Chen, A review of mitigating strategies to improve the thermal environment and thermal comfort in urban outdoor spaces. Sci. Total. Environ. **661**, 337–353 (2019)
41. S. Teshnehdel, E. Gatto, D. Li, R.D. Brown, Improving outdoor thermal comfort in a steppe climate: effect of water and trees in an Urban Park. Land **11**(3), 431 (2022)

# Practical Strategy to Generate a Cooling Effect for Mitigating Outdoor Heat Stress: Fanned Parasol

**Abstract** Sunshades are widely used to alleviate outdoor heat stress, yet conventional parasols often block airflow and trap heat, limiting their cooling performance. This chapter introduces an innovative "fanned parasol," which integrates a lightweight canopy with a compact fan to provide enhanced relief from thermal discomfort. The device is portable, affordable, and designed for practical outdoor use. Field experiments demonstrated that the cooling benefit of the fanned parasol is more pronounced in the initial stage of operation, when stored cooling energy within the canopy is most active. Compared with a traditional parasol, the fanned version strengthens thermal relief through both direct airflow and indirect cooling effects, effectively counteracting the drawbacks of airflow obstruction. Quantitatively, the indirect fan effect was identified as the dominant mechanism, contributing more than 60% of the overall enhancement in both operational stages. These findings highlight the potential of the fanned parasol as a practical strategy for mitigating outdoor heat stress and improving thermal comfort in hot urban environments.

## 1 Introduction

Outdoor heat stress poses a serious risk to human health, especially under the combined effects of climate change and the urban heat island phenomenon [1]. A thermally comfortable outdoor setting can encourage more outdoor participation [2]; for instance, outdoor activities may rise by 14% if the outdoor thermal environment is improved by 5 °C [3]. This enhancement supports human health and lowers building energy use [4, 5]. Conversely, excessive heat in outdoor conditions worsens health outcomes and elevates energy demand [6]. The strength of urban heat islands can be over 10 °C [7]. Such islands also drive cooling energy needs upward, e.g., by 34% [8], while reducing air-conditioning efficiency by about 25% [7]. Climate change further intensifies this issue, increasing urban heat island intensity by 1.4–5.8 °C and producing more extreme weather phenomena such as heat waves [9]. Overheated outdoor environments induce heat stress that seriously impairs health, manifesting as heat cramps, heat syncope, exhaustion, heat shock, diminished concentration,

 
Z. Fang et al., *Outdoor Thermal Comfort and Heat Stress Mitigation*, Environmental Science and Engineering, https://doi.org/10.1007/978-981-95-8899-2_16

fatigue, confusion, etc. [10]. In severe cases, heat stress may cause death [11]. For instance, the European heatwave of 2003 resulted in 40,000 fatalities [12], while in 2010, Japan recorded 1745 heat-related deaths [13].

Growing attention has been directed toward reducing outdoor heat stress [14, 15], which includes the development of evaluation approaches for outdoor thermal conditions (such as the Universal Thermal Climate Index—UTCI [16], the index combining solar exposure and wind conditions [17], and the adaptive-rational outdoor comfort model [18]), along with monitoring, assessing, and verifying mitigation strategies [19, 20]. Lai et al. [21] conducted a thorough review of outdoor heat stress mitigation strategies and concluded that adjusting urban geometry produced the strongest cooling benefit, followed by adding vegetation, applying reflective or cooling surfaces, and integrating water features. The main reason is that both optimizing city form and planting greenery can effectively provide shading, thereby reducing solar radiation, which remains the most critical factor influencing outdoor heat stress [17].

Shading measures are recognized as effective approaches to alleviate outdoor heat stress because of their ability to block sunlight [22, 23]. From a technical perspective, shade can lower direct solar radiation and surface temperatures, which in turn reduces both air temperature and mean radiant temperature, thereby lessening outdoor heat stress [24]. From a psychological perspective, people generally favor shaded areas as a way to cope with heat. Kim and Brown [25] observed that on hot days, individuals preferred shade along sidewalks whether they were sitting, standing, or walking. Similarly, Thorsson et al. [26] found that when air temperature exceeded 20 °C, about 80% of park visitors preferred to stay in shaded spots.

Using parasols is regarded as a convenient and efficient shading approach [27]. Although their cooling performance is slightly weaker than two other common shading forms—namely shade provided by buildings and trees—the parasol remains the most practical option [28]. For instance, Kusaka et al. [29] measured the parasol's cooling effect, reflected by a 4.4 °C reduction in UTCI, while trees achieved a 5.9 °C reduction. Dietrich [30] also noted that the cooling benefit of parasols was about 2 °C less compared with buildings and trees. Nevertheless, Huang et al. [4] revealed that parasols were more favored, especially among females, than seeking shade from trees or buildings. In Nagoya, Watanabe and Ishii [13] reported that approximately 26% of women used parasols to reduce outdoor heat stress, while in Taiwan, Lin [31] found the proportion to be 38%.

Due to the ease of use of parasols, researchers have attempted to strengthen their cooling performance to better alleviate outdoor heat stress. Golbabaei et al. [32] simulated six differently shaped parasols with COMSOL Multiphysics and observed that parasols could reduce radiant temperature by about 30%, with the conical type providing the greatest cooling benefit. Watanabe and Ishii [33] carried out experiments comparing parasols with various surface colors and materials for heat stress mitigation. Their results indicated that parasols were more effective under sunny weather than cloudy conditions, since the cooling mainly depended on reducing direct solar radiation. Moreover, by carefully selecting surface colors and materials, the cooling performance of parasols could be enhanced by approximately 2 °C [33].

This research presents an innovative design, termed the fanned parasol, aimed at strengthening the cooling performance for alleviating outdoor heat stress. Previous studies mainly examined the conventional parasol, which produces a cooling benefit solely by blocking direct solar radiation [28, 32, 33]. The fanned version incorporates a small fan into the traditional parasol, thus retaining its portability and practicality. In addition, the fan is low-cost and easy to use. Beyond reducing solar exposure, the fan promotes additional air circulation, thereby further improving the cooling impact. Airflow is recognized as a critical factor influencing thermal comfort [17, 34], and increasing it can enhance the cooling sensation [35, 36]. In other words, the fanned parasol integrates two strategies—minimizing direct solar gain and boosting air movement—to relieve outdoor heat stress. Since combining multiple mitigation approaches usually achieves stronger cooling effects than relying on a single measure [37], the fanned parasol is expected to outperform the conventional parasol in mitigating outdoor thermal stress.

In conclusion, the main contributions of this research are (1) to the proposal of a new type of parasol, namely the fanned parasol, which is low-cost, easy to use, and effective in alleviating outdoor heat stress and (2) to the clarification of its mechanisms that enhance cooling compared with a standard parasol. The structure of the remaining paper is arranged as follows. Section 2 introduces the experimental setup and methodology for evaluating the cooling performance of both the conventional and fanned parasols. Sections 3.1–3.5 present the collected meteorological parameters influencing outdoor thermal stress and the quantified cooling outcomes. Section 3.6 discusses the mechanisms explaining why the fanned parasol provides stronger cooling than the traditional parasol, and Sect. 3.7 outlines suggestions for future research. Finally, Section 4 summarizes the major conclusions of this work.

# 2 Methodology

## *2.1 Experiments*

The experiments were carried out on the iHarbour Campus of Xi'an Jiaotong University, located in the western part of Xi'an (Fig. 1). Xi'an lies in the central Guanzhong Plain of northwest China and features a temperate monsoon climate with hot summers. The city receives strong solar radiation, with a daily summer average of 414 W/m$^2$ [38]. Thus, reducing outdoor heat stress is an urgent challenge for Xi'an [39, 40]. Measurements took place in the open square in front of Building 19 on a typical hot day in August 2022. Three scenarios were designed: a reference case without parasols, one using a standard parasol, and one using a fanned parasol (Fig. 1b). The first prototype of the fanned parasol is displayed in Fig. 1d.The fanned parasol, which combines a portable umbrella with a small fan, remains easy to use and inexpensive, with the fan costing only about CN¥20–30. The fan is mounted inside the parasol at the upper section of the handle so that, during use, it is positioned above the user's head to improve air circulation in this region. Stronger airflow near the head effectively cools the body [41]. A perforated net separates the fan from the user, preventing accidental injury from the plastic blades while allowing airflow to pass through to strengthen ventilation. For accurate comparison, aside from the addition of the fan, the fanned parasol's design is identical to that of the regular parasol.

In the benchmark scenario, air temperature, wind speed, relative humidity, and globe temperature were recorded at the head height of a standing person, about 1.7 m above ground level. Watanabe and Ishii [33] recommended measuring at head height since the parasol's cooling influence is most evident in this region. In addition to these meteorological factors, for the two parasol scenarios, the inner surface temperatures of the parasols were also measured to help explain the mechanisms of enhanced cooling, which are discussed in Sect. 3.6. The observation period lasted 110 min, determined by the battery capacity of the fanned parasol. The measurement uncertainties were ± 0.3 °C for air temperature, globe temperature, and inner surface

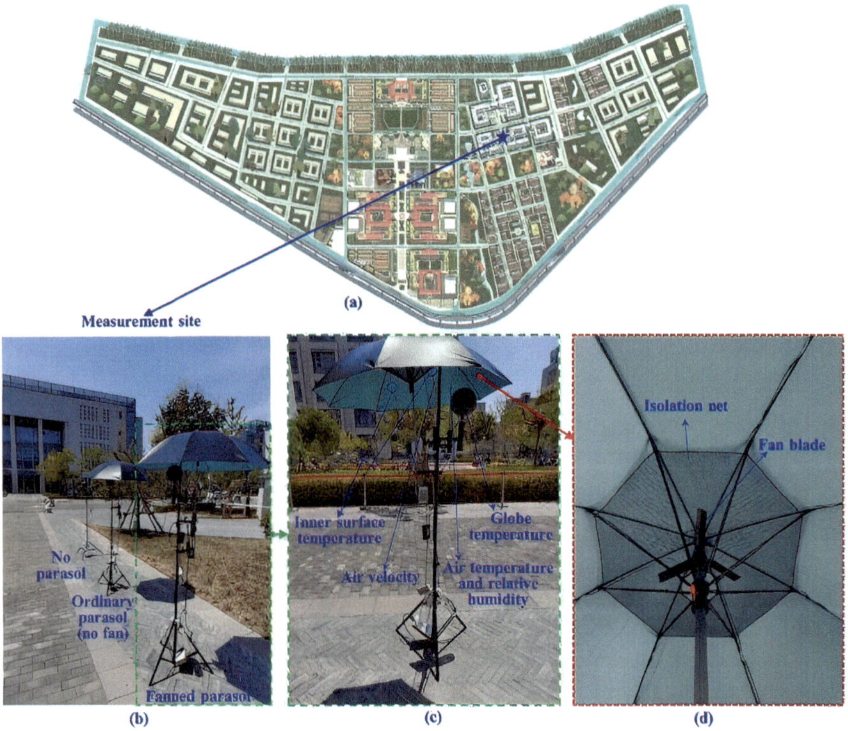


**Fig. 1** Experiment setup: **a** measurement site in iHarbour Campus of Xi'an Jiaotong University, **b** arrangement of parasols, **c** configuration of measurement devices, and **d** structure of fanned parasol

temperature; $\pm 5\% \pm 0.05$ m/s for air velocity; and $\pm 3\%$ for relative humidity. These values are consistent with those commonly reported in previous studies [34, 42].

It should be mentioned that two reasonable simplifications were adopted in the experiments. First, the parasol configuration was kept fixed, which simplifies real-world use but is appropriate for the comparative purpose of this chapter [32, 33]. In practice, people carrying parasols usually move outdoors, and such movement can raise the surrounding wind speed [43]; however, this effect applies equally to both the conventional parasol and the new fanned parasol. Since the objective here is to test the performance of the novel fanned parasol against the ordinary one, using a fixed setup without accounting for human movement does not significantly alter the comparison results, which is consistent with earlier research [32, 33]. For instance, Watanabe and Ishii [33] also employed a fixed setup similar to Fig. 1 when comparing different ordinary parasol designs. Second, the study did not include subjective surveys on occupants' thermal sensation and comfort. Outdoor heat stress involves two key dimensions: the physical aspect of heat exchange between people and the environment, and the subjective aspect of people's perception of that environment [1, 17]. This paper concentrates on the physical aspect by using the Universal Thermal Climate Index (UTCI) to highlight the advantages of the new fanned parasol over the traditional one. It is common for a single research article to emphasize only one aspect of outdoor thermal stress [44–46]; for example, Sadeghi et al. [44], Dhariwal et al. [45], and Ibrahim et al. [46] applied UTCI to evaluate outdoor heat stress objectively.

It should also be emphasized that supplementary experiments were carried out to further confirm the advantages of the novel fanned parasol compared with the conventional one in improving the cooling effect for reducing outdoor heat stress. The outdoor thermal conditions in these supplementary tests differed from those described earlier, but they likewise verified the stronger cooling performance of the fanned parasol over the ordinary parasol. To avoid redundancy and maintain clarity, the details of these supplementary experiments are provided in the Appendix.

## 2.2 *Quantification of Cooling Effect*

The Universal Thermal Climate Index (UTCI) is employed to reflect the combined influence of multiple meteorological factors on outdoor heat stress [47]. Along with the Physiological Equivalent Temperature (PET), UTCI is one of the two most commonly applied thermal indicators for evaluating outdoor thermal stress [48, 49]. In assessing the cooling benefits of shading strategies, Cheung and Jim [50] recommended UTCI over PET because UTCI could clearly distinguish the differences among various shading approaches for mitigating outdoor heat stress, whereas PET could not. Therefore, following previous studies [13, 50, 51], this research adopts UTCI as the primary index to assess outdoor heat stress. The cooling performance is expressed as the decrease in UTCI produced by the ordinary parasol or fanned parasol compared with the benchmark scenario (Eqs. (1, 2)). A greater reduction in UTCI indicates a stronger ability of the parasols to alleviate outdoor heat stress.

$$\mathrm{CE_{OP}} = \mathrm{UTCI_b} - \mathrm{TCI_{OP}} \tag{1}$$

$$\mathrm{CE_{FP}} = \mathrm{UTCI_b} - \mathrm{UTCI_{FP}} \tag{2}$$

Here, $\mathrm{CE_{FP}}$ and $\mathrm{CE_{OP}}$ represent the cooling effects of the fanned parasol and the ordinary parasol, respectively (°C). Meanwhile, UTCIb, UTCIFP, and UTCIOP denote the UTCI values of the benchmark scenario, the fanned parasol, and the ordinary parasol, respectively (°C).

The UTCI, introduced by Brode et al. [52], is expressed as a one-dimensional equivalent temperature that reflects human physiological responses to complex outdoor thermal environments. To derive these physiological reactions, such as core body temperature, skin temperature, skin wettedness, and sweat rate, the UTCI-Fiala multi-node model is applied based on the human body's heat balance [18, 53]. Computing UTCI requires several meteorological inputs, namely air temperature, wind speed, relative humidity, and mean radiant temperature [54]. The mean radiant temperature refers to the temperature of an idealized uniform black enclosure in which the radiant heat exchange of the body matches that of the actual surroundings [55]. It is derived from the measured air temperature, air velocity (Eq. (3) [13]), and globe temperature (Eq. (4) [56]). The UTCI values can be conveniently calculated using the RayMan software [48].

$$v_{a,10\mathrm{m}} = v_{a,h}\frac{\log\left(\frac{10}{0.01}\right)}{\log\left(\frac{h}{0.01}\right)} \tag{3}$$

$$T_{mrt} = \left[\left(T_g + 273\right)^4 + \frac{1.1\times 10^8 {v_{a,10\mathrm{m}}}^{0.5}}{\varepsilon_g D^{0.4}}\left(T_g - T_a\right)\right]^{0.25} - 273 \tag{4}$$

Where D is the diameter of the black globe thermometer, i.e., 0.150 m in this chapter; h is the height above the ground for the measurement of the air velocity (m); $T_a$ is the air temperature (°C); $T_g$ is the globe temperature (°C); $v_a$, 10 m is the air velocity at the height of 10 m above the ground (m/s); $v_a$, h is the measured air velocity at the height of hm above the ground (m/s); $\varepsilon_g$ is the emissivity of the black globe thermometer, i.e., 0.95.

# 3 Result

## *3.1 Measured Air Velocity*

Figure 2a illustrates the recorded air velocities for the benchmark case, the ordinary parasol, and the fanned parasol. The pronounced fluctuations in the data reveal the complexity of the outdoor wind environment. For the benchmark scenario, air velocity ranged from 0.29 to 2.62 m/s with an average of 1.09 m/s (Table 1). In contrast, the ordinary parasol reduced airflow, especially lowering the peak values, with velocities between 0.22 and 1.32 m/s and a mean of 0.59 m/s. This reduction can be attributed to the wind-blocking effect of the parasol. By integrating a fan, the fanned parasol promoted airflow and effectively counteracted this blocking effect. Its velocity values ranged from 0.68 to 1.99 m/s, averaging 1.34 m/s. Relative to the benchmark, the fanned parasol prevented excessively high wind speeds while improving the mean value, whereas the ordinary parasol reduced the average air velocity by 0.50 m/s and the fanned parasol increased it by 0.25 m/s.

## *3.2 Measured Air Temperature*

Figure. 2b presents the recorded air temperatures for the benchmark case, the ordinary parasol, and the fanned parasol. In the benchmark scenario, air temperature showed an upward trend since the tests were carried out in the morning, with minimum, maximum, and mean values of 37.3 °C, 41.2 °C, and 39.3 °C, respectively (Table 1). For the ordinary parasol, the temperature was initially lower than the benchmark but later exceeded it, reaching 36.9 and 41.2 °C as the minimum and maximum, and yielding a mean of 39.2 °C, which is nearly the same as the benchmark. The fanned parasol consistently maintained lower temperatures compared with both the benchmark and ordinary parasol, with minimum, maximum, and average values of 36.5 °C, 40.5 °C, and 38.4 °C, respectively. Relative to the benchmark case, the ordinary parasol lowered the mean air temperature by 0.12 °C, while the fanned parasol reduced it by 0.94 °C.

## *3.3 Measured Relative Humidity*

The changes in relative humidity for the three scenarios show an opposite trend to air temperature (Fig. 2c). In the benchmark case, relative humidity gradually declined, with minimum, maximum, and mean values of 43.0%, 51.5%, and 47.5%, respectively (Table 1). For the ordinary parasol, humidity started out higher than the benchmark and later dropped below it, with corresponding minimum and maximum values of 42.0 and 53.9%. Consequently, the mean relative humidity of 47.7% was

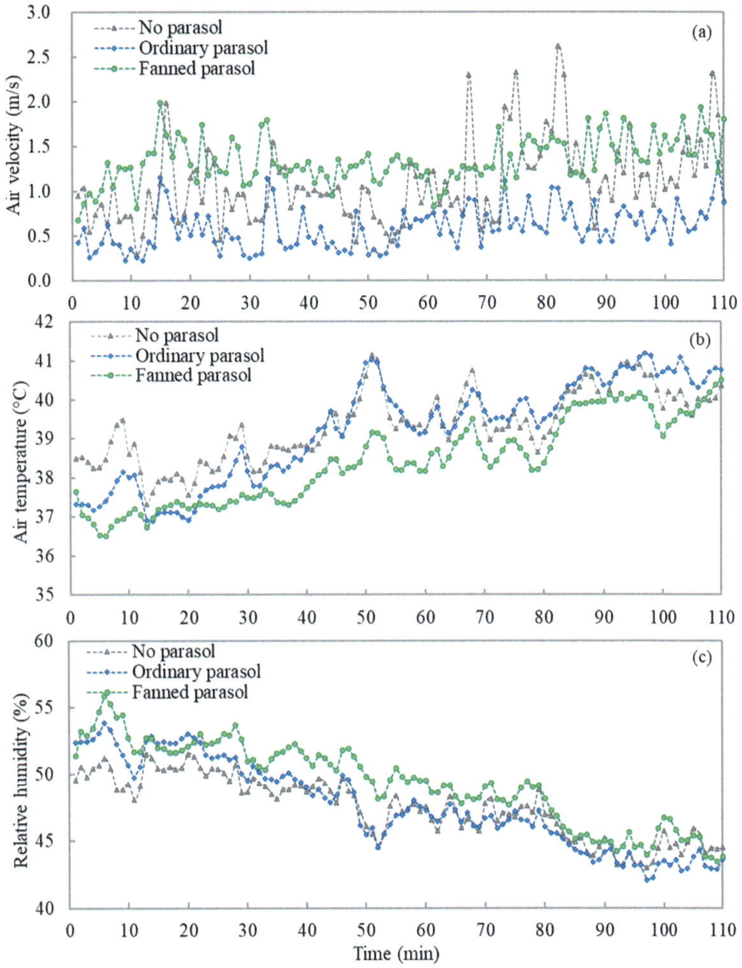


**Fig. 2** Measurements: **a** air velocity, **b** air temperature, and **c** relative humidity

nearly the same as that of the benchmark. In contrast, the fanned parasol generally maintained higher humidity compared with both the ordinary parasol and the benchmark, with minimum, maximum, and average values of 43.5%, 55.9%, and 49.3%, respectively. On average, the ordinary parasol increased relative humidity by 0.2% and the fanned parasol by 1.8%, compared with the benchmark case.

**Table 1** Measurements: uncertainty and mean value

| | Air velocity (m/s) | Air temperature (°C) | Relative humidity (%) | Globe temperature Stage I (°C) | Globe temperature Stage II (°C) |
|---|---|---|---|---|---|
| Measuring uncertainty | ± 5% ± 0.05 m/s | ± 0.3 °C | ± 3% | ± 0.3 °C | ± 0.3 °C |
| Mean value no parasol | 1.09 | 39.3 | 47.5 | 46.0 | 49.1 |
| Ordinary parasol | 0.59 | 39.2 | 47.7 | 35.0 | 43.6 |
| Fanned parasol | 1.34 | 38.4 | 49.3 | 33.3 | 41.1 |

## 3.4 Measured Globe Temperature

Figure 3 illustrates that the globe temperatures of the three cases generally rose because of the increasing solar radiation in the morning. The minimum values were 43.4 °C for the benchmark, 29.2 °C for the ordinary parasol, and 27.8 °C for the fanned parasol, while the maximum values reached 51.3 °C, 46.5 °C, and 43.4 °C respectively. Based on the magnitudes of reduction achieved by the parasols relative to the benchmark, the observation period was divided into Stage I with larger reductions and Stage II with smaller ones (Fig. 4). In Stage I (the first 10 min), the ordinary parasol and the fanned parasol reduced globe temperature by more than 8.5 °C and 10 °C, respectively, whereas in Stage II (the rest of the period), the reductions were generally below these thresholds. The distinction between Stage I and Stage II is attributed to the initial cooling energy stored in the parasols, since they had been placed in a 26 °C room prior to the test. The average globe temperatures were 46.0, 35.0, and 33.3 °C in Stage I for the benchmark, ordinary, and fanned parasols, and 49.1, 43.6, and 41.1 °C in Stage II (Table 1). Accordingly, compared with the benchmark, the ordinary parasol reduced the mean globe temperature by 11.1 and 5.5 °C in Stages I and II, while the fanned parasol achieved 12.7 and 8.0 °C reductions in the same stages.

## 3.5 Quantified Cooling Effect

Figure 5a indicates that the UTCIs derived from the measurements of the three cases generally increased. The minimum UTCI values were 44.1 °C for the benchmark, 35.7 °C for the ordinary parasol, and 33.8 °C for the fanned parasol, while the maximum values reached 50.2 °C, 46.8 °C, and 44.6 °C respectively. As shown in Fig. 5b, the UTCIs of both parasols can also be divided into two stages. In Stage I (the initial 10 min), the UTCI reductions relative to the benchmark exceeded 5 °C for the ordinary parasol and 7 °C for the fanned parasol, whereas in Stage II (the remaining time), the reductions were generally below 5 and 7 °C. These two UTCI

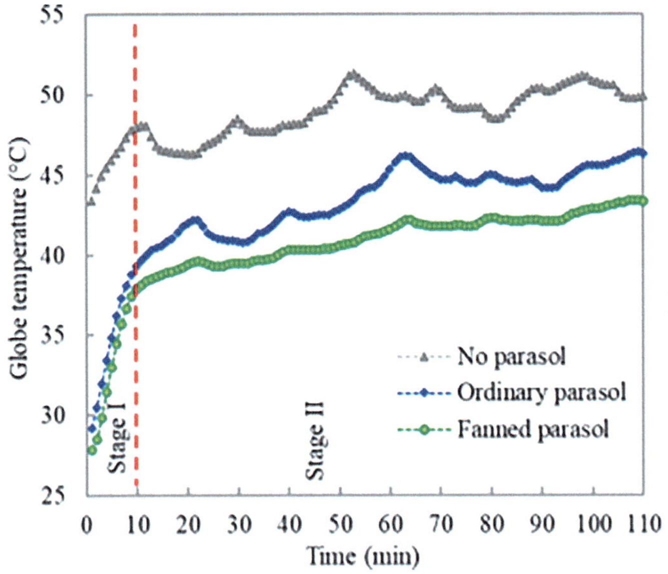


**Fig. 3** Measured globe temperature

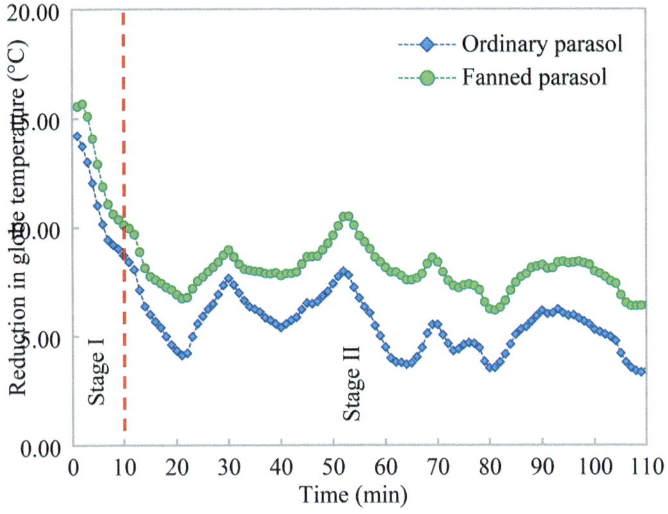


**Fig. 4** Globe temperature reductions by ordinary and fanned parasols compared with benchmark case

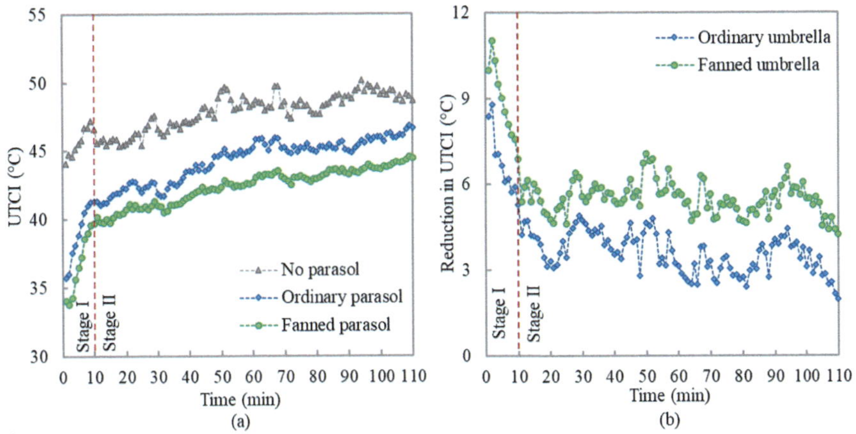


**Fig. 5** **a** Variations of Universal Thermal Comfort Index (UTCI) and **b** UTCI reductions by ordinary and fanned parasols compared with benchmark c

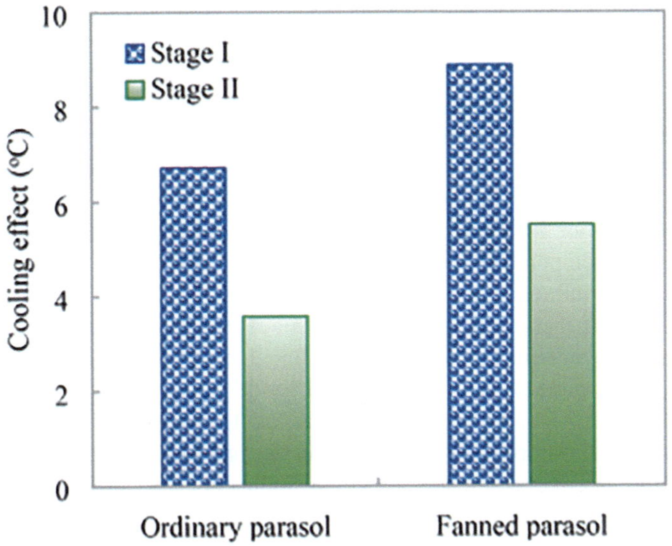


**Fig. 6** Cooling effects of ordinary and fanned parasols

stages correspond to the two stages of globe temperature, which is a critical input for UTCI calculation [17]. The average UTCIs during Stage I were 45.7, 39.0, and 36.8 °C for the benchmark, ordinary, and fanned parasols, while during Stage II the corresponding values were 47.9, 44.4, and 42.4 °C (Fig. 5a). Thus, compared with the benchmark, the ordinary parasol reduced UTCI by 6.7 °C and 3.6 °C in Stages I and II, and the fanned parasol achieved 8.9 °C and 5.5 °C reductions, respectively (Fig. 6). In other words, the cooling effects generated by the ordinary parasol were

6.7 and 3.6 °C, while those of the fanned parasol were 8.9 and 5.5 °C in the two stages.

### 3.6 Direct and Indirect Fan Cooling Effects

Figure 6 illustrates that relative to the ordinary parasol, which produced cooling effects of 6.7 °C in Stage I and 3.6 °C in Stage II, the fanned parasol achieved stronger cooling of 8.9 °C and 5.5 °C, respectively. This additional cooling arises from the fan and is termed the fan cooling effect, which is determined by the difference between the cooling impacts of the fanned and ordinary parasols. Accordingly, the fan cooling effect reached 2.2 °C in Stage I and 1.9 °C in Stage II. Moreover, the fanned parasol delivers this fan cooling effect through both direct and indirect mechanisms, allowing the effect to be categorized into direct and indirect components.

The direct fan cooling effect originates from the increased air circulation provided by the fanned parasol. As shown in Fig. 2a, the ordinary parasol produces a negative influence, namely wind-blocking, on the microenvironment beneath it. Comparable to the wind obstruction caused by artificial and tree shading noted by Ribeiro et al. [57], the ordinary parasol diminishes ambient airflow, reducing the mean air velocity from 1.09 m/s in the benchmark case to 0.59 m/s, a decline of 0.50 m/s. By contrast, the fan integrated into the fanned parasol successfully offsets this wind-blocking effect, boosting the airflow beneath the parasol by 0.75 m/s, increasing the mean air velocity from 0.59 m/s in the ordinary parasol case to 1.34 m/s.

The indirect fan cooling effect of the fanned parasol is partly attributed to the lowered air temperature, explained as follows. The ordinary parasol produces another negative influence, namely heat-trapping, within the shaded microenvironment. Because of thermal buoyancy, warm air rises, and the heat-trapping effect retains this warm air beneath the parasol, leading to heat accumulation and elevating the air temperature under the ordinary parasol. Evidence can be seen in Fig. 2b, where the air temperature of the ordinary parasol starts out lower than that of the benchmark case due to its initial stored cooling energy, but later surpasses the benchmark case as the heat-trapping effect dominates. By contrast, the enhanced airflow created by the fanned parasol alleviates this heat-trapping phenomenon. The increased circulation allows better air exchange between the shaded microenvironment and the surroundings, making the air temperature under the fanned parasol consistently lower than under the ordinary parasol, with the mean reduced by 0.8 °C (from 39.2 °C with the ordinary parasol to 38.4 °C with the fanned parasol) (Fig. 2b). Additionally, it is worth noting that fluctuations in air temperature also influence relative humidity, although the humidity ratios among the three cases remain broadly comparable.

On the other hand, the indirect fan cooling effect of the fanned parasol is linked to the decreased globe temperature, explained as follows. The parasol's surface material absorbs solar radiation, leading to a high inner surface temperature. The surrounding air beneath the parasol should normally dissipate this absorbed heat through convection with the inner surface. Yet, in the case of the ordinary parasol, the wind-blocking effect weakens air circulation, thereby reducing the heat transfer coefficient between the shaded air and the parasol's inner surface. In addition, the heat-trapping effect elevates the air temperature under the parasol, diminishing the temperature difference between the air and the inner surface. Consequently, the cooling capacity of the shaded air on the inner surface is restricted, resulting in a higher inner surface temperature. In contrast, the fanned parasol, by overcoming both wind-blocking and heat-trapping effects, lowers the inner surface temperature. This is confirmed by the supplementary data in Fig. 7, which shows that the inner surface temperature of the fanned parasol is consistently lower than that of the ordinary one. The reduced inner surface temperature further decreases the globe temperature. On average, compared with the ordinary parasol, the fanned parasol reduces the globe temperature by 1.7 °C (from 35.0 °C in the ordinary parasol case to 33.3 °C in the fanned parasol case) and 2.5 °C (from 43.6 °C in the ordinary parasol case to 41.1 °C in the fanned parasol case) during Stages I and II respectively.

Therefore, by counteracting the wind-blocking and heat-trapping effects of the ordinary parasol, the fanned parasol produces both direct and indirect fan cooling effects. The direct effect, stemming from the strengthened air circulation, and the indirect effect, arising from the lowered air and globe temperatures, are 0.29 °C and 1.88 °C during Stage I, and 0.68 °C and 1.25 °C during Stage II, respectively (Fig. 8a). In other words, the indirect effect accounts for 86.7 and 64.8% of the total enhanced cooling during Stages I and II. The effectiveness of the fanned parasol in boosting the cooling performance compared with the ordinary parasol is further evaluated using Eqs. (5, 6). Figure 8 shows that the contribution of the fan cooling

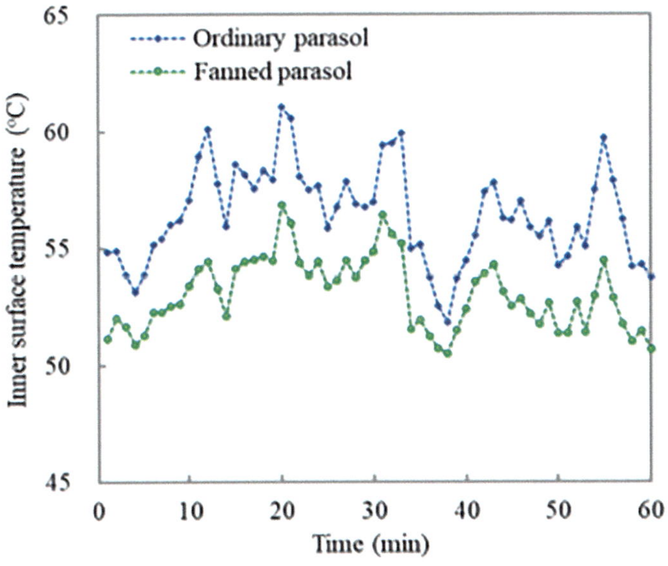


**Fig. 7** Inner surface temperatures of ordinary and fanned parasols

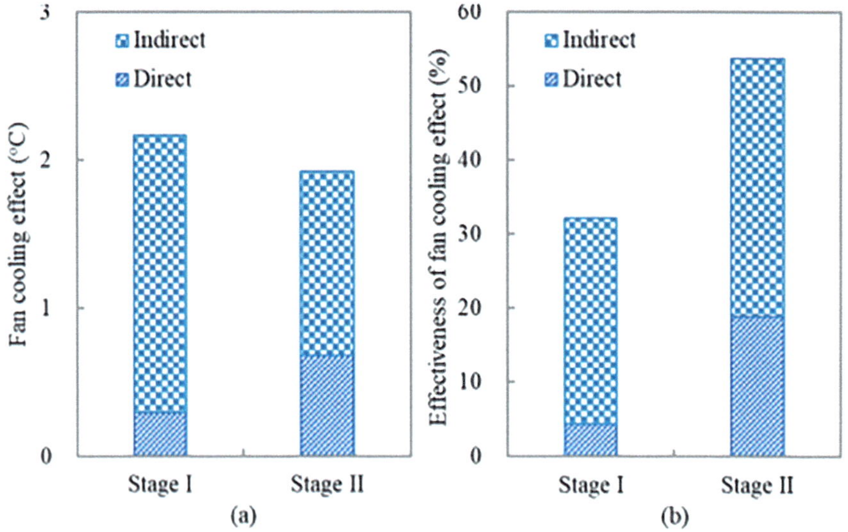


**Fig. 8** **a** Direct and indirect fan cooling effects of fanned parasol and **b** effectiveness of fan cooling effect of fanned parasol

effect reaches 32.3% in Stage I (comprising 4.3% from the direct effect and 28.0% from the indirect effect) and 53.7% in Stage II (comprising 18.9% from the direct effect and 34.8% from the indirect effect). Hence, the fanned parasol proves highly effective in mitigating outdoor heat stress, especially through its indirect fan cooling effect.

$$E_{direct,FP} = \frac{CE_{direct,FP}}{CE_{OP}} \times 100\% \tag{5}$$

$$E_{indirect,FP} = \frac{CE_{indirect,FP}}{CE_{OP}} \times 100\% \tag{6}$$

Where $CE_{direct}$, FP is the direct fan cooling effect of the fanned parasol (°C); $CE_{indirect}$, FP is the indirect fan cooling effect of the fanned parasol (°C); $CE_{OP}$ is the cooling effect of the ordinary parasol (°C); Edirect, FP is the effectiveness of the direct fan cooling effect of the fanned parasol (%); Eindirect, FP is the effectiveness of the indirect fan cooling effect the fanned parasol (%).

## *3.7 Future Studies*

This research proposes the innovative fanned parasol and experimentally verifies and interprets its role in improving the cooling performance for alleviating outdoor heat stress. In addition, four suggestions are provided for future investigations to support the broader application of the fanned parasol in outdoor heat stress mitigation.

- First, the design of the fan requires further optimization. Although a small fan is adopted and remains economically affordable (about CN¥ 20–30), its size should be adjusted to achieve a balance among stronger air circulation, energy use, and noise level. In addition, the installation position of the fan can be refined to improve the airflow distribution within the microenvironment beneath the parasol. In this way, air movement can be maximized to strengthen the direct fan cooling effect, while both the microenvironmental air temperature and the inner surface temperature of the fanned parasol can be reduced to enhance the indirect fan cooling effect.
- Second, Stage I warrants deeper investigation to better utilize the stored cooling energy within the surface material of the novel fanned parasol. Results of this chapter showed that Stage I exhibited a stronger cooling performance, primarily because the parasols were preconditioned in a 26 °C room before the experiments, allowing their surface materials to store cooling energy. That storage temperature could be further examined and optimized to maximize the cost-efficient use of this cooling capacity. In addition, deliberate selection of parasol surface materials with higher thermal mass may enhance the cooling intensity during Stage I and extend its duration.
- Third, it is advised to carry out subjective evaluations of the novel fanned parasol. In addition to the cooling effect achieved by objectively regulating heat exchange between occupants and the outdoor setting, the fanned parasol may also alleviate outdoor heat stress in a subjective way. Factors such as thermal comfort perception and thermal expectation are crucial to how individuals experience outdoor heat stress and can serve as pathways for mitigation [58, 59]. Based on the thermal adaptation theory, the fanned parasol enhances users' ability to regulate their surrounding thermal conditions, thereby potentially improving their subjective perception of the outdoor thermal environment [60].
- Fourth, further verification of the fanned parasol's effectiveness is required across various climate zones. While this chapter confirms that the device significantly improves the cooling effect for alleviating outdoor heat stress in hot and dry environments with low relative humidity, its performance in regions characterized by both high temperature and high humidity remains uncertain. Since thermal stress is usually more severe in such climates, additional testing is necessary to validate the applicability of the fanned parasol under these conditions.

# 4 Conclusions

This chapter presented the design and performance evaluation of a novel fanned parasol as a practical cooling strategy for reducing outdoor heat stress. By integrating a small fan into a lightweight canopy, the fanned parasol provides dual benefits of mobility and enhanced thermal relief compared with traditional designs. Field measurements carried out under hot summer conditions confirmed that the cooling effect is strongest in the early stage of use, when the parasol material releases stored cooling energy. Unlike conventional parasols, which can restrict airflow and intensify heat accumulation, the fanned parasol counteracts these limitations by delivering both direct airflow-driven cooling and substantial indirect thermal relief. Quantitative analysis showed that the indirect effect accounted for the majority of the enhancement, underscoring its role as the principal cooling mechanism. Overall, the fanned parasol offers a cost-effective and practical solution to improve outdoor comfort in hot climates, while future refinements in fan design and canopy materials may further expand its applicability across diverse urban settings.

**Acknowledgments** This study is financially supported by the High-Level Talent Pro-gramme of Shaanxi Province (Project No. 050700/71240000000113), the Fundamental Research Funds for the Central Universities (Project No. xzy012022078) and the Top Young Talent Programme of Xi'an Jiaotong University (Project No. 011900/71211222030705).

## Appendix

The following results (i.e., Figs. 9 and 10) are additional experiment measurements conducted on Sep 6, 2022. The experiment measurements focus on Stage II. Stage I has not been included because the cloudy outdoor condition during Stage I distorts the difference between the cooling effects of the ordinary parasol and the novel fanned parasol. The outdoor thermal conditions of the experiments in this appendix are largely different from those of the experiments in Sect. 2.1, with a lower mean air velocity (i.e., 0.36 m/s of the additional experiments verse 1.09 m/s of the experiments in Sect. 2.1), a lower mean air temperature (i.e., 37.8 °C of the additional experiments verse 39.3 °C of the experiments in Sect. 2.1), a higher mean relative humidity (i.e., 25.2% of the additional experiments verse 47.5% of the experiments in Sect. 2.1), and a lower mean globe temperature (i.e., 47.1 °C of the additional experiments verse 49.1 °C of the experiments in Sect. 2.1). The experiments in this appendix also demonstrate the superiority of the novel fanned parasol over the ordinary parasol. The cooling effect of the ordinary parasol is 4.7 °C. The novel fanned parasol further enhances the cooling effect to 7.2 °C.

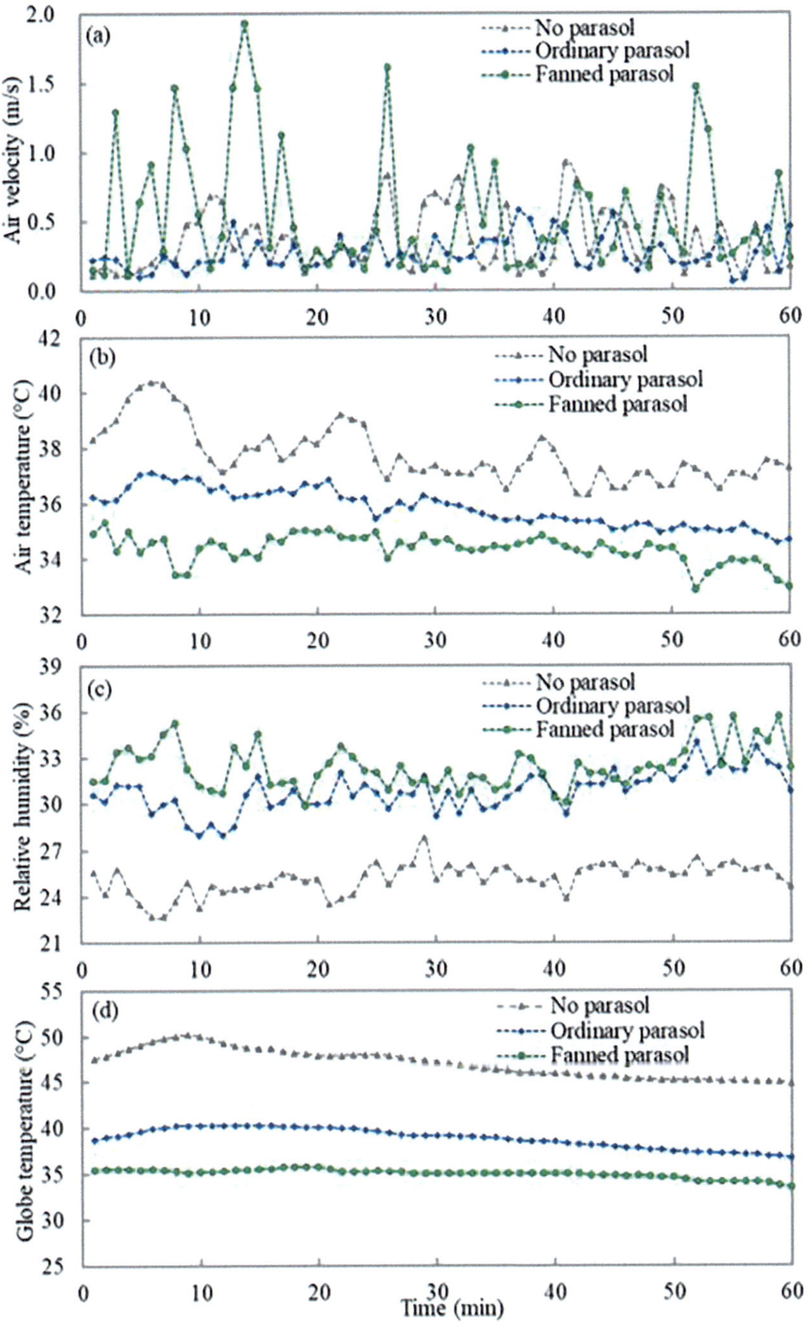


**Fig. 9** Measurements: **a** air velocity, **b** air temperature, **c** relative humidity, and **d** globe temperature

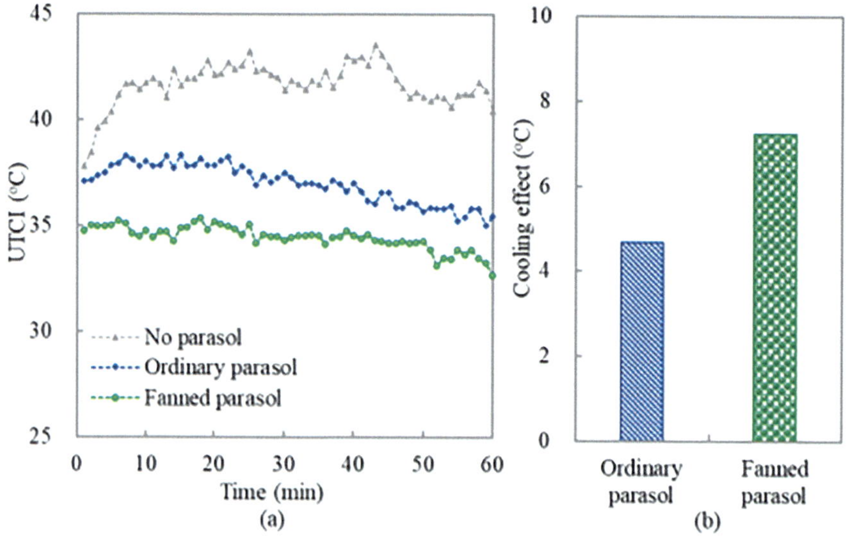


**Fig. 10** **a** Universal Thermal Comfort Index (UTCI) and **b** cooling effects of ordinary and fanned parasols

## References


1. S. Liu, Y. Xie, Y. Zhu, B. Lin, B. Cao, N.H. Wong, J. Niu, Z. Fang, D. Lai, W. Liu, J. Wen, Comparative analysis on indoor and outdoor thermal comfort in transitional seasons and summer based on multiple databases: lessons learnt from the outdoors. Sci. Total Environ. **848**, 157694 (2022)
2. D. Lai, B. Chen, K. Liu, Quantification of the influence of thermal comfort and life pattern on outdoor space activities. Build. Simul. **13**(1), 113–125 (2020)
3. L. de Montigny, R. Ling, J. Zacharias, The effects of weather on walking rates in nine cities. Environ. Behav. **44**(6), 821–840 (2012)
4. Z. Huang, B. Cheng, Z. Gou, F. Zhang, Outdoor thermal comfort and adaptive behaviors in a university campus in China's hot summer-cold winter climate region. Build. Environ. **165**, 106414 (2019)
5. S. Zhang, D. Niu, Z. Lin, Occupancy-aided ventilation for airborne infection risk control: continuously or intermittently reduced occupancies? Build. Simul. **16**(5), 733–747 (2023)
6. A. Sedaghat, M. Sharif, Mitigation of the impacts of heat islands on energy consumption in buildings: a case study of the city of Tehran. Iran. Sustain. Cities Soc. **76**, 103435 (2022)
7. E.J. Gago, J. Roldan, R. Pacheco-Torres, J. Ordóñez, The city and urban heat islands: a review of strategies to mitigate adverse effects. Renew. Sustain. Energy Rev. **25**, 749–758 (2013)
8. F.B. Errebai, D. Strebel, J. Carmeliet, D. Derome, Impact of urban heat island on cooling energy demand for residential building in Montreal using meteorological simulations and weather station observations. Energy Build. **273**, 112410 (2022)
9. J.A. Orosa, A.M. Costa, A. Rodríguez-Fernandez, G. Roshan, Effect of climate change on outdoor thermal comfort in humid climates. J. Environ. Health Sci. Eng. **12**(1), 1–9 (2014)
10. H. Kakaei, F. Omidi, R. Ghasemi, M.R. Sabet, F. Golbabaei, Changes of WBGT as a heat stress index over the time: A systematic review and meta-analysis. Urban Clim. **27**, 284–292 (2019)
11. K. Ward, S. Lauf, B. Kleinschmit, W. Endlicher, Heat waves and urban heat islands in Europe: a review of relevant drivers. Sci. Total Environ. **569**, 527–539 (2016)

12. N. Mazhar, R.D. Brown, N. Kenny, S. Lenzholzer, Thermal comfort of outdoor spaces in Lahore, Pakistan: lessons for bioclimatic urban design in the context of global climate change. Landsc. Urban Plan. **138**, 110–117 (2015)
13. S. Watanabe, J. Ishii, Effect of outdoor thermal environment on pedestrians' behavior selecting a shaded area in a humid subtropical region. Build. Environ. **95**, 32–41 (2016)
14. W. You, W. Ding, Effects of urban square entry layouts on spatial ventilation under different surrounding building conditions. Build. Simul. **14**(2), 377–390 (2021)
15. Y. Zhang, Z. Lin, Z. Fang, Z. Zheng, An improved algorithm of thermal index models based on ENVI-met. Urban Clim. **44**, 101190 (2022)
16. J. Li, J. Niu, T. Huang, C.M. Mak, Dynamic effects of frequent step changes in outdoor microclimate environments on thermal sensation and dissatisfaction of pedestrian during summer. Sustain. Cities Soc. **79**, 103670 (2022)
17. J. Li, J. Niu, C.M. Mak, T. Huang, Y. Xie, Exploration of applicability of UTCI and thermally comfortable sun and wind conditions outdoors in a subtropical city of Hong Kong. Sustain. Cities Soc. **52**, 101793 (2020)
18. S. Zhang, X. Zhang, D. Niu, Z. Fang, H. Chang, Z. Lin, Physiological equivalent temperature-based and universal thermal climate index-based adaptive-rational outdoor thermal comfort models. Build. Environ. **228**, 109900 (2023)
19. R.-L. Hwang, Y.-T. Weng, K.-T. Huang, Considering transient UTCI and thermal discomfort footprint simultaneously to develop dynamic thermal comfort models for pedestrians in a hot-and-humid climate. Build. Environ. **222** (2022)
20. S. Jia, Y. Wang, Effect of heat mitigation strategies on thermal environment, thermal comfort, and walkability: a case study in Hong Kong. Build. Environ. **201**, 107988 (2021)
21. D. Lai, W. Liu, T. Gan, K. Liu, Q. Chen, A review of mitigating strategies to improve the thermal environment and thermal comfort in urban outdoor spaces. Sci. Total Environ. **661**, 337–353 (2019)
22. A.S.H. Abdallah, R.M.A. Mahmoud, Urban morphology as an adaptation strategy to improve outdoor thermal comfort in urban residential community of new Assiut city, Egypt. Sustain. Cities Soc. **78**, 103648 (2022)
23. C.K.C. Lam, J. Weng, K. Liu, J. Hang, The effects of shading devices on outdoor thermal and visual comfort in Southern China during summer. Build. Environ., 109743 (2022)
24. Y. Li, B. Hong, Y. Wang, H. Bai, H. Chen, Assessing heat stress relief measures to enhance outdoor thermal comfort: a field study in China's cold region. Sustain. Cities Soc. **80**, 103813 (2022)
25. S.W. Kim, R.D. Brown, Pedestrians' behavior based on outdoor thermal comfort and micro-scale thermal environments, Austin, TX. Sci. Total Environ. **808**, 152143 (2022)
26. S. Thorsson, T. Honjo, F. Lindberg, I. Eliasson, E.M. Lim, Thermal comfort and outdoor activity in Japanese urban public places. Environ. Behav. **39**(5), 660–684 (2007)
27. M.M. Baruti, E. Johansson, J. Åstrand, Review of studies on outdoor thermal comfort in warm humid climates: challenges of informal urban fabric. Int. J. Biometeorol. **63**(10), 1449–1462 (2019)
28. I. Lee, J.A. Voogt, T.J. Gillespie, Analysis and comparison of shading strategies to increase human thermal comfort in urban areas. Atmos. **9**(3), 91 (2018)
29. H. Kusaka, Y. Nakamura, Y. Asano, UV parasol, dry-mist spraying, and street trees as tools for heat stress mitigation. J. Meteorol. Soc. Jpn. **100**(4), 677–685 (2022)
30. U. Dietrich, Urban street canyons–impact of different materials and colours of facades and ground and different positions of persons on outdoor thermal comfort. Int. J. Sustain. Dev. Plan. **13**, 582–593 (2018)
31. T.P. Lin, Thermal perception, adaptation and attendance in a public square in hot and humid regions. Build. Environ. **44**(10), 2017–2026 (2009)
32. F. Golbabaei, E. Karami, M. Shahi, Z. Safari, K. Azrah, Modeling and investigating the effect of parasol installation on solar radiant temperature reduction using COMSOL Multiphysics. Int. J. Occup. Saf. Ergon., 1–15 (2022)

33. S. Watanabe, J. Ishii, Mitigation of pedestrian heat stress using parasols in a humid subtropical region. Int. J. Biometeorol. **61**(11), 2009–2019 (2017)
34. S. Zhang, F. Cheng, X. Zhang, Z. Lin, Y. Cheng, Cooling effect of air movement of advanced air distribution under heating mode: experimental analyses of stratum ventilation. Build. Environ. **231** (2023)
35. Y. Yu, R. de Dear, *Thermal respite for pedestrians in overheated urban environments introduction of a dynamic analysis of outdoor thermal comfort* (Sustain. Cities Soc., 2022), p. 104149
36. S. Zhang, D. Niu, Z. Lin, Extending effective draft temperature to cover full range of air velocity. Build. Environ. **210**, 108738 (2022)
37. M. Santamouris, L. Ding, F. Fiorito, P. Oldfield, P. Osmond, R. Paolini, D. Prasad, A.J.S.E. Synnefa, Passive and active cooling for the outdoor built environment–analysis and assessment of the cooling potential of mitigation technologies using performance data from 220 large scale projects. Sol. Energy **154**, 14–33 (2017)
38. C. Hong, Y. Yang, S. Ge, G. Chai, P. Zhao, Q. Shui, Z. Gu, Is the design guidance of color and material for urban buildings a good choice in terms of thermal performance? Sustain. Cities Soc. **83**, 103927 (2022)
39. M. Gao, F. Chen, H. Shen, H. Li, A tale of two cities: different urban heat mitigation efficacy with the same strategies. Theor. Appl. Climatol. **142**(3), 1625–1640 (2020)
40. B. Yuan, L. Zhou, F. Hu, Q. Zhang, Diurnal dynamics of heat exposure in Xi'an: a perspective from local climate zone. Build. Environ. **222**, 109400 (2022)
41. S. Zhang, D. Niu, Z. Lin, Index of ventilation effectiveness regarding energy performance considering cooling effect of air movement: equivalent thermal utilization effectiveness. Build. Environ. **212** (2022)
42. Z. Fang, Z. Zheng, X. Feng, D. Shi, Z. Lin, Y. Gao, Investigation of outdoor thermal comfort prediction models in South China: a case study in Guangzhou. Build. Environ. **188**, 107424 (2021)
43. Y. Zhang, X. Zhou, Z. Zheng, M.O. Oladokun, Z. Fang, Experimental investigation into the effects of different metabolic rates of body movement on thermal comfort. Build. Environ. **168**, 106489 (2020)
44. M. Sadeghi, R. de Dear, G. Morgan, M. Santamouris, B. Jalaludin, Development of a heat stress exposure metric–Impact of intensity and duration of exposure to heat on physiological thermal regulation. Build. Environ. **200** (2021)
45. J. Dhariwal, P. Manandhar, L. Bande, P. Marpu, P. Armstrong, C.F. Reinhart, Evaluating the effectiveness of outdoor evaporative cooling in a hot, arid climate. Build. Environ. **150**, 281–288 (2019)
46. Y. Ibrahim, T. Kershaw, P. Shepherd, I. Elwy, A parametric optimisation study of urban geometry design to assess outdoor thermal comfort. Sustain. Cities Soc. **75**, 103352 (2021)
47. X. Xu, Y. Wu, W. Wang, T. Hong, N. Xu, Performance-driven optimization of urban open space configuration in the cold-winter and hot-summer region of China. Build. Simul. **12**(3), 411–424 (2019)
48. P. Kumar, A. Sharma, Study on importance, procedure, and scope of outdoor thermal comfort-a review. Sustain. Cities Soc. **61**, 102297 (2020)
49. O. Potchter, P. Cohen, T.P. Lin, A. Matzarakis, Outdoor human thermal perception in various climates: a comprehensive review of approaches, methods and quantification. Sci. Total Environ. **631**, 390–406 (2018)
50. P.K. Cheung, C.Y. Jim, Comparing the cooling effects of a tree and a concrete shelter using PET and UTCI. Build. Environ. **130**, 49–61 (2018)
51. S. Park, S.E. Tuller, M. Jo, Application of Universal Thermal Climate Index (UTCI) for microclimatic analysis in urban thermal environments. Landsc. Urban Plan. **125**, 146–155 (2014)
52. P. Bro"de, D. Fiala, K. Błazejczyk, I. Holmer, G. Jendritzky, B. Kampmann, B. Tinz, G. Havenith, Deriving the operational procedure for the universal thermal climate index (UTCI). Int. J. Biometeorol. **56** (3), 481–494 (2012)

53. A. Psikuta, D. Fiala, G. Laschewski, G. Jendritzky, M. Richards, K. Błazejczyk, I. Mekjavic, H. Rintama"ki, R. de Dear, G. Havenith, Validation of the Fiala multi-node thermophysiological model for UTCI application. Int. J. Biometeorol. **56**(3), 443–460 (2012)
54. G. Jendritzky, R. de Dear, G. Havenith, UTCI—why another thermal index? Int. J. Biometeorol. **56**(3), 421–428 (2012)
55. S. Zhang, D. Niu, Z. Lin, Mean radiant temperature calculated based on radiant heat dissipation of human body addressing effect of emissivity of inner surfaces of envelope. Sol. Energy **246**, 14–22 (2022)
56. W. Ouyang, Z. Liu, K. Lau, Y. Shi, E. Ng, Comparing different recalibrated methods for estimating mean radiant temperature in outdoor environment. Build. Environ. **216**, 109004 (2022)
57. K.F.A. Ribeiro, A.C.A. Justi, J.W.Z. Novais, F.M. de Moura Santos, M.C.D.J.A. Nogueira, S.A. Miranda, J.B. Marques, Calibration of the Physiological Equivalent Temperature (PET) index range for outside spaces in a tropical climate city. Urban Clim. **44**, 101196 (2022)
58. L. Chen, E. Ng, Outdoor thermal comfort and outdoor activities: a review of research in the past decade. Cities **29**(2), 118–125 (2012)
59. Y. Xie, J. Niu, H. Zhang, S. Liu, J. Liu, T. Huang, J. Li, C.M. Mak, Development of a multi-nodal thermal regulation and comfort model for the outdoor environment assessment. Build. Environ. **176**, 106809 (2020)
60. M. Luo, B. Cao, W. Ji, Q. Ouyang, B. Lin, Y. Zhu, The underlying linkage between personal control and thermal comfort: psychological or physical effects? Energy Build. **111**, 56–63 (2016)

# The Role of Building Overhead Layer in Improving Thermal Comfort in Urban Planning: A Comparative Study with Indoor Spaces

**Abstract** The overhead layer of buildings, often serving as semi-open transitional spaces, plays an important role in moderating outdoor microclimates and improving pedestrian comfort in urban planning. Thermal indices, particularly the Standard Effective Temperature (SET*), are commonly used to evaluate human thermal sensation; however, their applicability in dynamic outdoor and semi-open conditions remains limited. To address this, comparative experiments were conducted in an indoor badminton hall and on the overhead floor of a university building in Guangzhou. Thirty participants performed walking trials at different speeds followed by rest, during which environmental parameters and mean thermal sensation votes (MTSV) were recorded. Results demonstrated that the predictive performance of SET* was more reliable in transitional overhead spaces than indoors under dynamic walking conditions. Building on these findings, an enhanced SET* model (αSET*) and revised stress categories were proposed, extending the applicability of SET* to semi-open overhead environments. The refined model provides a thermal comfort range of 24–37.5 °C (MTSV within $\pm$ 0.5), offering practical guidance for integrating overhead structures into urban planning strategies to alleviate thermal discomfort.

## 1 Introduction

With the continuous improvement of living standards, people are paying more attention to thermal comfort in their environment. Research on indoor thermal comfort has long been emphasized [1–3], since individuals spend the majority of their time inside buildings [4]. As the pursuit of healthier lifestyles increases, outdoor activities have also grown, making outdoor thermal comfort an emerging research focus [5–7]. Meanwhile, the ongoing enhancement of urban development [8] and the growing construction of outdoor transitional spaces have created more favorable conditions for people's outdoor experiences. Consequently, in recent years, numerous studies have concentrated on these transitional environments [9–12].

Human thermal comfort is influenced by many elements, making it essential to integrate different environmental variables into a single composite index when

 
Z. Fang et al., *Outdoor Thermal Comfort and Heat Stress Mitigation*, Environmental Science and Engineering, https://doi.org/10.1007/978-981-95-8899-2_17

assessing indoor thermal conditions. Since the 1960s, research on thermal comfort has shifted from examining the effect of an individual parameter on humans to evaluating the combined influence of multiple factors, leading to the development of various indicators grounded in thermal balance theory. Among these, the standard effective temperature (SET)* introduced by Gagge [13] and the predicted mean vote (PMV) model developed by Fanger [14] are the most extensively applied. For indoor comfort assessment, both PMV and SET* are accepted as international benchmarks. However, experimental studies demonstrated that the PMV equation works reliably only when a person is wearing standard clothing and remaining still, but fails to predict sensations accurately under other attire or activity levels [15]. Nicol and Humphreys further observed that PMV is applicable only in relatively stable environments [16]. Therefore, its capacity to predict dynamic thermal changes is limited.

The SET* index integrates the impacts of varying activity intensities and clothing thermal resistance on human comfort, which makes it the most widely applied indicator today. Earlier research has shown that SET* can predict indoor thermal sensation with higher accuracy, especially for individuals sitting and working in enclosed spaces. Attempts have also been made to apply SET* in evaluating outdoor thermal perception; however, the outcomes were not ideal [17, 18]. For outdoor comfort assessment, the indices most frequently employed are PET (physiologically equivalent temperature) and UTCI (universal thermal climate index) [6, 19]. In our team's earlier investigations, the predictive performance of these two indices in transitional spaces was validated, and necessary adjustments were introduced [20].

Moreover, because the knowledge base regarding thermal comfort in transitional spaces is still limited, only a few studies have attempted to apply SET* in predicting human thermal perception. Physical activities not only modify surrounding environmental variables but also trigger sweating on the skin during movement, which further complicates the regulation of human thermal balance [21, 22]. Based on this analysis, using the current simplified SET* calculation makes it challenging to accurately capture the dynamic thermal sensation of the human body. Hence, to better evaluate the comfort of individuals in motion, the SET* model must be suitably adjusted according to the specific activity conditions.

In conclusion, since the applicability of the SET* model in evaluating outdoor thermal comfort remains debated, this study first examined its prediction accuracy for human thermal sensation under various indoor walking intensities, and then extended the tests to transitional environments. Subsequently, by incorporating both the environmental disturbances induced by walking and the metabolic rate variations at different walking stages, an improved version of the SET* model suitable for indoor and transitional spaces was proposed. Furthermore, using Probit probability analysis, the thermal sensation levels corresponding to αSET* (the modified SET*) at multiple walking speeds were determined, allowing the identification of comfortable walking conditions.

## 2 Methods

This study applies a combination of on-site experiments and questionnaire surveys. After completing the tests, the raw field data were entered and organized using office software. The processed data were first calculated with basic relevant formulas, and then further analyzed with thermal comfort evaluation tools or online platforms.

### *2.1 Testing Process*

In 2019, two experiments were carried out from March to April and from June to July. The test locations were the Guangzhou University badminton gym and overhead floor of a college building, representing the two locations of indoor space and transition space, respectively. The investigation site is shown in Fig. 1.

A thermal comfort data logger was installed at the experiment site to automatically monitor environmental parameters such as natural wind speed, ambient air temperature, relative humidity, and radiant temperature. Based on the recorded results, the indoor air temperature of the badminton hall was maintained at 22 ± 2 °C, with a relative humidity of 80% ± 10%. In contrast, the transitional space showed an air

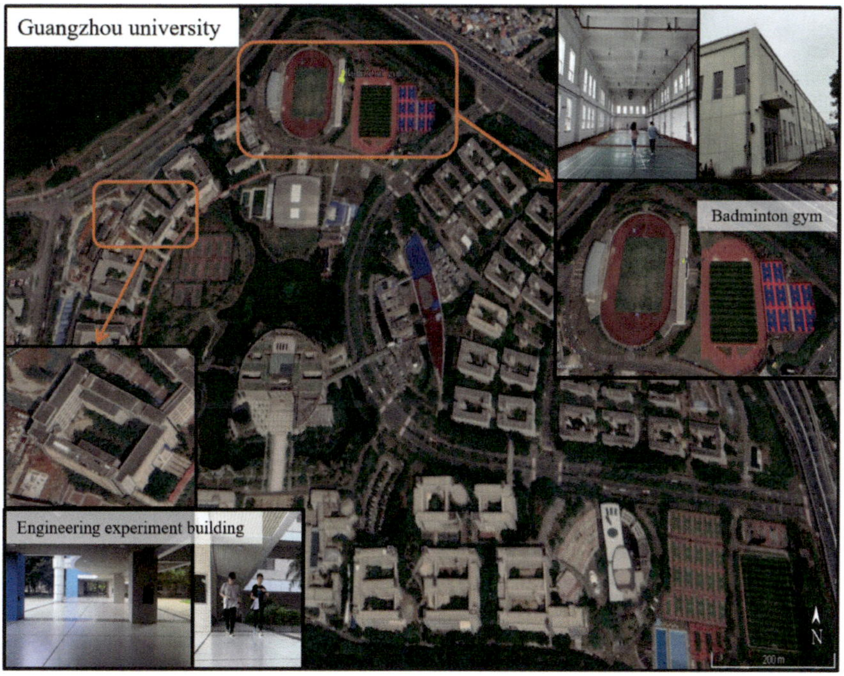


Fig. 1 Investigation site

temperature range of 34 ± 2.5 °C and a relative humidity of 85% ± 10%. Detailed information on the environmental parameters is provided in Table 1.

During the experiment, participants were instructed to walk at four fixed speeds of 1.2, 1.4, 1.6, and 2.0 m/s, which were controlled using a metronome. A portable wireless device for measuring air velocity and temperature was positioned at the height of the navel to automatically capture the wind speed caused by body movement. At the same time, participants orally responded to the thermal comfort questionnaire (recorded by an assigned assistant) and monitored their heart rate using a smart wristband. The whole procedure consisted of three stages: a preparation phase in which subjects sat quietly for 30 min, an exercise phase involving 20 min of continuous walking, and a recovery phase requiring 10 min of seated rest. The detailed measurements carried out in each stage are illustrated in Fig. 2.

**Table 1** Detail information of test condition

| Classification | | Air temperature $T_a$ (°C) | Relative humidity RH (%) | Black globe temperature $T_g$ (°C) | Air velocity $V_a$ (m/s) |
|---|---|---|---|---|---|
| Max. value | Indoor space | 24.14 | 84.15 | 24.20 | 1.22 |
| | Transition space | 35.70 | 95.00 | 36.10 | 7.70 |
| Min. value | Indoor space | 20.72 | 74.80 | 20.81 | 0.00 |
| | Transition space | 31.60 | 80.20 | 31.80 | 0.00 |
| Average value | Indoor space | 22.10 | 79.13 | 22.19 | 0.16 |
| | Transition space | 33.90 | 85.90 | 34.10 | 1.90 |
| Standard deviation | Indoor space | 1.00 | 2.60 | 1.00 | 0.60 |
| | Transition space | 1.10 | 3.80 | 1.20 | 1.20 |

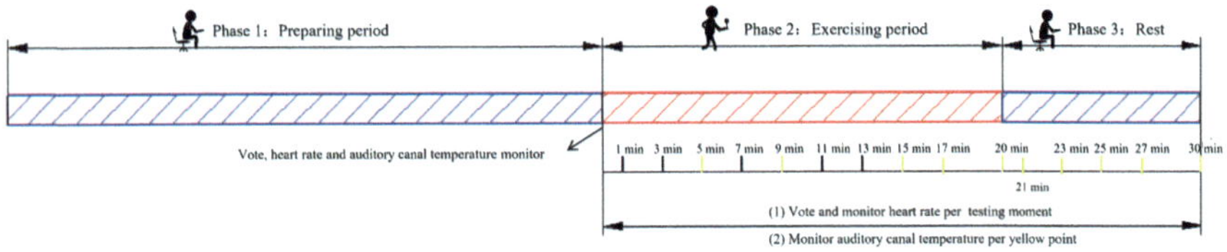


**Fig. 2** Experimental procedure

**Table 2** Details of instruments used in the experiment

| Instrument | Type | Parameter | Measuring range | Accuracy | Sampling rate (s) |
|---|---|---|---|---|---|
| Thermal comfort level recorder | SSDZY-1 | $T_a$ (°C) | – 20.0–+ 80.0 °C | ± 0.3 °C | 2 |
| | | RH (%) | 0.01–99.9% RH | ± 2.0% RH (10.0–90.0% RH) | 2 |
| | | $T_g$ (°C) | – 20.0–80.0 °C | ± 0.3 °C | 2 |
| Wireless universal wind speed and temperature recorder | WFWZY-1 | $V_a$ (m/s) | 0.05–5.00 m / s | 5% ± 0.05 m/s | 2 |
| Smart bracelet | 3XMSH05HM | HR (bpm) | 30–250 bpm | ± 3 bpm | 2 |

## 2.2 Measurement Instruments and Parameters

During the test, thermal environmental variables—air temperature ($T_a$), relative humidity (RH), globe temperature ($T_g$), and air velocity ($V_a$)—were monitored simultaneously. Meanwhile, participants' heart rate was also tracked using a smart wristband, which provided data for estimating the metabolic rate. The measuring devices applied in the experiment, along with their specifications, are summarized in Table 2.

## 2.3 Subjective Survey and Measurement

In both trials, a total of 30 healthy university students were recruited, with an average age of 22.3 years, comprising 18 males and 12 females. Participants were required to wear standardized clothing during testing. For the indoor experiment, the outfit consisted of long sleeved shirts, lightweight trousers, socks, and sneakers, giving a total clothing insulation of 0.61 clo. On the overhead floor, subjects wore short-sleeved shirts, light trousers, socks, and sneakers, with a total insulation of 0.57 clo. According to the participants, these clothing combinations represented the most comfortable attire for the season. Detailed demographic information is listed in Table 3.

The subjective questionnaire surveyed the participants' thermal sensation vote (TSV). The scale of subjective vote is shown in Table 4, which meets the thermal environment in ASHRAE standards 55 [23] and ISO7730 [24].

**Table 3** Detailed information of the participants

| Gender | Sample size | Age (years) | Height (cm) | Weight (kg) | BMI ($kg/m^2$) | BMR (kcal) |
|---|---|---|---|---|---|---|
| Male | 18.0 | 22.5 ± 1.5 | 175.3 ± 5 | 66.0 ± 8.0 | 19.3 ± 1.5 | 1480.2 ± 145.1 |
| Female | 12.0 | 21.5 ± 2.5 | 161.0 ± 3 | 50.2 ± 1.5 | 19.0 ± 1.3 | 1197.0 ± 20.0 |
| Average | – | 22.3 | 171.6 | 57.8 | 19.2 | 1338.4 |

**Table 4** Subjective vote scale

| Thermal sensation | Thermal preference | Humidity preference | Acceptability |
|---|---|---|---|
| – 3 cold<br>– 2 cool<br>– 1 slightly cool<br>0 neutral<br>+ 1 slightly warm<br>+ 2 warm<br>+ 3 hot | – 1 cooler<br>0 no change<br>+ 1 warmer | – 1 lower<br>0 no change<br>+ 1 higher | – 1 unacceptable<br>1 acceptable |

## 2.4 Data Analysis

In this research, the calculation of SET* was performed using the CBE Thermal Comfort Tool, accessible on the developer's website (http://comfort.cbe.berkeley.edu/). On this platform, the SET* value is obtained directly by entering the required parameters, including air temperature ($T_a$), mean radiant temperature ($T_{mrt}$), relative humidity (RH), surrounding air speed (Vs), metabolic rate (M), and clothing insulation.

The metabolic rate is typically evaluated through the measurement of inhaled oxygen and exhaled carbon dioxide [25]. In this experiment, however, heart rate was additionally recorded to indicate variations in the subjects' metabolic levels. A simplified estimation method based on heart rate [26] was applied in this study. The detailed computational formulas are provided in Eqs. (1–4):

$$M = \frac{HR - HR_0}{RM} + M_0, \tag{1}$$

$$RM = \frac{HR_{max} - HR_0}{MWC - M_0}, \tag{2}$$

$$MWC_{male} = (41.7 - 0.22 \times A) \times W^{0.666}\left(W/m^2\right) \tag{3}$$

$$MWC_{female} = (35.0 - 0.22 \times A) \times W^{0.666}\left(W/m^2\right) \tag{4}$$

# 3 Results

## *3.1 Adaptability of SET* in Evaluating Indoor Dynamic Thermal Comfort*

Previous research has shown that SET* is mostly applicable for assessing static human thermal comfort [27–29], while another study indicated its feasibility under dynamic situations [30]. In this study, the experiment was carried out indoors in the badminton hall, where participants were in motion; consequently, their metabolic rate exceeded 1 and the ambient air velocity was not zero. It was hypothesized that SET* could be adopted to estimate the subjects' thermal comfort during this process. Therefore, SET* was obtained from the online tool and compared with the mean thermal sensation vote (MTSV), as illustrated in Fig. 3. The findings revealed that the linear correlation between SET* and MTSV throughout the badminton hall test was relatively weak, with an $R^2$ of about 0.2.

As reported in a previous study [31], the metabolic rate generally becomes steady after 5 min of activity in the badminton hall. Accordingly, the experiment process was divided into two phases: a rising stage of metabolism (0–5 min) and a stable stage (7–20 min). The linear correlations between SET* and MTSV in these two phases are presented in Fig. 4. In Fig. 4a, it can be seen that during the initial rising stage, regardless of SET* fluctuations, MTSV values remained within ± 0.5, suggesting that participants perceived thermal neutrality. This phenomenon may be attributed to the effect of airflow, which dispersed part of the metabolic heat during this period. In contrast, during the stabilization phase shown in Fig. 4b, the metabolic rate had

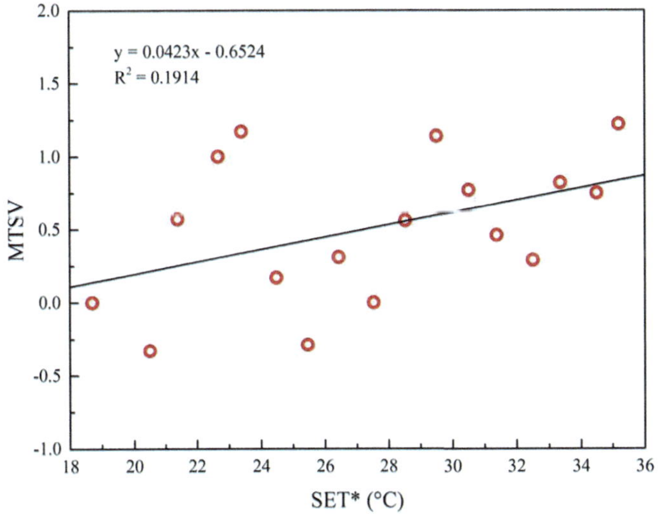


**Fig. 3** Correlation between SET* and MTSV during exercising period (badminton gym)

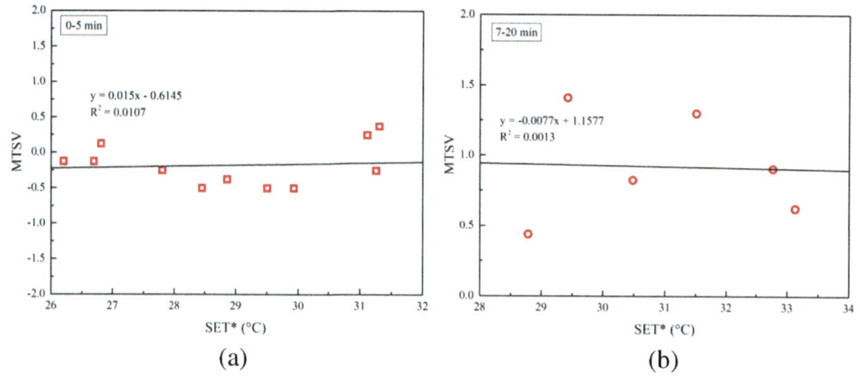


**Fig. 4** Relationship between SET* and MTSV at different metabolic stages (badminton gym)

already reached a high and stable level, which varied with walking speed. At this point, MTSV was mainly influenced by the airflow disturbance, and thus no clear linear correlation was found between SET* and MTSV.

## *3.2 Adaptability of SET* in Evaluating the Dynamic Thermal Comfort of Transition Space*

It is well known that the calculation of SET* requires multiple input parameters, including Ta, Tmrt, RH, Vs, M, and Iclo. Hence, the resulting values depend not only on environmental factors but also on clothing characteristics and human activity intensity. Although transitional spaces are generally not considered the primary application area for SET* as a thermal comfort index, Fig. 5 demonstrates that the linear correlation coefficient ($R^2$) between MTSV and SET* exceeded 0.6, suggesting a notable relationship between the two.

After confirming the potential correlation between MTSV and SET*, the metabolic rate variations of participants while walking in the transition space were examined, similar to the indoor study and with reference to our earlier work [20]. Specifically, minutes 0–9 corresponded to the ascending stage of metabolism, whereas minutes 11–20 represented the stabilized stage. Because the metabolic rate increased continuously during the first 9 min, dynamic equilibrium was achieved from the 11th minute onward. Accordingly, in the transition space experiment, the linear correlation between MTSV and SET* was analyzed for different stages, as illustrated in Fig. 6.

As illustrated in Fig. 6a, a strong linear correlation was observed between SET* and MTSV during 0–9 min ($R^2 = 0.9$), indicating that MTSV can be predicted with high accuracy from SET* in the metabolic rising stage. In this period, the metabolic rate (M) increased gradually over time, and since M is a key factor influencing

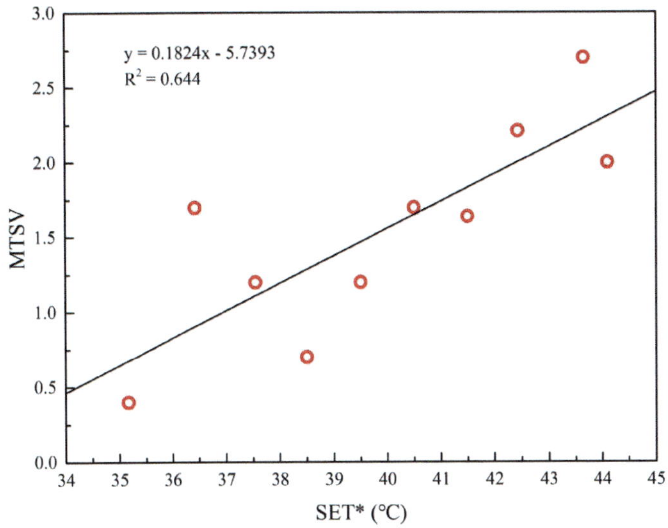


**Fig. 5** Correlation between MTSV and SET* (overhead floor)

SET*, the SET* value also rose accordingly. Moreover, consistent with our earlier study [29], both MTSV and M exhibited synchronous increases as time progressed. Therefore, the pronounced linear relationship between SET* and MTSV during the metabolic ascent phase was confirmed.

During the metabolic stabilization phase (11–20 min), the linear regression between MTSV and SET* revealed only a weak relationship, as indicated in Fig. 6b. In this period, participants' average metabolic rate (M) fluctuated around the level corresponding to each walking speed, showing that metabolism had entered a relatively steady state and no longer rose with time. With the external environment remaining stable, SET* values stabilized at about 40 °C (Fig. 7), while MTSV consistently ranged between 1.5 and 2.5 (Fig. 8). Consequently, the regression analysis showed a weak link between MTSV and SET* at this stage, suggesting that MTSV could not be reliably estimated solely through SET* in the stabilization period.

## 3.3 *Improvement of SET* Model*

### 3.3.1 Transition Space

The results indicate that in both indoor and transitional spaces, relying only on SET* calculations is insufficient to precisely predict the thermal comfort of people in motion throughout the walking process. To achieve reliable estimation of the mean thermal sensation (MTSV) for active individuals—whether indoors, in transition areas, or at different metabolic stages—the analysis began with the transition space.

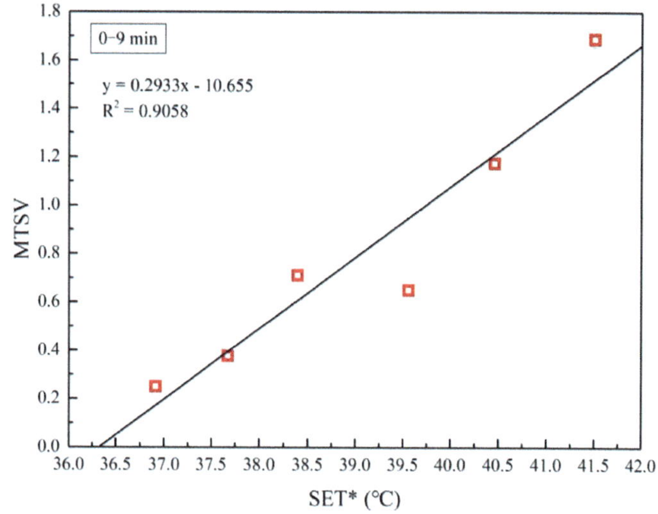


(a)

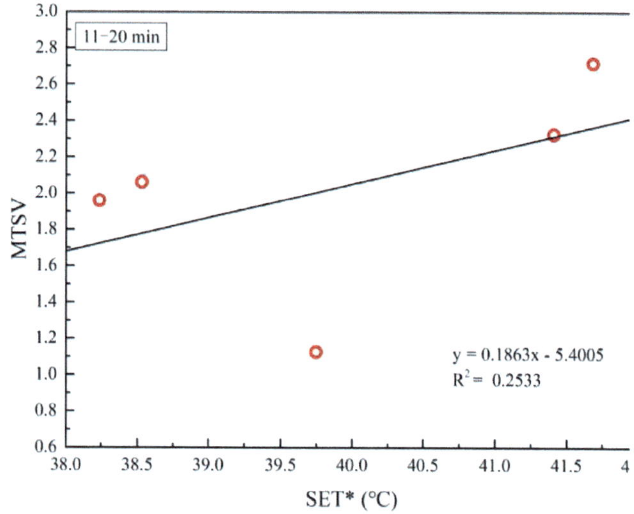


(b)

**Fig. 6** Correlation between SET* and MTSV at different metabolic stages (overhead floor)

In this environment, a clear linear relationship was observed between SET* and MTSV during the metabolic rising phase.

First, attention is given to the metabolic stabilization period (11–20 min) in the transition space. During this stage, among the six variables influencing SET*, Ta, Tmrt, and RH remained nearly unchanged, while clothing insulation ($I_{clo}$) was

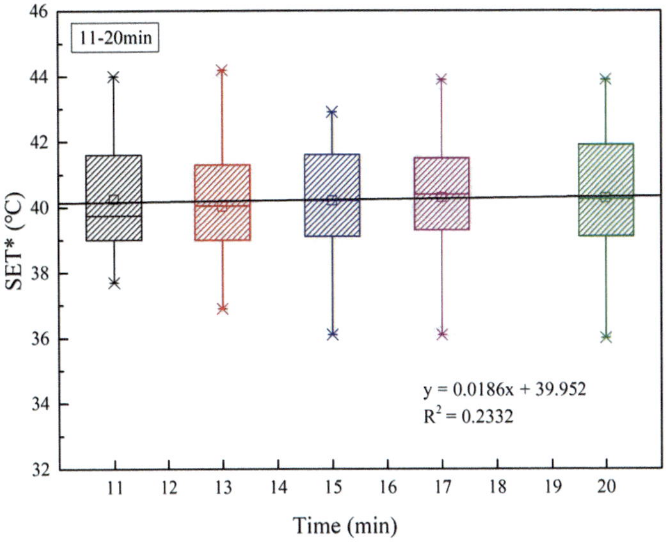


**Fig. 7** Time-dependent changes of SET* during the metabolic stabilization stage in the transition space

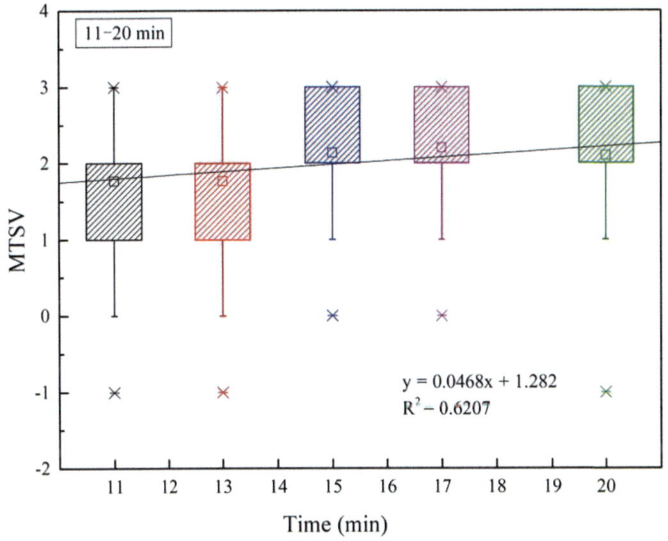


**Fig. 8** Time-dependent changes of MTSV during the metabolic stable stage in the transition space

constant and M stayed relatively stable. In contrast, the natural wind speed could not be controlled in the transitional environment, and walking at different speeds generated varying airflow disturbances, leading to noticeable fluctuations in local wind velocity. Thus, we infer that air velocity ($V_s$) together with metabolic rate (M) were the dominant factors influencing SET* during this phase.

By applying the stepwise regression function in IBM SPSS Statistics 25 (a software originally created in 1968 by Stanford graduate students Norman H. Nie, C. Hadlai (Tex) Hull, and Dale H. Bent), we constructed a linear model between SET* and the factors M and Vs, as expressed in Eq. (5). The regression output indicated that the correlation coefficient $R^2$ exceeded 0.8 with Sig. < 0.05, confirming a strong dependence between the response variable and predictors. Likewise, another regression formula was obtained for MTSV with M and V as inputs ($R^2 > 0.75$, Sig. < 0.05), as shown in Eq. (6).

$$SET^* = 1.057\ M + 2.247V_s + 32.429 \quad \left(R^2 = 0.83, sig. < 0.05\right) \tag{5}$$

$$MTSV = 0.602M - 1.057V_s + 1.612 \quad \left(R^2 = 0.76, sig. < 0.05\right) \tag{6}$$

From Eqs. (5) and (6), it can be observed that Vs shows a negative association with MTSV, meaning that higher Vs corresponds to lower MTSV. In contrast, Vs is positively related to SET*, indicating that an increase in Vs results in a higher SET*. The metabolic rate M demonstrates a positive correlation with both MTSV and SET*. Consequently, the inability of SET* to fully explain MTSV in the stabilization phase stems from neglecting the dual impact of Vs on these two variables. During this stage, the effect of disturbed airflow on thermal sensation surpassed that of metabolic rate, making Vs the dominant factor. Referring to the rising phase relationship (MTSV = 0.2933SET* − 10.655), by integrating Eqs. (5) and (6) and removing the variable M, Eq. (7) was derived:

$$\begin{aligned} MTSV &= 0.57SET^* - 2.337V_s - 16.873 \\ &= 0.2933\left(1.94SET^* - 7.968V_s - 21.19\right) - 10.655 \end{aligned} \tag{7}$$

The expression in brackets was defined as αSET*.

As illustrated in Fig. 9, after introducing αSET* to modify the SET* model during the stabilization phase, the correlation coefficient between MTSV and αSET* rises above 0.8. This outcome demonstrates that the revised model enables the prediction of MTSV throughout the entire walking process using a single unified equation.

### 3.3.2 Indoor Space

Figure 4a illustrates that during the first 0–5 min of the exercise stage, MTSV remained within the comfort interval (MTSV = ±0.5). However, from the seventh minute until the end of the activity, MTSV exhibited no distinct variation trend with

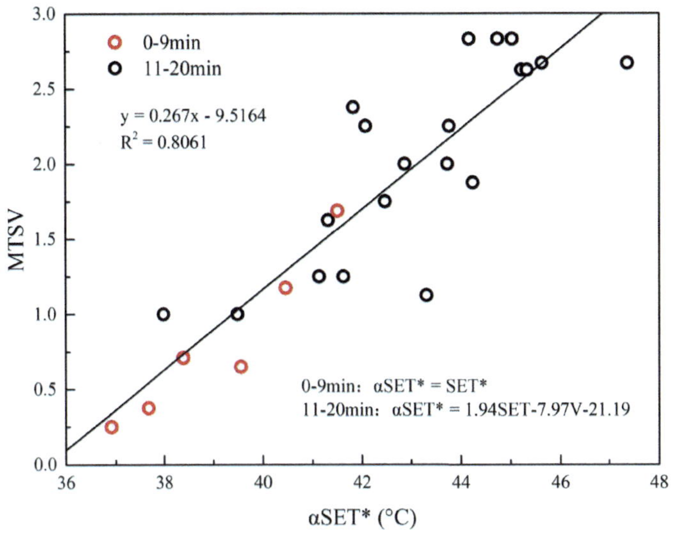


**Fig. 9** Correlation between MTSV and αSET* in transition space

time, as depicted in Fig. 4b. Following the analytical approach used for the transition space, regression analysis was conducted between SET* and M, Vs as well as between MTSV and M, Vs for the period of 7–20 min, leading to the derivation of Eqs. (8) and (9):

$$SET^* = 2.574M - 3.543V_s + 24.771R^2 \quad (8)$$

$$MTSV = 0.707M + 2.5V_s - 3.88R^2 = 0.80 \quad (9)$$

Similarly, based on the regression model (MTSV = 0.2933SET*−10.655) of the metabolic rising stage in the transition space experiment, the SET* regression equation obtained indoors can be written into the following equation:

$$\begin{aligned} MTSV &= 0.28SET^* + 3.47V_s - 10.69 \\ &= 0.2933(0.94SET^* + 11.83V_s - 0.12) - 10.655 \end{aligned} \quad (10)$$

The expression in brackets was defined as αSET*.

By integrating the improved models from both experiments, the outcomes shown in Fig. 10 were obtained. Apart from the 0–5 min data of indoor walking, the other revised datasets were analyzed using a unified regression method. It was observed that during the metabolic ascending phase, MTSV and αSET* exhibited a cubic correlation, while in the stabilization phase their relationship was linear. In both metabolic stages, the regression correlation coefficient exceeded 0.8. Furthermore, the intersections of the regression curves with different stress levels indicated the thermal sensation values corresponding to αSET* across various activity stages. The detailed results are presented in Table 5.

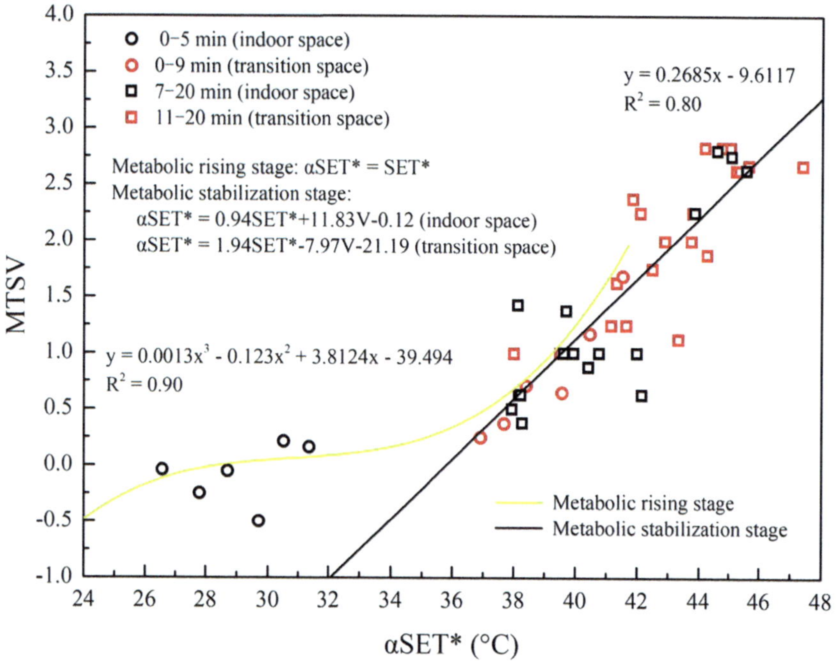


**Fig. 10** Correlation between MTSV and αSET* at different stages of metabolic rate during walking

**Table 5** Thermal sensation levels corresponding to αSET* at different metabolic rate stages

| Period | αSET* range (°C) | Thermal sensation for different stress categories | | | | | | | |
|---|---|---|---|---|---|---|---|---|---|
| | | – 0.5 | 0 | 0.5 | 1.0 | 1.5 | 2.0 | 2.5 | 3.0 |
| Metabolic rising stage | 26.6–41.5 | (24.0) | 28.7 | 37.1 | 39.2 | 40.9 | | | |
| Metabolic stabilization stage | 36.9–47.4 | | | 37.8 | 40.0 | 41.5 | 43.1 | 45.2 | 46.9 |
| Whole process | 26.6–47.4 | (24.0) | 28.7 | 37.5 | 39.6 | 41.2 | 43.1 | 45.2 | 46.9 |

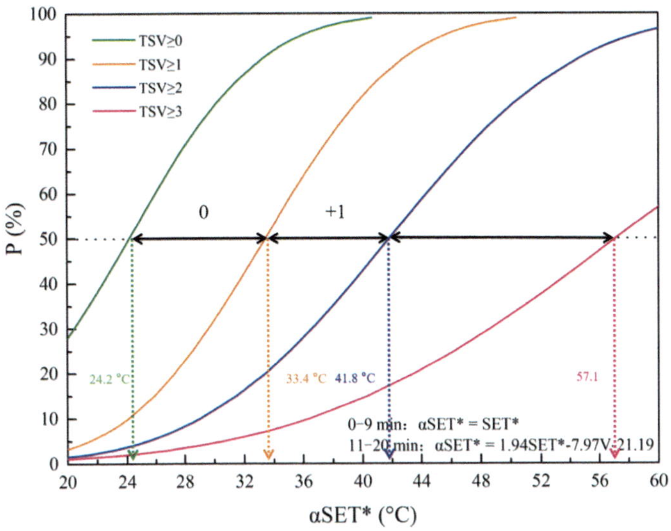


**Fig. 11** Probabilistic analysis of all walking speeds integrated in transition sp.

## *3.4 αSET* Thermal Interval*

The thermal sensation vote (TSV) values of participants at different walking speeds in both the badminton hall and the overhead floor were collected. Using Probit regression, cumulative probability curves for each TSV level were generated, as illustrated in Figs. 11, 12, 13 and 14. The point where each curve intersects the 50% probability line was regarded as the threshold for the corresponding thermal interval. In other words, the range of the n-level thermal interval can be expressed as: $SET^{**}TSV \geq n\ (p = 50\%) \leq TSV = n < SET^{**}TSV \geq n + 1\ (p = 50\%)$.

### 3.4.1 Probit Analysis in Transition Space

In the transition space, Probit regression was applied to obtain the probability curves of TSV levels at all walking speeds (Fig. 11) as well as at each individual speed (Fig. 12). A reference line at 50% probability was drawn, and the αSET* values corresponding to the intersections of the TSV curves with this line were determined. The outcomes are summarized in Table 6. As shown, for a fixed walking speed, thermal sensation levels rose with increasing αSET*. Conversely, at the same thermal sensation level, a higher walking speed was associated with a lower αSET* value, indicating that faster walking corresponds to a higher thermal level under reduced αSET*.

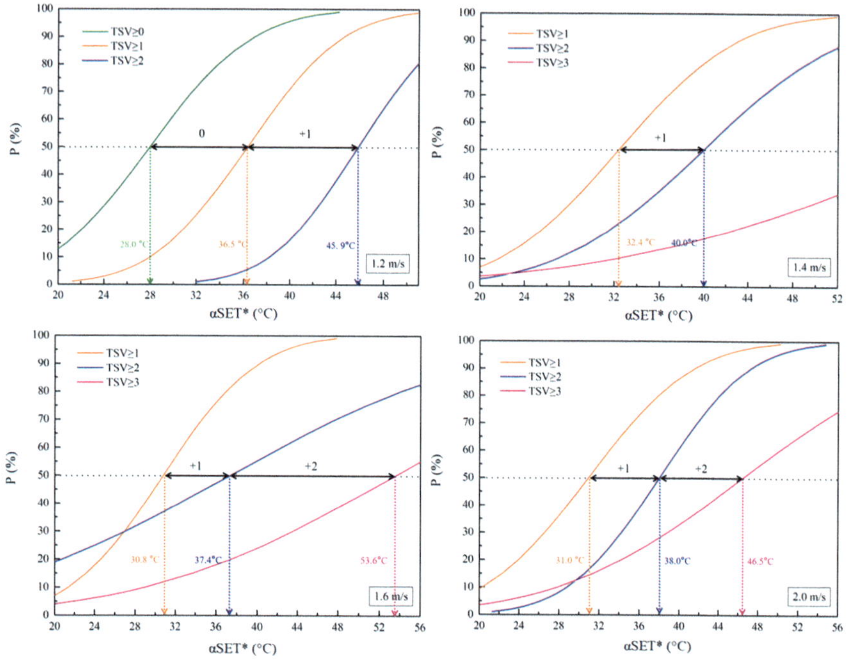


**Fig. 12** Probit analysis at different walking speeds in transition space

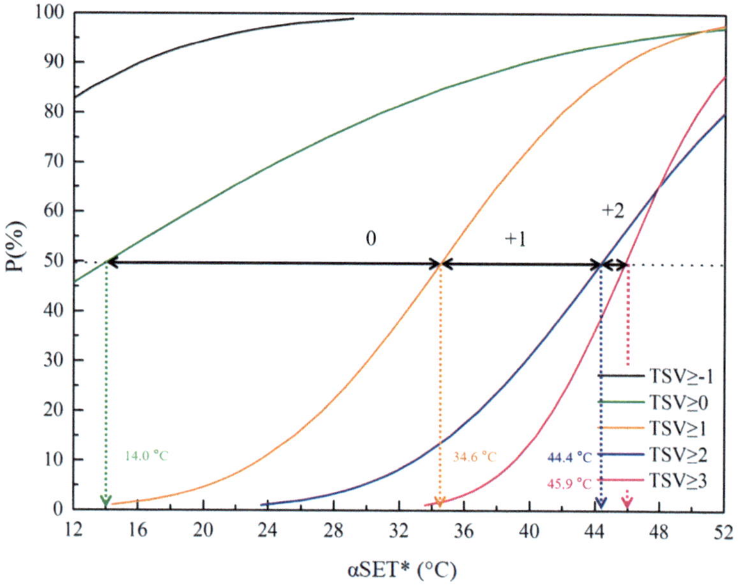


**Fig. 13** Probabilistic analysis of all walking speeds integrated in indoor space

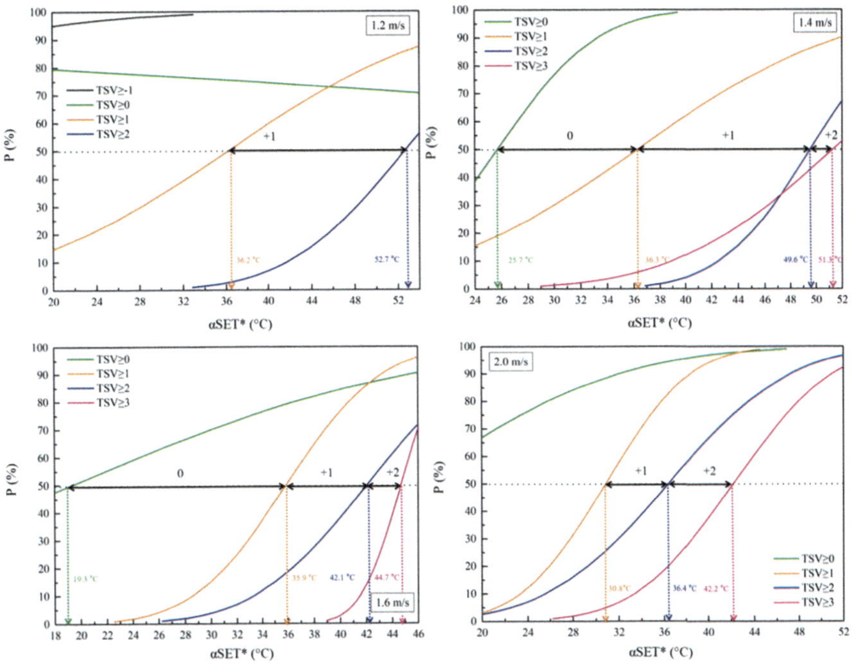


**Fig. 14** Probit analysis at different walking speeds in indoor space

**Table 6** Thermal sensation levels corresponding to αSET* in transition space

| Walking speed | αSET* range (°C) | Thermal sensation for different stress categories | | | |
|---|---|---|---|---|---|
| | | – 0.5 | 0.5 | 1.5 | 2.5 |
| 1.2 m/s | 30.4–51.4 | (28.0) | 36.5 | 45.9 | |
| 1.4 m/s | 33.6–49.6 | | (32.4) | 40.0 | |
| 1.6 m/s | 34.0–51.7 | | (30.8) | 37.4 | 53.6 |
| 2.0 m/s | 35.5–53.3 | | (31.0) | 38.0 | 46.5 |
| All speeds | 30.4–53.3 | (24.2) | 33.4 | 41.8 | (57.1) |

### 3.4.2 Probit Analysis in Indoor Space

Using the same approach, Probit regression was applied to the indoor space data, and the outcomes are presented in Figs. 13 and 14. Table 7 summarizes the thermal sensation levels corresponding to αSET* across various walking speeds. A clear difference was observed when compared with the transition space: the αSET* interval indoors was markedly narrower. Nevertheless, whether comparing αSET* values at the same sensation level or analyzing different sensation levels at identical walking speeds, the observed trend was consistent with that of the transition space.

**Table 7** Thermal sensation levels corresponding to αSET* in indoor space

| Walking speed | αSET* range (°C) | Thermal sensation for different stress categories | | | |
|---|---|---|---|---|---|
| | | − 0.5 | 0.5 | 1.5 | 2.5 |
| 1.2 m/s | 21.4–43.0 | | 36.2 | (52.7) | |
| 1.4 m/s | 27.7–46.4 | (25.7) | 36.3 | (49.6) | (51.3) |
| 1.6 m/s | 19.4–45.4 | (19.3) | 35.9 | 42.1 | 44.7 |
| 2.0 m/s | 20.8–51.5 | | 30.8 | 36.4 | 42.4 |
| All speeds | 19.4–51.5 | (14.0) | 34.6 | 44.4 | 45.9 |

## 4 Discussion

This study centers on the applicability and refinement of the SET* model in both indoor and transitional spaces of Guangzhou. After modification, the findings suggest that αSET* demonstrates strong predictive capacity for MTSV of moving individuals under the tested conditions. According to the measured results, the αSET* range lies between 26.6 and 47.4 °C, while the comfort interval is 24.0–37.5 °C. It is clear that the actual minimum αSET* is slightly higher than the lower comfort limit, implying that reducing the lower boundary by 2.6 °C would still keep it within the comfort zone.

Earlier research has also applied the SET* index to classify thermal intervals, though the majority of those experiments were carried out in outdoor environments. As shown in Table 8, when compared with prior studies, the present results reveal that the upper boundary of the αSET* range in both indoor and transitional spaces is higher than all outdoor cases. This difference arises because the participants in our experiment were engaged in walking, with the metabolic rate M surpassing 2 met. Moreover, this study refined the SET* calculation by incorporating the effects of wind speed and metabolic rate across different stages of activity. By contrast, in many other investigations, the participants remained standing still, and statistics indicated that over 90% exhibited a metabolic rate not exceeding 2 met, leading to lower SET* values.

The identified comfort zone of αSET* (corresponding to MTSV = ± 0.5 °C) under the combined indoor and outdoor walking experiments was 24.0–37.5 °C. When compared with the standard reference interval [32] (22.2–25.6 °C), the observed range is not only higher in temperature but also broader. This suggests that during the actual walking tests, participants demonstrated greater tolerance to the surrounding thermal environment. Furthermore, results from other studies reveal that their comfort ranges do not consistently align with the standard reference, implying that the values provided by the standard are mainly applicable to idealized situations.

In addition, results from the summer experiments were analyzed for comparison. This experiment, conducted in the transitional space, showed that the αSET* comfort range was 24.2–33.4 °C, which was closest to the comfort interval reported in the literature [36]. Relative to other research, the comfort zone in this case was slightly

Table 8 Summary of SET* thermal interval

| Category | | Location | Method | Season | SET*(αSET*) range/°C | Thermal sensation for different stress categories | | | | | |
|---|---|---|---|---|---|---|---|---|---|---|---|
| | | | | | | – 1.5 | –0.5 | 0.5 | 1.5 | 2.5 | 3.5 |
| This study | Both | Guangzhou, China | Linear regression of MTSV and αSET* | Spring/summer | 26.6–47.4 | | (24.0) | 37.5 | 41.2 | 45.2 | |
| | Indoor space | | Probit regression analysis | Spring | 19.4–51.5 | | (14.0) | 34.6 | 44.4 | 45.9 | |
| | Transition space | | Probit regression analysis | Summer | 30.4–53.3 | | (24.2) | 33.4 | 41.8 | (57.1) | |
| (ASHRAE standard) McIntyre [32] | – | USA | – | Annual | – | 17.5 | 22.2 | 25.6 | 30.0 | 34.5 | 37.5 |
| Feng [33] | Outdoor space | Guangzhou, China | Linear regression of MTSV and αSET* | Spring | 19–40 | | 19.2 | 25.1 | 31.0 | 36.9 | |
| | | | | Summer | 27–44 | | – | – | 31.6 | 38.1 | |
| | | | Probit regression analysis | Spring | 17–45 | | 18.3 | 28.9 | 30.3 | 33.6 | |
| | | | | Summer | 25–45 | | – | 27.2 | 30.7 | 34.6 | |
| Xi et al. [34] | Outdoor space | Guangzhou, China | Linear regression of MTSV and αSET* | Summer | – | | 20.6 | 27.8 | 35.1 | 42.3 | 49.5 |
| Chen et al. [35] | Outdoor space | Guangzhou, China | Linear regression of MTSV and αSET* | Summer | 20.5–38.6 | | 21.5 | 26.3 | 31.2 | 36.1 | |

(continued)

**Table 8** (continued)

| Category | | Location | Method | Season | SET*(αSET*) range/°C | Thermal sensation for different stress categories | | | | | |
|---|---|---|---|---|---|---|---|---|---|---|---|
| | | | | | | – 1.5 | −0.5 | 0.5 | 1.5 | 2.5 | 3.5 |
| Huang et al. [36] | Outdoor space | Taiwan, China | Linear regression of MTSV and αSET* | Annual | – | | 22.9 | 31.5 | 40.1 | 48.8 | 57.4 |
| Lin et al. [37] | Outdoor space | Beijing, China | – | Summer | | 13 | 20 | 30 | – | 35 | 40 |
| Tsitoura et al. [38] | Outdoor space | Greece | – | Annual | – | 13 | 18 | 23 | – | 30 | |

higher in temperature. In those other studies, the experimental locations were entirely outdoors, where the intensity of solar radiation was stronger, and human thermal sensation was more heavily influenced by external conditions. Thus, under similar outdoor climatic factors, when the SET* value exceeded the comfortable threshold for outdoor participants, individuals in transitional spaces still perceived a more comfortable environment.

## 5 Conclusions

This chapter examined the role of building overhead layers in improving thermal comfort by testing the applicability of the SET* index in both indoor and transitional environments. Field experiments revealed that while conventional SET* predictions cannot fully capture dynamic human thermal responses, the overhead floor provided a setting in which SET* showed greater predictive accuracy, particularly during the initial phase of elevated metabolic activity. By incorporating variations in metabolic rate, an enhanced model (αSET*) was developed, allowing for more reliable estimation of thermal sensation across both walking and resting stages. Furthermore, the conventional SET* comfort range was found to be unsuitable for outdoor and semi-open contexts, leading to the proposal of revised αSET* thresholds tailored to transitional spaces. These findings highlight the importance of overhead layers as effective microclimatic modifiers in urban planning, and demonstrate that refining thermal indices for such environments can significantly improve the evaluation and management of outdoor thermal comfort in cities.

**Acknowledgements** This work was supported by the National Natural Science Foundation of China (Project No. 51978180), Research Project of Guangzhou University (No. YG2020003). The authors thank all workers who participated in the surveys.

## References

1. H.J. Cho, J.W. Jeong, Evaluation of thermal comfort in an office building served by a liquid desiccant-assisted evaporative cooling air-conditioning system. Energy Build. **171**, 361–370 (2018)
2. C. Chun, A. Tamura, Thermal comfort in urban transitional spaces. Build. Environ. **40**, 633–639 (2005)
3. S.A. Damiati, S.A. Zaki, H.B. Rijal, S. Wonorahardjo, Field study on adaptive thermal comfort in office buildings in Malaysia, Indonesia, Singapore, and Japan during hot and humid season. Build. Environ. **109**, 208–223 (2016)
4. N.E. Klepeis, W.C. Nelson, W.R. Ott, J.P. Robinson, A.M. Tsang, J.P. Switzer, W.H. Engelmann, The National Human Activity Pattern Survey (NHAPS): a resource for assessing exposure to environmental pollutants. J. Expo. Anal. Environ. Epidemiol. **11**(3), 231 (2001)
5. S. Shooshrarian, C.K.C. Lam, I. Kenawy, Outdoor thermal comfort assessment: a review on thermal comfort research in Australia. Build. Environ. **177** (2020)

6. Z. Fang, Z. Lin, C.M. Mak, J. Niu, K.-T. Tse, Investigation into sensitivities of factors in outdoor thermal comfort indices. Build. Environ. **128**, 129–142 (2018)
7. Z. Fang, X. Feng, X. Xu, X. Zhou, Z. Lin, Y. Ji, Investigation into outdoor thermal comfort conditions by different seasonal field surveys in China, Guangzhou. Int. J. Biometeorol. **63**(10), 1357–1368 (2019)
8. I. Golasi, F. Salata, E.L. Vollaro, M. Coppi, Complying with the demand of standardization in outdoor thermal comfort: a first approach to the Global Outdoor Comfort Index (GOCI). Build. Environ. **130**, 104–119 (2018)
9. A.H. Abdullah, Q. Meng, L. Zhao, F. Wang, Field study on indoor thermal environment in an atrium in tropical climates. Build. Environ. **44**, 431–436 (2009)
10. S. Hussain, P.H. Oosthuizen, A. Kalendar, Evaluation of various turbulence models for the prediction of the airflow and temperature distributions in Atria. Energy Build. **48**, 18–28 (2012)
11. H. Liu, Z. Lian, Z. Gong, Y. Wang, G. Yu, Thermal comfort, vibration, and noise in Chinese ship cabin environment in winter time. Build. Environ. **135**, 104–111 (2018)
12. F. Wang, K. Pichatwatana, S. Roaf, L. Zhao, Z. Zhu, J. Li, Developing a weather responsive internal shading system for atrium spaces of a commercial building in tropical climates. Build. Environ. **71**, 259–274 (2014)
13. A.P. Gagge, Rational temperature indices of man's thermal environment and their use with a 2-node model of his temperature regulation. Fed. Proc. **32** (1973)
14. P.O. Fanger, Thermal comfort: analysis and applications in environmental engineering, in *Thermal Comfort Analysis and Applications Environmental Engineering*, (1972)
15. S. Gao, Y. Li, Y.A. Wang, X.Z. Meng, L.Y. Zhang, C. Yang, L.W. Jin, A human thermal balance based evaluation of thermal comfort subject to radiant cooling system and sedentary status. Appl. Therm. Eng. **122**(25), 461–472 (2017)
16. M.A. Humphreys, J.F. Nicol, The validity of ISO-PMV for predicting comfort votes in every-day thermal environments. Energy Build. **34**(6), 667–684 (2002)
17. Z. Fang, X. Feng, J. Liu, Z. Lin, C.M. Mak, J. Niu, K.T. Tse, X. Xu, Investigation into the differences among several outdoor thermal comfort indices against field survey in subtropics. Sustain. Cities Soc. **44**, 676–690 (2019)
18. T. Xi, Q. Li, A. Mochida, Q. Meng, Study on the outdoor thermal environment and thermal comfort around campus clusters in subtropical urban areas. Build. Environ. **52**, 162–170 (2012)
19. H.H. Ali, R. Al-Hashlamun, Assessment of indoor thermal environment in different prototypical school buildings in Jordan. Alex. Eng. J. **58**, 699–711 (2019)
20. Y. Zhang, J. Liu, Z. Zheng, Z. Fang, X. Zhang, Y. Gao, Y. Xie, Analysis of thermal comfort during movement in a semi-open transition space. Energy Build. **225**, 110312 (2020)
21. D. Schnorr, G. Dörner, F. Stahl, W. Rohde, H.M. Guddat, Human heat balance during post exercise recovery: separating metabolic and nonthermal effects. Am. J. Physiol. **294**(2), 1586–1592 (2008)
22. T. Sakoi, T. Mochida, Y. Kurazumi, K. Kuwabara, Y. Horiba, S. Sawada, Heat balance model for a human body in the form of wet bulb globe temperature indices. J. Therm. Biol. **71**, 1–9 (2018)
23. R.A. Engineers ASHRAE Handbook: Fundamentals (2017)
24. ISO, International Standard 7730, in *Ergonomics of the Thermal Environment-Analytical Determination and Interpretation of Thermal Comfort Using Calculation of the PMV and PPD Indices and Local Thermal Comfort Criteria*, International Standard Organization, Geneva, 2005
25. W. Li, Y. Zhang, L. Zhang, Z. Li, The research of the sport burden grade standard corresponding to the heart rate. Nat. Sci. J. Harbin Normal Univ. **24**(5), 99–102 (2008)
26. K. Fabbri, *Indoor Thermal Comfort perception: A questionnaire Approach Focusing On Children* (Springer, New York City, 2015)
27. A.T. Nguyen, M.K. Singh, S., Reiter an adaptive thermal comfort model for hot humid South-East Asia. Build. Environ. **56**, 291–300 (2012)
28. B. Li, C. Du, M. Tan, H. Liu, E. Essah, R. Yao, A modified method of evaluating the impact of air humidity on human acceptable air temperatures in hot-humid environments. Energy Build. **158**, 393–405 (2018)

29. Z.J. Yu, B. Yang, N. Zhu, T. Olofsson, G. Zhang, Utility of cooling overshoot for energy efficient thermal comfort in temporarily occupied space. Build. Environ. **109**, 199–207 (2016)
30. H. Du, C. Yang, Re-visitation of the thermal environment evaluation index standard effective temperature (SET**) based on the two-node model. **53** (2020)
31. Y. Zhang, X. Zhou, Z. Zheng, M.O. Oladokun, Z. Fang, Experimental investigation into the effects of different metabolic rates of body movement on thermal comfort. Build. Environ. **168** (2020)
32. D.A. McIntyre, *Indoor climate* (Applied Science Publishers Ltd, London, 1980)
33. X. Feng, Outdoor thermal comfort study for a whole year in Guangzhou, 2018
34. T. Xi, Q. Li, A. Mochida, et al., Study on the outdoor thermal environment and thermal comfort around campus clusters in subtropical urban areas. Build. Environ. **52**(6), 162–170 (2012)
35. R. Chen, Analysis and applicability of outdoor thermal environment evaluation index based on Guangzhou campus, 2015
36. R. Hang, T. Lin, Thermal comfort requirements for occupants of semi-outdoor and outdoor environments in hot-humid regions. Arch. Sci. Rev. **50**, 357–364 (2017)
37. B. Lin, X. Li, Y. Zhu, Y. Qin, Numerical simulation studies of the different vegetation patterns' effects on outdoor pedestrian thermal comfort. in *The Fourth International Symposium on Computational Wind Engeneering (CWE2006)*
38. M. Tsitoura, T. Tsoutsos, T. Daras, Evaluation of comfort conditions in urban open spaces. Application in the island of Crete. Energy Convers. Manag. **85**, 250–258 (2014)

GPSR Compliance

*The European Union's (EU) General Product Safety Regulation (GPSR) is a set of rules that requires consumer products to be safe and our obligations to ensure this.*

*If you have any concerns about our products, you can contact us on ProductSafety@springernature.com*

In case Publisher is established outside the EU, the EU authorized representative is:

Springer Nature Customer Service Center GmbH
Europaplatz 3
69115 Heidelberg, Germany

Batch number: 10732076

Printed by Printforce, the Netherlands